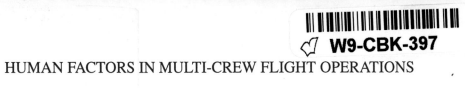

HUMAN FACTORS IN MULTI-CREW FLIGHT OPERATIONS

This book is dedicated to the memory of Col. Horace Orlady, father and grand-father of the authors. He soloed in 1916 and held Aviator's Certificate #28 of the Aero Club of America. Col. Orlady, or Hoy as we knew him, served as a pilot in the US Signal Corps as aviation was introduced into the US Army. He served his country in World War I and continued to serve as the Aviation Branch of the Signal Corps became the US Army Air Corps and then the US Air Force. He maintained a Reserve Commission during the interim between World War I and World War II, and served the US Air Force during World War II. Hoy was awarded the US Legion of Merit and the Officer of the Order of the British Empire for his services during World War II. It is a very real plea-sure to include memory of this remarkable aviator in our dedication.

In addition, and above all, this book is also dedicated to Ellen and to John (including Amy and Pumpkin) with a great deal of love and affection. We gratefully acknowledge the unfailing tolerance and understanding both displayed during the much longer than anticipated time that it took to write this book.

Human Factors in Multi-Crew Flight Operations

HARRY W. ORLADY
LINDA M. ORLADY

With a Foreword by
JOHN K. LAUBER

Ashgate

Aldershot • Brookfield USA • Singapore • Sydney

© Harry W. Orlady and Linda M. Orlady 1999

Published by
Ashgate Publishing Ltd
Gower House
Croft Road
Aldershot
Hants GU11 3HR
England

Ashgate Publishing Company
Old Post Road
Brookfield
Vermont 05036
USA

British Library Cataloguing in Publication Data
Orlady, Harry W.
    Human factors in multi-crew flight operations
    1. Flight crew 2. Aeronautics, Commercial
    I. Title II. Orlady, Linda M.
    387.7

Library of Congress Cataloging-in-Publication Data
Orlady, Harry W.
    Human factors in multi-crew flight operations / Harry W. Orlady,
Linda M. Orlady.
       p. cm.
    ISBN 0-291-39838-3 (hc) . -- ISBN 0-291-39839-1 (pbk.)
    1. Aeronautics--Human factors. 2. Flight crews. I. Orlady ,
Linda M. 1956- . II. Title.
TL553.6.O75 1999
629.132'5216--dc21   99-26346
                                        CIP

ISBN 0 291 39838 3 (HBK)
ISBN 0 291 39839 1 (PBK)

Printed in the United Kingdom at the University Press, Cambridge

Contents

List of Figures

List of Tables

Foreword

I am especially delighted to have been asked to write the Foreword for *Human Factors in Multi-Crew Flight Operations* written by my friends and colleagues, Harry and Linda Orlady. My association with Harry and Linda dates to the earliest days of my career in aviation human factors. I first met Harry, at that time a B-747 pilot for United Airlines, shortly after I went to NASA Ames Research Center in 1973. Captain Orlady had been previously associated with another colleague whose work is frequently referenced in the chapters that follow, Dr. Charles Billings, who entirely coincidentally, had also made a career change that took him to Ames that same year. It was through Dr. Billings that I first met Harry; and through Harry, that I first met his daughter, Linda.

I hold myself partially responsible for Linda's choice of career. During one of Harry's periodic visits to NASA, he brought Linda with him. Our conversation turned to flying, as it always did, and the next thing I knew, we were at Reid-Hillview airport making Linda's first flight in a light airplane. I was the pilot, Captain Orlady was in the rear seat of an airplane not quite as spacious as the B-747 he regularly flew, and Linda was in the copilot's seat.

Apparently, Linda enjoyed the flight, because shortly thereafter, she enrolled in the aviation program at Ohio State University. She submersed herself in the aviation program, became an instructor in both the Ohio State aviation science program and flight schools, and completed a masters program in business, concentrating on organizational behavior. Coincidentally, she also became my daughter, Sarah's flight instructor. None of us, Linda, Sarah, or I will ever forget the day when Linda soloed Sarah with me present to observe. I am not sure who was the more nervous—the fledgling pilot, the flight instructor, or the father! Lots of human factors were at play that day.

Although the Orladys have written this book primarily from the viewpoint of a pilot, it is not a book only *for* pilots, nor is it a book only *about* pilots. This is a book about the air transport industry and about human performance in the aviation system. It is a book that describes, in practical, useful terms, factors that affect the ability of humans to function efficiently and safely, regardless of their specific role or function within the aviation system. It is a book that will serve as a useful reference for anyone needing a basic understanding of operational air transport, of how system factors can affect the performance of people within the system, and also of how human performance can affect system performance, especially system safety.

There are many misconceptions about what 'human factors' is all about, and not infrequently, these misconceptions are fueled by the human factors community itself. In fact, in the early days at Ames, many of us (at least those of us with a sense of humor) used to refer to human factors as 'squishy stuff'. This was a reference to the fact that human performance research tends to be more subjective, qualitative and based on statistical inference, and less objective, quantitative and based on mechanistic models. Later developments, especially in the area of Cockpit Resource Management, as we called CRM in the early days, created opportunities for California style 'hot-tub' psychology to come to the fore. Some got the mistaken impression that human factors was about getting people to like each other so that they would then fly safely together. However, as you will see when you read this book, human factors in aviation is not about what people *think* or *feel* so much as it is about what people—pilots, flight attendants, mechanics, air traffic controllers, dispatchers, and others, *do*. Human factors is about observable, measurable human behavior—conducting timely crew briefings, making informed decisions, exercising leadership, actively managing fatigue and jet lag, monitoring, redistributing workload, maintaining professional discipline, effectively communicating, and so on.

Careful reading of this book will help dispel common misconceptions of what 'human factors' is about, and will provide the reader with practical information that will help achieve the desired levels of human performance in the aviation system. That this is important is amply reflected in the accident data: human performance and human error continue to be the leading cause of such mishaps.

John K. Lauber
Vice President – Training and Human Factors
Airbus Service Company
Miami

Acknowledgements

We are particularly pleased that Dr. John Lauber has written a Foreword for this book. Dr. Lauber is known throughout the aviation world as a foremost expert in aviation safety and in modern aviation human factors. His background as a NASA aviation psychologist and researcher, as an outstanding former Member of the US National Transportation Board (NTSB), as former Vice President of Safety and Compliance for Delta Airlines, and presently Vice President—Training and Human Factors, Airbus Service Company gives him a particularly unique and advantageous perspective to understand the multi-faceted problems of modern human factors in air transport operations.

We are also particularly pleased that we were able to secure the services of John Cirino. John, a consummate aviator and a US Airways captain, is responsible for all our illustrations and for the design, alterations and enhancements of the charts and diagrams we have used. Additionally, John also has provided us with valuable insight into the real world of day-to-day flight operations and considerably broadened our air transport perspective. He has made this a better book.

Finally, we would like to acknowledge the contributions that have been made to this book by Captain Lawson C. White and by Dr. Rolf Braune. Their support, their unfailing cheerfulness, and their willingness to spend a great deal of time reviewing our efforts have gone well beyond the call of friendship. The authors, while completely responsible for any mistakes, misstatements, or omissions in *Human Factors in Multi-Crew Operations,* are grateful considerably beyond the rewards of the deep and rewarding friendships we have with Captain White and Dr. Braune.

This book has been the product of discussions with many people over a great many years. Unfortunately, there are inadvertent omissions in the following list of university professors, aeromedical doctors, aviation psychologists, airline management personnel, aviation researchers, aviation lawyers and arbitrators, aviation human factors specialists, aviation safety experts, aeronautical engineers, pilot association representatives, and fellow pilot peers from US and foreign airlines. Many of them have been colleagues of both authors. The list illustrates the range of individuals and the disciplines that are concerned with the broad aspects of modern aviation human factors, flight crew performance, and aviation safety.

The following individuals have been important in both of our careers. Their names are listed to provide at least a meager acknowledgement of the contribution they have made to this book: Bob Alkov, Bob Barnes, Jerry Berlin, Charlie Billings, Nan Burnett, Stuart Bernstein, Gerry Bruggink, Bill Connor, Asaf Degani, Bill Dunkle, Earl Carter, Bill Edmunds, Del Fadden, Clay Foushee, Dick Gabriel, Curt Graeber, Richard Hackman, Rex Hardy, Dick Harper, Bob Helmreich, Dick Jensen, Neil Johnston, Barb Kanki, George Kidera, Les Lautman, Cliff Lawson, Jerry Lederer, Dan Maurino, Judith Orasanu, Bert Ruitenberg, Pete Siegel, Bill Traub, Earl Wiener, Leon Wollard, Tom Young, and the late Bill Ashe, Lloyd Buley, Frank Hawkins, Ted Linnert, and Clancy Sayen.

Introduction

Human Factors in Multi-Crew Flight Operations is an entirely new book. It is written from a particular point of view—the view of flight operations and from the viewpoint of the well-informed pilot. It is concerned with the application, and the effective implementation, of modern human factors in air transport operations. This book is concerned with the impact of new technology on the safety and efficiency of the air transport industry. The behavior and goals of the people who are a part of the air transport industry and the health and well being of the individuals who work within it are an important part of the industry's safety and efficiency.

The purpose of *Human Factors in Multi-Crew Flight Operations* is to provide a perspective of the air transport industry and an understanding of the human factors involved in current air transport operations at a level that should be of interest to anyone engaged in these operations. It supplies information with which professional transport pilots and other aviation professionals should be familiar. Therefore, this material also will be of value to serious students, engineers, scientists, managers, air traffic controllers, regulators, and any others who are concerned with transport flight operations. Despite the varying interests of this diverse group, they have common goals and a common need to be familiar with aviation human factors and the breadth of its interests. Finally, this level of aviation human factors should be of interest to general aviation pilots who wish to learn more of the background, the information, and the non-technical and behavioral skills and knowledge that are requirements for the career professional pilot.

The book should not be considered a text for academic researchers or a detailed text on specific subjects. This would require several books. The authors fully recognize that professional people within the aviation system have differing needs. We believe, however, that those professionals should be familiar with the material we will discuss. We know of no other single book that contains all of this varied information. References are supplied for those who wish to pursue further any of the individual subjects.

Safety is a Prime Consideration

Safety is given considerable emphasis in this book because practical contemporary aviation human factors research begins with a safety component. Operational management and economics, which may be considered outside of usual human factors boundaries, are frequently discussed. Ultimately these factors drive the implementation of human factors developments in real-world air transport operations.

The safety ramifications of the man-machine-environment interface (which is by no means restricted to advanced airplanes or to the pilot/airplane interface) are a major factor in the emphasis and expansion that is now given to human factors in the air transport industry. The considerations raised by that interface and their safety implications extend throughout this book. *Human Factors in Multi-Crew Flight Operations* covers both the human factors advances in air transport operations, which are a part of this challenging time, concepts such as the latent considerations within the system, and traditional operational human factors, which historically have not stressed behavioral factors.

Air transport is a large, dynamic, and complex industry. While both the size and status of the industry are essential considerations in any discussion of operational human factors, we should never forget that safety is air transport's prime consideration. Safety is considered throughout the book. We have a wrap-up discussion of air transport safety in Chapter 20 and a discussion of current safety problems in Chapter 21.

Human Factors is a 'Core Technology' and is Application Oriented

For a great many years, human factors in aviation has had an engineering orientation primarily because the industry has been, to a considerable extent, technology-driven. A point that is sometimes missed is that effective modern aviation human factors is also application oriented. Today's airplanes, the technological marvels that are a part of this business, must be operated safely and efficiently. Modern aviation human factors plays an important role in recognizing, establishing, and maintaining the factors and conditions which increase the safety and efficiency of operations in the airline world.

Human factors has become a 'core technology' in air transport operations. It is a core technology in much the same way that powerplants, meteorology, navigation, and communications have been, and still are, core technologies in aviation.

The Air Transport is a World Industry

Air transport has become a world industry. The International Civil Aviation Organization (ICAO), the aviation branch of the United Nations, has become a leader in achieving recognition that human factors is a basic part of aviation operations. Human factors has become a broadly based international concept. ICAO is discussed in Chapter 20—The Worldwide Safety Challenge and in Appendix D.

Anyone involved in the operations of air transport and its technology must now be sensitive to the human technology interface. He or she should also be aware of the cultural differences found in many aspects of global air transport operations. In an expansion of traditional thinking, modern human factors includes consideration of what we are calling the 'social environment'. Because air transport is a world industry, variations in and organizational cultures are important human factors. All of these are a part of the challenges and constraints of real-world day-to-day air transport operations.

Parts of the industry have a way to go. The late Frank Hawkins noted that as late as 1984, the representative of a large independent airline told an international audience that, 'Human factors is just an excuse for incompetence.'[1] Unfortunately, such a view represents too common thinking of earlier decades (see Chapter 3—A Brief History of Human Factors in Aviation). The earlier view presents an inaccurate view of aviation human factors. We hope to keep it in better perspective.

Human Factors in Multi-Crew Flight Operations has an essentially US orientation, because of the generally greater accessibility of US data, because of the US background of the authors, and because of the position of US aviation throughout the world. The prominence of US accident data in this book, and of references to these accidents, should not be considered an adverse reflection of air transport safety in the US. Neither should the emphasis given US aviation human factors activities create the impression that the US is the only source of such activities.

In some cases the data we discuss, particularly that of accidents, hull-loss accidents, fatal accidents, total fatalities etc., may appear inconsistent because they come from several different bases. The data can involve world airlines, or can involve only airlines that are members of IATA (the International Aviation Transport Association), or only airlines that are members of the Air

---

[1] Hawkins, Frank H., *Human Factors in Flight*, Second edition, Ashgate Publishing Limited, Aldershot, Hants, England.

Transport Association (US). Additionally, some data is from the US FAA, the NTSB, and ICAO. The problem of fully understanding ramifications of the data is not easy because the various groups do not use common definitions nor is the data that is available equally complete.

It should be clear that the human factors principles discussed apply to all transport airplanes regardless of their national origin. The prominence of Boeing data and the examples we use are due in principal part to the greater availability of that material and the authors' greater familiarity with its specifics. Worldwide statistics do not include data from the Republic of China or from the Commonwealth of Independent States (CIS and formerly the Republic of Russia) because only limited data are available from those entities. Appendix A—Safety in the CIS States and in the Republic of China—has additional information regarding air transport in these States.

Standardization and Certification Reciprocity

Increasingly, attempts are being made to achieve standardization and certification reciprocity among nations in order to increase safety and to facilitate the import and export of aviation products. Problems have been exacerbated with the global operation of air transport and with the growth of human factors issues in certification. There are a great many anthropological, language, and cultural differences among the countries of the world. Problems this creates in air transport safety will be discussed in Chapter 6—The Social Environment.

Air transport, of course, has no geographical boundaries. This book includes an Appendix O, which discusses multi-cultural operations and the increasingly important issue of multi–national certification. Appendix O is co-authored with Robert B. Barnes, an authority who has spent considerable time on this important subject. While there are aspects of the questions raised in Appendix O that go beyond the basic purposes of this book, advanced human factors issues have become increasingly important in this area. Today's professionals should be aware of the relevance of these issues in global air transport operations.

The Book

Chapter 1, 'Our Heritage in Air Transport,' begins with an overview of the heritage of air transport operations. We believe familiarity with the heritage of air transport helps better understand modern air transport human factors and the air transport industry itself. There have been many innovations in the past that are now part of our heritage. All parts of this industry, both technical and nontechnical, have thrived on them. There is no reason to think that we will not continue to have many innovations in the future.

Chapter 2, 'The Industry and its Safety Record,' gives a brief discussion of the industry's safety record and of its present size and status. This chapter occurs early in the book because we believe that the industry's size and status and its safety record are necessary background information that help keep human factors questions and challenges, and the air transport industry in perspective.

Chapter 3, 'A Brief History of Human Factors and its Development in Aviation,' briefly reports on the growth and development of human factors in industry generally, on human factors in aviation specifically, and on the growing consensual recognition of the importance of a broadened concept of human factors in air transport operations.

The remainder of *Human Factors in Multi-Crew Flight Operations* discusses the basic elements and considerations involved in the operation of air transports in Chapters 4-6, in operational considerations of particular importance in today's environment in Chapters 7-13, and in additional considerations meriting human factors attention in Chapters 14-17. Finally, in Chapters 18-22 we discuss non-punitive incident reporting, accident analysis, safety generally, current safety problems, and the authors' view of the future of this dynamic industry. The role of operational management and economics is discussed in Chapter 22 because they are keys to the implementation of the advances that are produced by human factors researchers and aviation engineers.

We have attempted to produce a balanced book. The chapters can be read in order, or the chapters read independently if there is immediate interest in that chapter's subject matter. The Appendices are somewhat different. They represent the considerable complexity that is an inherent part of the industry, present information that is not always easy to procure, and in at least one case present data and information that we believe is of interest but of insufficient maturity to be included in the main chapters. Appendix Q, which includes web site addresses that the authors have found particularly useful, is in a special category. The internet can be a marvelous information source although it is important to verify information. Some of these web sites are updated frequently or changed. Often, they provide valuable links to other useful sites.

General Considerations

The reader should recognize that in this book the word 'man' is used in its generic sense. Therefore, it includes both sexes unless it is specifically indicated otherwise. For example, 'man' should be understood to mean people or humans, including both males and females, and 'he' or 'she' and 'him' or 'her' should ordinarily be understood to mean either sex.

When it is found necessary to differentiate between flight deck and cabin crew and the inference is not clear from the context, the distinction will be made specifically. In all other cases, crew, flight crew, or aircrew refers to both cockpit and cabin crew members.

A major goal of the authors is to make *Human Factors in Multi-Crew Flight Operations* as user friendly as possible. Therefore, each chapter is reasonably self-sustaining, and we have placed the references we used at the end of each chapter. If the reader wants more information on a particular subject, a good source for that information will be easy to identify. In order to reduce the amount of cross-referencing necessary for the reader, several of the references and even the essence of the text will be repeated in related chapters.

1 Our Heritage in Air Transport

The Early Days of Powered Flight

This chapter is by no means a complete history of air transport. However, a brief understanding of air transport history and its heritage will help understand and appreciate the development and the application of today's aviation human factors.

Powered flight, as we know it, is a product of the 20th century. Most historians agree that it began with the historic flight of those two mechanical geniuses, Wilbur and Orville Wright.[1] Their epic flight took place at Kitty Hawk, North Carolina on 17 December 1903 and it has forever changed our world.

The first real pioneer in aviation was Leonardo da Vinci. His drawings were based on the way birds fly. Over 500 years ago he recognized, among other things, the importance of redundancy in air safety. His early drawings show duplicate wires for lift so that if one broke the other would hold (J. Lederer, personal communication, 2 February 1996). All biplanes still have lift wires in duplicate.

A New Undertaking

Over the years, men continued to envy the birds and to think about aviation. At the time of the Wrights, many others were also actively working on this new undertaking. Men such as Alberto Santos-Dumont and Gabriel Voisin of France, Sir George Cayley and Sir Hiram Maxim of England, Otto Lilienthal of Germany, Percy Pilcher of Scotland, Lawrence Hargrave of Australia, and Octave Chanute, Glenn Curtiss, and Samuel Langley in the US, all made major efforts to achieve powered flight. Each came close and each made valuable

---

[1] The Wrights were indeed geniuses. In addition to building the first airplane capable of powered flight, they built their aircraft engine in only 12 weeks, they were the first to conceive of propellers as rotary wings, they built the first aircraft simulator, they built the first flight data recorder (recording distance through the air, time, and engine rpm), and in 1913 they patented an automatic stabilizer that contained an angle-of-attack assessment vane.

1

contributions to a new and developing science.

Otto Lilienthal and George Cayley deserve special mention. In the early part of the 19th century, Cayley laid the foundation of modern aerodynamics and clearly envisioned a practical airplane. Unfortunately, he was unable to provide a suitable power plant (Moolman, 1980). The pioneering work of Lilienthal in early gliders had a big influence on the Wright Brothers. Wilbur Wright called him the 'greatest of the precursors'. Among other things, Lilienthal introduced the concept of crash survival. He firmly believed that gliding was a preliminary to mastering powered flight and met his death on 9 August 1896 in one of his many gliding experiments. Unfortunately, on the day he was killed, Lilienthal was not using the willow hoop (prellbügel) around his body which he had devised to absorb the energy of a possible crash (J. Lederer, personal communication, 2 February 1996 and Moolman, 1980).

The Wright's innovative world-shaking flight did not create an instant industry. There were few true believers, and they were scattered throughout the world. Most of the early faithful firmly believed that the primary use for this primitive device, called an airplane, would be to carry the mail and their efforts were mainly to get exploratory contracts from their governments. Initial attempts to prove that airmail was both feasible and desirable were made in Great Britain, France, Italy, Japan, Australia, and in the United States. However, Great Britain has the distinction of being the first nation to officially sanction an airmail trial. In February 1911, it staged a flight containing 6,500 letters and cards over a five-mile route at an Industrial and Agricultural Exhibition in Allahabad, India (Jackson, 1982).

The airplane's military uses in World War I provided considerable impetus to the growing interest in aviation because early airplanes were an effective supplement to the military ground forces. They were used successfully for observation and with limited success for the bombing of railroads, cities and other targets. Most of the bombs were dropped by hand. At this time people, especially investors, did not see an expanded civil use for the airplane. The aviation visionaries still believed that the primary civil use of an airplane was to carry the mail. European countries led the way in the carriage of mail both on the continent and internationally.

The First Scheduled Night Flight

In the US, an early requirement to get sufficient support and financing for this fledging industry was to prove that pilots could routinely fly at night. The first pilot to do this officially was Jack Knight who, on 23 February 1921, flew through the night from North Platte, Nebraska to Chicago. His flight was a crucial step towards completing the first scheduled continuous coast-to-coast airmail flight. Details of the flight indicate how trying flying was in those early times.

Knight had flown from Omaha to Cheyenne the day before his famous night flight. He then deadheaded (flew as a passenger) to North Platte where he was scheduled to fly back to Omaha. The incoming mail flight had a mechanical problem and it was 10:44 p.m. before the mechanics could ready the plane for the flight back to Omaha. A tired Jack Knight landed in Omaha at 1:10 a.m. the next morning.

In Omaha there was a crisis. The pilot, who was to continue the flight, was still in Chicago because the weather was judged unflyable. In Omaha, there was no one else to fly. Jack Knight, who not only had never before flown the route from Omaha to Chicago but was also fatigued, volunteered to continue. He borrowed a road map covering the Omaha to Chicago route from the Omaha station manager and using it flew on to make aviation history.

A raging snowstorm just east of Des Moines created one of Knight's most critical periods. The snowstorm was critical because Knight needed to find a set of railroad tracks to show him the way to his refueling stop at Iowa City and snow obliterated the railroad tracks.[2] Bonfires, at specific sites had been used to identify other cities enroute, but bonfires were of little help in a blinding snowstorm. Knight used his dead reckoning skills, and perhaps a bit of luck, to finally find the airport at Iowa City.

He arrived in Chicago at 8:40 a.m., the morning of 23 February 1921, after literally having flown all night. The flight made him an instant hero and one of aviation's immortals. A new pilot took over and the epoch flight continued until it reached Hazelhurst Field in New York 33 hours and 20 minutes after it had left San Francisco. Although this time seems lengthy by today's standards, in 1921 it was 65 hours (almost three days) faster than the fastest train. An enthusiastic Otto Praeger, the Second Assistant Postmaster, proclaimed that the coast-to-coast flight was a conclusive 'demonstration of the entire feasibility of commercial night flying'. The flight was a momentous step in civil aviation. Congress was so moved that the next day it approved the Air Mail Service's $1.25 million appropriations bill by a nearly 2-to-1 vote

---

[2] The early airmail pilots often followed railroad tracks to find their way. One of the landing lights in the early mail planes was slanted down to make it easier to follow a railroad at night.

(Jackson, 1982). This appropriation provided a much-needed monetary transfusion for the early US Air Mail Service and kept air transport alive in the United States.

Meanwhile, the Europeans pioneered airmail routes to Africa and South America as well as within Europe itself. Aviation progress was much more sporadic in the US where the Post Office led by Assistant Postmaster General Otto Praeger and the National Advisory Committee for Aeronautics,[3] vigorously advocated increased carriage of mail by air. World War I was over and the Congress, despite its passage of the Air Mail Services appropriation bill in early 1922, was still highly skeptical. The fledgling airlines needed more money, and Congress controlled the purse strings. In those early days passenger traffic was almost nonexistent. Airmail contracts with the government provided the only reliable money source.

US geography helped the new entrepreneurs. The sprawling and stretched-out United States provided a golden opportunity for the continued development of domestic airmail. Over 2,500 miles separated the principal cities and financial centers of the East and West Coasts. Even the fastest trains took nearly four days to span the continent. Reliable airmail could significantly reduce that time.

The 'Lighted Highway in the Sky'

Stimulated by Jack Knight's flight and led by Otto Praeger the Post Office made a major effort to prove that scheduled night mail flights were truly feasible. In 1922 an engineer named Joseph Magee was hired to study the problem of night airmail. After about a year of study, Magee's recommendation was for the Post Office to construct a system of beacons and emergency landing fields between Chicago and Cheyenne. This was the critical night stretch for a proposed transcontinental airway.

Magee planned major terminals at Chicago, Iowa City, Omaha, North Platte, and Cheyenne. Each of them would have a 36-inch revolving light mounted on a 50-foot tower. These lights would have revolving beams that swept the horizon three times a minute. Each would be visible for 100 miles on a clear night. In addition, emergency fields were recommended. These

---

[3] The National Advisory Committee for Aeronautics was founded in 1915 as an independent government body interested in aviation. It was later absorbed by the National Aeronautics and Space Administration (NASA). NASA was established in 1958.

were to be placed approximately 25 miles apart and each had an 18-inch beacon located on a 50-foot tower. The 902 mile illuminated airway that resulted was marked at three-mile intervals by gas lights that flashed 150 times per minute. The main airports were equipped with flood lights and field boundary lights. Landing lights were installed in the wings of the rebuilt British de Havilland D.H.4s, which were the airmail airplanes of that day.

There was nothing like the Post Office's lighted airway anywhere in the world. It made the airway a 'lighted highway in the sky'. This was a monumental technical achievement, and it made transcontinental airmail flights possible. The Post Office's lighted airways were completed under the regime of Col. Paul Henderson, who succeeded Otto Praeger. Henderson was not an aviation enthusiast. Earlier (1919) he had proclaimed that, '(Airmail was) an impractical sort of fad, and had no place in the serious job of postal transportation.'

When reviewing this era, the role of the Post Office, and of people like Otto Praeger, in pioneering the use of airway beacons should not be underestimated. This was true pioneering, made much more difficult because support from the congressional politicians in Washington was lukewarm at best. The US Army had an experimental lighted airway between Columbus and Dayton, Ohio which was used sporadically and the French had experimented with beacons to guide night flights but that was all.

The First Instrument Flight

The fledging industry still had its teething problems. During this formative period it became essential for pilots to learn to fly in clouds and in bad weather. They had to demonstrate that they could fly solely by reference to their instruments and a few of them were unable to do this. While the ability to fly solely by reference to instruments is commonplace among pilots today, it was not commonplace during the early years. Some pilots were unable to acquire these now necessary skills. The ability to adapt to a changing environment became an absolute requirement for pilots at a very early stage in this young industry. The need for flexibility and the ability to adapt has remained to this day.

The feasibility of flying solely by instruments, when the cockpit was entirely enveloped by clouds or fog, was proven by Lieutenant James H. Doolittle of the United States Air Corps. On 24 September 1929 he took off and landed at Mitchell Field, Long Island after flying in an entirely hooded cockpit over a measured course. Lieutenant Doolittle was guided only by his instruments, which included three innovations in aircraft instrumentation; a Kollsman precision altimeter, a Sperry Gyrocompass, and a Sperry artificial horizon. He also used special radio receivers. His flight, which lasted only 15 minutes, proved that flight without outside visual reference, was possible. It followed an intense year of research financed by the Daniel Guggenheim Fund for the Promotion of Aeronautics. The flight provided a positive milestone for the struggling airlines and Doolittle's airplane is now in the National Air and Space Museum.

The Fledgling Industry Makes Meaningful Progress

Many people believe that the world's first regularly scheduled passenger airline—the St. Petersburg-Tampa Line—had its inaugural run on 1 January 1914 (Jackson, 1982).

Figure 1.1   The First Airline
Source: *Illustrated Encyclopedia of Propeller Airplanes*, page 2, Gunston, 1980

Unfortunately, after a promising beginning, the brand new airline folded at the close of the spring tourist season and was not revived. At approximately the same time the airline SCADTA was flying its own airmail and passenger service between two large cities in Columbia, South America. SCADTA flew seaplanes up a river that cut the time between the cities from days on land to a few hours for the airplane flight. Some people claim that SCADTA actually flew the first scheduled passenger flights.

The Chosen Instrument

It seems somewhat ironic that meaningful progress for the fledging aviation industry did not happen in the United States but in Europe. This in spite of the fact that the US was the home of the first powered flight and has good claims that it had the first scheduled passenger airline. World War I had devastated much of the European rail network because the railroads had been ruined by artillery fire and by sporadic bombing. This laid waste to many heavily traveled land routes. Several European States also had routes to distant colonies. Combinations of rail and passenger ships took substantially more time than did air transport. With the War over in Europe, there was abundance of surplus aircraft and of trained pilots who needed jobs. It was a natural for aviation development.

The British led the trend with aviation advances during this period. Political leaders in Britain were convinced that the Empire should link its distant colonial outposts by air. In a far-reaching government decree, Great Britain combined all of its fledging and struggling airlines into 'the chosen instrument of the state for the development of air transport on a commercial basis'. They did this in 1924. The result was soon the world-renowned Imperial Airways that later became British Overseas Airways Corporation (BOAC). Much later, BOAC merged with British European Airways (BEA) and became the British Airways (BA) of today.

Other European Developments

Meanwhile France's Lignes Aériennes Latécoère was flying the mail across the Pyrenees to Barcelona, North Africa and eventually to South America where its pioneering continued. The airline changed its name to Aéropostale, which after consolidations similar to the consolidations being made in Great Britain, became a part of Air France. One of the early pilots of Aéropostale was the gifted writer, Antoine de Saint-Exupéry.

Other European countries were also active. KLM, the legendary and still highly successful airline, was founded in 1919. Ten years later KLM inaugurated the then longest airline route in the world—Amsterdam to far-away Djakarta.[4] The trip took 12 days and had a total flight time of 89 hours. Essentially the same development process was happening in Germany. The

---

[4] Djakarta was the capital of Indonesia, formerly known as the Dutch Indies and now known as Java. The 'D' in Djakarta has been lost and the city is now known as simply Jakarta.

postwar German government avidly promoted aviation and in 1926 consolidated all German airlines into a single airline—Luft Hansa—later known simply as Lufthansa. By the middle 1930s the Germans, while competing directly with Air France and Britain's Imperial Airways, had the largest commercial aviation system in Europe and were actively involved in South America.

Australia's QANTAS[5] pioneered routes from Australia to the Middle East and Europe. QANTAS was founded in 1921 and has some claim to being the second oldest airline in the world (Donaldson, 1997). QANTAS later operated from Australia to Europe in flying boats and recently celebrated its 76th anniversary. When QANTAS started their flying boat operation, the first two overnight stays on the flight to Europe were still in Australian territory. In the early days of the flying boats, the captain was responsible for virtually everything. He paid the accounts for fuel and hotel accommodation for crews and also for passengers. Passengers stayed at the same hotels as the crew on flights that often took a week or more.

There are conflicting claims regarding the order in which airlines were founded and in which they became operational. For example, Columbia's Avianca, which also claims to be the second-oldest airline in the world, states it began operating in 1919 and that the airline has been operating continuously from that date to the present. The important thing to remember is simply that air transport was a new and innovative method of transportation and that pioneering attempts to fly passengers and mail were made in many countries during these formative years.

The Early Pilots

This was a period when airmail pilots, like most pilots of that time, were a special breed. They were in every sense of the word, 'daredevils of the air'. These were the days of white scarves and goggles. It was the beginning of an era that continued through the early development of the airlines. Antoine Saint-Exupéry memorialized France's early pilots and entrepreneurs—men like Didier Daurat, Pierre Latécoère, Jean Mermoz, Henri Guillaumet and others who pioneered in Africa and South America. George Holt-Thomas, Sir William Sefton Branker, Bill Lawford, Jerry Shaw, Gordon Oley, and Sir Alan Cobham pioneered for Great Britain. Albert Presman was the guiding light for the early

---

[5] The acronym QANTAS stands for Queensland and Northern Territories Aerial Services.

KLM. The US had such pilots as Lincoln Beachey, Farr Nutter, Jack Knight, Charles Lindbergh, Bud Gurney, E. Hamilton (Ham) Lee, Max Miller, Randolph Page, William (Wild Bill) Hopson, and many others who earned their rightful place in aviation history.

One of the many stories of the early airmail period is of 'Wild Bill' Hopson, who once needed to get to New York for romantic reasons, but unfortunately was in Bellefonte, Pennsylvania. This didn't for a minute stop the enterprising Hopson. He simply hitched a ride with the pilot of the mail ship that was going through and because the mail compartment was full, 'got onto the wing, lay up against the fuselage, and held onto the guy wires all the way' to New York.

The same kinds of exploration and expansion were happening in Germany, France and in far away Australia. Germany was a major force on both the European continent and in South America. France pioneered in both Africa and in South America. For pilots, this was a worldwide era that has never been duplicated. A possible exception is in the time and lives of the World War II test pilots, who were portrayed so vividly in Tom Wolfe's *The Right Stuff*.

Rigid Airships and the Zeppelins

These early years were also the period of rigid airships and the glorious and eventually tragic Zeppelins. The Germans developed a rigid airship around the turn of the century and these very early aircraft behemoths were called Zeppelins. They were named after their inventor, a brigadier of cavalry, Count Ferdinand von Zeppelin. His name became virtually synonymous with all rigid aircraft of that era. While there were other rigid airships, for their relatively short life the Zeppelins were in an aviation class by themselves.

Zeppelins were the first truly commercial aircraft. They transported passengers for years before airplanes carried anyone but their crews. They proved their value in wartime by being responsible for the first aerial bombardment and bombed London, Paris, and other European cities during World War I. Later they were the first to cross the Atlantic Ocean and were the first aircraft to cross it against the prevailing winds.

In 1929, a Zeppelin circumnavigated the globe in 21 days. One of the most famous of these airships, the Graf Zeppelin, was completed in 1928. One year later it flew non-stop from Friedrichshaven to Tokyo, a distance of 6,980 miles. Prior to her retirement in 1937, the Graf Zeppelin flew over one million miles and carried 13,110 passengers without a mishap.

The tragic demise of these magnificent aircraft occurred in just 32 seconds in the horror of the fire of the Hindenburg at Lakehurst, New Jersey on 6 June 1937. The problem with these extraordinary aircraft was that they depended upon very flammable hydrogen gas to provide their lift capability. Helium, the only practical non-flammable lighter-than-air gas, was available only in the United States and not available in Germany. It took 32 seconds for fire to destroy the Hindenburg and a flourishing and growing industry. A reexamination of that accident is now suggesting that the painted external surface of the dirigible was first to catch fire and was actually responsible for this catastrophe. The paint on the Hindenburg was designed to protect it from the sunlight on its trips to and from Europe and had not been tested or used before on any other German dirigible.

In those early days, rigid airships were also being developed in other countries. One of these was 'Norge' which was built in Italy. Its Captain, Umberto Nobile, carried the Norwegian explorer Roald Amundsen across the frigid Arctic and the North Pole. The Norge's 3,180-mile arctic crossing in May of 1926 took 70 hours and 40 minutes. It flew from the frozen Kings Bay on the arctic island of Spitzbergen to a point near Point Barrow, Alaska. An American Navy Lieutenant, Richard E. Byrd, claimed to have reached the North Pole on a Fokker ski plane a few days before Amundsen. There is still controversy regarding that claim, although it has been generally accepted. Some maintain that Byrd's navigation must have been faulty because they believe the Fokker lacked the range required to have made the round trip, and that therefore Amundsen and the Norge were the first to cross the North Pole. Today, the dispute is of interest mainly to historians for airline trips from several countries routinely fly over the North Pole if it happens to be the shortest or the most desirable route for them.

The Airmail Days

During the 1920s and during much of the 1930s, America's struggling airlines barely survived by carrying the mail. The Air Mail Service Act of 1922 just kept them alive. These early airlines received another transfusion of money when Congress passed the Kelly Act. The Kelly Act, or Air Mail Act of 1925, instructed the Post Office to issue contracts to private operators to carry the mail transcontinentally. This gave the fledgling airlines the beginning of stability and permitted them to maintain at least the appearance of solvency.

Figure 1.2   The Graf Zeppelin

Their bread ticket was always a mail contract. Competition was limited but fierce.

On the bright side, passenger traffic gradually increased and far-seeing investors began to believe that the real future of the industry lay in developing a sound passenger business. By 1928, passenger traffic in the US had risen to 60,000 people. In 1929, it had risen to over 160,000. This surpassed even the well-established German airlines and came close to topping the total passenger counts of all other European countries combined. In the euphoric days leading up to the stockmarket crash of 1929 and the depression that followed, aviation stocks became one of the darlings of Wall Street and stock prices soared. Fortunes were made during this period. Meanwhile the major airlines were gradually brought under the control of the larger aviation companies (Allen, 1981) and affairs for the infant industry were finally beginning to look up.

Walter Folger Brown and the Watres-McNary Act of 1930

Herbert Clark Hoover was inaugurated as the 31st president of the United States in March of 1929 and shortly thereafter appointed Walter Folger Brown as his incoming postmaster general. This appointment was to have a major effect on the airlines of the United States. Postmaster Brown's duties included the duty to 'encourage commercial aviation…and to contract for the Air Mail Service'. It was a responsibility he took very seriously. Following the passage of the Watres-McNary Act on 29 April 1930, Brown became the virtual czar of the airlines.

One of the most important provisions of the Watres-McNary Act (which had been drafted by Brown) was a provision which empowered the postmaster general to bypass low bids for competitive mail contracts and instead allow them to the 'lowest responsible bidder' (Allen, 1981). The reason for that rather unusual language was that in the prior period it had been the practice of some of the more unscrupulous airlines to bid for mail contracts at a ridiculously low rate. Then, after being awarded the mail contract on the basis of the unrealistic bid, the slightly unscrupulous but successful airline would petition to the government for an adjustment and then be awarded a much higher rate.

As it evolved, Brown's basic plan was to have three transcontinental airlines and a fourth airline flying up and down the East coast (Allen, 1991). United was one of the airlines with its route from New York through Chicago to San Francisco. Another airline, Transcontinental and Western Airlines (TWA), was to fly from New York to Los Angeles through such cities as Pittsburgh and St. Louis. The third, American Airlines, was to take a southern route and fly from Washington, DC to Los Angeles via Atlanta, Oklahoma City and Dallas. The North-South airline was Eastern, which would fly up and down the East Coast between Miami, New York, and Boston.

Brown firmly believed that each of these routes should be flown by a large well-financed airline (close to the antithesis of today's deregulation). No longer would the mail routes be flown by an unstable amalgam of existing or newly formed carriers with connecting flights. His basic plan seemed to work. In spite of the fact that mail contracts were not necessarily awarded to the lowest bidder, mail costs to the Post Office, which had averaged $1.10 per mile in 1929 were lowered to 54 cents by 1933 (Allen, 1981).

Cancellation of the Mail Contracts

A major disruption to the developing industry occurred in 1934. It culminated in a disastrous decision by the newly-elected president, Franklin D. Roosevelt, to have the Army Air Corps fly the air mail while a major congressional investigation of the government's handling of the air mail contracts was being conducted.

It was a disastrous decision because the inadequately trained Army pilots were forced to fly without the training or navigational aids that had become standard among the airlines. They were compelled to fly over unfamiliar routes and also fly in some of the worst winter weather in years. Not surprisingly, almost immediately the inadequately trained and equipped pilots had a series of calamitous crashes. Five pilots were killed and six more were critically injured by the end of the first week. By the end of the first five weeks, twelve Army pilots were dead. The situation simply was not acceptable and new bids were called out. In the interim, the Army Air Corps flew the mail only in the daytime.

When things finally settled down, manufacturing and transportation operations were separated, and two new airlines had joined the then big four of American, United, TWA, and Eastern. The two new airlines were Braniff airlines, led by Thomas Braniff, which was awarded the lucrative Chicago-Dallas route, and Delta, which was led by C. E. Woolman. Delta was awarded a contract to fly mail from Charleston, South Carolina to Dallas and Ft. Worth. Today Braniff has disappeared through bankruptcy, and Delta is a major domestic carrier and a significant international airline.

The Airlines come into their Own

The middle and late 1930s provided two major innovations for the struggling airlines. The first innovation occurred when United Airlines added a registered nurse (called a stewardess) to each aircrew, the second innovation occurred when Donald Douglas designed and manufactured the DC-3.

The registered nurse stewardesses played an important role in increasing passenger acceptance of transportation by air. They were introduced by United Airlines on 18 May 1930 and soon became standard on airlines throughout the world. Stewardesses provided passengers with much needed reassurance because of their medical training. They also provided 'tender loving care' to often apprehensive passengers by taking care of a wide variety of passenger needs.

The DC-3 was a twin-engined airliner carrying 21 passengers. It featured a hydraulically operated retractable landing gear, improved shock absorbers, and adjustable propellers of the most advanced design available. Its strength and innovative aerodynamic design (which made it stable and much easier to fly than its predecessor DC-2) made the DC-3 the workhorse of the air and the standard for world airlines for the next two decades.

World War II reinforced the preeminence of the DC-3 in world air transport. The DC-3 provided yeoman service for the US, Canadian, British, and Russian Air Forces. A total of more than 11,000 were manufactured before this venerable airplane was superseded in the US by the 4-engined DC-4s. The DC-4s were followed by the pressurized DC-6s and Constellations, and later the DC-7s and the Super-Constellations.

International Airlines in the Early Years

Great Britain, Germany, France, the Netherlands, and Australia, as well as other nations were anxious to ensure that other countries did not obtain significant commercial advantages in the new industry. These States were also interested in taking full advantage of this new way to speed up the transportation of mail and passengers. In 1928, Great Britain pioneered with the first of the air transport flying boats. This was the Short Bros. Ltd. three-engined Calcutta. Britain's Imperial Airways used the Calcutta to open routes south to Africa and southeast towards India and Australasia. Multi-engine airplanes were needed to fly over hostile areas. Seaplanes were needed to fly over long stretches of water and to land in areas with no airfields. Australia continued to develop air routes in the other side of the world.

As early as 1926 the Germans flew what they claim were the first night passenger flights. By the early 1930s, their Zeppelins were flying around the world and flying regularly to South America. France's Lignes Aériennes Latécoère explored the West Coast of Africa and both the East and West coasts of South America. Its exploits have been well chronicled by Antoine de Saint-Exupéry.

The Germans also experimented with plane and ship combinations. They used a particularly innovative method of extending the range of their seaplanes. The extended range was needed to fly between Europe and South America. The German's technique was to land the seaplane astern of a specially equipped and strategically placed supply ship. The supply ship would then hoist the flying boat aboard, refuel it, and then catapult it into the air so the flying boat could continue its journey (Figure 1.3).

This was also a period in which the Europeans unsuccessfully experimented with very large airplanes. One of the first of these was the Italian Ca 60 (Figure 1.4). The Ca 60 was a gigantic eight-engined, nine-winged flying boat, which unfortunately crashed on its second test flight and was not rebuilt. Somewhat more successful was Germany's Dornier Do-X (Figure 1.5). The Do-X was a huge flying boat with twelve opposed engines (six with forward propellers and six with pushers). It was designed to carry 70 passengers

Figure 1.3   A Dornier 18 being Launched from a Ship at Sea
Source: Reproduced by permission of Lufthansa

Figure 1.4   The Ca 60
Source: The Museo Aeronautico Caproni di Taliedo

Figure 1.5   The Dornier Do-X

Source: *The Illustrated Encyclopedia of Propeller Airplanes,* page 64,
Jackson, 1982

more than 1,000 miles without refueling. Despite the fact that it did make one transatlantic flight to New York in 1931, the Do-X was never a commercial success. It was placed in a Berlin museum and finally destroyed by an air raid in World War II.

Juan Trippe and Pan American World Airways

Any review, albeit brief, of the heritage left by those who formed the airline industry in the United States must include something of the extraordinary history of Pan American World Airways (Pan Am). Its golden days—the days of its glorious 'Clippers'—were in the mid–1930s. Its demise in 1981, for economic reasons following the tragic bomb accident over Lockerbie, Scotland, caused nostalgic regret in a great number of aviation aficionados throughout the world. In 1996, the Pan American name and its famous logo were purchased in bankruptcy court by a start-up airline that has since gone into bankruptcy for itself and is now just reemerging.

Juan Trippe, whose name became synonymous with Pan American, was the son of a New York investment banker. He became a Navy bomber pilot and was entranced with aviation. He graduated from Yale University and in 1923 started with seven war-surplus Navy seaplanes that were being auctioned off at the Philadelphia Navy Yard. These he eventually turned into an airline that girdled the world, that represented the United States abroad and that became virtually an arm of the US State Department.

Trippe's original charter operation failed in 1924 but, with the help of influential friends, he managed to get a contract under the Kelly Act to fly the mail between New York and Boston. He was already thinking ahead. He left his original airline and with what has been called 'characteristic foresight' obtained permission from Cuban President Gerardo Machado to personally have exclusive landing rights as well as other concessions for flying into Cuba. Next, with wealthy friends, he formed a new company titled the Aviation Corporation of America and bought out the airlines that also wanted to fly into Cuba and later into South America. It is of some interest that one of the three companies that was also interested in this venture was formed by four Army Air Corps officers who were concerned about the potential threat to the Panama Canal of a German-run airline that operated in Columbia under the acronym SCADTA. One of the principal officers of this group was Major Henry H. (Hap) Arnold, then an intelligence officer, who later would lead the US Army Airforces in World War II (Allen, 1981).

The Kelly Foreign Air Mail Act and Pan Am

A crucial piece of legislation for Pan American was the Kelly Foreign Air Mail Act which was passed on 8 March 1928.[6] Trippe was a prime lobbyist for his airline during this period and actually helped Congressman Kelly write the bill. A critical clause empowered the Postmaster General to decide (irrespective of the bids submitted) to 'perform the services required to the best advantage of the government'.

Political thinking in Washington favored the concept of a 'chosen instrument' for international flying and Pan American World Airways was a natural choice. The rationale was that this was the only way that the US could compete with such state-backed foreign airlines as Britain's Imperial Airways, France's Aéropostale, the Netherlands's KLM, or Germany's Lufthansa. The 'chosen instrument' remained a basic part of US thinking until after World War II.

The early success of Trippe was phenomenal. In just three years he transformed one 90-mile route into an international airline that flew 20,308

---

[6] The Kelly Foreign Air Mail Act was a separate Act which supplemented the original Kelly Air Mail Act that had been passed in 1922. Representative Claude Kelly of Pennsylvania was the principal writer of both Acts.

miles in 20 countries. In terms of route miles, Pan American Airways was the largest airline in the world. Trippe accomplished all of this by the time he was 31 years old.

Lindbergh and Pan Am

In 1929, Trippe performed one of his masterstrokes. He secured the services of America's latest hero, Charles Lindbergh, to become Pan Am's Technical Advisor. While Lindbergh did not work exclusively for Pan Am, he was a technical advisor to the airline for three decades. During that period he surveyed the 2,000-mile Cuba–Panama route, both the East and West Coasts of South America, and the limits of the Pacific. Hiring Lindbergh was both a very sound technical move and a great public relations gesture. Lindbergh, who became an international celebrity when he became the first pilot to fly across the Atlantic, provided invaluable services. He was closely associated with the early pioneering of Pan Am.

Igor Sikorsky and the 'Pan American Clippers'

It soon became apparent that Pan Am needed a larger and amphibious airplane in order to fly its contemplated routes. To meet this need Trippe turned again to the brilliant Russian expatriate, Igor Sikorsky. Sikorsky had designed and built the first of the famous Pan American 'Clippers', the S-40s. Then to meet the need for a larger airplane, Pan-American, in 1932, issued specifications for a new flying boat. Sikorsky responded with the S-42, which made its first test flight two years later and in August of 1934 made its inaugural flight for Pan American.

The S-42 was an extraordinary airplane for its day. Sikorsky's S-42s were four-engined flying boats that held 50 passengers in a walnut paneled cabin and were the epitome of aircraft luxury. A basic requirement was that they be flying boats so that they would not be dependent upon airports. Airports were non-existent in many of the cities that Pan American wished to serve.

The longest and probably the most difficult route that Trippe contemplated was the route across the Pacific to Asia. Because international politics prevented the route to Asia over Alaska, Japan, and China which Lindbergh had recommended, the only other way to get there was to island hop all the way from San Francisco. Honolulu was no problem but the availability of bases stopped there. Pan Am built modern bases at Midway, Wake, and Guam, and later at Kingman Reef and Pago Pago. Still later it built bases at Canton Island

and in New Caledonia. The latter were needed for service to New Zealand and Australia. Construction crews were forced to erect radio masts, build maintenance facilities and housing for crews, and build specialized hotels and terminals for passengers. As vividly shown in Figure 1.6, it was frequently necessary to dynamite channels through the coral reefs that often surrounded the Pacific islands so that Pan Am's flying boats could land. It took five months and five tons of dynamite just to clear the coral heads and provide a safe landing area at Wake Island (Allen, 1981).

Figure 1.6   Dynamiting the Lagoon at Wake Island

A very real problem for the S-42s was their relatively limited range of only 1,200 miles. Juan Trippe's answer was to order three M-130s from the Glen L. Martin Company. These three seaplanes were named the 'China Clipper', the 'Philippine Clipper', and the 'Hawaii Clipper'. They had a range of 3,200 miles with a full passenger load of 41 passengers so the distance between San Francisco and Honolulu was never a problem for them. Their passenger interiors had the furnishings of a small luxury hotel. The S-42s were used primarily on South American routes, which had lesser distance requirements than the pacific routes.

The 1930s were a truly dynamic period. The Martin 130s quickly surpassed the S-42s, and then only three years later, the 130s were surpassed by the Boeing 314—the true queen of the skies. The Boeing-314 (Figure 1.7) was the finest flying boat ever produced and the largest commercial transport to fly until the arrival of the jumbo jets 30 years later. The Boeing-314 carried a crew of twelve, had sleeping quarters for the crew, and had so large a wing that a mechanic, who was part of the crew, could service the engines in flight. Its range was 3,500 miles, its cruising speed was 193 miles per hour, and its double decks seated 74 passengers or 40 passengers with berths. Pan American ordered six of the first B-314s, among them the famous 'Dixie, American, and Yankee Clippers'. It later ordered six of an improved version—the B-

Figure 1.7   A Pan American B-314 Lifting off San Francisco Bay

314A. Britain's BOAC, which was a principal rival, ordered three B-314s in order to compete on the prestigious North Atlantic routes.

World War II and the Post-War Years

World War II made several changes in airline history. Even before US direct involvement, many airline pilots, who had learned to fly in either the Army Air Corps (later the Air Force), the Navy, or the Marine Air Corps and retained their reserve commissions, were called back to active duty. Large numbers of the airlines' airplanes were requisitioned. Eventually national defense needs took approximately one-half of the airlines' airplanes and one-half of their pilots.

Because all transport airplanes manufactured during this period were immediately assigned to one of the Armed Services, there was no way for the airlines to replace their requisitioned fleet or to acquire new airplanes. After the Armistice, airplanes gradually became available to the airlines, new four–engined transports were built, and pilots returned to their civilian jobs. There was a substantial supply of trained pilots available for expansion.

Several innovative changes were made in the air transport industry following World War II. One of these was a growing recognition of the importance of human factors in air transport operations. First efforts were limited to revised selection procedures and very sporadic attempts to use this new science in the design, layout, and interpretation of displays and controls. Much of what is now known as modern human factors, and the subject of much of this book, happened years later.

This was a period of substantial evolutionary progress. Important, and mostly technological changes, saw the development of improved VHF radio navigation aids and communications, and the manufacture of such four-engined airplanes as the DC-4, the pressurized DC-6s, DC-7s, and the Lockheed Constellations.

By 1960, the demise of propeller airplanes on the major carriers began and the jet age, which began with the introduction of BOAC's ill-fated Comets, was upon us. The Comet's first flight was in 1949, its first scheduled flight was in 1952, and by 1953 there were firm orders for 50 airplanes with negotiations for an additional 100 in process. Unfortunately, the early Comets had a series of tragic accidents and they were voluntarily grounded. In the meantime the Boeing 707 was being designed and built. It had the big advantage of being able to benefit from the structural problems of the first Comets. In keeping with the finest traditions of air transport, information of the Comet's problems was made available by de Havilland to all others as soon as it was known. The B-707's first scheduled flight was on 10 December 1959 and it became the first of the truly successful commercial jets. In a major technological advance, the first supersonic jet—the British-French Concorde—flew its first scheduled flight in 1976. The Concorde has been operating in scheduled service for over 23 years without a passenger fatality, and has flown over 3,000,000 passengers supersonically.

In the US we now have deregulation—a phenomena which is gradually spreading worldwide. Manufacturers are building airplanes that carry up to 500 passengers, have 700 to 800 passenger airplanes on the drawing board, and are even discussing very large transports that will carry up to 1,000 passengers. In most parts of the world, capacity on both the ground and in the air are real problems. Expanded supersonic air travel is very much in our future. All of this has not been without a price. Since 1981, in the US alone 123 scheduled, charter, or all-cargo airlines have ceased to operate (Lampl et al., 1996). There can be little doubt that air transport is still an extraordinarily dynamic industry.

References

Allen, Oliver E. (1981). *The Airline Builders*, Time-Life Books, Chicago, Illinois.

Botting, Douglas. (1980). *The Giant Airships*, Time-Life Books, Chicago, Illinois.

Donaldson, Eric (1997). 'Aviation Medicine in Australia and New Zealand', *Aviation Space and Environmental Medicine*, May 1997, Aerospace Medical Association, Alexandria, Virginia.

Gilbert, James (1970). *The Great Planes*, Grosset & Dunlap and The Ridge Press, New York.

Gunston, Bill, ed. (1980). *The Illustrated Encyclopedia of Propeller Airliners*, Phoebus Publishing Co., London, England.

Jackson, Donald Dale (1982). *Flying the Mail*, Time-Life Books, Chicago, Illinois.

Josephy, Alvin M., Jr., ed. (1962). *The American Heritage History of Flight*, American Heritage Publishing Co., Simon & Schuster, New York.

Lampl, Richard and the eds. of Aerospace Daily and Aviation Daily (1996). *The Aviation and Aerospace Almanac*, McGraw Hill, New York.

Moolman, Valerie (1980). *The Road To Kitty Hawk*, Time-Life Books, Chicago, Illinois.

Nevin, David. (1980). *The Pathfinders*, Time-Life Books. Chicago, Illinois.

Prendergast, Curtis (1981). *The First Aviators*, Time-Life Books, Chicago, Illinois.

Saint-Exupéry, Antoine de (1932a). *Night Flight*, Harcourt Brace Jovanovich, Orlando, Florida.

Saint-Exupéry, Antoine de (1932b). *Wind, Sand and Stars*, Harcourt Brace Jovanovich, Orlando, Florida.

Saint-Exupéry, Antoine de (1942). *Flight to Arras*, Harcourt Brace Jovanovich, Orlando, Florida.
Saint-Exupéry, Antoine de (1976). *Southern Mail, Night Flight*, Penguin Books, Harmondsworth, England.
Serling, Robert J. (1982). *The Jet Age*, Time-Life Books, Chicago, Illinois.
Tryckare, Tre. (1970). *The Lore of Flight*, Cagner & Co., Gothenburg, Sweden.

2 The Industry and its Safety Record

Air Transport is a Dynamic Industry

The industry's very impressive growth and safety record did not happen overnight. Chapter 1—Our Heritage in Air Transport—revealed that the air transport industry was a struggling industry during its early and middle years. However, in an evolutionary and sometimes erratic process, the industry grew and operational safety continued to increase. The following excerpt from a paper given by Ray Gerber, a veteran Pan American pilot and pilot's representative, was reprinted in the July 1980 issue of the United Airlines' flight operations publication, *The COCKPIT.* It gives a picture of the industry's progress and helps develop a perspective that can be easily missed.

> The views expressed here reflect 35 years of airline flying as a crewmember. I have had considerable opportunity to observe and participate in the whole flight crew experience - its benefits, glories, frustrations, disappointments, achievements, and yes, its disasters. During that time I have been privileged, as many have, to be a part of the birth and startling growth of the airline industry from the days of 85-mph flying boats to the 1,400-mph supersonic all-weather era; from the 12-man operating crew to the three-man complement. Yes, you read me correctly-12 flight operations crewmembers. We marched in full uniform, double file, from the terminal, down the ramp, and aboard our flying boat in full view of the passengers and spectators, and marched off again at termination. Some of my first copilot training was in 'close-order drill'. Unbelievable today, and just one outward indication of the tremendous changes over the years in flight crew duties, attitudes and performance. (Note: Two-person flight crews are now the standard.)
>
> Even more startling has been the technological explosion which produced the superbly engineered aircraft we now fly. When I was hired in 1942, three Sikorsky S-42 flying boats were part of my airline's fleet. They were the survivors of 15 purchased less than 10 years before. Such an attrition rate would be unthinkable today. After transition from the Constellation to the DC-6, I was so impressed by the improved engine reliability-not a failure in the first six months-that I checked back in my log book and found I had experienced 42 engine failures in five years on the 'Connie'. That's over eight per year or almost one

for every month I actually flew. In 10 years on the B-707, I had only two shutdowns; one of those was a precautionary, not a failure. In seven years on the B-747, I had just one. That is progress. Add the improvements in ATC, airports, instrumentation, etc., and you have a steadily improving over-all safety record achieved in the past few decades. (Gerber, 1980)

Air Transport Today

The Major Airlines

In 1993, the world travel and tourism industry, of which air transport is a major part, had a cash flow of 3.5 billion US dollars per year. It is the largest industry in the world (Davis, 1993). International Civil Aviation Organization (ICAO) figures show that the domestic and international airlines of its contracting States carried over 1 billion 203 million passengers in 1994.

It is easy to forget how large the air transport industry really is. Each day in 1994 there were 5,800 airplanes in the US Air Traffic Control System (ATC) at all times. In 1995 US airlines carried over 543 million passengers, more than 495 million of them domestically and nearly 48 million on international flights. They made over 7 billion takeoffs and landings during that period. In the fourth quarter of 1995, US airline domestic revenue passenger miles (RPMs) rose by 2 billion over the same quarter one year ago. Available seat miles rose by 1.5 billion (Lampl, 1997). On 24 July 1996, Carol Hallett, President and Chief Executive Officer of the Air Transport Association (ATA), stated in a television interview with Jim Lehrer on the Lehrer News Report that US airlines were making 22,000 takeoffs and landings every day.

The industry's astounding growth is continuing. In 1996, US scheduled airlines surpassed all previous years in the number of hours flown, in flight hours, and in departures. In 1996, they logged 12.9 million flight hours, flew more than 5.4 billion miles and made 8.2 million departures. All of these were aviation records.

Demonstrating the extraordinary capability and flexibility of modern transport airplanes, Malaysia Airlines, flying the first Trent-powered B-777, set the record for the longest flight by an airliner—10,823 nautical miles non-stop from Seattle to Kuala Lumpur, Malaysia. Flying time was 21 hours and 15 minutes and the year was 1997. The former record was made in an Airbus A-340, which flew the 10,267 nautical miles from Auckland to Paris in 1993.

Air freight is also a big and growing business for both the primarily passenger airlines and for those that carry only cargo. This is a worldwide phenomenon. In 1997, the two largest US air freight operators, Federal Express (Fed Ex) and United Parcel Service (UPS), employed nearly 500,000 personnel worldwide, ran 196,000 vehicles, delivered 15 million documents and parcels daily, and had combined revenues of about $34 billion. Fed Ex has a fleet of more than 500 aircraft—including over 130 widebodies. New acquisitions consisting of 70 Boeing B-757Fs and 22 B767-300Fs make up just under half the UPS fleet. The rest of its heavy jet operations consist of B-727s, B-747s and DC8-70s. Every day and night the two carriers deploy nearly 1,000 aircraft ranging from Cessna Caravans to Boeing B-747s, and are well-established as among the largest airlines in the world.

From another perspective, US airlines hired 10,600 new pilots in 1996 and nearly 12,000 pilots in 1997. The 14 major US airlines were expected to hire more than 12,000 in 1998. These numbers are consistent with estimates made by both the University Aviation Association (UAA) and Aviation Resources Inc. The UAA estimates that an average of 8,000 pilots and 3,000 maintenance technicians will be hired by US carriers through 2010. British Airways will need to replace 1,500 pilots in the next ten years. It has been estimated that 20% of the 52,000 airline pilots in the US (10,400) will reach the mandatory retirement age of sixty in the next six years and that 47% (24,440) of them will reach retirement age by 2010. Of 51,853 jet pilots, 5,736 will retire at age 60 by 2000.

It is important to remember that hiring will not be limited to those categories. While the numbers sound high, one should remember both that the airlines are still growing and that every year large numbers of employees of all categories are retiring because of age. Those retiring have to be replaced by new employees. The distribution of US airline employees is shown in Table 2.1.

ICAO statistics show that in 1997 the balance sheets of the international scheduled airlines of ICAO's Contracting States showed total assets of over 277 billion US dollars. In 1994 these same airlines flew over 1 billion 203 million passengers. Three hundred and four million of these passengers were on international flights. ICAO also reported that these airlines employed a total of 305,253 individuals, including 20,709 pilots and copilots, 3,235 other cockpit personnel and over 42,000 flight attendants. The International Air Transport Association (IATA) has reported that by mid-1997, the worldwide passenger turbojet fleet included more than 10,300 aircraft. Seven manufacturers in various parts of the world are offering or planning to offer 24 turbojet aircraft in 47 seat sizes. Air transport is a large, dynamic, worldwide industry.

Table 2.1   US Scheduled Airline Employees in 1995

Pilots and Copilots	55,389
Other Flight Personnel	8,571
Flight Attendants	86,670
Mechanics	50,455
Aircraft and Traffic Service Personnel	251,056
Office Employees	41,851
All Other	52,995
Total Employees	**546,987**

Source: *The Aviation & Aerospace Almanac*, page 148, Lampl, 1997

Total operating revenues for US domestic and international carriers in 1995 were over 94 billion dollars. This figure does not include supplemental air carriers, commuters and air taxis. The Air Transport Association (ATA) reported that 11 US airlines had annual revenues of over one billion dollars. The revenues of 30 other airlines were between $100 million and one billion dollars. In addition, 54 regional airlines had annual revenues of up to $100 million (Lampl, 1997). If we look at other statistics, we find that in 1994, US certificated air carriers consumed 16,827,415 thousand gallons of fuel at a cost of 6.5 billion dollars. In 1995 the 95 airlines that were members of the Air Transport Association (ATA) employed a total of 546,987 employees and their total assets were 89,781 million dollars. In 1995, a major US airline (United Airlines) served 104 domestic and 39 international airports in 30 countries and 3 territories. Its United Express partners served 191 airports. Its total revenues were 14.9 billion dollars.

There is a wide variety in individual airlines. The larger US airlines operate from 400 to over 600 airplanes, have from 45,000 to over 85,000 employees, have over 8,000 pilots, and as many as 19,000 flight attendants. Regional airlines are different and, while some of them are quite large, others are relatively small.

Regional Airlines

Throughout the world, regional airlines are becoming more and more an integral and strategically important part of major airline networks. They are growing rapidly. Pressurized turboprops and pure jets are replacing the older piston-engine airplanes, and the industry has gone from the traditional 19-

seater to larger, faster, technology-driven aircraft. There is no reason not to expect the regional airlines to continue their impressive growth.

The Regional Airline Association reported that in the US there were 109 regional carriers in 1996. They served 782 airports, had 4.44 million departures, and averaged 566,311 revenue passengers per carrier. These 109 regional airlines had 61.8 million revenue passengers. They flew these passengers 12.2 billion revenue passenger miles, and had a healthy average load factor of 53.29 percent. In 1995, regional airlines operated 2,138 aircraft. Forty-six of its top regional carriers had code sharing[1] agreements with major airlines and they carried 96% of the industry's passengers. Regional airlines constitute a large and important industry. Regional airlines in Europe are also big business. In 1994, a regional airliner takes off or lands somewhere in Europe every ten seconds.[2]

Safety is Paramount

Air safety has been a prime consideration for the industry since almost the beginning of air transportation. It has been a prime consideration in the eyes of the regulators, the manufacturers, the operators, the line pilots, and certainly in the eyes of the public. The regulators, manufacturers, operators, and the line pilots are the four principal operating elements of the industry. Each has done its best to minimize the risk of an air transport accident. Unfortunately, the risk of a tragic accident can never be completely eliminated.

One of the more graphic examples of air transport's actual safety was given by Les Lautman when he was Safety Manager for the Boeing Commercial Airplane Company. He stated: 'If you were born on an airliner in the US in this decade and never got off you would encounter your first fatal accident when you were 2300 years of age and you still would have a 29% chance of being one of the survivors....' (Lautman, 1989). David Vinson, while he was Administrator of the FAA put it another way. He told the attendees of the ALPA Annual Air Safety Forum in August of 1996 that: 'you'd have to fly one flight a day for 27,000 years to be assured of being a fatal aviation statistic in the United States' (Vinson, 1996). This is safe transportation. There is not much doubt that the industry is doing a lot of things right.

---

[1] Codesharing is a process where an airline, known as the marketing airline, offers and sells services operated by another airline using the marketing airlines code. It is a simple and effective marketing tool.

[2] Flight International, 1-7 November 1995.

This chapter continues with a brief discussion of air transport safety because of the paramount importance of safety in all aspects of air transport operation. Chapter 20—The Worldwide Safety Challenge— discusses the subject of safety in considerably greater detail and Chapter 21—Current Safety Problems—discusses current safety problems.

The Safety Record

One of the difficulties in objectively looking at worldwide, regional, or even single country safety records is that while a great deal of information is available, it is difficult to keep such information in perspective. Much of the information can be misleading. For example, world airline statistics invariably lump large airlines and smaller airlines together and also lump together airlines from well-developed countries and those from developing countries. These differences can be important factors in a specific airline's safety record. Individual airlines can operate under different operating environments, under a very wide spectrum of different operating conditions, and can have quite different experience levels. A major US airline's annual operations can exceed the entire operational history of the airlines of many other countries.

Most developed countries do a reasonably good job of recording safety information. However there still are variations in as basic a concept as to what constitutes a reportable accident. ICAO has set official recording requirements, but until 1993 notification of accidents to ICAO was a recommended practice only, not even an ICAO standard. Compliance with the recommended practice was far from uniform and unfortunately, ICAO has no means to enforce its standards. ICAO worldwide data collected before 1994 is invariably incomplete. Accident/incident information in North America, in Great Britain, in most of Europe, and in some other developed countries is more accessible than one finds in several other parts of the world.

Airline statistics are frequently given in terms of airplane or passenger seat miles or of airplane or passenger seat hours flown. These are meaningful statistics for some purposes, but have little to do with airline safety. Benjamin Howard, an early airmail pilot, airline pilot, test pilot, and aviation safety expert, stated it well several years ago when he wrote, 'the risks associated with flying are some 95 percent the result of the flight having been made and 5 percent the result of the flight time and distance involved'. Fatal accidents or the number of fatalities are frequently cited by the media and are perceived by the traveling public as particularly significant. Two basic problems with the number of fatal accidents or the number of fatalities criterion are that these numbers are influenced by the size of the airplane and the type of operation and for fatalities, the particular load factor on the ill-fated day in which the accident occurred. In many accidents, survivability can be very much a matter of chance.

Some experts in airline safety believe that Total Losses and Major Partial Losses from the insurance industry furnish the most useful and reliable data that is available (Woodhouse and Woodhouse). Both measures are scrupulously defined in the insurance industry, but even then they are not always strictly comparable. Repair costs vary across airplane types and geographic regions, especially with major partial losses. ICAO asks for 'Substantial Damage' but many States neither investigate nor report 'Substantial Damage' accidents.

Frequently, efforts are made to create lists of those airlines with the 'best and the worst accident rates'. The differences between airlines in those lists are often statistically meaningless. Outside of the problem of defining what constitutes an accident, it is a basic principle that for a difference to be meaningful, the difference must be statistically significant. There is no single criterion that does not create problems. The important point is to be aware of other criteria and also to be aware of both the strengths and the limitations of any method of analysis being used.

World Air Transport Safety Today

Worldwide, there were 57 domestic and international fatal accidents with 1,840 fatalities in 1996 (*Flight International, 15-21 January 1997*). This is significantly greater than the previous decade's yearly average of 44 accidents and 1,084 fatalities for the previous decade. It again points up the problem that is immediately raised whenever the number of departures and passengers is increased, while essentially the same rate of fatal accidents per million departures is maintained. Under these conditions, the actual number of accidents and the number of passengers killed are bound to increase. When that happens, the politicians, the popular media, and the traveling public make it very clear that safety must be improved. Simply maintaining the existing very good safety record is not good enough.

Figure 2.1, which compares the Primary Cause Figures of Hull Loss Accidents for the Worldwide Jet Fleet from 1959 to 1996 with the Primary Causes for the last ten years, shows that the primary causes of hull losses have changed very little over the years. The investigating authorities have listed

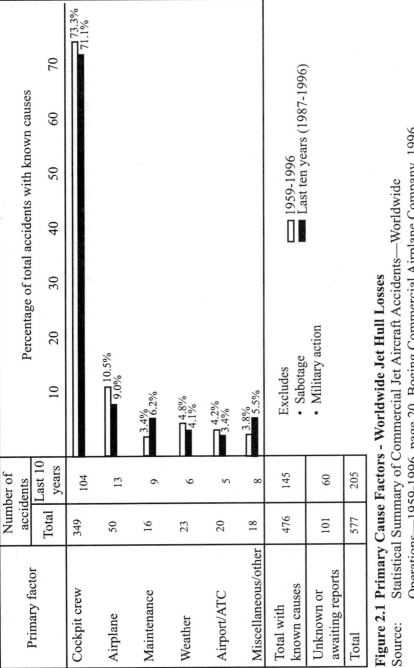

Figure 2.1 Primary Cause Factors - Worldwide Jet Hull Losses

Source: Statistical Summary of Commercial Jet Aircraft Accidents—Worldwide
Operations—1959-1996, page 20, Boeing Commercial Airplane Company, 1996

something associated with the flight crew as an overwhelming factor in each of these periods. The figure attributed to the flight crew is something in the area of 70%. Bruggink's study (1997) shows that at least in the US, this may be changing. Unfortunately, all of the underlying reasons for these tragic instances were seldom fully considered. The underlying reasons can range from factors such as the crew's failure to use proper procedures and the adequacy of training and supervision to organizational and regulatory problems and inadequate consideration of cultural and of basic inter- and intra-crew language differences. Much of the remainder of this book discusses factors involved in these difficult areas.

Figure 2.2 shows Hull Loss Accidents for the Worldwide Commercial Jet Fleet By Generic Group from 1959 (the beginning of the jet era) through 1996. It clearly shows that when new airplanes are introduced, they have had 'shake-down' problems, but also that they result in an increase in air safety.

US operators do a preponderance of jet air transport flying, and the US record has always been better than the non-US record if all non-US airlines are considered in one category. This is a statistic that should be viewed very

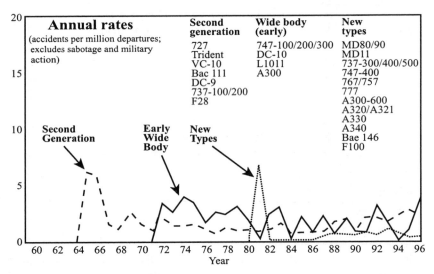

Figure 2.2   Hull Loss Accidents—Worldwide Commercial Jet Fleet by Generic Group

Source: Adapted from *Statistical Summary of Commercial Jet Aircraft Accidents—Worldwide Operations—1959-1996*, page 22, Boeing Commercial Airplane Company, 1996

carefully. It is really not fair to many non-US airlines for several have records that equal, or even exceed, the safety records of US airlines.

Generally speaking, the US has a significantly more advanced infrastructure than do many (but by no means all) of the countries that are included in the non-US operator group. Another reason that US transport operations generally have better records is that the US airlines are frequently more technologically advanced than those in the developing countries. Increased technology almost always means increased safety wherever it is found. Final advantages of the US airlines are that they operate in a relatively advanced regulatory and social environment It is a mistake not to recognize that the same statements could be made of several airlines of other developed countries.

It should always be remembered that many of the graphs, charts and figures we reproduce were designed for a US audience, and that many other advanced countries would look equally good if the graphs, charts, and figures had been designed for them. We know of no case, with the exception of departures and passengers enplaned, in which the Australians do not lead the rest of the world in safety. They have continually led the world's airlines, including US airlines, since the early propeller days. This is a considerable achievement, but the relatively small exposure of the Australian airlines is also needed to keep those comparisons in perspective.

Overall, between 1959 and 1995, the industry had 577 hull losses, with 154 of them occurring to US operators and 423 occurring to non-US operators. If we look at the 10-year period from 1987 to 1996, there were 205 world jet fleet hull losses. Of the 205, 41 jet fleet hull losses occurred to US operators and 164 to non-US operators. Aircraft manufactured in the Commonwealth of Independent States (CIS), the former USSR, are not included because complete operational data is not available for them although, as is indicated in Appendix A, that may be changing.

The graph in Figure 2.3 shows all non-US operators lumped into a single category. As we have suggested, it by no means tells us that all non-US operators have relatively poor records. Comparing the safety record of US airlines with the rest of the world should be done very carefully. However, as good as the over-all airline safety record really is, and as good as the safety record of an individual airline may be, we should always remember that all airlines have had close calls. The difference between an incident in which everyone escaped 'scot-free' and a tragic fatal accident is often little more than chance. There is absolutely no reason for any airline or any country to become com-

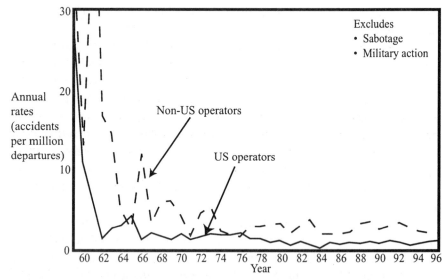

Figure 2.3   All Hull Loss Accidents in US and Non-US Worldwide Commercial Jet Fleet

Source: *Statistical Summary of Commercial Jet Aircraft Accidents— Worldwide Operations—1959-1996*, page 19, Boeing Commercial Airplane Company, 1996

placent about the safety of their operations in spite of sometimes superb individual operating records.

Safety in the United States

One of the low points in the safety record occurred in the first years that the Post Office started to carry the mail. Thirty-one of the 40 pilots hired in 1920 were killed in the first two years (Jerome Lederer, personal communication). Things have been getting better ever since. Looking at an early 50 years of experience, Figure 2.4 shows the fatal accident rates of US scheduled carriers per million departures from 1930 to 1990. The chart shows at least four things. First, while the early years were pretty rocky, the safety record's improvement trend was already well established in the propeller era and the improvement has continued.

Secondly, and unfortunately, while the safety record continued to improve from year to year, the curve of fatal accident rates per million departures became essentially asymptotic about the mid-seventies and it has continued to be essentially asymptotic ever since. While further improving the record has become increasingly difficult, the continued growth of the industry makes further improvement a requirement. This is a major challenge. If the fatal accidents per million departures stays essentially the same and if the industry continues to grow as forecast, the total number of fatal accidents is bound to increase to unacceptable levels.

A third point Figure 2.4 shows us is that during their first five years the jet transports, which then did not have a significant impact on the total commercial accident picture because they were just being introduced, had an accident rate per million departures that was considerably higher than that of the combined fleet. It took almost a decade for the jets to reach the safety rates that the well-established propeller transports had achieved. Unfortunately, it is very difficult to eliminate a 'break-in period' for radically new equipment.

A fourth point that Figure 2.4 illustrates is something one should always keep in mind. There can be fairly large year-to-year fluctuations in the yearly accident rate. Headlines telling us that a given year is much better or much worse than those of a preceding year should be carefully evaluated.

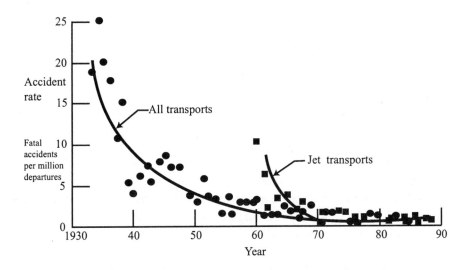

Figure 2.4   Fatal Accident Rates—US Scheduled Carriers
Source: *The Air Safety Story: Where We've Been and Where We're Headed*, figure 1, Lautman, 1989

When it is realized that a single large US airline makes more than 2,000 takeoffs and landings every day, again one is given the strong impression that a lot of people are doing things right. At least one of these airlines, which made over ten million departures during this period, has not had a flight crew related accident for over 16 years (Orlady, 1992). Other airlines have similar records. Until the unfortunate Cove Neck, New York mid-air explosion in 1996, Trans World Airlines (TWA) had not had a passenger fatality for 21 years, except for those involved in a single terrorism incident. While we should always remember that QANTAS, the Australian International Airline, has not had a passenger fatality in the entire history of its jet operations, we should also remember that the annual experience of the larger US airlines exceeds the total jet experience of QANTAS and of many other non-US airlines.

In the US, considerable concern about US air safety trends was raised in 1996 by two quite different but devastating crashes. The first was the May crash of a ValuJet DC-9 which crashed shortly after takeoff from Miami killing 110 people. The cause of that crash was an uncontrollable fire that started in a cargo compartment. The major cause of that crash seemed to have little to do with the flight crew or of the flight operations of the fatal flight itself.

However, during the investigation considerable concern was raised regarding the carriage of potentially dangerous cargo and the regulatory supervision of airlines that were created after deregulation, essentially the start-up airlines. The adverse publicity forced the financially strong ValuJet[3] (which up to this point had a good safety record) to merge with another airline. It had to give up its ValuJet name in hopes that it could erase the stigma caused by an accident that seems to have little to do with the flight operation of the fatal flight. It was a somewhat unusual accident in that it triggered an investigation that was only indirectly related to flight operations but nevertheless uncovered a number of other operational problems.

The second highly visible crash happened when a TWA B747-100 exploded off the coast of Long Island, NY in July of 1966 and killed 230 people. It is of interest that the apparent causes of each of these accidents did not involve either of the flight crews. Regardless of the connotations, such data is of little solace to the families of those who lost their lives in these tragedies.

Regional Air Safety Statistics

Safety among the regional airlines has been of concern to many people in spite of the fact that the regional airlines have steadily improved. Measured

---

[3] Actually, the ValuJet crash was the first fatal accident for a start-up airline since deregulation. Deregulation in the US began with the Airline Deregulation Act of 1978 (*US News & World Report*, 17 May 1996).

by fatal accidents, in 1995 the regional airlines had the safest record in their history. Regional airlines went from 0.11 fatal accidents per 100,000 departures in 1990 to 0.04 fatal accidents per 100,000 departures in 1995 (Regional Airline Association (RAA) Annual Report, 1996). The continued improvement of the regional airlines' safety performance will undoubtedly be reinforced by a recent action of the Federal Aviation Administration. In 1995 the FAA issued a final rule that required that the training, aircraft certification, and operating requirements which currently apply to Part 121 carriers (more than 30 seats) will also be applied in the future to any aircraft in scheduled service, that has from 10 to 30 seats. This change in the Federal Air Regulations was strongly supported by the RAA and its member airlines. Regional airline safety statistics from 1991 to 1995 by region are shown in Table 2.2.

Table 2.2   Regional Airline Safety Statistics

Year	Operations in the 48 Contiguous States, Hawaii, Puerto Rico and the U.S. Virgin Islands		Total U.S. Operations	
	Accidents	Fatal Accidents	Accidents	Fatal Accidents
1990	0.28	0.00	0.35	0.11
1991	0.50	0.17	0.68	0.05
1992	0.26	0.13	0.54	0.23
1993	0.34	0.05	0.48	0.11
1994	0.26	0.07	0.30	0.09
1995	0.19	0.02	0.32	0.04

Data source: derived NTSB statistics (Rates per 100,000 departures)

Source: *Annual Report*, page 22, Regional Airline Association, 1996

Critical statements regarding regional air safety statistics (like any others) should be considered carefully and critically. Walter Coleman, Regional Airlines Association President, in replying to a critic who stated that regional and commuter airlines are less safe to fly than the majors stated: '(the assertion) was not supported by the facts. In 1966, there were no fatal accidents involving larger regional aircraft, and the smaller aircraft had the lowest fatal accident rate ever' (Asker, 1997).

The European Regional Airlines Association says that concerns in the US about commuter safety are not relevant in Europe. It further states that, in spite of having different regulations, when the statistics are examined on a per cycle rather than on a per hour basis, European regional airline accident rates are comparable to those of the major European carriers. The US Regional

Airlines Association makes essentially the same point about US carriers.

Safety in the Rest of the World

While reviewing the 1995 safety record, *Flight International's* Annual Air Safety Review noted: that 'Once again Africa and South/Central America performed consistently the least well of all the world's regions' (Learmount, 1996). North America experienced a relatively large number of accidents in 1995, but it also had a small number of fatalities per accident. This was because most 1995 North American accidents happened to small cargo operators. The single exception was American Airline's disastrous Cali, Columbia crash. The Cali crash was unique. It took 163 lives, involved a B-757, and an airline with a good safety record. The B-757 had a 13 year operational history and had never before been involved in a fatal accident. For these reasons and despite the Cali crash, the 1995 relative safety record of North American airlines is actually much better than might appear. While North America had a relatively large number of accidents, it did so with about one-half of all the world's commercial transports.

1996 was a bad year for world transport operations because it included the largest number of deaths caused by air crashes. Despite this statistic, the average risk level for departures during the entire decade of the 1990s did not worsen. However, there was one more crash in 1996 than had occurred in the previous year and fatalities of 1,840 in 1996 exceeded the highest previous year total. In 1985, eleven years earlier, there had been 1,801 fatalities. Some observers believe that an adverse trend may have started in 1995. Whether or not these figures indicate a meaningful trend, they do rather strongly suggest that the industry has not been able to reduce accident rates fast enough to ensure that both crash numbers and fatalities do not increase as travel grows.

The Safety Record is Not Good Enough

To help keep the safety record in perspective, the US NTSB tells us that there were 42,860 passenger car fatalities in the years 1992 and 1993. During the years 1992 and 1993, there were 34 deaths among the US scheduled airlines, and 45 fatalities among the commuters. For another statistic, the Federal Railroad Administration tells us that there were 4,257 deaths at railroad crossings in the US in 1996. In spite of the fact that these are not exact comparisons because the populations involved may not be comparable and because of differences in exposure and many other variables, there is an obvious striking difference. Air transport is not one of our largest killers of people. Few would believe that the risk to their life is greater while riding in the family automobile than the risk to their life when they were flying on a scheduled airline. A big difference to the public is that an air transport crash kills large numbers of people all at once and acquires a great deal of long-lived publicity.

NTSB studies show that from 1993 through 1996, scheduled US carriers averaged only 0.02 fatal accidents per 100,000 flight hours, less than half the fatal-accident rate in the four-year period only one decade earlier (Carley, 1997). It accomplishes very little to state that scheduled air transportation is the safest of all transportation modes; that the really dangerous part of any scheduled flight is the drive to the airport; and that the daily carnage on our roads makes the consequences of our air crashes comparatively minor.

In the real world much of the public perception of airline safety is based upon absolute figures of world air transport crashes, not on statistical rates. As good as the safety record is in the US and in several other countries, it is not good enough, particularly when considering worldwide figures and the forecast growth of the industry. We should never forget that there are airlines with good safety records in every area of the world, and we should also realize that some areas have airlines with very poor records. Without question, a major objective of the world air transport industry is to improve the safety record of those airlines.

In 1996, the world's airlines killed 1,840 people in 57 accidents (Learmount, 1996). If the present fatality rate continues, a simple straight-line projection gives a fatality rate of something in the area of 3,680 airline deaths per year by 2015. Numbers like that are not acceptable. In the US, the FAA is estimating that by 2015, the industry will have 40 percent more discreet flights than it has today and it will carry 1 billion 200 million passengers per year. This is about double the number the industry carries today (Vinson, 1996).

There are at least three reasons the worldwide safety record is not good enough. The first is the extraordinary attention that any air crash attracts in the public media. Unfortunately, when an air crash occurs, it creates a great deal of publicity and in almost every case kills a lot of innocent people all at once. Any air transport crash in any part of the world is still big news. In the US, an air transport crash gets headline treatment for sometimes months. One reason is the very long time it takes to complete the accident investigation and to determine its primary or probable cause.

The essentially static accident rate per departure in the midst of considerable expansion is the second reason, and is an even more compelling reason that the present safety record is not good enough. The forecast growth of the industry predicts that without a further improvement in accident rates the actual crashes will reach a number that is completely unacceptable—something in the order of a major airline crash somewhere in the world every one to two weeks—depending upon the 'crystal ball' used in making the forecast. It is almost universally agreed that numbers such as these would not be acceptable to the general public. If the public perception is that the airline industry is unsafe because the public is reading weekly accounts of air crashes, the actual safety record is of little consequence.

The third, final, and inexorably overpowering reason that we cannot be satisfied with our present record was given us by the same Benny Howard, who in simple and straight-forward language reminded us that, 'Man's greatest sin is the unnecessary taking of human lives' (Howard, 1954). The industry must continue to work hard to improve.

As we have previously mentioned, accident rates are frequently misunderstood, whether given in terms of the actual number of departures or in terms of millions of passenger miles flown. Figure 2.5 shows us essentially what Benny Howard told us long ago. Of all the hull loss accidents of the world commercial jet fleet, 93.6% occurred in the takeoff, climb, descent, approach, and landing phases of a hypothetical flight of 1.5 hours. This, of course, is not surprising, for it is in these phases of flight that the airplane is close to the ground. It is very difficult to have a terrain accident at 35,000 feet.

In much of the media, safety is equated with the actual number of crashes, regardless of the number of departures or of the millions of passenger miles flown. These measures can be very misleading because if a rate deemed satisfactory remains equal but either departures or passenger miles increase, the actual number of crashes is bound to also increase. As air transportation grows, there will be an increase in the actual number of accidents in spite of the fact that the level of safety for a given flight has not changed an iota.

The general public, led by much of the media, also equates air safety with the number of crashes that occur. Unfortunately, the present record is not good enough, even if the carriers that have poor safety records could be brought up to the records of the airlines with good records. This is primarily because of the growth of the industry. Actually, no operations are so good that they cannot be improved and several of them must. It is clear that very few, if any, accidents are not preventable.

The need to increase the operational safety of the 'Third World' coun-

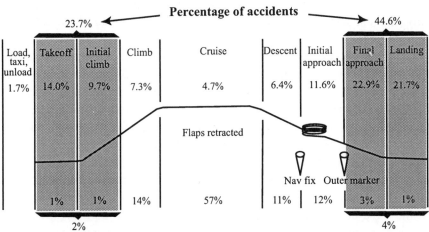

Figure 2.5   Percentage of Flight Involving Hull Loss Accidents

Source: Adapted from *Statistical Summary of Commercial Jet Aircraft Accidents—Worldwide Operations—1959-1996*, page 13, Boeing Commercial Airplane Company, 1996

tries or those countries with so-called 'underdeveloped economies' becomes increasingly apparent. An undeniable fact is that the industry sector, which carries about 12% of the world's air traffic, produced 80% of the fatal accidents in 1996. There is little doubt that future growth projections will be affected by the worldwide safety record.

At the 1995 Flight Safety Foundation's Annual International Airline Safety Seminar while exploring the industry's concern over accidents occurring in any part of the World, Earl Weener, Boeing's chief of systems engineering, provided figures that showed the disparity in safety standards between the world's safest regions and those consistently less safe is not only large, but growing. This is a problem for the entire industry. It is not restricted to only airlines in the regions with the poor economies and poor records.

Annual Results or Short Term Records can be Misleading

Accidents are not programmed by the calendar. They are obviously rare events and the number of aircraft accidents that occur in the short term or annually are too small for meaningful statistical analysis. In all accident data, the source, the completeness of the data, and the definitions used should be considered in context. ICAO tries very hard for consistency in these areas, but is dependent upon the statistics that are furnished to it from its 186 member States. The playing field on accident statistics is not always even.

The difficulty of relying on figures based on relatively short periods was demonstrated in 1995 when worldwide there were 23 fatal accidents in the first six months, but only 155 fatalities. This was because a fairly large number of the accidents (10) occurred with small aircraft on cargo flights. If the record for the first six months for 1995 had continued for the next six months, the industry would have achieved an all time safety record in terms of fatalities (*Flight International*, 'Airline Safety Review', 1996).

The industry was fortunate, for if more of the crashes that occurred in the first six months of 1995 had involved larger passenger airplanes fatalities would have been much higher. The total safety record of that year clearly shows that decreasing the number of fatal accidents during a period of significant growth is a major industry challenge. The air transport safety record, and the essentially asymptotic curve of accidents per million departures since about 1970 reinforces the basic message in no uncertain terms. The world air transport industry must lower the accident rate per departure.

Another difficulty in relying on short-term records is the misconception that can be reached by focusing on a single accident. This is well illustrated with the tragic accident that occurred in July of 1996 to Trans World Airlines. In this crash TWA experienced its first fatal accident in more than two decades. Excluding terrorist action, its fatality record had been completely clean for the preceding 21 years. This B747-100 accident, which killed 230 people on 17 July 1996 and spoiled TWA's very good record, was apparently caused by an explosion in the center fuel tank. The cause of the explosion has not yet been determined even after the most intensive and expensive aircraft accident investigation in US history. The investigation, which has already taken over one year and cost over 30 million dollars, is not complete at the time of this writing. However, the accident still gets considerable media attention. The crashed airplane had completed over 92,000 hours and about 17,000 flight cycles. Knowing that prior to this accident, the world fleet had recorded around 5.5 million flights since the last fatal accident involving passengers and that TWA had not had a passenger fatality (except for one terrorism incident) for 21 years helps keep that accident in perspective (*Flight International,* 'Headlines', 1996).

References

Asker, James R. (1997). 'Washington Outlook', *Aviation Week & Space Technology*, 7 April 1997, McGraw Hill, Inc., New York.

Aviation Week and Space Technology (1996). In 'News Breaks', 29 January 1996, McGraw Hill Companies, Inc., New York.

Aviation Week and Space Technology (1996). In 'News Breaks', 4 March 1996, McGraw Hill Companies, Inc., New York.

Billings, Charles E. (1997). *Aircraft Automation: The Search for a Human-Centered Approach*, Lawrence Erlbaum Associates, Inc., Mahwah, New Jersey.

Boeing Commercial Airplane Group (1996). *Statistical Summary of Commercial Jet Accidents: Worldwide Operations—1959-1996*, Boeing Commercial Airplane Group, Seattle, Washington.

Bruggink, Gerard, B. (1997). 'A Changing Accident Pattern', *Air Line Pilot*, May 1997, The Air Line Pilots Association, Herndon, Virginia.

Carley, William M. (1997). 'Final Approach, Landing Procedures Are Tied to Air Crashes: Safety Push is Begun', *The Wall Street Journal*, 11 August 1997, Dow Jones & Company, Inc., New York.

Davis, Robert (1993). 'Human Factors in the Global Marketplace', keynote address to the Human Factors and Ergonomic Society, Seattle, Washington.

Federal Aviation Administration (1990). *The National Plan for Aviation Human Factors*, Washington, D.C.

Federal Aviation Administration, (1995). *Administrator's Fact Book*, US Department of Transportation, Washington, D.C.

Flight International (1996). In 'Airline Safety Review', 17-19 January, 1996, Reed Business Publishing, Sutton, United Kingdom.

Flight International (1996). In 'Headlines', 24-30 July 1996, Reed Business Publishing, Sutton, United Kingdom.

Flight International (15-21 January 1997). Reed Business Publishing, Sutton, United Kingdom.

Howard, Benjamin (1954). 'The Attainment of Greater Safety', presented at the 1st Annual ALPA Air Safety Forum, and later reprinted for presentation at the Aircraft Accident Prevention Course, University of Southern California, July 1957.

Lauber, John K. (1991). 'Principles of Human-Centered Automation: Challenge and Overview', Presented at AIAA/NASA/FAA/HFS Conference Challenges in Aviation Human Factors: The National Plan, National Transportation Safety Board, Washington, D.C.

Lautman, Les (1989). *The Air Safety Story: Where We've Been And Where We're Headed*, Boeing Commercial Airplane Group, Seattle.

Lampl, Richard, ex. ed. (1997). *The Aviation & Aerospace Almanac 1997*, McGraw Hill Companies, Inc., New York.

Learmount, David (1996). 'Airline Safety Review, Off Target', *Flight International*, 17 January 1996, Reed Business Publishing, Sutton, United Kingdom.

Maurino, Daniel, E., Reason, James, Johnston, Neil, and Lee, Rob B. (1995). *Beyond Aviation Human Factors*, Avebury Aviation, Aldershot, Hants, England.

Norman, S.D. and Orlady, H.W., eds. (1988). *Flight Deck Automation: Promises and Realities*, p. 126 of final report of a NASA/FAA/Industry Workshop held at Carmel Valley, California.

Orlady, Linda M. (1992). 'C/L/R - How Do We Know It Works?', *Safetyliner*, Volume III, Issue 2, 1992, United Airlines, Chicago, Illinois.

Regional Airline Association Annual Report (1996). Regional Airline Association, Washington, D.C.

Shifrin, Carole A. (1996). 'Aviation Safety Takes Center Stage Worldwide', *Aviation Week & Space Technology*, 4 November 1996, McGraw Hill Companies, Inc., New York.

Thurber, Max (1996). 'Best Year For Pilot Hiring; Big Turnover In Regionals Seen', *Aviation International News*, 1 November 1996, Midland Park, New Jersey.

Vinson, David R. (1996). 'An Air Safety Challenge', presented at the Annual ALPA Air Safety Forum 21 August 1996, Federal Aviation Administration, Washington, D.C.

Woodhouse, Robert and Woodhouse, Rosamund A. (1997). 'Statistical Measures of Safety: An Objective Risk Indicator', *Proceedings of the Ninth International Symposium on Aviation Psychology*, The Ohio State University, Columbus, Ohio.

3 A Brief History of Human Factors and its Development in Aviation

The First Human Factors

There are those who maintain that the first prehistoric man to use a tool was really the founder of human factors thinking. The first US organized effort in this area was the 1898 work of Frederick W. Taylor at the Bethlehem Steel Works. His main concern was the restructuring of an ingot-loading task. Additional milestones were the work of Frank and Lillian Gilbreth in the early 1900s and the often-quoted work done by Hugo Munsterbers at Western Electric's Hawthorne Plant in the early 1920s. Little other early progress in human factors was made during this period with the exception of work done for the US Department of Defense (DOD). The DOD actively supported human factors in those early days and has continued its support of human factors to this day.

Many attempts have been made to achieve a better understanding of the role of humans in the manufacture of goods and services. Weimer (1995) noted that the literature contains at least 74 definitions of that role and includes such terms as human factors, ergonomics, human factors psychology, human factors engineering, applied ergonomics and industrial engineering. Gradually, the two terms 'ergonomics', used primarily in Europe, and 'human factors', used primarily in the United States, have reached the point where now they are considered essentially synonymous and interchangeable. There is still a relatively small school that restricts 'ergonomics' to considerations directly involving the machine or a component. They use the term 'human factors' in a broader sense that also considers the behavioral aspects that are involved in a total system approach. The word, 'ergonomics' (ergon= work and nomos= law), is usually attributed to K.F.H. Murrell, who, in June 1949, hosted a meeting in London which discussed the role of humans in accomplishing work. In the United States, a group of what can be called human factors (or ergonomics) professionals also recognized the need for better organization in this new profession. They held a meeting in 1955 at which they formed the Human Factors Society. In 1993, the Human Factors Society changed their organization title to The Human Factors and Ergonomics Society. This was an attempt: 'to resolve the continuing identification confusion arising from the use of both terms; to recognize the continuing contributions of Europeans and others; and to be recognized as a truly international organization.'

Human Factors in Aviation During and After World War I

Aeromedical physicians played an active role in pilot selection during World War I. Even the relatively simple medical examinations for pilots of that era effectively reduced the pilot initial training failure rate. One result has been that the US air regulations have required periodic physical examinations of pilots since the mid-1920s.

During the early development of aviation, aeromedical physicians found themselves increasingly involved in human factor activities. There were sound reasons for this involvement. The earliest human factors problems recognized in aviation were physiological in nature and it was natural that aviation physicians would become involved in aviation human factors activities. Man was entering a domain that was far different from the earthbound environment in which he had lived since the beginning of time. Problems associated with such factors as acceleration, altitude, vibration, noise, speed, temperature, and fatigue were naturals for the physiology-oriented aviation physician.

World War I provided an opportunity for both physicians and psychologists to demonstrate that they could make a meaningful contribution in the new aviation industry, particularly in selection and training. The result of a coordinated effort by aeromedical physicians and psychologists was responsible for the first organized attempt to deal scientifically with what they considered human factors.

The need for trying to understand the human element was dramatically demonstrated by the British. The British showed that, for every 100 aviators killed at the beginning of World War I, two met their deaths at the hands of the enemy, eight because of defective airplanes, and 90 were killed because of their own individual deficiencies (Air Service Medical, 1919, and Wilmer, 1979). Studies were undertaken on the effectiveness of pilot selection procedures in several countries, including Italy, France, Great Britain, and the United States. While these countries used widely varying criteria to measure effectiveness, the inevitable conclusion was that an overwhelming majority of casualties were due to human failure rather than structural failure or combat (McFarland, 1953). Psychologists of that era maintained that they could significantly improve the situation by better selection. They were partially successful.

In the period between World War I and World War II, most human factor activities concerned automobiles, which were a growing phenomenon (Weimer, 1995). Less prominent human factors activities involved the classic studies conducted at Western Electric's Hawthorne plant and the work performed in Army aviation. The work performed in Army aviation is of particular interest. A great deal of it centered on physiological investigations performed under the leadership of Captain (later General) Harry Armstrong, namesake of the USAF Armstrong Laboratory. In 1940, John Flanagan organized a comprehensive aviation psychology program for the US Army. During World War II, this program was greatly expanded and led to the US Army Air Forces Aviation Psychology Program.

Virtually the only human factors work in civil air transport before World War II was the pioneering work of Dr. Ross McFarland, a Harvard professor of physiology. He was interested in all aspects of aviation and became particularly interested in the effects of decreased oxygen at high altitude. Dr. McFarland also studied the effects of fatigue on pilots who were assigned the long flights that were necessary for Pan American World Airways to open new routes over the Pacific and Atlantic oceans and to Alaska. He became an official advisor to Pan American. His association with Pan American, interrupted by World War II was continued after the cessation of hostilities. In 1953, Dr. McFarland authored what proved to be a classic text—*Human Factors in Air Transportation*. Its publication followed another earlier and equally pioneering book by Dr. McFarland—*Human Factors in Air Transport Design*.

The Effect of World War II on Human Factors in Aviation

The Second World War provided considerable additional impetus to human factors progress in aviation. In England, a Climatic and Working Efficiency Research Unit was established at Oxford University, and at Cambridge, the Applied Psychology Laboratory of the University, under the leadership of Sir Frederick Bartlett, did outstanding work. Among other things the Applied Psychology Laboratory constructed an early simulator which became known as the 'Cambridge Cockpit' (Hawkins, 1987). Experiments in this simulator clearly demonstrated that for optimal flight crew performance the machine should be matched to the characteristics and limitations of the people that were going to fly it. While this point may seem to be simply common sense, it was an important statement at the time. It continues to have application today, especially in light of some of the developments in advanced technology cockpits. Other practical outcomes of the Cambridge program included contributions toward aircrew selection and training, the effects of sleep loss and fatigue, and increased understanding of the principles involving visual perception and instrument display.

In the United States following the War, aviation psychology centers were started at the University of Illinois, The Ohio State University, and shortly thereafter, at the University of Southern California and at several other universities. Today aviation human factors work continues at those aviation psychology centers and is continued routinely by the FAA's CAMI,[1] NASA, the Volpe National Transportation Systems Center, and the Armed Forces. Recently many others, including the Universities of Texas, Miami, Illinois, North Dakota, Embry-Riddle, Purdue and the University of Central Florida have become increasingly active. While the movement toward aviation human factors is more sporadic outside the United States, there is also a growing and considerable interest in many other countries. They have made valuable contributions in aviation human factors.

One of the leaders in the US during the period immediately following World War II was Lt. Col. Paul Fitts. In 1945 he was selected to head the new Psychology Branch of the Aero Medical Laboratory of the US Air Force. It has been a leading center of advanced human factors research and development ever since, although it is, of course, primarily concerned with military aviation. Today, the Psychology Branch is called the 'Human Engineering Division'. It recently celebrated 50 years of continuous research and development for the Air Force. In 1949, Col. Fitts left the US Air Force and headed The Ohio State University's Laboratory of Aviation Psychology (Roscoe et al., 1997).

On January 1946, Alex Williams opened the Aviation Psychology Laboratory at the University of Illinois, and another giant of that period, Alphonse Chapanis, joined the Psychology Department at Johns Hopkins University. At least one of Chapanis' many publications of that era is *Research Techniques in Human Engineering*. It is still very good reading nearly four decades later.

Human Factors After World War II

A large proportion of human factors activity had been limited to considerations involving the decreased air density found at higher altitudes. In the days

---

[1] CAMI is the Civil Aeromedical Institute. It is a branch of the US FAA and is located in Oklahoma City, Oklahoma.

immediately following World War II, aviation human factors also became interested in the placement of cockpit instruments and other kinds of technical advancements.

Later, during this same period, many improvements were made in aviation. They included the development of the 'Basic T' in primary flight instruments. The Basic T principle required that primary instruments for all transport airplanes should consist of an airspeed indicator, an attitude indicator, an altimeter, and direction indicator, all arranged in a standard T formation.

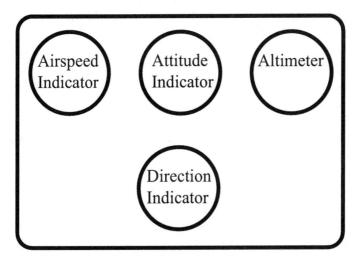

Figure 3.1   Illustration of Basic T

Other innovations that had human factors implications and that followed World War II included such items as: the development of VHF (very high frequency) radios for navigation and voice communication; increasingly sophisticated autopilots which were easier to operate and which later included altitude hold; and the increased acceptance of the use of flight simulators for training. Recently 'glass cockpits' with digital flight and engine instruments, cathode ray tube (CRT) and flat panel displays have made available a plethora of information that represented significant technological advancements. They have created new and challenging human factors problems.

With a history of demonstrated adaptability on the part of airline personnel generally, and of pilots in particular, and with the industry's technical engineering orientation, it is not surprising that human factors in air transport was not given a great deal of attention, particularly during the early years following World War II. This changed slightly when shortly after its introduction, one of the new transports experienced a series of disastrous cabin fires. This was the piston-engine powered and pressurized DC-6. The fires led to a US requirement that a third crewmember be required on all transport airplanes with a gross weight over 80,000 pounds. These fire incidents were responsible for the famous (or infamous) 80,000-lb. rule. The now defunct US Civil Aeronautics Board (CAB), which then had certification responsibilities, promulgated the Rule. The fact that the fires were the result of a freak design problem in which under certain conditions gasoline vapors flowed into a cabin heater air intake and later caught fire did not change the requirement for a third crewmember. Few today would call promulgation of the 80,000-lb. rule an example of the results of an optimal human factors evaluation.

The logic of the 80,000-lb. rule was frequently challenged. Ultimately, all of the parties involved, the manufacturers, the operators, the unions representing the pilots, the union representing the flight engineers, and the regulating authority (the FAA) agreed that simple weight was not a valid criteria for establishment of the need for a third crewmember.

The 80,000-lb. rule was rescinded in 1964 when the FAA promulgated FAR 25.153. It stated that workload would be the primary criteria in establishing crew complement. Appendix D of FAR Part 25 listed six workload functions and ten workload factors that should be considered in determining the workload that would be imposed on the flight crew of any new airplane. These factors are discussed in detail in Chapter 10—Workload.

Physiological Problems in Pressurized Cabins and Early Jet Operations

The introduction of pressurized cabins, and later, of jet engines represented significant engineering advances. They also created many day-to-day operational problems that fully occupied top operating, manufacturing, and regulatory executives. Pressurized cabins required an additional system—the engine bleed air pressurization system—to maintain cabin pressurization. Heating and cooling created additional demands and required additional flexibility. Both heating and cooling came from a new source—the new air conditioning packs that are powered by air bled off from the engines themselves. Pilots were required to know and understand the new systems and the radically different jet engines.

The industry made considerable progress with the physiological problems raised during this period. Today, most of the physiological problems that arose have been identified and, in most instances, seem reasonably well controlled. Problems identified include those associated with the general quality, humidity, and oxygen requirements of increased recirculated cockpit and cabin air and higher radiation and ozone levels at high altitudes. Others include the effect (which can be particularly serious) of ground-based advertising laser beams on crews and passengers, increased circadian arrhythmia because of the increased range of the latest transports, and the fatigue associated with difficult scheduling on both long- and short-haul flights.

The solution to several of these problems remains controversial. While some, like the fatigue problem, are independent of the power plant, it seems clear that many of these problems are exacerbated by jet operations. Medical, air ambulance, and other emergency flights have fatigue and specific problems that are unique to their own operations. Helicopter and short takeoff and landing (STOL) operations have special and important human factors problems of their own.

New and additional normal and emergency operating procedures, including those for emergency descents, were required. Jet engines, while basically simpler than the old piston engines, were larger, faster, and were more fuel critical than their piston predecessors. The jet engines generally also had longer ranges, and with the increased ranges, significantly greater operational flexibility. All of these required new rules and new procedures. The jet era, which had just begun, had its teething problems.

One of the reasons it took so many years to significantly expand human factors' horizons in aviation was that while pilots were always considered to be key players in industry operations, both operators and engineers considered pilots adaptable and individually expendable if they could not adapt. Problems and advances were technological and were believed to be engineering problems. Therefore, it is not surprising that problems were considered in what were then standard engineering terms.

Increased safety and efficiency were clearly the industry's objectives. Technology was believed to be the principle that could lead to increased safety and efficiency. If technical improvements were made, pilots were expected to adapt in order to make the technical improvements work in day-to-day operations. Operators and financiers felt that obvious problems could be dealt simply with common sense. Many did not see any need for an expanded human factors effort in air transport operation. The situation was not helped by many human factors 'experts' who were woefully weak both in understanding operational problems and in understanding day-to-day operations.

Psychological Areas Involving Behavior and Crew Performance

During the years following World War II, and particularly in the later years, aviation human factors and the operational industry became more and more involved with issues related to the psychological areas involving human behavior and performance. This started with the promulgation and then the withdrawal of the 80, 000-lb. rule. In the last two decades, team coordination (both within and without the cockpit), decision-making and flight crew cognitive processes have received increasing emphasis, much of it from the implementation of Crew Resource Management (CRM) programs. The effective use of equipment in all operating environments, the optimum use and design of multi-purpose controls and displays, and the design of advanced cockpit and cabin layouts required, and still require, considerable attention. It is now recognized that other psychological areas, including the effects of social cultures on performance, need a great deal more consideration than they had previously received.

A basic reason for the continued and increased interest in aviation human factors is that despite the overall safety record in air transport, aviation safety experts had become increasingly disturbed with the continuing prominence of the human factor in air transport accidents. It was a problem that could not be ignored. The continuing inability of the industry to do anything effective about this obvious problem was particularly frustrating. This concern was eloquently stated by the late Hugh Gordon-Burge when, more than two decades ago, he addressed IATA's Safety Advisory Committee regarding approach phase accidents with these words:[2]

> When looking at all the approach phase accidents, it is really possible to do no more than to remark once again on their seeming inevitability year after year; on their almost exact similarity year after year; on the airlines apparent inability to prevent their numbers increasing, let alone to reduce them; and on the continuing prominence of the human factor in the causal chain of events. It therefore seems pointless to go on repeating what has been said before....(regarding)...this most pressing—and depressing—matter.

The Impact of the IATA 20th Technical Conference

A watershed in the long process of dealing effectively with this 'most pressing—and depressing—matter' occurred in 1975 when the International

---

[2] Hugh Gordon-Burge as Chairman of the IATA Safety Committee in his Opening Address to the Safety Committee (about 1977).

Air Transportation Association (IATA) devoted its entire 20th Technical Conference to human factors. The Conference was opened with a thought provoking address by IATA's Secretary General, Knut Hammarskjöld. He stated, among other things:[3]

> Analysis of our accident data clearly indicates that our historic trends towards reducing accident rates can only be resumed if we address and solve the problem of why critical human beings fail to fulfill expectations. These accidents fifteen or twenty years ago would have been superficially dismissed as due to 'pilot error' or 'controller error.' We now know beyond a shadow of doubt that these descriptions of accident causes are at best misleading and at worst irresponsible.
>
> The issues we now call 'human factors'...are the critical issues...I believe that at this point in time in the history of civil aviation there is enough knowledge available to launch a serious and effective attack on this — the 'Last Frontier'— of the airline safety problem.

Despite this clear and eloquent language and despite a growing consensus of the importance of human factors among the operators of the world airlines, the Conference was forced to conclude, after five days of sometimes conflicting lectures and discussions, that 'the wider nature of Human Factors and its application to aviation seem still to be relatively little appreciated'. An outstanding exception was KLM, the Royal Dutch Airline, which in 1977 instituted its KHUFAC[4] course. The purpose of the course was to provide a systematic educational program to increase an awareness of human factors among its operational staff.

Human Factors Today

Today, modern technology has raised new challenges. A major change for many has been the change from three-person to two-person cockpit crews regardless of gross weight or number of engines. Other new challenges include increasingly complex airplanes, the increasingly complex environment in which they must be operated, and the modern airplane's embedded computers

---

[3] Knut Hammarskjöld, Secretary-General of IATA in his Opening Address to IATA's 20th Technical Conference, Istanbul, November 10-14, 1975.

[4] The acronym KHUFAC stands for the KLM Human Factors Awareness Course. Additional information regarding it can be found in *Human Factors in Flight*, Second Edition, especially page 331.

and their associated flight deck displays and documentation.

The use of digital instrumentation, which has many advantages over the well-tried and proven analogue instruments, has caused problems for many. Still other human factor challenges involve new documentation that includes maps, charts, and their display on CRTs or LCDs. New display and operational problems have been created by GPWS and the advanced EGPWS, and by TCAS and WSAS. Navigation innovations include such things as GPS, and the concept of 'free flight'. Head Up (HUD) and enhanced vision displays for minimum visibility landings raise additional valid human factors questions. These innovations are discussed in Chapter 22—The Air Transport Future. National boundaries are having less and less meaning to the industry, although the political ramifications, overflights, landing rights, and cabotage still remain. The US has become the major air transport country in the world.

Major efforts are being made to design a new SST (supersonic transport) without the droop nose that is now required for taxi-out, takeoff, approach and landing, and taxi-in. Because of the fuselage angle required for efficient flight, the forward visibility from the cockpit in the Concorde (the only commercial present SST) does not cover the intended flight path during these flight phases without drooping the nose. If the associated problems are solved and the droop nose is not required, pilots would taxi-out, takeoff, approach and land, and taxi-in entirely by instruments and without any actual forward visibility. This would obviously involve a radical change from the operation of present aircraft.

Considerable concern is being evidenced in determining refinements for staff selection, and for the training and checking of, not only pilots, but check pilots and instructors as well as all other personnel involved in airline operations. This concern also includes the selection and training of personnel in the regulatory agencies. A growing consensus is beginning to realize that the social environment of transport operations can no longer be neglected. The social environment can serve as a vital factor in understanding the human factors involved and increasing the safety and efficiency of air transport operations. Recognition of the social environment is perhaps the latest of the many changes in human factors orientation. The social environment is discussed in Chapter 6—The Social Environment.

Today it looks very much as if the entire air transport industry is on the right operational track. Human factors, including its psychologically-based ramifications and the importance of the total social environment, are recognized as a critical part of operations by nearly all segments of the industry. The formerly conventional, and often rather narrow, use of the term 'human factors' can be a critical oversimplification. Thanks to a growing consensus in the industry and helped considerably by the recent leadership exhibited by the International Civil Aviation Organization (ICAO), broadly based human factors is now recognized as a core aviation technology in most progressive parts of the world. It is becoming as much a core technology as are such traditional areas as meteorology, power plants, aerodynamics, navigation, and communications.

Organizations such as IATA and IFALPA fought long battles in forum after forum before progressive parts of the industry were able to influence ICAO leadership and to encourage meaningful action by ICAO. The IATA 20th Technical Conference which was held in 1975 and was devoted to human factors is considered by many to have been a major turning point in reaching full recognition of the importance of human factors in the operation of transport airplanes. The IATA *Airline Guide to Human Factors* (1981) is still a very readable and useful publication.

ICAO is a political institution. The promotion to a position of ICAO leadership in worldwide human factors was actively urged by the Canadian, the French, the British, the Russian, the US and some other State's government representatives. ICAO's leadership could not have been as successful without that political support. It is the only organization that can influence all of the countries involved without raising questions of ulterior motives and sovereignty. ICAO is performing a very useful leadership function and coordination function that could not be handled by any other organization.

The need for increased safety and efficiency and the need for the development of large long-range airplanes in a changing international social order are all important parts of the environment in which today's and tomorrow's airplanes must operate. Economics and operational efficiency have made the two-person crew the new standard crew complement for all future airplanes. One person simply cannot routinely fly today's airplanes alone. There is also the special problem created by a pilot incapacitation. All of this has led to an expanded role for the copilot and the 'team approach' in the operation of these airplanes. The universality and inevitability of human error, as well as recognition that even a captain can make an inadvertent mistake, reinforces these concepts.

Clearly aviation human factors has become multi-disciplinary and requires special and very broad qualifications. Many of these considerations require qualifications well outside the realm of an individual specializing only in medicine or psychology. In fact, David Meister, an experienced human factors practitioner, has stated, 'It is possible, in view of the fragmentation of specialty areas, that there is no such animal as a general purpose HF scientist.' (See Appendix B.)

The human factors field continues to evolve, in part because of contributions from individuals with a wide range of background, experience and education. It has become incumbent upon any individual with operational responsibility to be familiar with human factors and also to recognize when specialized expertise is required. It follows that in those cases it may be important to utilize outside sources to secure the needed knowledge. Manufacturers, operators, and regulators need human factors expertise. Each of them not only have common human factors needs, but also can have needs that are unique to themselves.

Dan Maurino, John Lauber, David Woods, Neil Johnston, Clay Foushee, Jim Reason, Earl Wiener, Bob Helmreich, and Charles Billings are current and long-time contributors to the expansion of the traditional view of aviation human factors. Today, such aviation human factors researchers as Barb Kanki, Judith Orasanu, Nadine Sarter, Asaf Degani, Mica Endsley, and Dick Jensen reinforce their efforts. This is only a partial list and should in no way be considered complete. It is simply an example of the number and variety of people working in aviation human factors who are making a substantial contribution to the never-ending quest for increased air transport safety. In addition to these, the industry has its own operational professionals who work with human and other factors on a day-to-day basis, and whose contributions are, to a considerable extent, responsible for the industry's very fine overall safety record.

The Continuing Prominence of Human Factors

As discussed in Chapter 2—The Industry and its Safety Record, one of the great truths of air transport operations is that air safety, and even the perception of air safety, is a paramount consideration. Human factors has several definitions, among them: 'The study of the physical, psychological, psychosocial, and pathological variables which affect human's performance;'[5] and 'the field of science concerned with the optimization of the relationship between people and the machines they operate through the systematic application of human sciences integrated within the framework of systems

---

[5] *Aerospace Glossary for Human Factors Engineers*, Society of Automotive Engineers (SAE) ARP 4107.

engineering.'[6] Human factors is definitely involved in air transport safety and is one of the reasons for air transport's overall excellent safety record. While human factors is not the only area that should be carefully examined in aircraft accidents, inevitably aviation human factors is directly involved at some level in virtually all of these tragic events.

References

Air Service Medical (1919). US Government Printing Office, Washington, D.C.

Chapanis, A., (1959). *Research Techniques in Human Engineering*, The Johns Hopkins Press, Baltimore, Maryland.

Christensen, J.M. (1987). 'The Human Factors Profession', In *Handbook of Human Factors*, edited by Salvendy, G., John Wiley & Sons, New York.

Hawkins, Frank H. (1993). *Human Factors in Flight*, Second ed. by Orlady, Harry W., Ashgate Publishing Limited, Aldershot, England.

IATA (1981). *Airline Guide to Human Factors*, International Air Transport Association, Montreal, Canada.

Jensen, Richard S. (1997). 'A Treatise on the Boundaries of Aviation Psychology, Human Factors, ADM, and CRM', in program for *The Ninth International Symposium on Aviation Psychology*, Columbus, Ohio.

McFarland, R.A. (1953). *Human Factors in Air Transportation*, McGraw-Hill Book Company, Inc., New York.

McLucas, J.L., Drinkwater, F.J. III, and Leaf, H.W. (1981). *Report of the President's Task Force on Aircraft Crew Complement*, Government Printing Office, Washington, D.C.

Orlady, Harry W. (1995). 'The Evolution of Minimum Crew Certification', in *Proceedings of Eighth International Symposium on Aviation Psychology*, The Aviation Psychology Laboratory, The Ohio State University, Columbus, Ohio.

Society of Automotive Engineers (1988). *Aerospace Glossary for Human Factors Engineers*, ARP 4107, Society of Automation Engineers, Warrendale, Pennsylvania.

Stanley N., Cori, Louis, Carl, and LaRoche, Jean (1997). *Predicting Human Performance*, Helio Press, Pierrefonds, Quebec, Canada.

Weimer, J. (1995). *Research Techniques in Human Engineering*, Prentice Hall PTR, Englewood Cliffs, New Jersey.

Wilmer, W. H. (1979). 'The Early Development of Aviation Medicine in the United States', *Aviation, Space, and Environmental Medicine*, May, 1979, Alexandria, Virginia.

---

[6] Richard S. Jensen, 'A Treatise on the Boundaries of Aviation Psychology, Human Factors, ADM, and CRM'.

4 The Physical Environment and the Physiology of Flight

The External Environment

From the time of the legends of Icarus and Daedalus, people have recognized that the external environment is a critical consideration for flight. The legend tells us that Icarus, who was the son of Daedalus (the master Athenian craftsman and inventor), escaped, with his father, from Minos the king of Crete. They fastened feathered wings to their arms with a 'pliant' wax so they could make their escape by flying away from their Cretan prison. Unfortunately, Icarus disregarded the instructions from his father. He was told to 'wing (his) course along the middle air', but instead flew too close to the sun. The 'pliant' wax, which was holding his feathered wings together, melted. Then as the Croxall translation put it, 'His feathers gone...Down to the sea he tumbled from on high, and found his fate...'

There is another version of that story. A group of British scientists, who were members of the British accident investigation board staff, decided to investigate the legend utilizing all their modern skills and knowledge. Their conclusion was that Daedalus and Ovid had it all wrong. The pliant wax that held Icarus's feathered wings together did not melt because he flew too close to the sun, but failed because Icarus flew too high, the temperature at altitude was too cold, the wax became brittle and cracked and would no longer hold the feathers. Icarus then plunged to his death into the Ionian Sea.

Today in aviation, we are concerned with several aspects of the physical environment. Nearly all of these aspects vary with altitude. The physical combinations and range of atmospheric elements found at sea level are not found at the altitudes that are routine for normal flight. Temperature, atmospheric pressure, vibration and turbulence, noise, ozone concentrations, high altitude ultraviolet radiation, and ionizing radiation are all of concern.

One perhaps tangential aspect of the environment in which we fly that has generated concern recently is the use of ground-based lasers aimed into the sky. These lasers are used primarily for advertising and operate with little regulatory guidance. They are a particular problem during approaches and inadvertently, have caused serious injury to the vision of pilots blinded by the beams. Strictly speaking ground-based lasers are not a part of the natural environment, but they become part of the environment in which pilots fly and are therefore of concern. Efforts are being made to promulgate effective regulatory guidance for ground-based lasers. Unfortunately, designing effective regulatory guidance can be a very slow process. This is particularly true since ground-based lasers are also an international problem and providing effective regulatory guidance is a challenge for virtually every country that has an air transport operation.

The Atmosphere [1]

The earth's atmosphere is divided into many layers or zones. These layers are of interest to pilots primarily because of the temperature and pressure changes that occur. At higher altitudes, there are also problems of increased ozone concentrations and of ultraviolet and ionizing radiation.

The first layer of the atmosphere is the troposphere which according to the International Standard Atmosphere (ISA) begins at sea level and extends from sea level to about 26,000 feet at the poles, to 36,000 feet in temperate latitudes, and to 52,000 feet over the equator. Within the troposphere air cools at the dry adiabatic rate of 5 1/2 degrees Fahrenheit per thousand feet in clear air and at the wet adiabatic rate of 3 degrees per thousand feet within clouds. The tropopause is the transition zone between the troposphere and the next layer, the stratosphere. The tropopause is of interest because the decrease in temperature that occurs as altitude increases in the troposphere ceases there. The temperature remains essentially constant in the stratosphere although with an increase in altitude pressure continues to fall. The ISA assumes that the temperature of the tropopause in the vicinity of the equator will be -56.5 degrees Celsius (-70 degrees Fahrenheit). The height of the tropopause varies with altitude from about five miles at the poles to about eleven miles at the equator. It also varies seasonally.

Air transport flight takes place in the troposphere, the tropopause and in parts of the stratosphere. Above the stratosphere, the temperature increases in the mesosphere. Above the mesosphere are the ionosphere, the Kennelly-Heaviside layer, and the Appleton layer. These latter four layers are of interest only to space travelers.

---

[1] Much of this section is unabashedly based on Green et al.'s *Human Factors for Pilots* and the Second Edition of Hawkins' *Human Factors in Flight*. They are highly recommended texts for any one who wishes more information on basic physiology and its effect on flight.

The air that we breathe and in which we travel has weight. Air pressure is caused by the weight of air from the top of the atmosphere as it presses down upon the layers of air below. For this reason, atmospheric pressure decreases with increasing altitude, as does air density. There is simply less air above. The barometric (or aneroid) altimeter measures air pressure to determine altitude. Relatively small increases in height at low altitude cause considerably greater change in air pressure than an equal change of height at a higher altitude because a given volume of air is less dense and weighs less at altitude.

A very practical consequence of the differences of pressure change with altitude is that a given pressure differential, which air traffic control uses to separate aircraft, gives less actual separation at high altitudes than it does at lower altitudes. Current and well-established ATC (air traffic control) rules have made the 1,000 foot separation the worldwide standard. Currently the ATC system uses 2,000 feet (as measured by conventional barometric altimeters) for altitudes above 29,000 feet. Many believe that this separation figure creates larger than necessary actual separations at these altitudes. This results in a wasteful use of already crowded airspace.

In an attempt to resolve this problem, ICAO has recommended new reduced vertical-separation minima (RVSM) rules. The rules were introduced above the North Atlantic in early 1997, and are planned for introduction later in all parts of the world. Minimum Aircraft System Performance Specifications (MASPS) criteria, requiring digital air data computers and associated systems that enable altitude keeping within 50 feet, are required for operation in RVSM airspace. This would double the airspace available above 29,000 feet by reducing separation levels above that altitude to 1,000 feet determined by pressure differential. This, of course, is less than 1,000 actual feet of separation. The subject is still being debated for worldwide use in part because of the equipment costs that would be required by non-airline aircraft operators. The use of 1,000 feet above 29,000 feet for aircraft separation was approved by an ICAO panel charged with reviewing the general concept of separation and has now been approved by most operating organizations.

The altitude/pressure relationship in the atmosphere is shown in Figure 4.1.

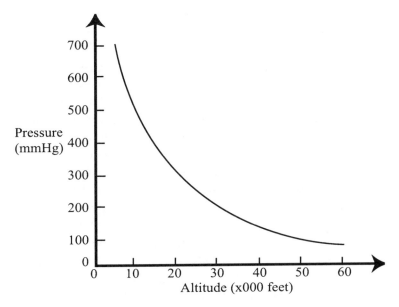

Figure 4.1   Altitude/Pressure Relationship in the Atmosphere

Man's Oxygen Needs and Limitations

In human factors we are concerned with the needs and limitations of man who must routinely travel and operate throughout the limits of the troposphere and in parts of the stratosphere. Air, in a remarkably consistent fashion through all altitudes, is composed of 21% oxygen, 78% nitrogen, and one percent of other gases. The other gases are primarily argon, water vapor, and carbon dioxide. For our purposes, the atmosphere can be considered a mixture of 21% oxygen and 79% nitrogen. Carbon dioxide does have important physiological considerations that are discussed later.

We are particularly concerned with oxygen because every living cell of the body requires oxygen. Oxygen is produced both by the chemical oxidation of food products in the tissues and by the respiratory process which delivers the oxygen to each cell and carries away waste products in the form of carbon dioxide. Respiration (our normal breathing) exchanges oxygen and carbon dioxide between the environment and the tissues. It does this by using the existing pressure differential between the capillary surface of the alveoli (very small sac-like structures in the lungs) and the outside atmosphere.

Oxygen and carbon dioxide waste products are carried within the body by the blood. Unfortunately, the body has only a small store of oxygen so there is a constant demand for replenishment of oxygen. The pressure that forces oxygen into the blood is called oxygen partial pressure. It decreases with altitude, in accordance with Boyle's law, with the normal decrease of atmospheric pressure with altitude. Other laws which deal with the amount of oxygen in the blood are Charles' law, which relates pressure to volume, and Dalton's law, which deals with the relationship between the partial pressure of any component gas of a mixture to the fractional concentration of that gas.

Man's need for oxygen is the main reason that we have pressurization systems for any airplanes that fly above ten or twelve thousand feet. Today's jets maintain the equivalent of an eight thousand foot cockpit and cabin altitude while they are flying at higher altitudes. Supersonic transports, which have much higher cruising altitudes, require higher cabin pressures to maintain the desired equivalent cabin altitudes. The pressurization system of the supersonic Concorde maintains a cabin level of the equivalent of 5,000 feet while at cruise altitude. This is lower than the cabin altitudes usually found in present day jet transports.

Oxygen Balance Within the Body

The body maintains an appropriate oxygen balance through a complicated process. Air, which enters the nose and mouth, passes down the bronchial tree and terminates in the alveoli of the lungs. The alveoli diffuse oxygen across the capillary membranes and into the blood. This process is illustrated in Figure 4.2.

Blood is the transport medium that carries oxygen from the lungs throughout the body by carrying a protein molecule, hemoglobin, that holds oxygen in the blood and carries it to the tissues. In addition to general health reasons, a very good reason for pilots not to smoke in flight is that tobacco smoke has a negative affect on the ability of hemoglobin to carry oxygen. This negative effect is caused because smoking increases the amount of carbon monoxide in the hemoglobin molecule. The hemoglobin molecule has a much greater affinity for carbon monoxide than it does for oxygen. Because the hemoglobin molecule cannot carry both gases at the same time, carbon monoxide is carried instead of oxygen. This leads to an increased susceptibility to hypoxia. Hypoxia is a deficiency of oxygen in the tissues sufficient to impair function or performance. Under certain conditions, the hypoxia can become severe enough to deprive the individual of oxygen. The condition of

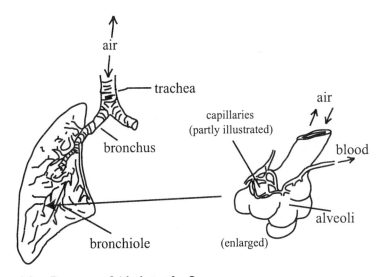

Figure 4.2   Passage of Air into the Lungs
Source: Adapted from *Human Factors for Pilots*, page 5, Green et al., 1991 and used with permission of Ashgate Publishing Limited

no oxygen is called anoxia. Anoxia can lead to permanent damage to a person's brain or to one or more of his/her organs.

To understand better the mechanism that transports oxygen to our tissues, it is helpful to understand at least the rudiments of our circulatory process. A block diagram of that process (Green et al., 1991) is reproduced below in Figure 4.3. A diagram of the heart is shown in Figure 4.4.

Red blood cells, which are richly laden with oxygen-carrying hemoglobin, pass from alveolar capillaries into the pulmonary veins enroute to the left atrium of the heart. It then passes to the left ventricle of the heart. From there, the richly oxygenated blood passes from the left ventricle through the aorta and branches to all parts of the body. Finally, the oxygenated blood reaches capillaries that transfer the oxygen to the tissues.

The capillaries are very thin-walled vessels. In the capillaries, oxygen is diffused down a pressure gradient across the cell walls and is exchanged for carbon dioxide. The now carbon dioxide laden capillary blood is then transferred to the veins. Blood in the venous system is forced back to the heart through the veins and a series of one-way check valves. The purpose of these check valves in the venous system is to prevent wrong direction flow. When the venous blood reaches the right ventricle of the heart, it is pumped to the lungs and the whole process begins again.

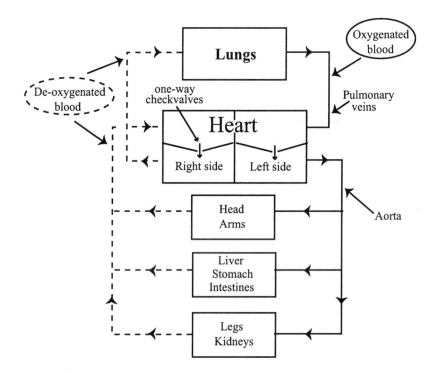

Figure 4.3   Block Diagram of the Human Circulation
Source: Adapted from *Human Factors for Pilots*, page 7, Green et al.,
                1991 and used with permission of Ashgate Publishing Limited

We have already seen how pressure, and therefore the way that the amount of oxygen in the air decreases with altitude. The actual concentration of oxygen remains at about 21% and does not diminish with altitude, but remains as 21% of an ever decreasing quantity of air expressed as diminished air pressure and density. A given volume of air at sea level has greater air pressure and air density than an identical volume of air has at altitude. Because of Boyle's, Charles', and Dalton's laws, the identical volume—still containing 21% oxygen—contains less actual oxygen at altitude than it does at sea level. The body needs a specific amount of O_2.

In order to compensate for the decreased amount of oxygen, oxygen masks can be used to increase the proportion of oxygen in the airmix the pilot breathes until he/she is breathing 100% oxygen. The equivalent of sea level oxygen can be maintained to 33,700 feet if the pilot is on 100% oxygen and reasonable performance can then be maintained until about 40,000 feet. Above

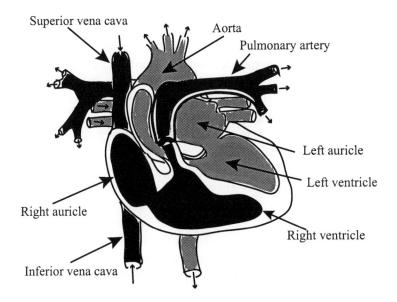

Superior vena cava

Aorta

Pulmonary artery

Left auricle

Left ventricle

Right auricle

Right ventricle

Inferior vena cava

Figure 4.4   The Heart

40,000 feet, pressure breathing, where oxygen is delivered under positive pressure, is required. In order to assure that oxygen is available to flight crews in an emergency decompression, pressure breathing is a requirement for flight crews on supersonic transports because they cruise well-above 40,000 feet.

Pressure breathing is a very tiring process to perform and requires practice to perfect. It is much too rigorous for routine transport operations. Pressure breathing is used in supersonic transports only in an emergency involving a complete loss of pressure, and is used only until altitudes are reached that do not require it. Even the best oxygen masks are cumbersome and uncomfortable. The routine use of any sort of oxygen masks for passengers or flight crew is not acceptable.

The Federal Aviation Regulations are quite specific about the requirements for supplemental oxygen. For example, on pressurized cabin aircraft with two pilots at the controls, FARs require each flight crewmember to have 'a quick donning type of oxygen mask that can be placed on the face with one hand from the ready position within five seconds, supplying oxygen and properly secured and sealed when flying above 35,000 feet' (FAR 91.211). At altitudes at and above 41,000 feet, a pilot at the controls is required by regulation to wear and use an oxygen mask at all times. The supersonic Concorde has a waiver against this provision.

Hypoxia and Hyperventilation

Hypoxia occurs when there is insufficient oxygen to meet tissue needs. In aviation, it is usually associated with the fall in ambient pressure that accompanies a climb to altitude. Hypoxia first affects the higher mental functions because brain tissues are most sensitive to a lack of oxygen. Hypoxic effects happen more quickly at higher altitudes. The time for useful consciousness (TUC), which is usually defined as the time in which a person can be expected to take effective preventive measures, can be as little as 40 to 74 seconds at 30,000 feet (Sells and Berry, 1961). Actual TUC varies with individuals, the aircraft altitude, the rate at which pressure falls, and the physical activity of the individual at the time of the incident.

A second time interval that is important in decompression measures the time of safe unconsciousness (TSU). This is the time a person can remain unconscious without risk of brain damage from lack of oxygen. Studies have shown TSU to be about two minutes (Gaume, 1970).

Likely symptoms for hypoxia are obvious personality changes, impaired judgment, and muscular impairment (especially if finely coordinated movements or decision-making is involved). Other symptoms are short-term memory impairment, sensory loss (especially color identification, and then later touch, orientation, and hearing), and finally loss of consciousness and even death. If the hypoxia occurs at high altitudes, death can occur within a very few minutes. One of the most obvious physiological signs is that a hypoxic individual usually becomes cyanotic, with the lips and fingertips developing a blue tinge because so much of the hemoglobin is in the deoxygenated state.

In an effort to get more oxygen, the hypoxic individual may also hyperventilate. Hyperventilation is simply over-breathing, or breathing more than is required in an environment of low ambient pressure to remove excess carbon dioxide in the blood. Over-breathing is of little use in alleviating the deprived oxygen state in an environment of low ambient pressure. Sufficient oxygen is simply not available.

Hyperventilation changes the acid-base balance of the body and can result in symptoms such as tingling, visual disturbances (particularly tunneling or clouding), hot or cold feelings, anxiety, impaired performance, and even loss of consciousness. Hyperventilation is a process in which too much carbon dioxide is eliminated from the blood stream by the over-breathing. This renders the blood excessively alkaline and produces the symptoms mentioned. While the loss of consciousness sounds severe, it does allow the respiration to return to normal. The individual recovers quickly assuming that there are no other problems.

A seeming paradox occurs when over-breathing in the cockpit causes severe symptoms there, but causes no difficulty when jogging or indulging in other exercise and using the same amount of respiratory ventilation. The reason, of course, is that when jogging the pilot's exertion generates extra CO_2, which must be eliminated from the blood. This is done by the extra breathing caused by the jogging. The process is entirely normal.

If the symptoms occur in flight, it is often difficult to distinguish between hyperventilation and hypoxia. If the cabin or cockpit altitude is below 10,000 feet, hyperventilation is the most likely cause. Above 10,000 feet, hypoxia should be considered first. It can be the most serious case. Fortunately, hyperventilation—frequently associated with anxiety or high stress levels— occurs much more frequently with passengers than with flight crew. It very rarely causes serious problems for passengers but obviously is more serious in those rare cases that pilots are involved.

Unwanted cabin decompression can occur in flight. This is a problem if the cabin altitude is above 10,000 feet for it can cause hypoxia in anyone exposed to that environment. While cabin decompression is rare, it is a very real hazard in transport aircraft and usually requires a rapid emergency descent to a lower and friendlier altitude. Airline flight crews are checked routinely for their ability to handle rapid cabin decompressions (including explosive decompressions) efficiently and to perform emergency descents.

Decompression Sickness

Decompression sickness is reasonably rare but should be understood by all professional aviators. It can occur with exposure to the reduced atmospheric pressure at altitude when bubbles of nitrogen that are normally in the fluids in body tissues come out of solution. Instances of decompression sickness have been reported with ascents to altitudes above 18,000 feet. Decompression sickness occurs because the body is normally saturated with all of the nitrogen it will hold. The actual amount of nitrogen is relatively small because the body will absorb only a small amount of nitrogen When ambient pressure is abruptly reduced, particularly with steep ascents, some of the nitrogen in the tissue fluids comes out of solution as bubbles. Symptoms vary depending upon the body site involved. Nitrogen bubbles in shoulders, elbows, wrists, knees, ankles, and other joints cause an aching pain known as 'the bends'. Nitrogen bubbles occurring in the skin cause the 'creeps', in the respiratory system the 'chokes', and in the brain the 'staggers'. Ultimately, the individual may collapse and in rare cases death may result—all from the unwanted formation of nitrogen bubbles.

While decompression sickness is quite rare, the incidence is greatly increased if the individuals have been scuba diving shortly before the flight. This is because when breathing air under pressure, as in scuba diving, the body's store of nitrogen is increased and on a subsequent ascent, the nitrogen may come out causing decompression sickness. A good general rule is that one should not fly within 12 hours of swimming or diving using compressed air, and avoid flying for 24 hours if a depth of 300 feet has been exceeded.

Entrapped Gas and Barotrauma

Because body temperature remains essentially constant and pressure drops with altitude, the volume of any gas entrapped within the body cavities increases. Gas can be entrapped in the middle ear if the soft walls of the top of the Eustachian tube become swollen, as in a cold, so that air in the middle ear cannot be vented via the Eustachian tube to the throat, mouth and ambient air. The air (gas) in the middle air then becomes trapped, and because of the decreased pressure on the outside of the ear drum, the air in the middle ear expands resulting in a distortion of the eardrum. This can cause extreme pain and even injury to the drum. The condition is known as otic barotrauma. Figure 4.5 shows the anatomy of the ear and illustrates the structures, which can be involved.

The sinuses are cavities within the skull that are located above the eyes, in the cheeks, and at the back of the nose. The air, or gas, in these cavities vents much easier on ascent than on descent and vent in much the same way that the air in the middle ear vents. If an individual is suffering from a cold, flu, or a nasal allergy, the soft tissues around the sinuses and the Eustachian tube expand and obstruct or close the normally open-air passageways. The result is discomfort and pain that increases as the descent is increased. For this reason, aircrew should avoid flying with any of these symptoms. Failure to avoid flying with these symptoms can result in a barotrauma serious enough to make the individual unable to fly for considerable periods.

A lesser, but on occasion, painful example of an entrapped gas can occur in the small bowel when gas, usually formed because of the fermentation of certain foodstuffs such as beans or some highly spiced foods, cannot be vented either through the mouth or the anus. The result can be a very painful stretching of the bowel wall. Prevention is simply to avoid such foods before any flight in which the individual is likely to be exposed to large changes of pressure.

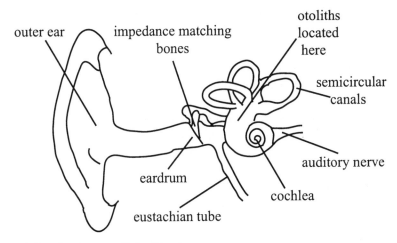

Figure 4.5   Anatomy of the Ear

Source: Adapted from *Human Factors For Pilots*, Second Edition, page 12, Green et al., 1991, and used with permission of Ashgate Publishing Limited

Finally, poor teeth or poor dental hygiene can create problems. Gas pockets can occur as a result of poor fillings or from a dental abscess. The result can be extreme pain under some flying conditions. Prevention is easy, as it simply requires good dental care and hygiene.

Hearing and the Physiology of the Ear

The ear provides two functions that are important in flying. The first function is hearing and the second function is balance. When we hear sound, the sound waves are passed down the external ear to the eardrum. The eardrum vibrates and these vibrations are passed across the middle ear by a series of small bones. The small bones of the inner ear are attached to the eardrum on one side and to the cochlea on the other. They thus transmit vibrations of the eardrum occasioned by the sound waves striking the eardrum. The cochlea then converts these vibrations to nerve impulses that are sent to the brain where they are recognized as sounds. (See Figure 4.5.)

The inner ear also detects angular and linear accelerations of the head to provide the body with the information it needs to maintain balance. The semicircular canals, which detect angular accelerations, and the otoliths, which detect linear accelerations perform this function. The otoliths are minute calcareous particles sometimes called the ear stones. Jointly, the semicircular canals and the otoliths make up the vestibular apparatus, whose purpose is to provide data to the brain that enables it to maintain spatial orientation and to control other systems that need that information. For example, eye movements are controlled so that a stable picture of the world is maintained even when the head is moved.

In order to maintain an accurate model of orientation, the brain also needs visual information that it gets from the eyes. An unfortunate effect of the motion stimuli the brain receives is that it can also lead to nausea, vomiting, hyperventilation, pallor and cold sweating. These symptoms can occur whenever a person is exposed to real or apparent motion of an unfamiliar form. While motion sickness, which is caused by inappropriate stimulation of the cochlea, can be extremely incapacitating, it is an entirely normal response to perceived stimuli. Under sufficient provocation, anyone with a normal sense of balance can suffer motion sickness.

Most people rapidly adapt to motion or to perceived and modified motion—as they find in simulator flight. Motion sickness can be a very real problem in flight training, particularly in the military. Flight training is a new experience for many and if individuals are not exposed to this new stimulation for a few weeks because they are on leave or for any other reasons, the adaptation can wear off and the motion sickness can appear again. Motion sickness is experienced by as many as one third of military trainees, some of whom lose their flying status as a result. Motion sickness is a very real phenomenon. It has even occurred in a fixed-based simulator that had no motion at all but did have very strong visual cues.

In transport aviation, motion sickness is much more of a problem for passengers than for flight crew where it is very rare. This is probably due to two reasons. The first is that transport flying does not involve the sort of maneuvers one finds in stunt or military flying, and the second reason is that flight crews become well-adapted to the range of motion that occurs in their environment.

Vision

Vision is the most commonly used of our primary five (some maintain we have eight) senses, and good vision for aviators is essential. It is also essential that pilots have a good understanding about the fundamentals of the visual process. Our discussion starts with the basic physiology of the eye and its functioning. It includes problems associated with the wide ranges of light intensity that pilots face, and concludes with the fascinating subject of visual illusions.

The Physiology of the Eye

The basic structure or anatomy of the eye is shown in Figure 4.6. It is relatively simple and straightforward although a detailed examination of the eye and its functioning can become extraordinarily complex.

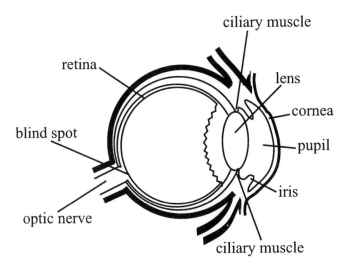

Figure 4.6   The Anatomy of the Eye
Source: *Human Factors in Flight*, Second Edition, page 109, Hawkins, 1993 and used with permission of Ashgate Publishing Limited

The analogy of a camera is often used to describe the basic functioning of the eye. This is because as rays of light enter the eye, the eye's lens focuses the rays onto the retina in the same way that a camera lens focuses rays of light onto film. For reasons that are well beyond the scope of this discussion, the image that is thus transmitted to the retina is transmitted in an upside-down fashion. The retina converts the light rays into nervous impulses which are transmitted to the brain via the optic nerve, and there an accurate real-world visual picture is constructed.

A basic problem with brightness levels occurs at the retina. This is because the retina is made up of light sensitive cells called rods and cones. The center of the retina is called the fovea and is composed entirely of cones. Moving outward, some of the cones are replaced by rods until at the periphery of the retina there are no cones but only rods. The rods are more sensitive to lower levels of illumination than the cones, but they are not sensitive to color. At low levels, we see only in monochrome. The rods do not see the color. This means that color intensity must be maintained at higher levels if color is used to furnish information.

The eye adapts to changes in light levels in two ways. One is simply by changing the size of the pupil and therefore the amount of light that reaches the retina. The size of the pupil opening contracts under strong light and expands under limited light conditions. The pupil can expand 16 times under very low levels of illumination. The second adaptation involves the interaction of the rods and cones. As light decreases, the visual-sensing task transfers from the cones to the rods. Unfortunately, this transfer does not take place instantly. The complete transfer can take nearly 40 minutes, with the time required increasing with age. The process is called dark adaptation. Most have seen an application of dark adaptation when going from a well-lighted lobby into a darkened theater.

Visual Acuity

Visual acuity, or the capacity of the eye to resolve detail, is usually expressed in terms of the smallest letter that can be read at 20 feet compared to normal vision. For example, if vision is stated as 20/20, it means that the individual can see at 20 feet what one with normal vision can see at that distance. If the vision is stated as 20/40, the individual can see at 20 feet what a person with normal vision can see at 40 feet.

The level of acuity is not distributed evenly across the retina. The highest level where detail can best be resolved is located on the fovea. It is used when concentrating or reading. This high degree of resolving power drops when just a few degrees from the fovea. Whenever we encounter anything that must be interrogated in detail, the object is automatically brought to fixation on the fovea. This leads to a common problem for many aviators, especially those just beginning instrument instruction. It is called 'tunnel vision' or sometimes 'fixation', and is manifested by focusing all attention on a single instrument and ignoring others which may be equally important. Actually, tunnel vision or fixation is a cognitive issue, not a foveal issue. The eye will move wherever the brain tells it to go. Therefore training and experience minimize this very real problem which is a natural characteristic of man.

Visual acuity in the real world is affected by such factors as brightness, the contrast between ambient lighting and the subject, the contrast within the displayed object, glare (including the angular relation of the glare to the subject), and finally the time allowed to view the object. Wearing appropriate sunglasses provides protection against many problems including the very high light levels found at upper altitudes. Good glasses also provide protection against the damaging blue and ultra-violet light that is not absorbed by the ozone layer. Both occur in a higher proportion at altitudes than occur in light at lower levels. Some ultra-violet radiation, which is found in sunlight, is necessary for the body to produce vitamin D. Normally harmful levels are absorbed by ozone in the atmosphere. Harmful effects of excess ultra-violet include sunburn, heat exhaustion or heat stroke, premature aging of the skin, wart like growths called solar keratoses, and most serious, a considerable later risk of skin cancer.

Accommodation

In order to see clearly at different distances, very sophisticated muscles in the eye change the refractive power of the lens. This process is called accommodation, and it utilizes what is called binocular convergence. A very real problem occurs in accommodation when visual cues are weak, such as with empty field visual conditions. An empty visual field means a field with the absence of discriminating visual cues. These can occur in either day or night conditions. In an empty field, the muscles that control accommodation take up a testing or relaxed state. Research has shown that a majority of individuals focus at about one meter or at slightly more than one yard when in the relaxed focus condition, not at infinity as was first believed. This condition is sometimes called empty field myopia or the dark focus. It can be significant because if the eye is searching for distant objects, such as another airplane, the eye will not be adjusted to detect them. Tanker pilots have to deal with the dark focus regularly for empty field myopia occurs frequently at high altitudes. The physiological phenomenon is known as the Mandelbaum or screen porch effect.

The Mandelbaum effect happens when a visual stimulus coincident with the dark focus traps the focus at that distance. Acuity beyond that distance is then degraded. Whenever the distance to the dark focus coincides with the distance between the eye and the windscreen, any visual stimulus on the windscreen such as dirt, moisture, crazing, sun-glare, reflections, or bug-spatter can trap the focus. Visual acuity beyond the windscreen is then impaired. In order to minimize this effect, it is a good practice for pilots to deliberately focus on far-away clouds, on stars at night, or on the wing tips of the airplane, if they are visible.

Color Vision

Light perceived by the human eye consists of electromagnetic radiation (energy waves) with wavelengths between 400 and 700 nanometers. Different wavelengths produce the sensations of violet, blue, green, yellow, orange, and red—the colors of the rainbow. White light is a complex combination of all these colors. Sensations of color are created when the wavelengths impinge upon the retina and stimulate nerve signals that are processed in the rear part of the brain. Within the retina, those cones that are concentrated in the central fovea area of the eye play the major role in receiving the color signals. One result is that colors are perceived better if they are viewed directly. Color discrimination is poor at the periphery of vision. It is also poor at low levels of illumination (Clayman, 1989). The reason, of course, is that the rods are more sensitive to low levels of illumination than are the cones. Unfortunately, the rods do not see color.

The use of color in aviation should be closely correlated with pilot licensing standards for color blindness. It has been estimated that 5-10% of men suffers from color blindness, which is simply the inability to perceive differences in color. It is rare in women although they can carry an apparently incurable congenital gene which is transmitted to their offspring (Hawkins, 1993). The prevalence of this disorder is lower in Asians or American Indians and somewhat lower among some other races. The most common form of color blindness affects red and green discrimination, which affects about 4% of Caucasian males. Unfortunately, people with this affliction see these important colors as shades of yellow, yellowish-brown or gray. Misidentification of red and green has definite safety ramifications, as for example, with interpreting a cockpit warning light or interpreting light gun signals from an air traffic controller. A perhaps apocryphal story is of a series of aircraft crashes during World War II that were finally traced to a worker who was unable to discriminate the coding colors used in aircraft wiring and consequently made several improper wiring connections.

There are many variations of the color blindness deficiency, which have positive advantages if used carefully. The use of color in documentation or cockpit instrument design is an example. Another consideration for designers in aviation products is that many individuals directly involved with air transport operations do not have physical licensing requirements or any type of relevant color testing.

Depth Perception

Depth perception, or the judgment of distance, has been important to man from his early existence. It was important to early man so he could better leap from branch to branch in a tree, or to throw rocks at animals more accurately. Judgment of distance is a critical skill for modern man in the current-day driving of an automobile and certainly in the landing of an airplane. At least nine cues to judging distance have been identified, although scientists still are unable to estimate the relative criticality of each. Depth perception is a complex activity.

Most people believe that stereo vision is the most important depth cue. The stereo cue results from the fact that we normally have two eyes sensing the same object. The retinal images formed by the two eyes are not focused precisely on the same part of the retinal mosaic and, while there is no doubt that this information provides very valuable knowledge, there is great uncertainty about just how the brain processes this data. Despite the acknowledged value of stereo vision in depth perception, there is no longer any question that people with only one eye (monocular vision) also are able to function effectively in tasks that require a high degree of depth perception. Wiley Post, a pioneer US high-altitude pilot and the first man to fly around the world alone, was an outstanding example for he had only one good eye. He wore a patch over the other.

Additional cues to depth perception include such things as the cues of superposition and motion parallax. Superposition includes differences in the relative size and height of the object in a plane, in texture gradient, the reduction of brightness, and in color contrasts. Motion parallax is found when the object or the observer is moving. Moving distant objects can appear stationary while stationary near objects seem to move.

When some of these cues are lacking, impoverished, or in conflict, the ability to perceive height and distance may be seriously degraded. If this occurs, errors can result. An example of this sort of error is haze or fog interfering with vision. In very clear weather objects such as hills appear near and dwarfed; in misty weather, they seem remote but look large. The dimming of an image as a function of distance is known as aerial perspective and is another depth cue (Gabriel, 1975).

Photic Stimulation

In a completely different area, exposure to flashing lights having certain characteristics such as critical frequencies, high brightness, contrast, and even some kinds of diffusion, can induce dangerous reactions even for apparently normal people. This is known as intermittent photic stimulation. Effects range from drowsiness, nausea, and disorientation to convulsions and trances. Frequencies which approximate the alpha rhythm of the brain (9-12 Hz) or a multiple thereof (e.g., 18-24 Hz) are particularly dangerous and must be considered by any designer who is using flashing lights for any reason. This is an old, old phenomenon. It is said that one of the tests used to discover soldiers for Caesar's legions was to expose them to light interrupted by the spokes of a chariot wheel. Individuals deemed vulnerable to this fairly gross example of photic stimulation were disqualified.

The Pilot's Visual Task

Pilots have a complex visual problem in modern cockpits. Traditionally, cockpit design was based on the pilot using one correction for looking at charts, manuals and the instruments and, if necessary, another correction for looking outside for distance vision. Today, the increasing complexity of both the aviation system and modern airplanes has changed pilot visual needs. The increasing complexity of many aspects of today's airspace and aircraft can require more reading of and attention to charts and manuals in the cockpit. For many pilots, this requires glasses with a near- correction based upon the ability to read sometimes fine print under less than optimum reading conditions. A second need is to be able to see flight instruments clearly. This is a middle distance requirement, which fortunately also takes care of a need to see easily and clearly data and input devices in the center pedestal.[2] A third need is the distance vision required for good vision outside the airplane, and a fourth is again, a near to middle vision requirement to be able to see and read controls that are in the upper panels above normal eye level.

As it is impractical to have a pair of glasses to address every type of vision, compromises are obviously required. In order to compensate for the effect of the aging eye on accommodation, pilots in today's airplanes, commonly use bifocal, trifocal, and varifocal correcting glasses. Generally

---

[2] Current US Medical Standards require 20/40 or better in each eye separately with or without correction at age 50 and over as measured at 32 inches for a Third Class Medical. (Also see Appendix P—FAA Medical Standards.)

speaking, the glasses work very well, but consultation with a well-informed aviation-oriented ophthalmologist is strongly recommended to take full advantage of modern technology.

Design Eye Position in Today's Aircraft

A very important consideration in cockpit design is determining the design eye position. When the pilot's seat is adjusted properly and his/her eye is in the design eye position, the pilot should have adequate visibility of the outside world and be able to see all important displays within the flight deck without further head movement.

If the eye position is too low, the pilot loses part of the exterior visual segment. The upper perimeter of the glare shield and the nose of the airplane block it. Inside the airplane, an eye position that is too low can also mean that the control wheel (or yoke) obscures some of the instrument displays, which may be required for an instrument approach. Obviously, this is not a problem with an airplane controlled by a side stick, although eye position is important for other reasons. A general rule is that the design eye position should be located to allow the pilot during low visibility landings to see the length of approach or touchdown zone lights that would be covered in three seconds at the final approach speed. If the eye position is too high, outside vision is impaired by the upper panel.

Present day airplanes indicate the design eye position by means of an indicator dot or stripe, usually on a window post. Adjusting the pilot seat so that it coincides with the design eye position has become a routine and important step in the cockpit set-up procedure, especially during low visibility landings and category II and III approaches (defined in Glossary). Older airplanes do not have an eye position indicator and there sometimes is a wide vertical variety in the actual eye position used. While training has proven very helpful in minimizing this problem, many pilots who are flying older airplanes may unnecessarily restrict themselves by using an improper eye position.

Illusions Commonly Encountered in Flight

Man generally takes the accuracy of perception for granted. This has been part of our heritage. In folk wisdom, it is attested to by the old phrase that 'seeing is believing'. Unfortunately, this is not always true.

Visual Illusions

Visual illusions are a fascinating subject and one, which all professional pilots encounter in their careers. There are several types. Common types include geometric optical illusions and depth and distance illusions. In spite of the fact that these illusions have been known for many years, and many theories have been proposed to explain them, very little is available that is truly definitive. A selection of classic geometrical optical illusions is shown in Figure 4.7.

A list showing the complexity of illusions is given in the US Air Force's 'Human Factors Glossary of Terms'. The list is given in Appendix C.

While low visibility approaches undoubtedly create the most critical conditions for pilots, most of the accidents or incidents arising from visual illusion problems seem to occur in relatively good weather. They frequently involve non-precision approaches when the pilot is in visual flight conditions. Many of the illusions already shown are common to several stages of flight.

A visual illusion that is common to night flight occurs when a stationary single light in a dark field often appears to move. This is referred to as the autokinetic illusion. Another common illusion is associated with rain on the windshield. Unfortunately, rain on the windshield involves many variables, just a few of which are the exact airflow over the windshield, the thickness of the water, and the effect of windshield wipers and various rain removal and repellent systems. These variables make exact determination of the illusory effect of rain very difficult, but there is no question that it is a real phenomenon.

Mist and fog, as well as atmospheric pollution can influence the judgment of distance. It is a common experience for pilots who are used to flying in the relatively polluted air of most urban areas to grossly misread their altitude in a clear atmosphere because distant objects appear much closer than they actually are. Other common illusions are the mistaking of a sloping cloudbank (either base or top) for a level horizon or the mistaking of ground lights for stars.

Velocity and Height

Nearly everyone has seen instances of the effect of height on apparent velocity. As height increases, apparent velocity decreases. That is the reason that the apparent speed of an airplane flying at altitude seems to be moving much slower across the sky than it actually is.

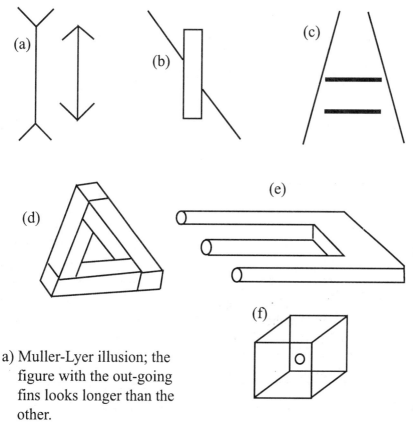

a) Muller-Lyer illusion; the
   figure with the out-going
   fins looks longer than the
   other.

b) Poggendorf figure; the crossing line looks displaced.

c) Ponzo illusion; the horizontal bar in the narrower part of the
   converging lines looks longer than the other.

d) Penrose triangle; impossible to perceive as a real object.

e) Schuster illusion; three arms at one end becoming two
   at the other.

f) Necker cube; the small circle appears sometimes at the
   back and sometimes at the front of the cube.

Figure 4.7   Illusions Involving Ambiguous Perceptions of Depth
Source: *Human Factors in Flight,* Second Edition, pages 118-119,
            Hawkins, 1993 and used with permission of Ashgate Publishing
            Limited

An important effect of this phenomenon is the velocity illusion during taxi that is created with different cockpit heights. For example, in airplanes with large pilot eye to wheel heights such as the Boeing B-747, the pilot is at the height of a third-story window. Most pilots are used to judging ground speed from driving a car or from the considerably lower height of the cockpits of other airplanes. It has become an absolute requirement to control taxi speed in the larger airplanes by reference to a cockpit readout of the actual ground speed. Otherwise, the temptation to taxi at excessive speeds is almost irresistible creating real problems for making comfortable turnoffs, turns, and gate approaches. Interestingly enough, this lesson was learned in part during the certification of the B-747 when blown tires resulted from excessive turnoff speeds after landing.

Another common problem occurs when a pilot has to land at an unfamiliar airport. While perspective and size cues are important for establishing the proper glide slope, not all runways have the same dimensions. On approach the runway that is 100 feet wide looks quite different than does one that has a width of 150 feet. Different spacing for runway lights can also cause erroneous height and velocity judgments. These are seldom problems for airline pilots because the width of the runways is standardized at 150 feet by international agreement through Annex 14 to the Chicago Convention. Even approach lights, runway lights, and taxiway lights are controlled by Annex 14. A good example is at London's Heathrow airport where the runway lights are located inside the actual width of its runways.

Other Visual Illusions

Illusions are simply a part of flight. A very common one occurs even before leaving the gate when as the loading bridge pulls away one has the impression that the airplane itself is moving. Blowing sand or snow during taxiing can create the same illusion. In admittedly rare cases, aircraft have been allowed to creep slowly into an obstruction because the airplane was thought to be stationary. In other instances, brakes have been suddenly applied unnecessarily while taxiing causing very real problems for the cabin staff and sometimes even injury.

Just after takeoff the impression of a false horizon can occur when surface lights are mistaken for stars or when sloping cloud levels are mistaken for a level horizon. As a result, the airplane can be placed in a critical attitude or position. Pilots must rely upon the aircraft's instruments.

Visual illusion problems are relatively benign during cruise but nevertheless a nuisance if they are not understood. One type of illusion concerns mountain tops in the distance which may appear to be above an aircraft's cruising altitude, but are actually well below. Charts and the airplane's altimeters, if set correctly, provide the best and only reliable guidance.

Another type of illusion occurring at cruise involves discrimination of the position and movement of other aircraft. For example, most pilots flying jet transports have seen approaching aircraft that initially appeared to be at a higher altitude but which slowly descended as they approached and finally pass below them. Pilots have been known to have 1,000 foot separation from opposite traffic but, when one or both of the airplanes are maneuvered, have inappropriately, actually increased the chance of a mid-air collision.

In another scenario, if a potentially conflicting airplane appears on the same spot in the windshield and does not move the two airplanes are on a collision course. Therefore the blind spot present in the normal eye can be a very real problem for the single pilot if the conflicting airplane happens to be in the blind spot. Windshield posts can be a problem, but are less in a two-person airplane. However, it is a mistake to assume that both pilots will always be looking outside. Figure 4.8 shows a very easy way to demonstrate the existence of the blind spot, which is present in each eye.

It is generally agreed that the approach and landing are the most critical phases of flight and there are at least three very real illusion possibilities when making a visual approach. The first involves either sloping terrain or a sloping runway. When an approach is made over terrain that slopes down toward the runway, the impression is created that the airplane is too low, even if the airplane altitude is entirely normal for the runway. If the approach is made toward a runway that slopes downward from the threshold, the pilot is again given the false impression of too much height. Usually the terrain and the runway slope in the same direction so that the perception of too much height is cumulative.

The reason for this illusion is that the pilot normally makes the approach at an angle of about three degrees to the terrain. If the terrain and or the runway slope downward at about one degree, the perceived approach angle to the pilot is only two degrees and hence gives the pilot the appearance of being too low. Exactly the reverse happens with an upslope.

A second illusion involves the 'black hole' phenomenon. 'Black hole' illusions occur most frequently when approaching over the sea, jungle, or desert, and there are no other intervening lights. When there are no lights other than the runway or airport lights and the approach is made over what appears to be simply a black void or hole, a very strong impression of extra height is created.

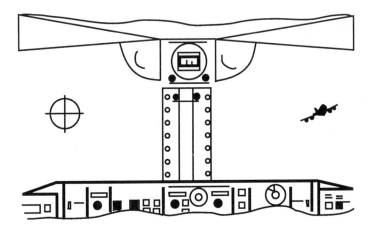

For a demonstration of the blind spot, hold this figure at arm's length, close the left eye and keep the right eye open. Now move the picture towards the face, keeping the right eye focused on the cross. The aircraft will disappear and then reappear as it gets closer.

Figure 4.8   Proving the Blind Spot

Source: Adapted from *Human Factors in Flight,* Second Edition, page 115, Hawkins, 1993 and used with permission of Ashgate Publishing Limited

Shortly after the introduction of the Boeing B-727, Dr. Conrad Kraft of the Boeing Company conducted basic research that identified the 'black hole' illusion and discovered the 'black hole' hazard. His research was prompted by the occurrence of crashes in that new airplane. Dr. Kraft and several others were given the task of finding out if there was anything basically wrong with the airplane.

Dr. Kraft examined the accident reports of the B-727 and of other transports occurring during that period and found that many of them occurred during approaches at night when the approaches were made over dark areas near cities that had an upward tilt. After several trial efforts, he constructed a model city on a moving table. The Model City was constructed from a large photograph of a city and airport. Simulated light sources were provided by puncturing pinholes through a soft wood backing to give slightly raised holes that were in effect small spherical light sources. The ultimate visual-scene simulation was convincing to pilots who flew in a stationary transport cockpit simulator with movement provided from the moving table.

Dr. Kraft identified a basic problem and received many awards for his simulator-based research. His principal findings were that pilots perceived an illusion of excessively high altitude when making a night straight-in approach when the airport was at the edge of a city and when the city had an upward tilt of from one to three degrees from the airport to the horizon. There was nothing wrong with the B-727, which has had an exemplary safety record ever since. From a human factors research standpoint, an interesting comment made by Dr. Kraft was that if a full-motion simulator had been made available for the study, he probably would not have discovered the basic problem (Orlady et al., 1988).

The 'Whiteout' Phenomena

Understandably, pilots tend to rely on their visual perception and judgments to determine glide slope, speed, and altitude. These can generally be made with an adequate degree of accuracy. However, when the external visual environment is lacking in cues or offers cues that can be misleading because they differ from those most often experienced, difficulties can arise. Visual approaches in the Arctic or Antarctic are notoriously difficult because of the homogeneity of terrain and the resultant lack of depth perception cues. The dangerous and notorious 'white outs' are an example. The US Navy Weather Research Facility at Scott Base, McMurdo Station in the Antarctica has written: '...in a whiteout condition a dark object is visible for many miles while a snow-covered object, even a mountain, next to the observer is invisible' (Vette, 1983). While, a full discussion of the 'whiteout phenomenon' is beyond the purview of this book, it is strongly suggested that anyone who is likely to be exposed to 'whiteout conditions' seriously study the appropriate literature. Whiteout is generally believed to have been a major cause of Air New Zealand's tragic crash on Mt. Erebus in the Antarctic on 28 November 1979.

Non-visual Illusions

Misinterpreted linear accelerations can be particularly important. Acceleration is detected in the semicircular canals of the middle ear. Unfortunately, forward acceleration and pitch-up create a similar sensation. It is very easy to mistake acceleration for a pitching sensation in an airplane. This is the somatogravic illusion and is particularly evident during the initial part of a go-around when the airplane both accelerates and changes to a climb attitude. A very common go-around failure during training is the failure to change to the go-around attitude after adding power. Some trainees, under the somatogravic illusion, feel the pitch-up sensation because of the acceleration when power is added and therefore do not raise the nose to the go-around attitude. In more than one accident, it was found that the gear was going up and full power was added as the airplane continued its downward course into the ground. Figure 4.9 illustrates the forces that create the somatogravic illusion during a go-around.

The only force experienced by a pilot in level flight at constant speed is his/her own weight. If the pilot accelerates forwards (as when power is added for a go-around), an inertial force is produced by the acceleration that acts as is shown in Figure 4.9. The resultant of these two forces in an accelerating aircraft operates similarly to the weight force in a climbing aircraft and gives the pilot a somotogravic illusion that he/she is climbing (Green et al., 1991).

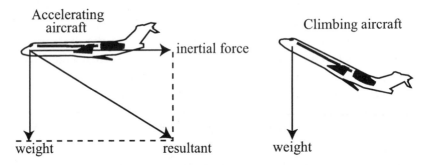

Figure 4.9   The Somatogravic Illusion
Source: Adapted from *Human Factors for Pilots*, page 49, Green et al., 1996 and used with permission of Ashgate Publishing Limited

The somatogravic illusion is used in motion-based simulators to simulate acceleration and deceleration of the airplane with pitching up to simulate acceleration and pitching down to simulate deceleration. A slow washout of the simulator motion is used to return the simulator to a neutral position.

Another type of balance illusion is known as the 'leans'. For example, if the pilot rolls the airplane into a shallow bank at a rate that is below the threshold of detection in the semicircular canals and then returns to straight and level flight at a slightly higher roll rate, he/she may feel that the airplane is not level but banked the opposite way. This feeling occurs because the resultant of the turning and weight vectors is aligned with the head and foot axis of the pilot. In a shallow and coordinated turn, the airplane is perceived as being in a wings-level attitude. Then, if the airplane is returned to an actual wings-level attitude at a rate that stimulates the semicircular canals, the pilot's perception is that he/she has banked the other way. The feeling of 'leans' may persist for a considerable time.

Training for Illusions

Captain J.A. Brown of American Airlines gave an example of an airline approach to the problem of illusions when he told the attendees at IATA's 20th Technical Conference:

> It is...futile to attempt to train crewmen to cope with visual illusions by training them under all types of conditions which may produce illusions. The specific illusion that may some day cause...a serious problem may or may not be duplicated. More importantly, such experience provides no assurance that the crewman could recognize his problem much less cope with it during an actual operation. The answer lies in procedures that will assure safe flight regardless of what one crewman thinks his eyes are telling him. Training that will convince...that illusions can be expected is important, but the key to safe operations lies in compliance with protective procedures.   (Brown, 1975)

It seems very clear that from a pilot's or a training executive's standpoint, the soundest solution to vestibular and other illusion problems is to trust the most reliable information available—the aircraft instruments—rather than to rely upon either specific illusion training or upon individual sensations. This tenet is a basic first lesson in instrument flying and is essentially the only sound operational advice that can be given on the subject of illusions.

We believe it is also important for the professional pilot to understand the hazards implicit in illusions. It is inevitable that sometime in the career of the professional pilot, he/she, perhaps suddenly, will be faced with these phenomena. It is highly probable, and certainly possible, that the performance of this pilot will already be degraded by fatigue, circadian arrhythmia, perhaps some form of additional stress, or any combination of these and other potentially degrading factors. One cannot expect thinking individuals to follow rules such as suggested in the preceding paragraph if they do not understand the reason for such rules.

Acceleration

Man is adapted to live on earth under the influence of the force of gravity. This normally creates a 1G environment. G is recognized as the basic unit of acceleration. It is frequently expressed as a formula, $1G = 32.2$ ft/sec^2. In aviation, we are concerned with long duration and short duration acceleration. Short duration acceleration lasts under one second and almost always is related to impact acceleration forces. Long duration accelerations are perceived as increases in body weight so that arms and legs become harder to move, the head becomes heavy and organs are displaced from their normal position. There are also changes in blood circulation because of the apparent increase in weight. In a 2G situation, the individual has effectively doubled in weight. This also affects the blood and has the same effect as if the heart has to push blood of increased weight to the head. The effect is a decrease in the amount of blood that reaches the brain and this results in the symptoms described in the next paragraph.

The effect of increased G forces reduces blood pressure at the head level to the point where first the eyes and then the brain may not receive the blood (and therefore the oxygen) they require for normal functioning. The result is graying out and eventually unconsciousness. While these phenomena usually occur at about 3.5G, the use of anti-G straining maneuvers and special anti-G suit can delay the onset of loss of vision and consciousness to about 7.5 or 8G. Such procedures require training, are very tiring, and are of relatively short term. They are important for specific kinds of military flying.

Negative G occurs during some kinds of aircraft maneuvers such as pushovers, inverted flight, outside loops, and some kinds of spins. Here, gravitational forces force blood to the head and the heart is unable to get the excess blood away from the brain. The results are the opposite of positive G and even more uncomfortable. Symptoms include facial pain, the bursting of small blood vessels in the face and eyes, the pushing up of the lower eyelid, 'redout', and the slowing of the heart because of the upward rush of blood from the lower body. The maximum tolerable level is -3G and this can be tolerated only for a relatively short period.

The body's tolerance for impact Gs varies with the parts of the body. For example, man can tolerate up to 25G in the vertical axis and up to 45G in the fore to aft axis. Accelerations that produce G forces in excess of these limits result in injury or death. These limits are of obvious concern to cabin and cockpit designers. Pilots and flight attendants are provided more protection than are passengers, largely because of more effective lap and shoulder harnesses and higher rated 'G' tolerant seats. Protective harnesses must be comfortable, easy to use, and cause a minimum of restriction to normal body movement if crewmembers can be expected to use them routinely.

Selected Physical Aspects of Today's Environment

One of the earliest interfaces recognized in flying was between liveware[3] and the physical environment. Early-day pilots were given helmets to protect them against noise and cold, flying suits to protect their bodies against the cold, oxygen masks to protect them against the more rarefied air at altitudes, and goggles to protect eyes against the airstream. All of these measures were aimed at enabling man to function in a changed physical environment. Later developments include enclosed cabins, pressurization, sophisticated air conditioning systems, and modern soundproofing.

Newer physical environmental problems include the possible effects of prolonged exposure to ozone concentrations at subsonic jet levels and to high-altitude ultraviolet and ionizing radiation. Supersonic flight exacerbates most of them. Another new physical environmental problem, this time affecting all airplanes, is the effect of ground-based laser light advertising on flight crews and passengers who may be inadvertently exposed to potentially very hazardous laser rays. Each of these new problems may require specific regulations, better measuring, the establishment of realistic conservative exposure limits, monitoring of the personnel involved, and additional training for flight and cabin crewmembers and for the crews operating ground-based lasers.

An old problem, often ignored, but which is getting increased attention is fatigue. The speed of transmeridian travel and the economic need to keep aircraft and their human payload flying 24 hours a day increases problems associated with disturbed biological rhythms and related sleep disturbance and sleep deprivation. (See Chapter 14—Fatigue and Stress.)

Still another part of the environment that is becoming increasingly critical is directly related to the effects of transport aviation's phenomenal growth. Air transport must function safely and efficiently in teeming passenger terminals while utilizing congested gates, taxiways, and runways. It must then operate in increasingly crowded skies. The economic and political costs of improving the infrastructure of the air transport industry are considerable.

---

[3] Liveware signifies the persons involved in the operation.

References

Boeing Commercial Aviation Co. (1988). *A Glossary of Terms: Human Errors in Aviation*, Boeing Commercial Aviation Co., Seattle, Washington.

Clayman, Charles B., ed. (1989). *Encyclopedia of Medicine*, American Medical Association, Random House, Inc., New York.

Gabriel, R. F. (1977). 'Some Potential Errors in Human Information Processing During Approach and Landing', Douglas Paper 6587, Douglas Aircraft Company (now Boeing), Long Beach, California.

Gaume, J. G. (1970). 'Factors influencing the time of safe unconsciousness (TSU) for commercial jet passengers following cabin decompression', *Aviation, Space, and Environmental Medicine,* 41:382-385.

Green, Roger G, James, Helen, Gradwell, David, Green, Roger L, (1996). *Human Factors for Pilots*, Avebury Technical, Aldershot, England.

Gunston, Bill (1986). *Jane's Aerospace Dictionary*, Jane's Publishing Company, Ltd., London.

Hawkins, Frank H. (1993). *Human Factors in Flight*, Second Edition, ed. by Orlady, Harry W., Ashgate Publishing Limited, Aldershot, England.

Josephy, Alvin M. Jr, ed. (1962). *The American History of Flight*, The American Heritage Publishing Co., Inc., New York.

Koonce, Jefferson M. (1988). *Aerospace Glossary for Human Factors Engineers*, SAE Aerospace Research Publication 4107, Society of Aeronautical Engineers, Warrendale, Pennsylvania.

McFarland, Ross A., (1953). *Human Factors in Air Transportation*, McGraw Hill Book Company, New York.

Orlady, Harry W., Hennessy, Robert T., Obermayer, Richard W., Vreuls, Donald, and Murphy, Miles R. (1988). *Using Full-Mission Simulation for Human Factors Research in Air Transport Operations*, NASA Technical Memorandum 88330, Ames Research Center, Moffett Field, California.

Sells, S.B. and Berry, Charles A, eds. (1961). *Human Factors in Jet and Space Travel*, The Roland Press Company, New York.

5   Those Magnificent Flying Machines and Their Internal Environment

Brief History of the Development of Those 'Magnificent Flying Machines'

Many transport airplanes were manufactured in the period following World War I. Most of the early ones carried only mail. Nearly all of these airplanes were redesigned bombers. They were redesigned in hopes that their countries could take advantage of this new method of transport and were or became the transports of their day. Human factors, as we know it today, was simply not considered. In Europe, the new transports included airplanes like Handley Page's DH.4A and W.10, the Farman F. 60 of the Goliath Series, the Vickers Commercial Vimy, the Aeromarine F-5L, de Havilland's DH.34, Dornier's Wal, the Messerschmitt M.20, Tupolev's ANT-3, Short's Calcutta flying boats, and there were many others.

Junkers and Fokker were the main European builders of transport aircraft in the first years after World War I. Their success was fueled by the energy and success of Luft Hansa, Germany's premier airline—the Lufthansa of today, of the Dutch airline, KLM, and of the political support these airlines got in their home countries. KLM ordered two of the early Fokkers in 1920. The Junkers W34 made its first flight in 1925 and helped open up air transport in several remote areas of the world. Fokker's pride and joy was the tri-motored F.VII/3m. It was a modification of the single-engine F.VII that had been designed to compete in the United States in the Ford Reliability Trial of September 1925. The Ford Reliability Trials were intended to promote the reliability of aircraft that were intended for airline use

During the middle and late 1920s, the Fokker F.VIIs became some of the most famous airplanes in the world. One of the best known was Charles Kingsford Smith's *Southern Cross*, which, laden with fuel, flew from Oakland, California to Brisbane, Australia. The *Southern Cross* arrived in Brisbane on 9 June 1928. It had crossed the entire Pacific Ocean after 83 hours and 11 minutes in the air.

Across the Atlantic in the United States the beginnings of viable transport aviation were also evident. The early airmail pilots operated a series of biplanes designed mainly to be flown for the mail. The Ford Tri-Motor, affectionately known as the 'Tin Lizzie', was a product of this era and has even been called by some the first advanced US transport. The first Ford, model 4-AT Tri-Motor, was a high wing open cockpit monoplane with an enclosed cabin for eight passengers (Gunston, 1980). It first flew on August 2, 1926 and set the pattern for a series that culminated in transports that were powered by three 300-hp Whirlwind radial engines, carried up to 15 passengers, and also enclosed the cockpit. Development of the Fords ended in 1932 with the advent of the Boeing 247 and the Douglas DC-2. There are still Ford Tri-Motors flying both in the US and in South America.

Figure 5.1   The Ford Tri-Motor
Source: *The Illustrated Encyclopedia of Propeller Airplanes*, page 42,
               Gunston, 1980 and used with permission

Other airplanes that were designed during this formative period of air transport development included Lockheed's famous Vegas and Boeing's 40As and 40Bs that led to the highly successful Boeing 80s. A first for the Boeing 80 was a jump seat at the rear of the cabin for a stewardess. Ellen Church is considered to be the first airline stewardess. She worked a flight in a Boeing 80 on 15 May 1930 for United Airlines and forever changed airline history. The Boeing 80As were the first Boeings that had an enclosed cockpit and were designed primarily to carry passengers. Mail and cargo were a secondary consideration.

The French, Germans, English, Italians, and the Russians continued to manufacture new, and for that period, advanced airplanes including larger Short flying boats, Fokker F.XIIs, Sikorsky S-38s, the Latécoère 28, Tupolev's ANT-9, the famous Dornier Do X, and Italy's three-engined twin-hulled Savoia-Marchetti S.66 flying boat. The Fokker F.XIIs were constructed entirely by wood. They were used by several airlines but were doomed after a 1931 airliner crash in which one came apart in an Indiana thunderstorm and killed the famous Notre Dame University football coach, Knute Rockne.

In 1933 Boeing flew a series of B-247s, which may have been the first of the truly modern airliners. It began US domination in this field. The B-247s were low-winged cantilevered monoplanes powered by 550-hp geared Pratt and Whitney Wasp engines which had the then new controllable pitch propellers. The 247s were soon passed by the new Douglas DC-2s and shortly thereafter by the DC-3. Air transport has always been a dynamic industry.

The Douglas DC-3, whose inaugural flight was flown in 1933, was unquestionably one of the most important transports ever built. Except for the flying boats of Pan American and Great Britain's Imperial Airways, the DC-3 soon became the premier airliner throughout the world. The DC-3 was a low-wing cantilevered monoplane with a retractable undercarriage and a rotatable tail wheel that could be fixed for takeoff and landing. It was simply the best air transport being built at that time. By 1939 the DC-3 was carrying 90 percent of the world's airline passengers (Serling, 1982).

The Effects of World War II

Large-scale production of the DC-3 began with the entry of the US into World War II and before the war was over 11,000 of them had been built. The Air Force called them C-47s or C-53s while the Navy designated them R4Ds. They were still DC-3s. Nearly 2,500 more were built by the Russians under license from Douglas. The Russians called them PS-84s and later L2Ds. In those days human factors of virtually any sort were given little consideration.

The Douglas C-54 was developed for the Air Force during World War II. It had 4 engines and was much larger than the DC-3. After the war, the unpressurized C-54s were purchased by the airlines and were known as DC-4s. Pressurization was developed just before the war in the Boeing 307, which was called the Stratoliner. The Stratoliner first flew in 1938 and was the first production airplane ever manufactured with a pressurized cabin. The human factors involved in flying at higher altitudes with pressurized cabins had to be considered.

Figure 5.2   The Douglas DC-3
Source: Courtesy of United Airlines

World War II prevented further development of pressurized airplanes. The airlines had to wait for peace before they could begin replacing their DC-3s. Once peace was declared, the military DC-4s were released to the airlines, and manufacturers began building the pressurized DC-6s, DC-7s and Lockheed L-749s and L-1049s. The latter were perhaps better known as the Constellations and the Super Constellations.

The Jet Age

The jet age began with Britain's ill-fated Comets, which had their first scheduled flight in 1952. Unfortunately, the Comets, which doubled the speed and altitude of existing transports, had a fatal design flaw. While the Comets met or surpassed all existing and predicted structural standards, its square windows developed small fatigue cracks at its corners. Round or oval shapes (like an egg) provide a stronger structure subject to less fatigue and stress cracking. The square windows on the Comet eventually gave way under the rapid changes in altitude to catastrophic fatigue failures that resulted from the Comet's unusually fast climbs and descents. The result was a series of disastrous decompressions and the airplane was doomed. The US waited for the Boeing 707, which was first flown in 1958 and had its first scheduled trip in 1959. About a year later, the Douglas DC-8 flew its first trip. Both Boeing and Douglas had learned from the pioneering Comet's problem, details of which, in the best tradition of air transport, had been provided to them by de Havilland. In this era, human factors was still very much secondary to engineering considerations. Pilots were simply expected to adapt to this new technological wonder. Those who could not adapt were restricted to propeller airplanes or eliminated.

During this first half century of transport aviation, transports had grown from small open cockpit airplanes which weighed in the neighborhood of 1,200 pounds and were powered with 300 to 400 hp engines, and carried either no or as little as two passengers to today's giant jets. Today's largest jets carry over 400 passengers, have engines that generate many times the power of those early engines, and have a non-stop range that connects virtually any two population centers in the entire world. Today, manufacturers have on their drawing boards supersonic airplanes and planes that can carry from 700 or 800 to over 1,000 passengers.

Some present transports are very large airplanes. For example, if a B747-400, one of the latest transport aircraft, was placed in a standard US football field, whose dimensions are 160 feet by 300 feet, the nose could be placed on one goal line and the tail would reach the 20 yard line at the far end of the field. The wing tips would extend 26 feet over each of the sidelines. The Wright Brothers' entire first flight could be completed entirely in the B-747's cabin interior. Airbuses of similar size are also built by the European consortium that manufactures these airplanes. Stretched models of these airplanes are now being manufactured on both sides of the Atlantic.

In the meantime we should remember that we have had a supersonic transport for some time. The Anglo-French Concorde—the world's first successful supersonic transport—flew its first passengers on 21 January 1976. The Concorde has flown 23 years, carried over 2 million passengers in over 44,000 flights, and continues to fly without having had a single passenger fatality. It is an impressive record, but unfortunately, the Concorde is only marginally economic to operate.

Presently, the largest airliner in the world is Russia's Antonov An-225. It has six engines and is able to carry a payload of up to 250 metric tons with gross weights approaching 1,323,000 lbs. The first and only An-225 was built in 1988 and is still flying. It would be a gross mistake to assume that we will not continue to make bigger and better airplanes in the future.

Anthropometry in Air Transport

As shown in more detail in Chapter 6—The Social Environment, crewmembers vary within their own populations, and by ethnic group, and gender. Typical body measurements and strength of movement covering the range found in a very diverse potential pilot population must be considered in the aircraft design stage. All required flight deck controls must be within the reach of all of the projected pilot population and require only normal strength. It can be inordinately expensive and nearly impossible to make later modifications.

Anthropometric data is carefully considered in each new airplane. A disadvantage of present data is that virtually all of the limited research on variations in size, gender, and race has been performed in the US or in Europe, much of it involving personnel in the armed forces. For example, little data is available that gives anthropometric data for a potential flight crew population from the Far East. Despite all this, the thoughtful application of anthropometric considerations by the manufacturers and, to a lesser extent by the original certification authorities, has minimized problems in this area. To accommodate the wide variation in potential crew member differences, the cockpits of today's airplanes usually are configured to accommodate individuals whose stature ranges from 5'2" to 6'3", and to consider the strength requirements of both genders and the reach potential of hopefully all prospective pilot populations.

The sizes and shapes of pilots can vary tremendously. Somewhat ironically, in the early days and through all of the piston era, an early airmail pilot and the most senior airline pilot in the US, the late E. Hamilton (Ham) Lee, was so short that he was forced to carry a special cushion that enabled him to reach the rudder pedals of a conventional DC-3. Ham finished a full career with an unblemished record, all with his special cushion. In the early days airplanes were built with minimum concern regarding the anthropometric needs of the pilots who might fly them.

The importance of sound and relevant anthropometric principles has grown in the recent years with the common use of female pilots, who added greatly to the range of the pilot population, and by the extension of the world air transport market to virtually all parts of the world. The size, shape, and strength of people varies by gender and by race well beyond the distribution found in individual homogenous populations in the US and in Europe.

Today, all major manufactures use computer-generated models to study anthropometric problems. For example, in one case, a three-dimensional wireframe mannequin that can be altered to fit an infinite number of height and body proportions is seated at the design reference point. In a simulated fashion the computer-designed mannequin can be seen operating all of the controls, seeing the displays, and performing any task requiring external vision. Unfortunately, because there is still a paucity of relevant worldwide anthropometric data, designers must still use considerable judgment in reaching their final conclusions.

The Cockpit's Physical Environment

The cockpit provides the internal physical environment for the flight crew. Design considerations for the cockpit require considerable advance planning because space in a modern airplane is fixed in design and always at a premium. The competition for passenger seat and baggage space, cargo space, passenger hand-held baggage, galley space, rest rooms, other passenger service items, flight attendant seats, and flight attendant modules makes compromises in the cabin inevitable. These items also compete for space on the flight deck. Many items, which range from flight bags to smoke goggles and fire axes, need to be accessible in flight.

Flight Crew Compartment

The flight crew compartment must be concerned with front instrument panels, with circuit breaker panels, airplane and system controls, and with communication modules. The design engineers must also be concerned with space for pilot seats, with storage for crew baggage, crew hand-held baggage, coats and uniform caps, with required emergency equipment, and with rest facilities on long-range airplanes. Seats in the cockpit area for observers, commonly referred to as 'jumpseats', are an additional requirement. They are required both for observation of flight crew performance by the FAA or company and for crew personnel movements. One of the latest airplanes, the Boeing 747-400, has three cockpit seats, in addition to those for the captain and first officer. The flight deck of this long range aircraft has a dedicated flight deck crew rest area, which consists of two bunk beds in an enclosed area located aft of the cockpit.

Shoulder harnesses are required for crews and cannot be allowed to restrict reach or impede movement of the controls. Cockpit seats must be comfortable and adjustable fore-and-aft and vertically. They must fit a wide variety of crewmembers and must have an indentation in the front of the seat for a sophisticated crotch strap fitting to prevent the submarining of a pilot under the harness during an accident, and also to permit full travel of the yoke. While it is not fully accepted in some circles, a side-arm controller eases this problem. Comfortable crew seats are an important item because US flight crews are required to spend all of their flight time in the cockpit with seat belts fastened (other than for physiological comfort or scheduled rest periods). Latest transports have fully adjustable pilot seats with sheepskin covers to provide extra comfort over a long period. These seats include lumbar adjustments, and adjustments that firm up the cushions underneath the back of the thighs.

Ergonomically, the Airbus side-stick controller provides considerable extra flight-deck space by replacing the conventional wheel and yoke found in most other airplanes. US manufacturers have not gone to side-stick controls in their fly-by-wire aircraft because of a philosophical desire to maintain a high degree of commonality with earlier generation aircraft and a possible lack of sufficient feedback from one side-stick to the other. This point is briefly discussed at the end of this chapter.

A fire-resistant door, which is normally locked but also can be unlocked reasonably easily, enables cabin crew access to the cockpit and crew access to the cabin and to some rest rooms and crew rest facilities. For security reasons, the unlocking of the cabin/cockpit door must be restricted. Access to the cockpit is controlled by the cockpit crew and supplemented with special keys or a coded knock for the cabin crew devised on a trip-by-trip basis.

Internal Environmental Considerations

Man is constituted to function efficiently under a reasonably narrow set of conditions (See Chapter 9—Man's Limitations, Human Errors, and Information Processing). Cabin temperature, pressure, humidity and noise are important passenger and crew comfort considerations. Temperature and pressure are controlled by the airplane's air conditioning and pressurization systems. Insulation and other soundproofing control cabin and cockpit interior noise to a considerable extent.

The cockpit has an additional problem caused by the noise created by the rush of air over the windscreens. Some airplanes, and most helicopters, get additional noise from their powerplants and propellers or rotors. The cockpits of earlier airplanes were so noisy that they prohibited normal conversation without using the intercommunication system and uncomfortable headsets. Fortunately, these problems are seldom found in the latest jet aircraft although noise levels remain a problem for most turboprop airplanes.

Higher Altitudes

The higher altitudes flown and the time spent at altitude that are part of long range flights have galactic (solar and cosmic) radiation complications that require serious consideration from pilots, flight attendants, the airlines, and regulatory authorities. Both low levels of radiation, as well as shorter-term high peaks, may be harmful to humans. An insidious characteristic of ionizing radiation (the type that is of biological concern) is that it is invisible and that it can be applied to the human body without sensation or obvious effect. Neither ordinary glass, aircraft structures, treated windshields or clothing provide protection against ionizing radiation. In spite of the fact that the basic radiation problem has been recognized for some time, its effects are still highly controversial.

US FAA Advisory Circular 120-52 (AC 120-52) provides excellent background material for this difficult subject. For example, there are generally accepted radiation rules and limits for pregnant women, although there is not complete agreement on some of the specific rules and limits. Another excellent source of background material is the 1992 FAA Office of Aviation Medicine Report, *Radiation Exposure of Air Carrier Crewmembers II,* by Frieberg et al. The FAA also has developed a computer software program (CARI-5E) for calculating the dose of galactic radiation that would be received on any given flight. CARI-5E can calculate the data from galactic radiation received on any user-entered flight profile at any geographic location for altitudes up to 87,000 feet and for dates as far back as January 1958. The CARI-5E software can be downloaded from CAMI's Radiobiology Research Team's website at http://www.cami.jccbi.gov/aam-600/610/600radio.html.

There is little question that exposure to galactic radiation increases with altitude and that the atmosphere provides less protection in northern latitudes than in latitudes near the equator. At slightly over 5,000 feet, residents of Denver receive more normal radiation than do residents of such relatively low level cities as Chicago or New York. The effects of ionizing radiation on humans are cumulative, and a great deal of it is absorbed by humans in very small units as a part of ordinary daily living. Sources can be as mundane as computer terminals, microwave ovens, high voltage electric transmission lines, television sets, medical and dental x-rays, some medical treatments, and even living in a home with high radon levels.

CAMI scientists, with the aid of colleagues at the National Oceanic and

Atmospheric Administration (NOAA) and the Bartol Research Institute which is located at the University of Delaware routinely monitor galactic radiation levels. Other research groups worldwide studying the effects of galactic radiation include a group at the University of South Carolina that is working in cooperation with the ALPA's Aeromedical Department. Dr. Donald Hudson, Head of the ALPA Aeromedical Department, agrees wholeheartedly that galactic radiation is an area that needs to be continually watched.

The jet age began in 1959 and ten years later the operations of several airlines was entirely by jets. This means that flightcrews have had nearly three decades of exposure to jet altitudes. This is a period long enough to provide a reliable longitudinal study. Additional, but obviously limited, data should also be available from Concorde crews as the Concorde has operated at supersonic altitudes since 1976. Some of the fears that periodically arise among flight crewmembers and even passengers are an increased liability to various kinds of cancer as well as premature death.

Presently, CAMI, various airlines throughout the world, and flight crew unions cooperate with interested universities in conducting studies on the effects of long-term exposure to the levels of exposure found in jet transportation. Both the long- and short-term effects of galactic radiation remain serious concerns for flight crewmembers. A documented longitudinal study by recognized and neutral experts of the effects, if any, of galactic radiation is needed to allay fears that are very real and are compounded by the lack of such a study. A longitudinal study is both difficult and expensive. Very little hard evidence, pro or con, regarding the effect of flight crew exposure to the levels of galactic radiation that flight crewmembers encounter is available today.

A continuing difficulty is that occasionally the radiation area, which is difficult, complex, and highly technical, has not been reported accurately by a not always fully-informed media. The fact that collective bargaining issues may be involved does not make the problem easier. Fortunately, in this area all aspects of the industry have a common goal. It is however, extremely important to get information from recognized experts. While further treatment of radiation and its effects is well beyond the scope of this book, the authors, as well as many radiation experts, believe that it is a subject that deserves continued attention from pilot or flight attendant unions, the airlines, and the regulatory agencies involved. An effective longitudinal study by recognized experts should be given a very high priority.

Another problem at higher altitudes is ozone and its effect on lungs and breathing. Ozone is a variant of oxygen that occurs naturally and is caused by the action of the sun on oxygen in the air. Ozone provides a very useful function in that it absorbs ultraviolet rays that come from the sun and protects life on earth. Unfortunately, ozone is also a toxic gas. Adverse effects seem far more related to concentration than to periods of exposure. The first adverse symptom of excessive ozone is usually dryness of the nose and throat followed by chest discomfort that can lead to more serious lung irritation and emphysema. Eye irritation has sometimes been reported. Individual responses vary widely and usually appear in people most physically active in flight. Flight attendants are often the first to notice excessive concentrations of ozone.

Ozone levels of concentration increase rapidly above the tropopause, increase in winter and spring months, and increase with latitude and altitude. Strong winds prevailing during certain times of the year occasionally bring ozone concentrations south and to altitudes routinely flown by many jet transports.

Ozone is partly destroyed as it passes through the pressurization system. The effectiveness of this method of protection varies by airplane type. Many airplanes that do not provide sufficient ozone protection through pressurization use carbon filters and catalytic materials to control the ozone hazard. Airplanes that do not provide ozone protection are advised to select routes and altitudes that do not have excessive levels of ozone (Hawkins, 1993).

The FAA has set limits of 0.25 ppmv (parts per million by volume) (CFR 121.578) and it appears that for healthy people indefinite exposure to levels below that maximum are harmless. However, levels as high as 0.57 ppmv have been reported on polar routes (Preston, 1979).

Humidity

Humidity is a real problem, particularly in pressurized airplanes. The problem is greater for pilots and cabin crew because they fly many more trips than the average passenger and therefore have a considerably increased exposure to the dry air in most pressurized airplanes. The outside air at altitudes is very dry and it becomes even drier when it is heated as it passes through the aircraft engine compressor and is then used for the cockpit and cabin. Symptoms caused by exposure to dry air include throat and nasal irritations and an increased susceptibility to subjective fatigue. These symptoms become mainly an annoyance to the passengers. They are more serious to the flight and cabin crews because of their repetitive exposure.

Most of the unpleasantness caused by low humidity is caused because the upper respiratory tract and the skin become dehydrated. The result is irritation of the throat and nasal passages, a general feeling of unpleasantness, and an increased susceptibility to minor infections in the breathing apparatus. Another frequent complaint arises when the conjunctiva (the mucous membrane that lines the inner surface of the eyelid and the exposed surface of the eyeball except for the cornea) and the cornea becomes dry causing unpleasant, scratchy, dry-feeling eyes.

Attempts at increasing the humidity levels, especially for long flights have been only marginally successful and the economic penalties have been very high. Not only are the required extra water tanks and additional plumbing expensive to install and maintain, but they take away revenue producing weight and require additional expensive space. If increased humidity is provided for more comfort at altitudes, the additional humidity condenses into water vapor and water droplets when the cabin returns to the increased pressure found at lower altitudes. This can have an adverse affect on electrical fixtures and equipment and result in saturated airplane insulation. The latter can cause dripping at low altitudes and frozen insulation at higher cruise altitudes.

Later airplanes (the Boeing B747-400 is an example) have substantially increased the humidity in the cockpits only. Unfortunately, the cockpit air, which is comfortably humidified during the long cruise period, becomes saturated during descent because of the decreased cockpit and cabin air pressure. Unwanted moisture condensation can then occur in the cockpit.

Recirculated Air

There is a continued controversy over the use of cabin and cockpit air for heating and cooling rather than using suitably treated fresh air from the outside. Maintaining the quality of cabin and cockpit air is a complex issue because pressurization, ventilation, contaminants, humidity, and temperature are all elements that should be considered.

There is general agreement that tobacco smoke and recirculated tobacco smoke is the single most important air pollutant. An ICAO resolution states that all airline flights should be smoke free—a position strongly supported by the Aerospace Medical Association. All US domestic flights have been required to be smoke free for several years. In addition, the number of US and other International flights that are now non-smoking flights have increased dramatically.

Cabin Air Studies

Studies of particulates in cabin air and of cabin air quality generally continue to be controversial even though the only two reasonably comprehensive studies that are available suggest very strongly that the problem has been overstated in the media and in some legal circles. The first was a study done by the US Department of Transportation (DOT) in 1989. This study monitored 92 US air carrier flights and revealed that respiratory particulates, carbon monoxide (CO), and ozone levels were well below the standards of the FAA, the American Society of Heating, Refrigerating, and Air Conditioning Engineers (ASHRAE), and of the Occupational Safety and Health Administration (OSHA). In addition, cultures for bacteria and fungi revealed colony counts less than those found in public buildings.

In 1984 the Air Transport Association (ATA) published a study in which the cabin air of 35 flights of eight large airlines was monitored using state-of-the-art instruments that provided direct and continuous readouts. This study included older aircraft such as the B-727, which had a 100% cabin air turnover, and newer aircraft such as the B-757 and the DC-8, which had a 50% cabin air turnover. Respiratory contaminants, CO levels, ozone, and volatile organic compounds were well below established standards. Like the DOT study, there was no evidence of significant air pollution. Bacteria and fungi colony counts were well below National Institute for Occupational Safety and Health (NIOSH) recommendation (Rayman, 1997).

One reason for the findings may be the use of High Efficiency Particulate Airfilters (HEPAs) in the most recent airplanes. A HEPA filters out 99% of anything greater than 0.3 μm (micro x 10^{-6}). Most airborne particles are much larger than this, which explains why the counts were so low in the DOT and ATA studies. It has been recommended that 'all commercial aircraft should have HEPAs installed and maintained' (Rayman, 1997).

A recent study on the microbiological composition of airliner cabin air is discussed in Chapter 17—The Challenging Role of the Flight Attendant. The study found that levels of microbiological concentrations found in the airliners studied were lower than typically found in locations associated with common daily activities and that therefore 'the risk of disease transmission as a result of microbial concentrations in airline cabin air is correspondingly low'.

The Problem of Recirculated Air

The problem of recirculated air is not simple. The most recent report of the Subcommittee on Passenger Health of the Aerospace Medical Association notes that airlines get some of the same complaints of the quality of cabin air from passengers on non-smoking flights as they get from passengers on flights that permit smoking. In a sentence that emphasizes the complexity of this issue, the Subcommittee's report states: 'The so-called problem of cabin air quality is most likely multifactorial (hypoxia, decreased barometric pressure, crowding, inactivity, temperature control, jet lag, noise, three dimensional motion, fear, stress, individual health, alcohol consumption, etc.) and we need to look at all possible causes before discarding any' (Thibeault, 1997). There seems little question that cabin air quality raises valid issues that need further examination.

Recirculated air also raises an economic issue for it takes considerably less power from the engines to recirculate already processed air from the engine's compressor than it does to take air from the outside of the airplane, bleed it off from the compressors, send it through an air cycle machine for processing, and finally deliver it to the cabin or cockpit. The manufacturers and the airlines use as much recirculated air as they think is reasonable for economic reasons of fuel cost savings.

Human Factors in Transport Aircraft

The outstanding technical achievements the industry has achieved in recent years have been accompanied by recognition that considerations of all of the factors that affect the people that fly transport airplanes and work with them must be considered. While a broadened view of aviation human factors has come into its own, the primary goal of the air transport industry is still safety.

In the past, cockpits were formed in traditional flight deck design from an assortment of instrument displays, knobs, switches, controls, and warnings of various sizes, shapes and colors that, for all practical purposes, were simply selected from a particular manufacturer's catalogue. Historically, the primary job of the design team was to see that all of the desired equipment could be squeezed into the allotted space. Excellent work was sometimes done in individual instances. Human factors concern did not extend to the whole aviation system.

Today, a coordinated flight deck design concept takes into account human capabilities and limitations in the broadest sense. Designers and certification officials are acutely aware of the basic differences between genders and between nations. People in each category are a part of the total aviation market. Today's systems approach considers the final product as a part of an integrated human-machine complex. The airplane is simply an important part of the aviation system. All of the people that will be involved in the operation of the airplane must be considered.

An Integrated Systems Approach

Today, in order to accomplish a truly systems approach, a major manufacturer in the US has a human factors staff that consists of more than 30 graduate level human factors professionals, including 12 Ph.Ds. The stated goal of this human factors staff is 'to apply human factors principles in the design of airplanes and support products so they can be produced, operated, and maintained safely and efficiently.' This group of human factors professionals includes expertise in human error, cognition, perception, decision making, ergonomics, and procedures development (Graeber, 1997). Major European manufacturers have similar staffs. An integrated system approach for the manufacture of transport airplanes is accepted on a worldwide basis.

Basically, the flight deck is a workplace. It has consoles, controls, and displays, which must be operated in a complex and often changing environment. A basic principle is that all of the consoles, controls and displays must be operated efficiently and safely. This must be done while always keeping in mind the basic limitations of the individuals who will operate and support the airplane and its systems.

In the cockpit, consideration must be given to such things as reach, the fit and function of seats and controls, and internal and external vision requirements. Accommodation for the use and storage of maps, flight manuals, and charts, and such comfort items as coffee cup holders, footrests, provisions for food service, and the storage of personal items such as suitcases and clothing are all considerations.

The increased sophistication of today's flight compartments is the result of continued effort by joint teams of airplane manufacturers, airlines, pilot groups, and instrument manufacturers. The result is an obvious improvement over the old cockpits, and this work continues as the industry progresses. While this discussion is primarily based upon the flight deck, it is obvious that there are human factors considerations throughout the airplane and throughout the other parts of the aviation system. Consideration must also be given to the needs and limitations of the variety of people who will be called upon to operate and maintain the system components.

General Design Considerations

The human being in the cockpit is the critical and the most flexible component of the aviation system. His/her performance in the cockpit is the last opportunity to compensate for any combination of weakened system defenses that can lead to accidents or incidents. There are wide variations in individual size, individual limitations, and in individual performance. It is not surprising that a great deal of effort is made to maximize the performance of all crewmembers.

Gerald Stone has divided, what he has called, ergonomic design factors into two main groups: behavioral (i.e., dependent on the operator), and those items that are design-dependent and therefore dependent upon the manufacturer (Stone, 1989). His model of ergonomic efficiency considerations is shown in Figure 5.3, and his model of human characteristics is shown in Figure 5.4. These are not simple concepts. It is obvious that professionals and careful planning are required to satisfy sometimes conflicting needs.

A major international problem in the cockpit involves the use of color. One of the reasons is that there are no international rules regarding the use of colors, although there is a growing recognition that this is a problem area that is rife with international complications. A major difficulty is that not all cultures view some specific colors in the same light.

In the US, the principal aviation colors of red, amber, and green are now specified in FAR 25.1322 as follows:

- Red for warning lights (lights indicating a hazard which may require immediate corrective action);

- Amber, for caution lights (lights indicating the possible need for future corrective action);

- Green, for safe operation lights; and

- Any other color, including white for lights not described in paragraphs (a) through (c) of this section, provided the color differs sufficiently from the colors prescribed in paragraphs (a) through (c) of this section to avoid possible confusion.

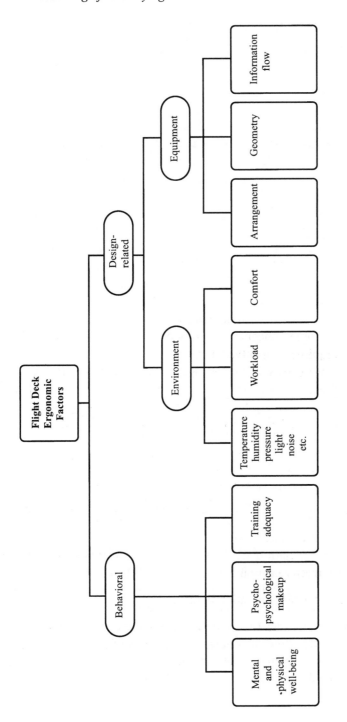

Figure 5.3 **Ergonomic Efficiency Considerations**
Source: 'The Ergonomic Integrated Flight Deck', page 2, Stone, 1989

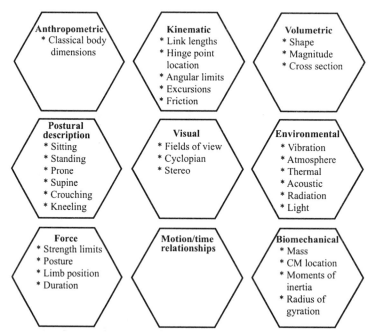

Figure 5.4   Human Characteristics

Source: Adapted from 'The Ergonomic Integrated Flight Deck', page 3, Stone, 1989

Figure 5.5 shows the number and complexity of the items that must be considered in the design of a modern transport. Clearly these items should be considered as early as possible in the design stage as changes or modifications made during later stages of the manufacturing process can be both time-consuming and inordinately expensive.

Three elements are involved in developing the crew station for a new airplane. These three elements are (1) the size and shape, (2) the reach, and (3) the vision of the prospective user population. The size and shape of the prospective user population can become very involved both because of the number of combinations possible and because of the dearth of data for other than US and European males. A typical computer model utilizes 14 external body dimensions of the 5th, 50th, and 95th percentile groups of specific groups of potential pilots in order to ensure that all crewmembers are able to reach their controls from various positions in the flight compartment. The designer is forced to utilize all available data and then use informed judgment to cover gender and racial differences.

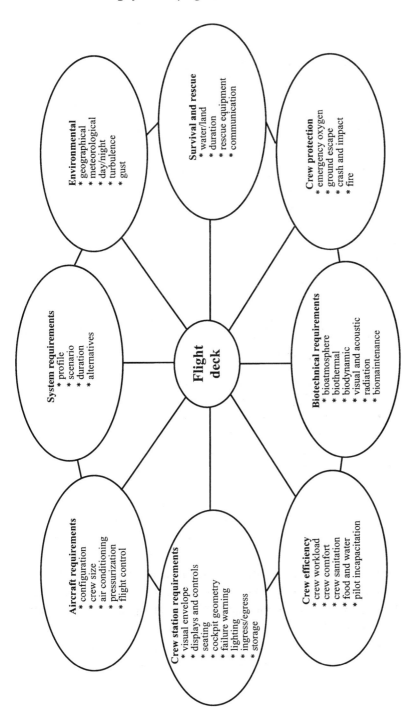

Figure 5.5   Integrated Flight Deck System
Source: 'The Ergonomic Integrated Flight Deck', page 5, Stone, 1989

A second element is reach. Reach has three variables: restraint configuration (utilizing three positions of the shoulder harness), type of grip (fingertip, pinch, or grasp), and type of clothing. Twenty-seven combinations of reach can be simulated. One manufacturer has stated that it examined the sizes and shapes of 3,600 individuals in determining limits that were acceptable (Stone, 1989).

The third element involved in developing the crew station for a new airplane is vision. Here consideration must be given for both displays within the cockpit and for the external environment. External vision requirements can define the configuration of the windshield. Requirements consider, among other things, the blocking obscuration of such items as fire handles and control wheels. External visual requirements may change drastically in a future SST (supersonic transport) which may be designed without a droop nose and have quite different visual external requirements. (See Chapter 3—A Brief History of Human Factors and its Development in Aviation and Chapter 22—The Air Transport Future.)

An interesting concept in the visual element involves the visual cone. It is well known that the field of vision narrows under periods of high concentration or stress. In order to recognize this phenomena, information necessary during critical periods of flight is displayed within what is known as the primary vision cone. This has been defined as the limits of vision associated with eye motion alone. The secondary vision cone includes the area which can be visually covered with a combination of both head and eye motion.

Recently, increased attention has involved crash injury protection. The National Highway Transportation Safety Board developed a computer program to analyze the reaction of passengers in automobile crash situations. The program was adapted by the Air Force to analyze aircraft crash scenarios and the program was later modified to extend to commercial transports. It has been used in recent times to specify acceptable locations for head-up displays in present aircraft.

To illustrate the complexity of these issues, Table 5.1 show 38 items which must be considered in the development of the flight deck of a new transport (Stone, 1989).

While even this list may not include all of the items that may be required in future air transports, there is little question that the human factors involved are important and that the issues raised should be raised at an early stage.

Table 5.1   Flight Deck Equipment Requirements

* Primary flight	* Closets
* Secondary flight	* Suitcase storage
* Navigation	* Briefcase storage
* Communication	* Coffee cup holder
* Power	* Pencil/pen holder
* Systems (hydraulic,	* Trash container
electrical, environmental)	* Eating facilities
* Landing gear	
* Emergency	
* Door	Emergency equipment

Operational items

* Fire axe
* escape equipment
* Headsets
* smoke goggles/hood/PBE
* Microphones
* Flotation devices
* Cables
* Fire extinguishers
* Oxygen masks
* Writing tables
* Map holders
* Flight manuals   Hand holds
* Navigation charts
* Lights   Windscreen

Seats   Crew members

Foot rests   Observers

Source: 'The Ergonomic Integrated Flight Deck', page 10, Stone, 1989

The 'Sidestick' Issue

There have been many innovations in air transport cockpits with human factors implications in the past few years, but none which was more dramatic
and initially controversial than the Airbus Industries sidestick. The following
excerpt from an article by Ron Rogers[1] in the June/July 1998 *Air Line Pilot*
titled, 'Flying the B-777-300', illustrates the operational and human factors
complexity of the sidestick issue. The initial controversy about the sidestick

---

[1] At the time of writing, Captain Rogers is Chair of the ALPA Airworthiness, Performance, Evaluation and Certification Committee.

Figure 5.6   Airbus Flight Deck
Source: Courtesy of Airbus Industrie

is subsiding as experience with it has increased. Figure 5.6 displays the inside of an Airbus cockpit.

Why no Sidestick?

Why did Boeing elect not to use a sidestick? A sidestick has many advantages. It is comfortable, and it frees up a lot of real estate in front of the pilot, but it also brings a few problems. One is that the pilot not flying cannot monitor the pilot flying's inputs through control movement. Also, the autopilot's input can be monitored more easily through observing the movement of the control yokes. The use of a "small displacement controller" also makes exerting high control forces to indicated edges of the flight envelope nearly impossible. As a result, Boeing would have had to effect hard limits on the flight controls rather than the desired soft limits. (This is a basis difference in operational philosophy.) Boeing made a detailed analysis using an in-flight simulator and found the control yoke to work better. The B-777-300 uses a somewhat smaller yoke than other Boeing models. Whether one prefers a sidestick or yoke, few can argue that anyone produces a better-handling aircraft than Boeing. (Rogers, 1998)

It should be clearly understood that there are many pilots who do prefer the sidestick controller, and in general, Airbus aircraft. Certainly both manufacturers have made significant contributions with their design elements and both produce excellent aircraft. Some of the differences between their automation philosophies are discussed in Chapter 11—Automation.

Summary

This has been an obviously highly selective and brief history of the development of air transport airplanes. Bill Gunston's *The Illustrated Encyclopedia of Propeller Airliners*, Time-Life and Oliver E. Allen's *The Airline Builders*, and *The American Heritage of Flight* which is edited by Alvin M. Josephy, Jr. are recommended texts for any reader wishing more complete versions of this history. *The American Heritage of Flight* goes up to the Korean War and the Boeing 707s.

References

Bryan, C.D. B. (1979). *The National Air and Space Museum*, Harry N. Abrams, Inc., New York.

Caidin, Martin (1992). 'High-altitude radiation risk', *Professional Pilot*, January 1992, Alexandria.Virginia.

FAA, Civil Aeromedical Institute (1998). CARI-5E (computer program available at http://www.cami.jccbi.gov/aam-600/610/600radio.html.), Oklahoma City, Oklahoma.

Gilbert, James (1970). *The Great Planes*, Grosset and Dunlap, Inc., New York.

Graeber, Curtis L. (1997). 'Enhancing Safety Through Human Factors', The Boeing Commercial Airplane Company, Seattle, Washington.

Green, Roger G, James, Helen, Gradwell, David, Green, Roger L, (1991). *Human Factors for Pilots*, Avebury Technical, Aldershot, England.

Gunston, Bill, (1980). *The Illustrated Encyclopedia of Propeller Airplanes*, Phoebus Publishing Co., London, England.

Hawkins, Frank H. (1993). *Human Factors in Flight,* Second Edition ed. Harry W. Orlady, Ashgate Publishing Co., Ltd., Aldershot, England.

Preston, F.S. (1979). 'Aircrew Stress', in *Symposium on Human Factors in Civil Aviation*, 3-7 September 1979, The VNV Dutch Airline Pilots Association, The Hague, Netherlands.

Rayman, Russell B. (1997). 'Passenger Safety, Health, and Comfort, A Review', *Aviation, Space, and Environmental Medicine*, Aerospace Medical Association, Alexandria, Virginia.

Rogers, Ron (1998). 'Flying the B-777-300', *Airline Pilot*, June-July 1998, Herndon, Virginia.

Sells, S. B. and Berry, Charles A.(1961). *Human Factors in Jet and Space Travel*, The Ronald Press Company, New York.

Serling, Robert J. (1982). *The Jet Age*, Time-Life Books, Chicago, Illinois.

Stone, Gerald, (1989). 'The Integrated Flight Deck', presented to SAE, Aviation Research and Education Foundation Human Error Avoidance Techniques Conference, Douglas Aircraft Company (now Boeing), Long Beach, California.

Wagstaff, Bill (1996). 'High-altitude radiation: How much is too much?, *Aviation International News*, 1 October 1996, Midland Park, New Jersey.

6 The Social Environment

To ignore the setting is to enfeeble the application.   (Charles Perrow, 1982)

The last defense against a transport aircraft accident or incident resides in the flight deck of the aircraft. Therefore, it is not surprising that from 60% to 80% of transport accidents are found to have significant cockpit crew involvement with those unfortunate occurrences. This statistic has not varied over the past half-century. In the past, most accident investigations have attributed the principal cause of the accident to pilot error. Jens Rasmussen has given us one reason that so many investigations attributed the accident to pilot error when he noted that, '...the identification of an event as a human error depends entirely upon the stop rule[1] applied for the explanatory search after the fact' (Rasmussen, 1987). This chapter (and other chapters in this book) goes beyond that first level of investigation.

Today, we know that human involvement is not restricted to the cockpit inhabitants alone. Other factors may well have been involved and many traditional investigations have stopped too soon. Concerned observers believe that current human factors provide an opportunity to make a demonstrable improvement in air transport safety. A growing number of experts believe that the 'social environment'[2] is either ignored, or that in some cases it is defined too narrowly (Maurino, et al., 1995).

Today's aviation human factors is concerned with a total system approach. While it is particularly concerned with behavioral issues and the social environment, it is important to recognize that the other more traditional elements of human factors in aviation continue to play a significant role in air transport operations. Present day aviation human factors is concerned with meeting the identified needs of a very dynamic industry. Human factors in air transport operations first expanded from its early physiological concerns to

---

[1] The 'stop rule' tells us that many accident investigations stop when one or more possible causes are found which are familiar and to which something can be done for correction. Therefore, they are acceptable as an explanation. The explanation then may tell us what happened, but not why.

[2] As we are using the term, social environment includes the national culture, the regulatory culture, and the organizational, corporate, or airline culture.

those involving man-machine interface with dials and switches, and it has continued to expand until today's multidisciplinary aviation human factors also addresses behavioral and many other aspects of the man-machine-aviation system.

The Social Environment in Today's Operations

Aviation human factors tells us that in order to understand the 'liveware' in today's air transport operations, it is necessary to understand not only the individuals directly involved with the operation, but also to be aware of the social environment—the national, regulatory, and organizational cultures—that can affect their behavior. Individual countries have cultural differences that distinguish them from other countries.

These differences can affect aircraft operations. There are also major, though sometimes less obvious, culture differences even among airlines in the same country. These differences also can affect aircraft operations. Ashleigh Merritt (1993) has defined culture 'as the values and practices that we share with others that help define us as a group, especially in relation to other groups.' It is a helpful definition.

The national culture is the broadest of the factors involving the social environment. It affects the behavior of virtually all who have responsibilities. The national culture can be quite different in different nations. Air transport safety, which is the primary concern for everyone, involves the behavior of all individuals involved in air transport operations. If modification of any of the values or practices that are a part of the national culture is required, special attention to individual needs may be necessary.

In many areas, regulatory culture, which is the second of the factors affecting operational behavior, can be considered a subset of the national culture. Regulatory effectiveness, which to a considerable degree is affected by the regulatory culture, is believed by some to be a major problem in many Third World States (Faizi, 1997). Obviously, the quality of a given State's regulatory system is limited by the availability of fully qualified personnel. While this is a minimal problem for some fortunate States, for others it presents a considerable challenge.

The third factor is the organizational culture. It can also be called the airline or corporate culture. But whatever it is called, safety should be its central element. Whether or not an airplane or system is good, mediocre, or even poor, once that airplane or system is purchased, it is the absolute responsibility of the airline that purchased it to operate that new airplane or system safely and efficiently over a long period of time. The organizational or corporate culture is a major factor in determining whether or not the airline can do this. Safety, which was always important, has now become a requirement for an airline's survival.

The National Culture

National cultures can be a significant consideration in air transport operations for many reasons. National cultures can affect an individual's response to regulatory rules and pronouncements, and also can affect the individual's response to his or her organization and its policies and procedures.

Important motivational factors can vary considerably as a function of nationality and cultural heritage. Even perceptions can differ depending upon the person's cultural background (Gabriel, 1975). Well-embedded beliefs and values are very resistant to change and when they do change, they change very slowly. Beliefs and values are part of each individual's cultural background and inevitably are an important part of the national culture.

Pilots from different countries have different cultures. While the operation of identical airplanes under routine conditions in different cultures will not appear very different to the casual observer, even then, cultural difference can affect the behavior of the crew. The impact and underlying influence of culture may become significantly more apparent if the crew is challenged with unanticipated operational problems.

This is important to recognize for although cultural differences can create subtle differences under normal conditions, they may become even more important under flight conditions that are not normal. Cultural difference problems are exacerbated with adverse operational conditions. The reason is that all people tend to revert to ingrained well-established behavior under stressful conditions. Inappropriate operational behavior can be a part of a national culture and entirely normal within that context. In other cases, normal behavior can be entirely inappropriate for some non-normal or emergency conditions and sometimes even for conditions that can be considered routine and normal.

National culture was almost certainly a factor in the Japan Air Lines flight that under normal conditions in good weather landed in the San Francisco Bay short of Runway 27. The copilot, unwilling to challenge his superior, was well aware of the impending water landing but took no corrective action. Since that time, the airline has actively pursued CRM training that is responsive to the cultural needs of its pilots.

Individualism, Collectivism, and Power Distance

The national cultural environment can, for the purposes of this discussion, be characterized in terms of individualism, collectivism, and power distance. The Dutch psychologist Hofstede who is truly a pioneer in cross-cultural psychology, coined these terms (Hofstede, 1980). Considered within the flight deck context, individualism refers to a high degree of independence among crewmembers, even if it is sometimes sublimated by regulatory and organizational demands. Collectivism refers to greater interdependence and to group-oriented efforts that are inherent in the society. Power distance can be defined as the extent to which the less powerful person in a society accepts inequality in power or influence and considers it normal, justified, and usually completely rational.

The three dimensions of individualism, collectivism, and power distance are important parts of aviation human factors in air transport operations. These dimensions can vary between and, to a lesser degree, within nations (Johnston, 1993, and Redding and Ogilvie, 1984). They are particularly relevant because there appears to be a correlation between individualism, power distance, and accident rates (Weener and Russell, 1993). However, it is an oversimplification to ignore such factors as infrastructure, training, maintenance, equipment and the regulatory and organizational cultures. Each of them is a necessary element that should be considered in evaluating the safety and efficiency of air transport operations.

Power distance is a particularly important measure. The relationship between captain and copilot, which involves power distance, has been called the trans-cockpit authority gradient. The term was described by Professor Elwyn Edwards in 1975 (Edwards, 1975), and until recently has been given little organized attention.

A society, which has a low power distance and high individualism, has a relatively easier time accepting many of the concepts that are required to have a good cockpit team and a desirable trans-authority cockpit gradient. However, even a society that has a very good power gradient, cannot guarantee it will have no problems. Every society has a wide variety of individuals. Some of them adapt to change very nicely, while others find change much more difficult. Developing a good cockpit operating team requires major philosophical changes for those individuals whose normal behavior is not tolerant of the position of other members of the team. There is no room for an inappropriate cockpit authority gradient in today's flight operations.

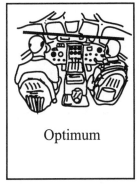
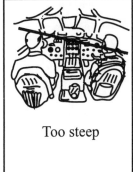
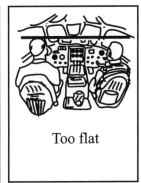

Optimum   Too steep   Too flat

Figure 6.1   The Trans-Authority Gradient
Source: *Human Factors in Flight,* Second Edition, page 35, Hawkins, 1993 and reprinted with permission of Ashgate Publishing Limited

Societies with a high power distance and low individualism usually find it more difficult to develop a good team operation. Safe and efficient air transport requires a relatively low power distance and relatively high individualism in flight crews. They are important qualities in flight crews because a high power distance, between captain and copilot, especially when combined with a low level of individualism on the part of the copilot, is incompatible with the team concept. The team concept is necessary to safely operate today's transport airplanes. This by no means suggests that such adaptation is not possible but it does suggest that flight crews that come from a high power distance and low individualism culture have a greater problem. They have to change fundamental personal characteristics, at least while they are flying a transport airplane.

According to Hofstede and others, individualism is high and power distance is relatively low in developed countries such as those in most of Europe, North America, Australia, and in New Zealand. Perhaps coincidentally, these countries have good safety records. One reason may be that those countries that have a large power distance have a greater problem in effectively developing many of the new human factor concepts. One of the most important and difficult of these concepts is for the copilot to effectively monitor and, if necessary, challenge the performance of the captain because their national culture reinforces the concept that the 'captain is king' and cannot make a mistake.

Safety in air transport is a complex phenomenon. As we have noted, the infrastructure, training, equipment, regulatory, corporate philosophies, and in some cases the available skill levels of the operating crews are important. National culture often plays a significant role in all of them. However, we should never forget that there are good safe operations in all regions of the world. Correlations are often easier to establish than underlying causes.

The operational heritage of air transport operations makes determination of the level of assertiveness that is appropriate in subordinate crewmembers a difficult problem in virtually all national cultures. It is a mistake to assume that any problems associated with captain—copilot operational relationships are restricted to countries with poor safety records. For example, the US NTSB has listed the failure of the copilot to effectively monitor or to actively intervene if required as a causal factor in several recent accidents (NTSB, 1994). Actually, this phenomenon is not new. It has been a problem since the days that a copilot was added to the flight crew. What has changed are the methods used to address the problem.

It has been stated that 'facts do not do well out of context'. Unfortunately, in air transport operations safety facts are too often taken out of context. There are, and always have been, obvious differences in relevant factors as they exist in several parts of the world. Despite clear differences in power distance, collectivism, and individualism in geographical areas and between countries, there are airlines with good records in every region in the world. Regional generalizations of air transport safety can thus be very misleading. This point is important for it shows that some airlines have been able to operate effectively despite inherent safety problems in the social environment in which they operate. Researchers have not always recognized these differences. The result has been an oversimplification and a series of too broad generalizations that have been stated as universal facts.

Other Factors

In addition to measures like power distance and individualism, two other human factors that can modify the national culture may be important in aviation safety. They are (1) the individual's specific background, i.e., military or civilian, the type and level of his education, and other parts of his national culture, and (2) a familiarity with computers. This latter can be important in the operation of advanced technology airplanes. Both factors can influence behavior. When considering pilots, it is a mistake to insist that familiarity and competence with computers is age related. The important point is that pilots must be flexible and be able to adapt to new conditions. The evidence regarding the importance of a familiarity with computers is sketchy because there is simply little comparative data available. Nevertheless, many trainers of pilots that are transitioning to advanced cockpit airplanes hold that belief. This concept is further discussed in Chapter 16—Selection and Training.

Whatever the name and importance of these subcultures, there is little question that individual behavior is affected by the social background of the individuals. In any culture, younger pilots, especially initially, seem at greater ease with the considerably higher level of computerization in new technology airplanes and are less resistant to changing operating concepts than are some of their older compatriots. Both groups are important parts of the aviation system. A somewhat cynical and imaginary observer might suggest that the older pilots in our cultures sometimes could use some of the flexibility and imagination suggested by the younger pilots, while the younger pilots would do well to have some of the conservatism and skepticism that experience has given their elders. One experienced observer believes that while younger pilots frequently learn new concepts easier than the older pilots in transition training courses, the older pilots make up their seeming deficiencies by the end of the course, and at completion are in the forefront of the class (B. S. Grieve, personal communication).

The human factors characteristics that are a part of the national culture must be recognized if we are to increase the safety and efficiency of air transport operations. Some of them may need special attention. They can involve fundamental structures of social life that are difficult to change and which normally change only slowly through time. Not surprisingly, Redding and Ogilvie (1984) have found that flight crews carry with them cultural conditioning, which influences their operational behavior. Unfortunately, the need to further increase the safety record in modern air transport operation does not permit the luxury of allowing a slow adaptation to critical behavioral modifications. Special consideration may be required to implement desirable operational safety concepts such as Crew Resource Management. Chapter 13—Crew Resource Management (CRM) and the Team Approach—is devoted to this topic.

General acceptance of new behavioral concepts and operational strategies has evolved during the past decade. New concepts include CRM and many changes that are an inherent part of CRM. These include the acceptance of such items as increasing the responsibility and authority of the 2nd in command, the total team concept in aircraft operations, and stressing the importance of the PNF's (pilot-not-flying's) effective monitoring of the performance of the PF (pilot-flying) regardless of cockpit status of either. Equally important is recognizing that errors will occur, that any individual can make an error, and recognizing that it is essential to both minimize and then control the inevitable operational errors. These concepts and attitudinal changes have become an inherent part of modern transport aviation. Establishing these concepts and the allied supervisory attitudes seems to be more difficult to accomplish in some cultures than in others.

The Regulatory Culture

Regulatory responsibilities are an important national function. They are a crucial consideration in the certification and operation of airplanes, systems, and personnel. Regulatory cultures may vary considerably between different countries. A principal factor is that individual countries can be quite proprietary regarding their own regulatory rights and responsibilities. It would be a major mistake to underestimate the importance of the regulatory culture.

Because many safety measures are expensive for the operators to implement, it frequently has been necessary to make desirable safety measures a new regulatory recommendation or requirement in order to insure uniform implementation. This is part of the regulatory culture. In the US, examples of regulatory safety requirements are those for Ground Proximity Warning Systems (GPWS), Traffic Collision Avoidance Systems (TCAS), and Wind Shear Advisory Systems (WSAS). An example of a safety recommendation in the US is the FAA's strong support and promotion of the CRM movement (Advisory Circular 120-51B). The extent of regulatory safety measures varies between nations. Safety regulations have become even more important since economic deregulation of the airlines is occurring throughout the world.

Economic Deregulation

Economic deregulation of the airlines in the United States, and in many other countries, is a fact of life. One of the consequences of economic deregulation is that airline companies can no longer depend upon often quite substantial subsidies in order to meet their operating costs. It also means that even well established and well-run airlines have to face hard competition from 'bare bones' start-up companies on routes that historically have been non-competitive. Within each airline there is internal competition among departments for scarce dollars. Unfortunately, safety and other operational considerations must compete for these scarce dollars. A strong, capable, and responsible regulatory agency that recognizes current human factors minimizes the possibility that relevant safety items become part of the economic competition that is inherent in a deregulated society. It is extremely doubtful that such safety provisions as GPWS, TCAS, and WSAS, or their equivalents, each of which costs money would be installed on all US airlines if they were not required by regulation.

However there are exceptions. There are cases in which individual operators without a mandatory regulatory requirement have taken safety initiatives. An example is the recent announcement that a major cargo airline in the US is installing collision avoidance systems in all of its aircraft in spite of the fact that they are not required in cargo-only airplanes. Another is the $400 million dollar commitment that 15 US airlines made to install smoke detectors in the cargo holds of 3,700 airplanes. This commitment was made in advance of any regulation. It was made after the 11 May 1996 Florida Everglades crash of a ValuJet DC-9. During the investigation it became reasonably sure that fire in a cargo hold was responsible for the crash that killed 110 people. Another major exception to the rationale in the previous paragraph has been the recent purchase of enhanced ground proximity warning systems (EGPWS) by major airlines throughout the world. This followed the crash of the American Airlines B-757 at Cali, Colombia.

It is believed that the Cali accident may well have been prevented by EGPWS, which was not available before the crash. At least one major manufacturer has already made EGPWS standard equipment for all the transports it manufactures in the future, and it is also expected that there will be retrofit requirements in many countries. The US air transport industry recognized that EGPWS should alleviate a valid safety problem and took action without waiting for a regulation. As this chapter is being written, 175 planes have been equipped with EGPWS that requires navigation systems that use global-positioning satellites and digital maps to alert pilots of dangerous terrain. By 2003, 6,300 other transport airplanes will have similar installations. The cost of the new equipment is at least $600 million (Flight International, 6 January 1998). The FAA also moved with alacrity in this area. It started regulations requiring future installation of EGPWS in US airline airplanes that will become official as soon as regulatory processing requirements are completed.

International Regulatory Standards

In an action, which reinforces the importance of ICAO SARPS (Standards and Recommended Practices—see Appendix D), the US FAA in the early 1990s developed a program called the International Aviation Safety Assessment Program (IASAP). This program monitors the effectiveness of the compliance of other States with international aviation safety oversight rules—essentially ICAO SARPS. Airlines from a State that does not provide the stipulated safety oversight are prohibited from entering the lucrative US market. The FAA has made it clear it is not assessing whether or not an individual airline is safe or less safe. Its concern is whether or not the State has a civil aviation authority in place and whether or not that civil authority ensures that accepted operational and safety procedures are maintained by its air carrier(s). In a very real sense, the IASAP evaluates the effectiveness of the State's regulatory culture.

Several years later in an effort to increase safety by expanding ICAO's role in international air safety, Dr. Assad Kotaite, President of ICAO, told the ICAO Council on 24 February 1997 that audits of national air transport standards performed by ICAO should be the accepted norm. He then called on ICAO's 185 member States to give ICAO the necessary powers to perform these audits. Dr. Kotaite suggested that the means to accomplish this: 'might well be found in the introduction of international technical inspections, or safety and security audits, which call on states to rectify disclosed deficiencies'. He states further that, 'ICAO, as an international body, should be empowered to check closely the implementation of safety and security standards, and to carry out regular inspections' (Kotaite, 1997). This obviously would be an expansion of the ICAO role in international air safety and could well take the place of the IASAP movement that was started by the US FAA.

Very few ICAO observers expected rapid movement on any proposal that has such far-reaching consequences as would these ICAO audits. Traditionally, approval of similar projects has been a slow process. However, in this case a meeting held by representatives of 145 countries on 10-12 November 1997 has accelerated the audit program. This meeting supported the use of selected ICAO audits among ICAO's 185 member States and careful implementation of the audit results. The US FAA and the European Civil Aviation Conference (ECAC), which represents 36 European nations, are among the outside organizations that have supported the ICAO program.

ICAO audits may help the regulatory authorities of certain Third World, or what have also been called 'developing countries'. These countries can have special problems. Many of them do not have a well-developed infrastructure or individuals with the expertise that is required to develop a sound regulatory culture. The head of IATA's Safety Committee, Pakistan International Airline's corporate safety chief, Captain Amjad Faizi, recently very bluntly told an international conference that regulatory slackness is at the top of the list of Third-World airline safety problems (Faizi, 1997).

Budgets are inevitably tight in these countries. This increases an undesirable tendency to have the person who is head of the national airline also to be the chief of the regulatory authority. While this is understandable, it creates an obvious potential conflict of interest. A second problem is that with a shortage of qualified people, the temptation is very real to make the head of the regulatory authority simply the beneficiary of a 'political plum'.

Another organizational problem for developing countries is that frequently the head of the regulatory authority is either military or from an exclusively military background. Some of these individuals are not sensitive to the differences between the military and the civilian safety culture or to the ways that the military organization varies from a civilian business organization. In some cases the military actually runs at least part of the civil airline structure. This problem can be very closely related to problems associated with a lack of fully qualified personnel.

Finally a problem in some States is that the regulatory authority also is the responsible investigative authority. On occasion, this can create a very real conflict of interest. Too often, because of the lack of functional separation and the lack of expertise in the regulatory authority, the regulatory authority can become little more than a rubber-stamping agency. Overall safety is not enhanced (Faizi, 1997).

The Organizational Culture

The organizational or corporate culture has a direct influence on the safety of operations. This is particularly true for flight crews because constant supervision is impractical, probably impossible, and certainly undesirable from the pilot's perspective. It is one reason that Professor J. Richard Hackman of Harvard University referred to cockpit crews as self-managing teams. A powerful value system that is part of a corporate and professional culture is crucial to ensure that unacceptable deviations from standard operating procedures and from other important safety practices are not an accepted part of day-to-day operations. Corporate culture has a powerful influence on employee behavior.

As others have noted, 'organizational culture is a widely used but variously-defined term' (Maurino et al., 1995). However, regardless of definition, a given and successful manner of addressing safety deficiencies in one culture may be entirely inappropriate (and therefore ineffective) in another

(Maurino, 1994). Behavior, which appears benignly paternalistic and encouraging in one culture, can seem invasive, rude and insensitive in another. This is an important consideration, for the organizational culture must be considered carefully. It is a critical element in the air transport system. As J. Richard Hackman (1986) has noted, 'the overall task of completing a flying mission is always a team task, and 'cockpit crews always operate in an *organizational context'* (italics supplied).

Making that same point, John Lauber has noted that: 'An individual's performance never takes place in a vacuum, but always occurs within an organizational and cultural context' (Lauber, 1989). Professor Charles Perrow of Yale University further emphasized that point when he wrote: 'Human factors work takes place in an organizational setting....To ignore the setting is to enfeeble the application' (Perrow, 1982).

An airline's safety culture is, of course, a critical component in the quality of its operation. In spite of the fact that some very good airlines have been able to keep their operations separated from other corporate activities, we believe that the airline's safety culture should be an intrinsic part of the corporate culture. There is little doubt that the safety culture is much stronger if it is a recognized central part of the corporate culture. It is of some interest that well before there was today's interest in the effect of corporate culture on air safety, Lautman and Gallimore noted that: 'In the broad context of management, these operators (that had superior safety records) characterize safety as beginning at the top of the organization, with a strong emphasis on safety, *that permeates the entire operation'* (Lautman and Gallimore, 1987, italics supplied).

There can be real differences among organizations within a national culture and, because of the variations, analyzing cultural differences among organizations can involve some very gray areas. Within each airline, the organizational culture also is affected by the individual cultures that distinguish the other personnel who are involved in day-to-day-operation. Each individual is part of a specific culture that is a part of the safety and organizational culture, and eventually, of the social environment in which operations are conducted. For example, ramp personnel, pilots, or flight attendants are not always like each other or even like their unions, nor do they always behave like other members of their group. The norms of behavior for these groups may not be very similar to the behavioral norms of doctors, accountants, passenger agents, or plumbers. One of the important purposes of an airline culture is to embrace and sublimate the diverse individual cultures toward the common goals of increasing operational safety and efficiency.

Many aspects of the corporate or organizational culture are unwritten. 'Unwritten rules of behavior can have a profound influence on both perception and actions.' Captain Neil Johnston has written eloquently of the real-world importance of both written and unwritten organizational rules and in their interpretation and implementation. In 'Organizational and Motivational Aspects of Air Carrier Pilot Decision Making,' Captain Johnston has noted:

> There is always an organizational climate; and clearly this can be both good and bad. There are unwritten rules of behavior, and perceptions as to the consequences of certain types of behavior and decision-making. When you find these out, you find out what the reality of an organizational decision making climate is like.

> So what does all of this mean in practice? Well, on one hand, an organization can be very supportive of its pilots, give good guidance and support, and enjoy the respect and confidence of its pilots. But, on the other hand, an organization can also be a source of stress and inconsistent management. There can be considerable role ambiguity and psychological distrust and uncertainty; for instance, fear of disproportionate or arbitrary discipline, or uncertainty associated with inconsistent standards. In addition, pilots may have perceptions in these matters which do not fully accord with the facts. However irrational and unfounded such perceptions might be, once they exist they are a tangible and significant influence on operational decision making and pilot performance in general. They must consequently be viewed as tangible and potentially significant influences on pilot behavior.

> As I have argued elsewhere:

> In an environment where everything from a bird strike or tyre burst to an unruly passenger not only requires a report, but also carries the possibility of a detailed inquiry, an individual can perceive himself to be under continuous background pressure and stress. Supervisory procedures ...set the background to the work environment.... The link between such factors and stress as well as aviation safety, has been [well] identified...   (Johnston, 1986)

The corporate culture must always work within the regulatory culture and the contribution of the corporate culture to safety is critical. As we have previously noted: 'Whether or not an airplane or system is good, mediocre, or even poor, once that airplane or system is purchased, it is the absolute responsibility of the airline that operates it, to operate that new airplane or system safely and efficiently.' In many ways, the corporate culture is the most important culture of them all. It is where we find the 'bottom line'.

Making a Sound Corporate Culture Work in an Airline Environment

The Four Ps

Implementation of the basic principles that are inherent in a good corporate and good safety culture is not easily accomplished. NASA's Asaf Degani, the University of Miami's Earl Wiener, Robert Mudge, and many others have studied this problem. In an attempt to identify and organize this task, Degani and Wiener have combined the most relevant issues into the Four Ps (Degani and Wiener, 1994). The Four Ps are: Philosophy, Policies, Procedures, and Practices.

Philosophy

The first of the Four Ps is philosophy. Philosophy, as it is manifested in the corporate culture, forms the basic foundation. In order to effectively implement basic philosophic principles, corporate policies and procedures must be consistent with the corporate philosophy. Assurance that the routine practices on the airline conform to established policies and procedures is also required.

The basic operational concern in an airline operation is assure that it is safe and efficient. It is important to recognize that all factors involved in safety and efficiency are important and that the Four Ps are but one facet of a complex system. Other issues involving safety include such elements as decision making, ergonomics, infrastructures, regulations, automation, management, etc. and they are all relevant. All of these factors should be considered in any attempt to better understand and improve safety and efficiency in air transport operations.

Airline philosophy is often confused with airline culture.[3] While the terms are indeed different, as they are used in aviation, they can be very similar. A dictionary definition of philosophy states that among other things, philosophy can be considered 'a system of principles for guidance in practical affairs'.[4]

Every airline has a corporate culture and that culture should reflect its corporate philosophy. Both should indicate a strong safety policy that perme-

---

[3] *Webster's New American Dictionary* (1995), Smithmark Publishers, Inc., New York, defines culture as 'the customary beliefs, social forms, and material traits of a racial, religious, or social group.'

[4] *The Random House Dictionary* (1980). Ballantine Books, Random House, Inc., New York.

ates the entire organization. Unfortunately, the safety policy is often not always clearly stated nor is it always effectively implemented even if clearly stated. For example, the British Report of the Investigation into the Clapham Junction Railway Accident states, among other things: 'The railway company management commitment to safety is unequivocal. The accident and its causes have shown that bad workmanship, poor supervision and poor management combined to undermine that commitment.'

It obviously takes more than just high-sounding phrases in company bylaws if they are to be effectively implemented. The corporate safety policy must be aggressively, practically and effectively promoted at all levels, as it obviously was not in the Clapham Junction Railway case. The aggressive, practical, and effective promotion of a sound corporate policy at all levels is increasingly the cost of economic survival in the air transport industry. It is not enough simply to have a good philosophy written somewhere that has little impact on day-to-day operations.

Policies

Corporate policies obviously should reflect corporate philosophy. The purpose of corporate operational policies is to state the manner in which the management of the airline expects its operations to be conducted. Operational policies cover many areas including flying, training, maintenance, the exercise of authority, personal conduct, and punitive actions. They may also include operational items such as fitness for duty, the responsibilities of individual crewmembers, adherence to the principle of the fail-safe crew,[5] and support for CRM principles—within the cockpit, between the cockpit and the cabin, and with other operating personnel both within and outside of the company

Company policies have more flexibility and are more dynamic than company philosophy because policies must adjust to changes in the operating environment, while the basic principles in overall corporate philosophy remain unchanged. Policies implement company philosophy and involve specific areas. They should remain flexible in spite of the fact that some policies may have matured. For example, policies about such areas as the role (not the specific duties) of the captain and the evolutionary role of the copilot have been modified over a period of years. Others such as the policy toward the use of

---

[5] While the fail-safe crew is usually associated with pilot incapacitation, the concept is much broader. A fail-safe crew is one in which the failure or degradation of any part of the crew for any reason does not impair the basic safety of the flight.

automation, implementation of CRM principles, the team concept, the role and responsibilities of the cabin crew, and the interface and relationship of the cabin crew with the flight crew can be maturing policies. An example of a policy that has changed over time, and is now solidified for most carriers, is the no smoking policy for both passengers and crewmembers.

The development of policies is a dynamic process. Policies can be modified or changed as they mature. In virtually all cases, it is important that the rationale for individual policies be well understood by the people who must implement them. Good communication is an extremely important part of this process (see Chapter 7—Basic Communication). Policies should be consistent with the policies of other departments and with overall company philosophy. As Nicole Svátek has aptly said:

> It is difficult if not impossible to form a clear picture of "the life" of an organization when each area is working without really knowing what the others are doing. The department's culture then becomes more meaningful than the overall corporate culture and competing rather than complementary goals become the norm. (Svátek, 1997)

Procedures

It should come as no surprise that procedures must also be consistent with both corporate philosophy and its policies. Degani and Wiener have stated that if philosophy and policies are specified in writing, then:

- A logical and consistent set of cockpit procedures that are in accord with the policies and philosophies can be generated,

- Discrepancies and conflicting procedures will be easily detected, and

- Flight crews will be aware of the logic behind every Standard Operating Procedure. (Degani and Wiener, 1994)

While the third point may be considered optimistic, it is highly desirable. There is no question that the three reasons given by Degani and Wiener are sound.

Operating procedures involve normal and non-normal procedures. The most serious of the non-normal procedures are sometimes called emergency procedures, but that term is slowly being eliminated by a major manufacturer and by some airlines. Their rationale is that the former emergency procedures should be included with the non-normal procedures for three reasons. First, it is very difficult to consistently differentiate between what have been called irregular procedures and emergency procedures. Second, the items included in emergency procedures are trained and checked for routinely. And third, pilots must be equipped to handle the so-called emergency procedures at any time that an emergency might happen, so that they are in effect simply non-normal procedures.

The proper execution of non-normal or emergency procedures is strengthened by having them include a minimum number of memory items and then completing them with a 'read, do, and verify' checklist. Examples of such items are engine failures, engine fires, and rapid decompressions. There are still wide differences in these areas among airlines. In this chapter and in this book, the term non-normal also includes emergencies. If we use the term emergency, it does not include all non-normals.

Practices

The final P refers to practices—what actually happens on the line in day-to-day operations. This is the real benchmark for routine, day-to-day operational behavior. The decisive test is whether or not established procedures for both normal and non-normal operations are routinely followed in line operations. In the instances where established procedures are not followed, the differences are often referred to as the 'book way' or 'training department way' versus 'how things are really done on the line'.

There can be understandable reasons for not always following Standard Operating Procedures (SOPs). On occasion policies and procedures seemed sound when written in a comfortable and well-lighted office or conference room, but are impractical in line operations. Therefore some published SOPs are modified or ignored by the people who have to use them. This happens under both normal and non-normal conditions. It is important to note that an airline that has done a good job with the first 3 Ps will have an internal system to solicit feedback about practices that need to be improved and also an internal system to incorporate desirable changes formally into new procedures and perhaps even a new policy. This is one of the ways that procedures can be dynamic. It is one way that the creators of procedures learn that what pilots are doing out on the line is the same as or different from stated procedures. Chapter 9—Man's Limitations, Human Errors, and Information Processing includes a discussion of some other reasons for non-adherence to SOPs.

References

Bolman, L. (1980). 'Aviation Accidents and the Theory of the Situation', in *Resource Management on the Flight Deck*, NASA Conference Publication 2120, Ames Research Center, Moffett Field, California.

Degani, Asaf and Wiener, Earl L. (1994). *On the Design of Flight-Deck Procedures*, NASA Contractor Report 177642, June 1994, Ames Research Center, Moffett Field, California.

Edwards, Elwyn (1975). 'Stress and the Airline Pilot', paper given at BALPA Medical Symposium, London, England.

Faizi, Amjad, (1997). Quoted in 'The Last Challenge', *Flight International*, 8 - 14 January 1997, Reed Business Publishing, Sutton, Surrey, United Kingdom.

Gabriel, Richard F. (1975). 'A Review of Some Universal Psychological Characteristics Related to Human Error', presented to International Air Transport Association 20th Annual Technical Meeting, Istanbul, Turkey, Douglas Paper 6401, Douglas Aircraft Company, (now Boeing), Long Beach, California.

Gunston, Bill (1986). *Jane's Aerospace Dictionary*, Jane's Publishing Company Limited, London, England.

Hackman, J. Richard (1986). 'Group-Level Issues in the Design and Training of Cockpit Crews', in *Cockpit Resource Management Training*, NASA Conference Publication 2455, Ames Research Center, Moffett Field, California.

Hawkins, Frank H. (1993). *Human Factors in Flight*, Second Edition, ed. by Harry W. Orlady, Ashgate Publishing Ltd., Aldershot, Hants, England.

Hofstede, G. (1980). 'Culture's Consequence', Sage Publications, London, England.

Johnston, Neil, (1993). 'CRM: Cross-Cultural Perspectives', *Cockpit Resource Management*, eds., Wiener, Earl L., Kanki, Barbara G., and Helmreich, Robert L., Academic Press, Inc., San Diego, California.

Johnston, A.N. (1986). 'Organizational and Motivational Aspect of Air Carrier Pilot Decision Making', in *Proceedings of the Third International Pilot Decision Making Conference*, Transport Canada, Ottawa, Canada.

Kotaite, Assad, (1997). Quoted in Air Transport, *Flight International*, 5 - 11 March 1997, Reed Business Publishing, Sutton, Surrey, United Kingdom.

Lauber, John K. (1989). 'Human Performance Issues in Air Traffic Control', *The Air Line Pilot*, June, 1989, Air Line Pilots Association, Herndon, Virginia.

Lautman, L.G. and Gallimore, P.L. (1987). 'Control of the Crew-Caused Accident', *FSF Flight Safety Digest*, June, 1987, Flight Safety Foundation, New York, and Boeing Commercial Airplane Company, Seattle, Washington.

Maurino, Daniel E. (1994). 'Cross Cultural Perspectives', *The International Journal of Aviation Psychology*, 2 November 1994, Lawrence Erlbaum Associates, Hillsdale, New Jersey.

Maurino, Daniel E., Reason, James, Johnston, Neil, and Lee, Rob B. (1995). *Beyond Aviation Human Factors*, Avebury Aviation, Ashgate Publishing Ltd, Aldershot, England.

McKenna, James T. Slack (1997). 'Commitment Hinders Safety Gains', *Aviation Week & Space Technology*, 12 May 1997, New York.

Merritt, Ashleigh (1993). 'The Influence of National & Organizational Culture on Human Performance', in *The CRM Advocate*, Charlotte, North Carolina.

NTSB (1994). *Safety Study-94/01*, National Transportation Safety Board, Washington, D.C.

Perrow, Charles (1982). 'The Organizational Context of Human Factors', essay presented through grant provided by Grant Number SEs–08014723, Sociology Division, National Science Foundation, Department of Sociology, Yale University, New Haven, Connecticut.

Rasmussen, Jens (1987). 'The Definition of Human Error and a Taxonomy for Technical System Design', in *New Technology and Human Error*, edited by Rasmussen, Jens, Duncan, Keith, and Leplat, Jacques. John Wiley & Sons Ltd., Chichester, England.

Redding, S.G. and Ogilvie, J.G. (1984). 'Cultural Effects on Cockpit Communications', University of Hong Kong, Presented at Flight Safety Foundation, Inc. Conference, Zurich.

Svátek, Nicole (1997). 'Human Factors Training for Flight Crew, Cabin Crew, and Ground Maintenance', *Focus on Commercial Aviation Safety*, The United Kingdom Flight Safety Committee, Choham, Woking, United Kingdom.

Taylor, Laurie (1997). *Air Travel: How Safe Is It?*, 2nd edition, Blackwell Science Ltd, Oxford, England.

Weener, Earl F. and Russell, Paul D. (1993). 'Crew Factor Accidents: Regional Perspective', Presented at the 22nd International Air Transport Association Technical Conference, 6-8 April 1993, Montreal, Canada.

Yamamori, Hisaaki (1986). 'Optimum Culture in the Cockpit', *Proceedings of the NASA/MAC Workshop Cockpit Resource Management Training*, NASA Conference Publication 2455, ed. by Orlady, Harry W. and Foushee, H. Clayton, Ames Research Center, Moffett Field, California.

Zeller, Anchard (1966). 'Summary of Human Factors Session', *Proceedings of FSF's 19th Annual International Air Safety Seminar,* Flight Safety Foundation, Inc., Arlington, Virginia.

7   Basic Communication

The Five Types of Communication

Communication of all sorts is an essential part of an air transport operation. Social, economic, and technological efficiency all depend on effective communication. There is one-way communication, e.g., from cockpit instruments and warning systems to the pilot, and there is person to person communication. This includes communication between individuals on the flight deck, individuals in the cabin, and other individuals who are involved in the operation, including management, and the regulatory authorities. Person to person communication is usually two-way communication but can also be only one-way as in a rule or order.

There are four kinds of person to person message types, each of which are important. In advanced airplanes, there is another type of communication with and between the airplane computers. In this book it is listed as type five.

- Type one is verbal as between crewmembers.

- Type two is non-verbal as might be appropriate with hand signals from the ground to the cockpit, between crewmembers during certain routine operations, during or after an unusual occurrence, or between cabin crew and passengers during or after an encounter with clear air turbulence.

- Type three is written communications which can be from a manufacturer wishing to guide the airline's maintenance personnel in best methods of servicing an engine, from a regulatory agency as it issues regulations, or from the management of an airline as it issues manuals, technical bulletins, SOPs, checklists, etc.

- Type four, which could be considered a modification of type three, is usually quite simple and consists of both written and considerably more graphic communication than is ordinarily appropriate. It usually is aimed at airline passengers.

- Type five, which mostly involves advanced technology airplanes, is becoming more and more important in air transport operations. It is communication with and between computers in the airplane, and is still another type of communication.

While a detailed discussion of computer communication is beyond the scope of this book, other than as it is discussed later in this chapter and in Chapter 11—Automation, all aviation professionals should be aware of the characteristics and of the reliability implications of a computer program. Greene and Richmond have warned about the practical problems involved in thoroughly testing computer software (especially computer codes) in the flight test environment and in fully understanding the intermediate failure states that can be involved in day-to-day line operations (Greene and Richmond, 1993).

The basic difference between types three and four is that four is aimed at a very special and different audience. It is designed to communicate safety information to a passenger audience with different native tongues and mixed language abilities. Combinations of both symbols and simplistic written text are commonly used to communicate to a passenger audience whose characteristics vary widely. Because there are many passengers who may not be familiar with the language used, the most effective communication with this group can well be printed symbols, diagrams, or pictorial messages given on movie screens. Unfortunately, too often the passenger audience has only minimal interest in the content of the safety message.

Recently communication with the passengers has utilized the video screens that are now a part of the passenger entertainment systems found in longer-range aircraft. Safety information is given verbally and demonstrated on the screen. It is reinforced with pictorial demonstrations of individuals using the safety equipment. Such demonstrations do not entirely solve the language difference problem. However, their pictorial demonstrations are a communication improvement to the passenger audience. On selected flights, safety information is given verbally in more than one language by bi- or multi-lingual flight attendants. A major US airline has recently announced that 50% of its new hire flight attendants are bi-lingual.

Written and Voice Communications

Both written and voice communications are extensively used in air transport operations. Each has special characteristics and principles, some of which apply to both.

Standardization is one of those characteristics important in both written and voice communication. Standardization is particularly important in critical voice communication and in such operational documentation as checklists and SOPs. Written communications seldom have the time pressure that can be involved in some voice communications.

Official communications are invariably written communications. They are reasonably permanent and, if they are well written, minimize misunderstandings. It is important that they be written with a clear knowledge of the characteristics of the intended audience. This may require different versions of the same manual or bulletin. In this case, it is essential that each version convey the same information. This is an important and a joint responsibility of the manufacturer and the airline. An airline that buys a transport aircraft manufactured in another country must ensure that all of the operating and maintenance manuals it receives are clearly understood by the people that will use them.

A manufacturer who is interested in selling its aircraft to airlines in all parts of the world has to consider many factors that may not be a consideration for an airline in the country in which the aircraft is manufactured. A US manufacturer expecting to sell its transports in countries which have different languages and cultures, e.g., India, China, Russia, France, or Brazil, has to consider the effective meaning or sometimes the misunderstandings of the words it uses in the equipment and operating manuals it sends to such countries. If the market were only a US airline, the chances of misunderstanding, or the failure to transmit specific information accurately, would be considerably less. European, Russian, Southeast Asian, or Latin American manufacturers, of course, have an identical problem when they sell their products to other countries.

Word intelligibility, or the extent to which a word is understood by the reader, is a critical factor. Words that are perfectly clear to a native of Omaha, may not be clear at all to the native of St. Petersburg, Buenos Aires, Paris, Berlin, Beijing, or other cities where English is not the mother tongue. Aviation professionals of non-English speaking countries have an obligation to learn more than the ICAO approved list of air traffic control procedures, if they are to become fully cognizant of the operating nuances of the complex machines they operate.

Although this practice unfortunately happens in nearly all countries, it is not satisfactory for airlines to expect their pilots to do significant parts of their operational learning while flying the new airplane and to also adapt its operating characteristics to the cultural environment in which it will be used. If the new airplane is designed certificated, and operated in a country whose language and culture are different from that of the airplane's mother country, effective communication and line problems are more difficult.

In quite another area, considerable unnecessary confusion is caused by the practice of individual manufacturers calling an instrument having the same function by different names. It increases learning and transition difficulties. Too often using different terminology for units with the same function is done only for a perceived commercial identification advantage. Because of its potentially adverse effects, there seems little excuse for having different terminology simply for a perceived commercial advantage. If the function, system, or instrument is essentially the same but is not the same in all aspects, the communication problem is a different one.

Other interpretation problems can occur depending upon whether the communication is by voice or is written. Sentences, which may be ambiguous when written, can be made perfectly clear if they are spoken with appropriate stress, pitch, and timing. Even common words that have more than one meaning, such as 'tear 'or 'lead', can be clearly differentiated when spoken. If written, they can only be correctly identified through context. On the other side of the coin are such words as 'wait' and 'weight' which sound the same and can only be identified by context when they are spoken. There is no problem regarding their meaning when they are written. Still another problem is words that can be used as either nouns or adverbs, or as directive verbs. Such words are 'hold', 'arm', 'pack', and 'clear'. Their meaning depends upon their use and they are a problem whether used in written or verbal communication. Another problem is words containing 'ough'. 'Ough' can be pronounced in nine different ways in English.[1] While English is the international language of aviation, it is not an easy language for many for whom English is not their mother tongue.

Voice communications are used between flight crewmembers and between crewmembers and flight operations management—especially when associated with supervision on check rides and the like. It is important that the content of this communication from management is consistent with management's official written comments on the same subject. Few things are more disruptive than a check pilot who says, 'I know that is not what the book says, but I want you to do it this way'.

---

[1] The following sentence contains all nine ways in which 'ough' can be pronounced: 'A rough-coated, dough-faced, thoughtful ploughman strode through the streets of Scarborough; after falling into a slough, he coughed and hiccoughed.'

Voice Communications Under Stress

In voice communication, researchers have found that frequently used words or phrases are better understood if fully articulated, as opposed to using shortened or truncated versions. Longer words are frequently better understood and more effective. Such words, or sometimes phrases, are better recognized because they are understood after clearly hearing only parts of a word or phrase that is frequently used.

For the same reason, in noisy conditions phrases can be better understood than single words as long as they are phrases that are used often and are well known. This is a particularly important consideration in the design of checklists and procedures, for often checklists and procedures must be utilized with considerable background noise. The choice of words or phrases can be a complex and complicated issue. While brevity is usually desirable, the critical point is that under less than optimum conditions, the word or phrase utilized should be so well known that if only a part of it is heard clearly, the message can still be clearly understood.

In most cases, the frequency with which a word or phrase is used in everyday life can be an important factor in determining the accuracy of the communication. While there are exceptions, and in spite of the fact that shorter words are usually most often used, a longer word may be more effective, particularly under difficult communication conditions. This is particularly true in a tense situation. A good example is to compare the effectiveness of the two words 'no' and 'negative'. 'Negative' is more reliably understood in stressful voice communications than the more frequently used 'no' (Hawkins 1993).

Repetition is effective in voice communication under stressful conditions. For example, repeating the words, 'hurry, hurry, hurry', frequently in a passenger evacuation is more effective than simply exclaiming the word 'hurry' just once. Other evacuation examples are the commands to be used by a flight attendant standing near a usable exit. 'Come this way, come this way', followed by 'Jump, jump' for the passengers at the exit are more effective if repeated. The repetition reinforces the importance of the simple command. The repetition must be done with discretion, for if overdone it can become annoying and distracting. In a real emergency, the passengers can be expected to be in a far from optimum nervous state and cannot be expected to be paying careful attention to subtleties in evacuation commands. For example, one airline uses the command 'release your seat belt' instead of 'unbuckle your seat belt', because it believes that the word, 'unbuckle', could easily be confused with the word, 'buckle'.

The ICAO Alphabet

Standard alphabetical words have been devised to increase intelligibility, particularly when communications conditions are poor. They also reduce the risk of misunderstanding at other times. These specific standard words can be of particular help when the interpretation is complicated with multi-lingual problems. The standard word alphabet was originally designed to assure high intelligibility when used by nationals of the member nations in NATO (North Atlantic Treaty Organization). In 1955, ICAO adopted it for universal application. Words with Latin roots were given preference because it was found that the Latin roots created maximum clarity across the wide spectrum of accents and languages in use. The standard ICAO alphabet is shown in Appendix E.

Problems still exist in this area as is evidenced by the language used in such accidents as the 1990 crash of an Avianca (Colombian) jet in Cove Neck, New York. This crash might well have been averted but for a poor, or at least questionable, choice of English words by the copilot whose native language was Spanish and by the lack of sensitivity by a rushed and very busy approach controller. Other crashes in which language is probably involved include the 1977 collision between a Pan American and a KLM jet at a foggy airport at Tenerife, Spain and the recent collision between a Saudi airliner taking off from New Delhi, India and an inbound Kazakhstan (a CIS State) cargo plane. The PAA/KLM disaster was the deadliest accident in aviation history and killed 582 people. Despite the prominence of the US in aviation and its promotion of the use of English as the language of aviation, the US was one of the last countries to accept the ICAO alphabet.

Speech—The Most Common Communication Medium

Speech is our most common communication medium. It is used both on and off the job and is preferred as a method of communication by most because it is by far the easiest to perform. Speech is used often, it is depended upon, and it is important. It has been stated that, 'Perhaps no other essential activity in aircraft operations is as vulnerable to failure through human error and performance limitation as spoken communication' (Monan, 1986).

Speech Characteristics

Speech is the means of communication most often used on the flight deck, excluding communication from the cockpit instruments. Speech is used frequently by every member of the flight crew. The conventional four primary characteristics of speech are intensity (decibels), frequency (Hertz or Hz), harmonic composition or quality, and the time or the speed with which words are spoken. We will discuss each one.

Intensity

Intensity is a first characteristic of speech. Intensity is measured in decibels (dB), and results in the sensation of loudness. The lower threshold for normal hearing is 0 decibels. Decibels are measured on a logarithmic scale and each rise of 10 dB represents a tenfold increase in its physical energy. Sounds generally become annoying at about 80 to 90 dBs and become potentially damaging at 85 to 90. The following examples show a comparison of decibels in everyday terms:

Sound-proofed room	0 dB
Library	30 dB
Average office	50 dB
Speaking voice	60 dB
Street corner	70 dB
Vacuum cleaner	70 dB
Large jet taking off nearby	100 – 120 dB

Frequency

Frequency is a second characteristic of speech. Frequency is measured in Hertz or cycles per second and gives rise to the sensation of pitch. Healthy humans can detect frequencies from 16 Hz up to about 20,000 Hz, with children sensitive to considerably higher frequencies. Voice frequencies normally range from about 1,000 to 9,000 Hz. A low frequency tone will not sound as loud as a high frequency tone of the same intensity because people are usually less sensitive to frequencies below 1000 Hz than to higher frequencies. Continued exposure to loud noises always results in a hearing loss.

A gradual frequency hearing loss is an inevitable accompaniment of advancing years. It is usually not critically debilitating. Pilots who have spent several years in noisy cockpits develop a characteristic hearing loss of higher frequencies because of long exposure to the noisy environment.

Harmonic Composition

Harmonic composition is a third characteristic of speech. It is often spoken of in terms of quality. A change in the harmonic composition of speech can change a sympathetic phrase into a sarcastic one, or an important operational phrase into one that is dull and seemingly routine and pedestrian. The word 'hello' when spoken by a friend has a different quality than the same word spoken by the same person when answering a telephone. This quality is important in aviation. Whether or not public address (PA) announcements from either the cabin or the cockpit crew convey an impression of authority, competence, and friendliness depends on their harmonic composition as well as their information content. The harmonic composition of speech either from a pilot or from an air traffic controller on the ground regarding an ATC message can convey urgency or even an emergency as well as just conveying a routine communication.

Time

The final and fourth characteristic of speech is related to time—or the rate at which the words are spoken, the length of the pauses, and the time spent on different sounds. Frequently standard PA announcements made by a harried cabin attendant reflect the time compression that is a part of many of the cabin attendant duties on short flights. Such announcements lose much of their effectiveness simply because of the time factor. This speech characteristic, involving time, rate, and the timing of sounds is important in ATC communications. Unfortunately, the necessary pace of operations sometimes forces controllers to speak at virtually unacceptable rates. For example, at rare times traffic at a busy US airport has been so heavy that controllers were forced to speak very rapidly, to suspend all readbacks, and simply to urge all pilots on their frequency to listen carefully. The importance of readbacks[2] is discussed in the next paragraphs.

---

[2] Readbacks are the aural repeat transmission back to the controller of a clearance or other ATC instruction as heard by the pilots to ensure that accurate transmission of the original message had occurred. A hearback is the act of a controller's active listening to a pilot's readback of an ATC clearance or other instruction. The first use of the term in this sense was probably in May, 1990 in Number 11 of the ASRS's monthly bulletin, *CALLBACK*. The ATC Handbook has now made ensuring correct pilot acknowledgement of information, clearances, or instructions by the controller a regulatory requirement.

Readbacks and Hearbacks

A continuing problem in speech communication is the problem of expectations or anticipations arising from the understanding received from what the originator believed was a clear English word or phrase. If an answer from an air traffic controller is expected to be 'approved', it may be heard as 'approved' when in fact the answer was 'not approved'. The more of the speech content which is lost through the quality of the transmission or through noise impediments, the greater is the risk of expectation playing a role—possibly a disastrous role—in the interpretation of the aural message. Some specific problems of noise and other impediments to communication are discussed at the end of this chapter.

Readbacks are extremely important in pilot/controller communications. Their purpose is to be sure that clearances are fully understood and that the receiving party will comply with the clearance. Readbacks of clearances delivered by ATC are a standard procedure required by the US FAA. They are also required by an ICAO Annex and the regulations of most sovereign States.

There are many examples of problems in voice communications. For example, early in the history of the US Aviation Safety Reporting System (ASRS) a report was received stating that the five-letter fix designator, TOUTU, has a phonetic sound which is the same as the number, '22'. TOUTU was a navigational fix on an approach to runway 27 at Elkhart Municipal Airport in Elkhart, Indiana. Pilots being cleared to TOUTU frequently simply turned to a heading of 22 degrees resulting in location, altitude, and distance confusion between pilots and controllers, in addition to the obvious flight hazard such confusion creates.

In another country, Hawkins reported that: 'In the UK airway system, there is a holding point called Eastwood. On a number of occasions a pilot, whose natural language was not English, received the expected instructions from ATC to take up a different heading during vectoring. When instructed to turn to Eastwood, he interpreted this as an instruction to turn eastward and so altered course onto a heading of 90°' (Hawkins, 1993). Frequently it is only through exposure to such situations that procedural and nomenclature corrections are made. Too often, an incident must occur before the problem is recognized. In the meantime, all aviation professionals should be aware of the problem.

To make the pilot-controller communication process even tighter, the controller is required to evaluate pilot readbacks for discrepancies and make corrections if they are needed. This process is called hearback and confirms the clearance or instruction as repeated by the pilot. The controller, being human, is vulnerable to expectations in the same fashion as the pilot. On occasions the controller hears only what he or she expected to hear which, of course, was the clearance or instruction he/she has just given. Readbacks and hearbacks are subject to clipping, masking, blocking or distortion, any of which can contribute to the problem.

Figure 7.1 illustrates the basic communication process between pilots and controllers. The process usually begins with the controller's clearance, instructions, or information. The pilot's readback and then the controller's hearback follow it if clarification or a misunderstanding occurred. In some cases, the process begins with a request from the pilot. In this case, while the pilot's request is the initial communication, the remainder of the procedure is the same.

The importance of the hearback process is well-illustrated by the report an ATC supervisor sent to the ASRS in which he reported a readback error that occurred while he was supervising an ATC trainee. The report is given below:

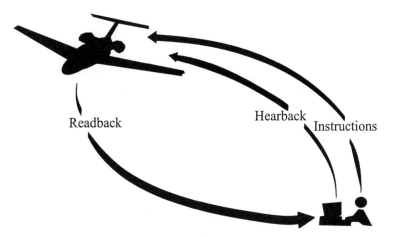

Figure 7.1   Pilot-Controller Communication
Source: *Methods and Metrics of Voice Communications*, page 21, edited by Kanki and Prinzo, 1996

Aircraft A was given a descent from 8,000 feet to only 7,000 feet (6,000 feet would be the norm on this route). Pilot read back 6,000 feet, which was not caught by either of us. We tried to get him back to 7,000 feet, but he went to 6,500 feet before he climbed back. Aircraft B was one mile in trail at 6,000 feet at the same speed.

A contributing factor was my over-reliance on the trainee, who is fairly well along in his training. I was assuming he would catch the problem, so I was not listening as intently. Also, the [typical] descent from 8,000 to 6,000 feet probably had the pilot expecting to hear 6,000. Only goes to prove the importance of readbacks being heard and understood.

(CALLBACK, No. 220, October 1997)

Unfortunately, these are not rare occurrences. It is not unusual for a pilot climbing out to readback a clearance reporting at 16,000 ft. when he/she was really cleared to 14,000 ft., and not have the discrepancy noted by the controller. In one example, the aircraft was cleared to an altitude of 14,000, and told to expect a later altitude of 16,000 ft. The pilot expected to hear 16,000 and reported going to 16,000 ft. Because 14,000 was the altitude assigned by the controller, this altitude was the controller's expected altitude and this was the altitude he thought he heard. Both the pilot and the controller were mistaken. In another typical example, a flight was cleared to 12,000 ft. and told to expect later clearance to 27,000 ft. The pilot's readback was correct but the flight climbed directly to 27,000 ft., entirely forgetting the 12,000 ft. restriction. In this case anticipation and a limitation of short-term memory may have been the cause. However, the result was a near miss with an airplane that was descending to 6,000 ft. in an approach to a busy midwestern airport. One veteran air traffic controller has made it a personal rule never to mention an altitude in a clearance that was not the altitude at which he wished the pilot to stop.

In many cases, the hearback controller does not follow the pilot's readback with an aural message because he/she gets the necessary confirmation from a proper readback. In those cases, such as a clearance for the pilot to land on runway 25 Left, or to takeoff on runway 27, if a proper readback is received further communication would be redundant, unnecessary, and only clutter up frequencies that are frequently too busy.

The numerical magnitude of the problems associated with what is basically a simple process is shown in an ASRS study in which 1652 citations of controller or pilot communication techniques were reported to the ASRS in the year 1993. These incidents included 96 percent of the 11 categories of communication problems that were reported (Connell, 1996). In an earlier study on pilot-controller communications, Monan found that:

Of the 417 occurrences, 328 featured airman listening/response deficiencies. Mishearing the numbers took place in 174 deviations, not hearing amended clearances was reported in 38, and inadequate "Roger", "Okay", and similar shortcuts in acknowledgment resulted in 46 "non-adherences to ATC instructions". In 41 additional controller/pilot dialogues, airmen insisted that they had read the numbers correctly: the controller had erred in transmission. "I may have goofed" read a typical narrative "but I don't think so". Finally 71 transmission errors were attributed to intracockpit mismanagement of the clearance information.   (Monan, 1986)

In the meantime, good communication practice can minimize the adverse operational consequences inherent in dependence on voice. Flight deck crewmembers should be aware of clearances to their aircraft and the intentions of the pilot flying. Whenever there is any doubt about the details of a clearance, such doubt should be clarified. In most cases, this will require confirmation with air traffic control. The importance of good readbacks and hearbacks should be stressed, and while good readbacks and hearbacks do not guarantee that there will be no clearance misunderstandings, they considerably increase the probability that misunderstandings will be detected. This is an example of a fundamental approach to good operations and air safety. The approach is first to minimize the number of inevitable errors, and secondly to detect them promptly and minimize their adverse operational consequences.

US vs. ICAO Phraseology and Procedures

Three very big issues in aviation communication involve terms and phraseology. One issue involves the common use of US (English) terms and phraseology because standard terms and phraseology are not consistently used between pilots and controllers. Controllers are held to standards by regulations and they do better than do pilots who are not subject to any sort of regulation. The result is considerable variation among pilots and confusion and frequently less than optimum communication among Part 121 pilots, Part 135 pilots, general aviation pilots, military pilots, non-English pilots, and controllers. One reason for the confusion between US pilots, non-English pilots, and controllers is that the US ATC glossary determined by the FAA and used by US controllers is different than ICAO's ATC glossary which is used by controllers and pilots throughout the rest of the world. The second issue arises from the difference between US communication standards and ICAO communication standards. A third problem is that there are procedural differences between the US and ICAO procedures. While some of the differences may be of little consequence, others are significant and have definite safety ramifications.

Within the US and to a considerable extent because pilots are not held to any uniform standards, pilot/controller communication discipline is very lax. Unfortunately, it is almost a tradition among some pilots that being inventive, non-standard, and using pet phrases or shortcuts in communication to controllers is an indication that the pilot is an 'old pro'. Fortunately, many of today's pilots know better. However, non-standardization in pilot/controller communication is still a very real problem.

As we have noted, the US was one of the last ICAO States to approve the ICAO alphabet and it makes limited official efforts to utilize the ICAO alphabet effectively. The International Air Line Pilots Association (IFALPA) has been forced by its members to complain to the US FAA regarding the undisciplined pilot/controller communication that it finds being used by US pilots. The lack of communication discipline among US pilots makes communication in aviation English difficult for many foreign international pilots flying in the US. It also increases problems for foreign controllers when US crews fly into other countries.

Although there is little disagreement regarding the magnitude of the problem, limited efforts have been made to correct it. Improvements have been largely due to the efforts of individual pilots. Some US airlines are much better than others are, and it would not be fair to paint all US pilots with the 'undisciplined communications' brush.

Because of the critical importance of good communication in aviation, the very real need for terms and procedures to have identical meaning where transport airplanes are flown is a definite problem. The existence of a large number of factors that exacerbate communication problems and the need for common communication and procedural standards is obvious. Stress, language and cultural differences for both pilots and controllers, and less than optimum communication conditions and equipment are an inherent part of aviation. So are occasional adverse environmental conditions. All of these increase the difficulty of ensuring good and effective voice communications.

Misunderstandings Between Pilots and Controllers

The SAE's G-10 subcommittee on Operational Communications has identified another communication problem area. Recognizing that for communication to be effective it must be thoroughly collaborative and that each participant's actions influence the other, the subcommittee believes that pilots and controllers need to better understand each others task requirements, problems, areas of flexibility, and responsibilities in order to minimize operational misunderstandings and misinterpretations. The subcommittee's proposed Aviation Resource Document (ARD) 50045, which is entitled 'Operational Communication: Cross Operational Education and Training', deals with this problem and the cross training that will be required.

Speech in Accident Investigations

Speech is becoming an increasingly important factor in accident investigations (Brenner, et al, 1996). Leaders in this movement have been the Russians who recently have cooperated with and exchanged considerable research with US and other scientists in a fashion that formerly was not politically acceptable. The speech measures that are of most interest to accident investigations are firstly, acoustic measures, which contain frequencies, including the fundamental frequency range, amplitude, and the relative energy distribution among the characteristics being examined. Secondly timing measures, which include such things as speaking rate, relative speaking/silence time, and latency to respond. Thirdly, contour measures, which relate to the shape of the speech energy waveform when it is plotted over time. Finally, the speech researchers are interested in psycholinguistic measures that include such things as changes in the articulation of words and also determination of whether or not the communication is appropriate and effective given the demands of the flight situation.

This sort of data is secured from cockpit voice recorders (CVRs)[3] and from the recorded voices of controllers who may have been involved in the

---

[3] At present two kinds of inflight recorders, the Cockpit Voice Recorder (CVR), which also records other sounds, and the Flight Data Recorder (FDR) are required for all airplanes in US air transports and in the air transports of many other countries. A third, the Digital Flight Data Recorder (DFDR), which is an advanced FDR, is now being required. These are the famous black boxes that play a prominent part in each accident investigation. Actually, none of them are black but, rather, are orange-colored.

Modern DFDRs can store a minimum of 88 details of the operation of the airplane for every second of flight. Generally, aircraft built before 18 October 1991 are required to monitor only 11 parameters. Aircraft built after that date are required to have at least 16 to 29 depending upon whether or not they are equipped with a flight data acquisition unit (FDAU). Airplanes built later are required to monitor and store from 29 to 88 parameters (FAR Part 121).

accident. In the US, these recordings have become an integral part of NTSB investigations and they are used in most other countries. The information secured by CVRs helps explain what happened in an accident. Researchers hope that in the future, with better recorders and more in depth analyses of flight and controller recordings, it will be easier to identify additional factors involved in the accident. The early CVRs store all cockpit sounds for 30 minutes on a revolving metal tape before the crash or a power interruption terminates the recording. There has been general recognition that the thirty-minute period is inadequate.

Noise and Other Impediments to Communication

Noise has been called simply 'unwanted sound'. It has also been defined as 'a sound of any kind, especially when loud, confused, indistinct, or disagreeable'.[4] Unfortunately, many airplanes create noise that is both loud and disagreeable. While the judgement as to the level of disagreeability is quite subjective, aircraft noise is seldom confused or indistinct. All aviation professionals should be sensitive to the fact that the tolerance of the general public has its limits regarding noise of any sort. Those limits seem to be decreasing. While a complete discussion of noise and its potentially disturbing consequences is beyond the purview of this book, the following paragraphs cover some areas of particular interest to transport aviation.

From the viewpoint of most people exposed to it, noise is not only generally unwanted sound but it interferes with all communication, especially speech communication. Whether or not it interferes with communication or is just an annoyance, noise is a factor that must be considered by any professional interested in operational communication or who is interested in air transport operations generally. Another factor is that protracted loud noise has health ramifications.

Considerable research has been done on the subject of noise. Much of it has been stimulated by community protection organizations that have been concerned with noise in the vicinity of airports. They have stimulated the development of quieter aircraft engines, operational noise-abatement flight procedures (which sometimes reduce flight safety margins), and the soundproofing of residential and work premises.

---

[4] *The American Heritage Dictionary of the English Language* (1978).

The development of ear protectors for those who cannot avoid exposure to noise has been fostered by industrial health concerns. Effective control of the noise problem involves applying measures at the noise source, or if that is not practical, noise control at the location of the receiver. A pair of good quality ear-defenders of the 'muff' type is essential equipment for persons exposed to a noisy environment, including those associated with noisy sports or hobbies. A new technology headset that has electronic noise-cancelling capability shows great promise for some situations.

Aviation professionals should be aware that there is another kind of noise other than the noise that is just loud and disagreeable. Frank Hawkins has given us a definition of this latter type when he called noise, 'sound which has no relationship to the completion of the immediate task' (Hawkins, 1993). One example might be extraneous conversation between crewmembers or extraneous conversation with a flight attendant that serves as a distraction to the task at hand.

There is frequently a great deal of redundancy in spoken language, and the redundancy can add considerably to the intelligibility of the basic message. This is a valuable characteristic when having to communicate under unfavorable and noisy conditions. Even if the message is distorted or surrounded by noise, with redundancy sufficient information can get through to convey its essential meaning. Without redundancy, a potential problem especially under noisy conditions is that the listener will interpret the things he/she does hear based on previous experience, learning, and expectation. Particularly if part of the message is obscured, there is a very real probability that the receiver may develop a false hypothesis. The false hypothesis that is based on an incomplete communication, an incorrect expectation, or an inappropriately applied past experience is a very real hazard in aviation.

The Signal-to-Noise Ratio

The relationship between the loudness of the signal and that of the background noise is called the signal-to-noise ratio. In determining intelligibility this factor is usually more important than the absolute level of the signal or noise. Either increasing or decreasing the volume and thereby changing the noise and signal together is seldom helpful. A danger in simply increasing volume is that auditory fatigue increases and at higher levels so can distortion. Both also can effect hearing performance.

Masking, Clipping, and Blocking

A technical aspect of aural communication that is highly relevant in noise control is called masking. Unwanted noise from the environment often masks speech or other important sounds. It masks them simply by covering them up. The noise may come from an aircraft engine, road traffic, typewriters in an office, teletype machines, or people talking. It may even come from electromagnetic interference from a number of sources including radiotelephony or, in some cases, from the telephone system itself.

The best way of protecting speech and other aural communication from the masking effects of noise is to control or isolate the masking noise at the point it originates. However, this is not always feasible. The use of standard phraseology and other good communication practices is always helpful. Caution must be used to avoid misinterpretation of partially masked signals with erroneous expectations, even with the use of standard phraseology.

Another technical aspect of aural communication is clipping. Clipping is not an uncommon occurrence and occurs when a part of the message is either inadvertently or consciously clipped off or omitted. Clipping can occur because of an improper microphone technique such as speaking before the microphone is keyed, or simply because the transmitted message was not complete. The best way of avoiding clipping problems is to use good communication techniques.

A final problem occurs when a transmission is blocked or 'stepped on' by another who is transmitting on the same frequency. If this occurs, the transmission can be simply blocked by the first one on the frequency. This is a frequent problem with partially heard or misunderstood communications. If a microphone button becomes inadvertently stuck leaving an entirely open mike, it can block an entire frequency.

Data Link

The introduction of data link systems, with clearances automatically printed on a cockpit printer or displayed on a CRT or flat panel, should help minimize some communication problems. However, the industry has already learned that while data link may well minimize clearance misinterpretations (and data link has other advantages), it can also create new difficulties and some very real problems of its own. This is an almost inevitable consequence of any innovation.

The major objective of data link is to ensure that clearances, and certain other information, are received and promptly seen and understood. A disadvantage of data link is the loss of the so-called party line, where pilots hear messages to other planes in the area that tell them of the location of the other airplanes, the instrument approaches and departures in use at that time, the specific weather involved, etc. While most would agree that the party line is helpful in day-to-day operations, there is far from a consensus as to its total value. If implementation problems are solved, a remaining question will be whether or not the advantages of sending clearances and other information by data link outweigh the advantages of the party line and being forced to live with the problems associated with transitioning to the new communications system.

A considerable amount of research in several countries involves the optimum way to use data link, including the advantages and disadvantages of having hard-copy clearances sent directly from air traffic control to the cockpit via a cockpit printer, or a cockpit CRT or flat panel display. If data link is implemented, compromises and tradeoffs as well as provisions for an intermediate transition period will have to be made.

Worldwide Operational Communication

Problems in communication are of critical importance to air transportation. Because air transport is an international concept, the language and cultural differences that are inherent in the international aviation community cannot be ignored. There is very little excuse for the sorry state of much pilot controller communications in the US, although many US pilots do very well. Neither does there seem any excuse for basic operational communication differences between US and the rest of the world as represented by ICAO. We can only hope that it does not take an aeronautical catastrophe to generate meaningful improvement in these areas.

A very fundamental problem in worldwide operational communication involves the use of English, for those for whom English is not the 'mother tongue'. While English is not an easy language, if any language is an easy one to learn, English is the international language of aviation and familiarity with aviation English and beyond that a workable understanding of English is becoming a requirement for anyone involved in international flying. This includes the ground personnel who must communicate with flights from countries other than their own. The next section of this chapter discusses some aspects of this problem.

How and Why English was Established as the Official Language of Aviation

In 1944, the government of the United States invited 55 allied and neutral states (World War II was still in existence) to meet in Chicago, and 52 of the invited states attended. The purpose of the meeting was to consider the international problems of civil aviation. The outcome of the meeting was the Chicago Convention on Civil Aviation and the formation of the International Civil Aviation Organization (ICAO), which is further explained in Appendix D.

The Chicago Convention includes 96 articles which '...establish the privileges and restrictions of all contracting states, provide for the adoption of international standards and recommended practices affecting air navigation, recommend the installation of navigation facilities by member states and suggest the facilitation of air transport by the reduction of customs and immigration formalities' (Taylor, 1988).

Meanwhile by international agreement, English was established as the international language of aviation. This was probably because English speaking nations had dominated the design, manufacture, and operation of civil aircraft. Additionally air transport has developed to a great extent in the US and other English speaking nations, and the unilingual US has played a major part in the growth and development of air transportation throughout the world.

This dominance of the English speaking nations, although now challenged in some circles, seems likely to continue. It has made the English language itself a part of human factors in aviation. Air transport does not recognize territorial boundaries. The continued need for communication in a single universal language for international aviation is very real. Miscommunication adversely affects air safety.

When English is Not your Native Tongue

The use of aviation English can create special communication problems for those whose mother tongue is not English. Most basic rules and equipment manuals are written in English and equally important, the language for international air traffic control is English. This means the non-native English speaking pilot has to learn basic English phraseology to fly across the boundaries of his/her native state. In many of these states, a substantial portion of air traffic is international. The task of these pilots is made somewhat easier by the adoption of an internationally agreed phraseology, but the problem is not simple for several reasons.

One reason is that people 'think and reason' best in their native tongue. If emergencies or partial emergencies arise, and they are inevitable, and the participants (on the ground or in the air) do not have English as a mother tongue, they, and the aviation system, can be at a distinct disadvantage. For example, this can be a problem for the marginally adept pilot flying into the US and unable to take full advantage of the strengths provided by the air traffic control system. Similarly, the US pilot flying into a foreign country where the air traffic controllers understand and use standard ICAO phraseology but are not fluent enough in English to go beyond it can encounter the same problems. The lack of proficiency in English has nothing to do with the technical proficiency of the pilots, or of the controllers, for that matter.

Language in Air Transport Crashes

Language difficulties may have been a factor in such crashes as the 1996 crash of an American Airlines flight at Cali, Columbia, the 1990 crash at Cove Neck, NY of an Avianca jet during an approach to JFK airport in New York, the Birgenair B-757 which crashed on 6 February 1996 near Puerto Plata, Dominican Republic, and the mid-air collision between a Saudi Arabian Airline's B-747 and a Kazakhstan Airline's Ilyushin I-76 which occurred near Charkhi Dadri, India on 12 November 1996.

The Saudi airliner had just taken off from New Delhi and the Kazakhstan cargo plane was on its approach. When they collided and crashed, it killed 351 people, was the world's worst midair collision, and the third worst aviation disaster of any kind. The official report of that accident is not available as this is being written but it is of interest that a native Indian controller and Russian and Saudi Arabian flight crews were involved. All used English as the international language of aviation. Another example is the 1977 ground collision of a PAA jet and a KLM jet that was taking off from a foggy airport in Tenerife, Spain, killed 582 people and was the deadliest accident in aviation history. It may well have involved English, Dutch, and Spanish language problems.

Another very real problem arises when a pilot minimally proficient in English flies into a very busy US airport such as Chicago, JFK in New York, Miami, Atlanta, Los Angeles, or San Francisco. There a busy controller, already forced to speak faster than normal, may sometimes slip into US idiomatic phrases or use local geographic references that are not known to the foreign pilot. It is somewhat ironical that the international pilot representing body, IFALPA, has found it necessary to complain about the poor quality of the radio telephony phraseology at certain US facilities to the US FAA. Serious misunderstandings of spoken communications between pilots and between pilots and controllers still exist (Taylor, 1988).

One of the most graphic examples of a language (and we believe also a training) problem was demonstrated when a trained and scheduled Asian copilot, who spoke only a minimum of English seriously asked his non-English speaking captain what 'pull-up, pull-up' meant when he heard it from the GPWS system that was installed in the airplane they were flying. The airplane crashed shortly afterward (at Urumqi, a city in northwest China, in 1993) while the captain and copilot were discussing the meaning of this English phrase in their native language. This is a true story. It is not a reflection on the basic abilities of the captain or the copilot, but does point out some of the real difficulties of pilots for whom English is not their mother tongue.

Is the Specialized Vocabulary Enough?

Language training is inadequate if it is limited to the specialized vocabulary of aviation and also does not include the linguistic confidence that comes from competency in conversational English. Routine ground-air communication uses a restricted language that is not difficult to master, however nonroutine situations have no official phraseology. Linguistic competency may be critically important in non-routine situations.

Another reason that good general English is desirable is that effective operational communication is made very difficult if the pilots cannot communicate easily with each other. This is particularly true with multicultural cockpits and the teamwork that is required for their effective operation. There is considerably more to the smooth operation of a modern transport than simply reeling off checklists, that too often are essentially given by rote.

A quite different type of language problem can occur when an ATC controller speaking English as a second language gives an inbound crew an altimeter setting of '992'. If the crew sets in 29.92 inches of mercury in their altimeter (in the US altimeter settings are always given in inches of mercury) and starts a descent not knowing that the controller meant 992 hectopascals (which is a unit of pressure used in the controller's country), the error would put the aircraft 600 feet too low. In minimum weather this could result in a disaster (*Aviation Week and Space Technology*, 4 November 1996). This was an actual occurrence and the problem may well have been inadequate training for either or both the controller and the flight crew.

Despite the existence of these and similar language and cultural problems, a very high percentage of pilots and controllers recognize them and controllers provide good and safe air traffic control services for a wide variety of aircraft that are flown by pilots with minimal competence in aviation English. Unfortunately, there are also a relatively few pilots and controllers whose training and cultural indoctrination provides very little real help or protection.

References

Aviation Week and Space Technology (1996). *Aviation Week and Space Technology*, November 1996, McGraw and Hill, New York.

Brenner, Malcolm, Mayer, David, and Cash, David (1996). 'Speech Analysis in Russia', in a report entitled *Methods and Metrics of Voice Communications*, the proceedings of a workshop organized by the Federal Aviation Administration, National Aeronautics and Space Administration, and Department of Defense, Kanki, Barbara G., editor, and Prinzo, O. Veronica, co-editor. DOT/FAA/AM-96/10, FAA Civil Aeromedical Institute, Oklahoma City, Oklahoma.

Billings, C.E. and Cheaney, E.S. eds. (1981). *Information Problems in the Aviation System*, NASA Technical Paper 1875, Ames Research Center, Moffett Field, California.

CALLBACK (1997). From NASA's Aviation Safety Reporting System, Ames Research Center, Moffett Field, California.

Connell, Linda (1996). 'Pilot and Controller Issues', in a report entitled *Methods and Metrics of Voice Communications*, the proceedings of a workshop organized by the Federal Aviation Administration, National Aeronautics and Space Administration, and Department of Defense, Kanki, Barbara G., editor, and Prinzo, O. Veronica, co-editor. DOT/FAA/AM-96/10, FAA Civil Aeromedical Institute, Oklahoma City, Oklahoma.

Greene, Berk and Richmond, Jim (1993). 'Human Factors in Workload Certification', presented at SAE Aerotech '93, Federal Aviation Administration, Seattle, Washington.

Hawkins, Frank H. (1993). *Human Factors in Flight*, Second Edition, edited by Orlady, Harry W., Ashgate Publishing Ltd., Aldershot, England.

ICAO (1984). *Manual on Radiotelephony*, Document 9432-AN/925, International Civil Aviation Organization, Montreal, Canada.

Kanki and Prinzo (1996). *Methods and Metrics of Voice Communications*, the proceedings of a workshop organized by the Federal Aviation Administration, National Aeronautics and Space Administration, and Department of Defense, Kanki, Barbara G., editor, and Prinzo, O. Veronica, co-editor. DOT/FAA/AM-96/10, FAA Civil Aeromedical Institute, Oklahoma City, Oklahoma.

Monan, William P. (1986). *Human Factors in Aviation Problems: The Hearback Problem*, NASA Contractor Report 177398, Ames Research Center, Moffett Field, California.

Taylor, Laurie (1988). *Air Travel: How Safe Is It*, BSP Professional Books, London.

8 Documentation, including Checklists and Information Management

Documentation for the Cockpit Crew

Many types of documentation are used in flight operations. Much of that documentation must be used on the flight deck and under a variety of operating conditions. Flight deck documentation ranges from large manuals, checklists, operational data cards, and flight operations bulletins to maps, charts, and approach plates. Equipment manuals, flight operations manuals, checklists, operational data cards, maps, charts, and approach plates are all a part of daily operations.

Pilots have long complained about the excessive documentation required in their job. The late Ruffell Smith, in 'A Simulator Study of the Interaction of Pilot Workload With Errors, Vigilance, and Decisions' noted that 'Without the aircraft and operations manuals, the amount of paperwork needed for this two-sector scenario (Washington Dulles to New York JFK to London Heathrow) has a single side area of 20m².' (Twenty square meters is the equivalent of a wall that is nine feet high and approximately 24 feet in length.)

Ruffell Smith's landmark study also notes that four loose-leaf volumes of operating manuals (two relating to the aircraft and two to the general operating policy of the airline) as well as printouts of flight plans, copies of weight and balance sheets, and weather maps and other weather information were also required for this flight. This documentation was, of course, in addition to the maps and charts mentioned in the previous paragraph.

Other airlines have similar requirements but may have the material organized in a different fashion. One major airline had so many required documents that they used to joke about 'a tree being killed for every flight'. Few would quarrel with the observation that this mass of paper information is a major problem for the flight crew and for the flight operations and documentation departments. Keeping all of this data current is a formidable task. It is not easy, but extremely important, for users of the data to ensure that they have the latest revision.

In another example, Phillip Brooks found that pilots are so buried in alpha-numerics that a pilot for a specific major US airline may have to consider

Figure 8.1   Paperwork Needed to Fly from Washington, D.C. to London
Source: *A Simulator Study of the Interaction of Pilot Workload With Errors, Vigilance, and Decisions*, page 5, Ruffell Smith, 1979

up to at least 28 'separate volumes, manuals, charts, bulletins, documents, logbooks, reports, electronic messages or videos—all of this before a single checklist has been called for'. He noted that: 'just one of those texts—the manual which contains airport-specific charts and procedures for the "North American Division" including Canada, the US and Mexico....*contains more than 3,100 pages and roughly 2,000,000 bits of alpha-numeric or symbolic information.*' While much of this information is not utilized on each trip, any of it may be required. Truly, the information glut is a valid problem.

It should be noted that parts of the problem of too much required paper in the cockpit can be alleviated with the CRTs or 'flat panels' of 'glass cockpits', but this by no means suggests that the problem of a plethora of information would be eliminated.[1] Many hope that this problem will be alleviated with the advent of the 'electronic flight bag' and the 'paperless airplane'. With an

---

[1] Presently, even if approach plate information is available on a CRT or flat panel, pilots are required to check the validity of the data with the paper version.

electronic flight bag, all required and verified data can be called up by the flight crew on a CRT or a flat panel and pilots presumably will not have to carry approach plates and be responsible for keeping the revisions current. However, electronic flight bags are still very much in the formative period and a great many design and implementation problems must be solved before they can be a standard part of airline operations. While electronic flight bags are not here now, they seem definitely a part of the future for at least parts of the industry. However, even after electronic flight bags have become standard equipment for the major airlines, it seems doubtful that all transport airplanes will have them in the foreseeable future. Electronic flight bags are discussed in more depth in Chapter 22—The Air Transport Future.

Typography and Standardized Nomenclature

The typographical quality and the consistency of document terminology are significant issues in all types of documents. There is little question that much present flight operations documentation can be improved. Areas such as typefaces, character height, the use of upper- and lower-case characters, and line length and spacing are all important. Paper quality and such graphical aspects as layout, color-coding, fonts and character contrast are major considerations. Although it is not a part of documentation, assuring that adequate cockpit reading conditions prevail in the airplane is important for all documentation. This includes provisions to control lighting, glare, and the angular alignment of the light source.

Ensuring that standard nomenclature is used, that information is indexed logically so that cross-references are easily found, and that operating tables, to the extent they are required, are easy to find and read under operating conditions are all very real problems. Asaf Degani's 1992 NASA Contractor Report on typography is an excellent reference. The importance of ensuring that standard nomenclature in flight deck documentation is used consistently in all references is emphasized in the reports of American Airlines' tragic 1995 accident at Cali, Columbia. Both Romeo (a non-directional beacon, or NDB, about 130 nautical miles and 120 degrees off the desired path and Rozo, another NDB, have the same radio frequency. Both are identified on the radio with an 'R'. Unfortunately, when 'R' was selected on the FMC (flight management computer) it picked Romeo as its next destination because Romeo was the closest FMC choice with an identification of 'R'. It was a fatal mistake. There was also a programming mistake. The FMC database identification for the desired Rozo listed was ROZO, not the single letter 'R'. As Earl Wiener has so aptly put it, computers are 'dumb and dutiful'. The ill-fated FMC computer did exactly what it was programmed to do. It led the tragic airplane directly toward Romeo, which was 130 miles away and 120 degrees off-course. It, of course, is an oversimplification to assume that this was the sole or primary cause of this tragic accident. There were several other factors, including other nomenclature and standardized ATC language problems, involved in the crash.

Company Written Communications and Operational Documentation

A basic requirement for any flight operations management is to be able to communicate effectively with its pilots. It is particularly important when major changes such as the FAA's AQP are being developed or when a CRM program is being instituted. To communicate effectively, constant liaison between operational management and the pilots is necessary. There are important reasons for such an approach. It can help ensure that pilots understand the reasons for their procedures and the reasons for any changes. Secondly, it almost certainly secures pilot interest and perhaps positive involvement in the consistent utilization of the process and procedures being developed. Aer Lingus's Captain E. A. Jackson put it nicely when he told IATA's 25th Technical Symposium:

> ...there is an absolute requirement to consult, discuss, justify and defend all procedures, changes and rules where the pilot is professionally and personally involved. This makes constant liaison with working groups, association councils and overall grouping necessary on many levels...with the maximum exchange of information, views and policies absolutely vital. (Jackson, 1975)

Flight manuals, equipment manuals, checklists, and operational bulletins are all important communications media. They should be useful documents that reflect the character and operating philosophy of the airline and reflect the airline's operational procedures and its training. It takes more than simply the issuance of manuals or directives from the top of an operational 'Mount Olympus' to communicate effectively with pilots.

Manuals and Flight Operations Bulletins

Manuals furnish a reasonably permanent guide to the way that flight operations departments want their flight operations conducted and the ways they want specific kinds of equipment operated. On the other hand, flight operations bulletins, which may be given other names, are usually used for a specific purpose or condition. They frequently are of a temporary nature. However, flight operations bulletins have essentially the same readability requirements of manuals and should adhere to the same typographic principles. Both of them should use the same terms and be carefully indexed in a common fashion. Both should be carefully written with high standards of technical writing and with appropriate illustrations.

Documentation Material in the Flight Management Computers (FMCs)

At least some of the documentation material that is in various manuals, that are usually kept in flight bags and not always convenient to use in the cockpit, are available in advanced technology airplanes via the flight management computers (FMCs) and control display units (CDUs) that are part of the overall Flight Management System (FMS). The FMCs provide a wealth of information. They include material for several performance aspects such as 'V speeds' for go or no-go speeds on takeoff, rotation, and climb-out; and for flap and approach speeds for all approaches and landings. The FMCs also provide power settings, optimum altitudes and vertical and lateral navigation information.

Additionally, in glass cockpit airplanes pilots can select navigation displays on their CRTs or flat panels which not only provide the moving position of the airplane but which can also display all surrounding navaids, airports, and waypoints. In addition, the CRTs or flat panels can display other items such as a particular instrument approach, the required holding pattern or a moving performance half-circle which displays the aircraft's climb or descent capabilities. This map display capability has been called by some the greatest of the innovations in glass cockpits. Ironically enough, on some models, it is possible to select so much information to be displayed that the message, 'excess data' will be generated. One wonders if there is also a similar message generated in the pilot's brain.

Computerized and Manually Created Flight Plans

Most airline flight plans, particularly for the larger airlines, are generated by computers picking the optimum and preferred routes, the most favorable altitudes, and the fuel required for the specific trip. The computer does this by looking at the planned time to destination, and comparing the alternatives available given the performance of the airplane and the projected payload.

The computed flight plans, which have been prepared by a dispatcher, then require signatures of the captain and the appropriate dispatcher to indicate their approval of the computer plan. This is a far cry from the days when one of the copilot's main responsibilities was to manually make out a flight plan under the captain's direction for the captain's and the dispatcher's final approval and signature. This method is still used in the aviation system. Although the need may not occur often, all pilots should be familiar with the types of information required and the intricacies of making out a manual flight plan.

Basic Checklists

Checklists are also operational documentation and a part of procedures. Checklists are considered by many to be the foundation for the standardization of operating procedures and a key to increased operational safety. Because of their importance, checklists are considered separately and we will devote considerable discussion to them.

Checklists are used for two quite different purposes. One purpose is to set-up or configure a system or even the whole airplane so that it is ready for operation, or for a particular part of an operation such as for a take-off or a landing. A second purpose is to ensure that an engine or a system's normal or non-normal (including emergency) operations will be handled properly and to verify that these procedures have been performed correctly.

The manufacturer's checklists, which are furnished with the airplane, usually consist of only those items that are necessary to fly the airplane safely and efficiently. They are frequently increased by additional items that are considered important to the individual airline and that reinforce the way the airline wants its airplanes flown. In these cases, the later items furnish another, but subservient purpose, for the checklist.

Methods or Philosophies for Designing Checklists

There are four basic methods or philosophies used in the design of checklists, although many airlines have further modifications. All of them require verification of the checklist items, usually by a crewmember who was not responsible for the required action. One checklist method is the 'read, do, and verify' method. Checklists designed using this method do not have items that are time critical and need immediate performance. There is no need to commit these checklists to memory although pilots should be intimately familiar with them. On some airlines, the initial challenge is omitted in a 'read, do, and verify' list. Items, such as those before start-up or after engine shut down, are completed with a cockpit flow control method and then verified. This second method sometimes called the 'do, challenge, and verify' checklist is very close to the first checklist method. The 'do, challenge, and verify' checklist does not have the preliminary 'read' step.

The third method omits the challenge and is simply a 'do and verify' checklist, where a flight crewmember completes a given item and then another crewmember reads the checklist and verifies that the required item is correctly completed. It assures a double check in that one crewmember does the required action and another crewmember verifies that it has been done. This method allows the flight crew to use flow patterns of checklist items in the cockpit from memory quickly and efficiently. 'Do and verify' has been a preferred checklist method for normal routine items such as preparing for a landing. The fourth method is a combination method, where the 'do and verify' checklists are used in non-normal or emergency checklists in conjunction with 'immediate action' items, which must be memorized.

There are few items in the fourth category and they are important. In this fourth method, after the first immediate action items are completed, 'read, do, and verify' items, which need not be committed to memory, can be completed at the convenience of the flight crew. There are only a few situations that require an immediate response from the flight crew. Examples are an explosive decompression, stall warning, GPWS or EGPWS pull-up warning, TCAS resolution (TR), or a rejected takeoff at close to V_1. Obviously, the first priority for the crew is to fly the airplane and to then handle the emergency or irregular situation. For quite different reasons, the 'read, do, and verify' or the 'do, challenge, and verify' checklists are also used by many airlines under those conditions that have a large number of required actions but with little time pressure, such as those for cockpit set-up, pre-starts, normal shut-downs, and flight terminations.

There is no way that one can depend upon accurately remembering all required steps in stressful and demanding circumstances, especially if they are rarely encountered. Non-normal or emergency procedures are often complex and their correct performance can be critical. To meet this problem, the combination method is used for critical and emergency situations. It first consists of critical actions requiring a very few immediate action items that must be committed to memory. These items are then followed by a verify checklist and then by a read, do, and verify checklist to take care of the 'clean up' items.

The Three Kinds of Checklists

There are the three different kinds of checklists—the normal checklist, the non-normal or abnormal checklist, which includes what are sometimes called emergency items, and an expanded checklist. Each is designed according to the manufacturer's and the airline's checklist philosophy. With most airlines, non-normal checklists are included in a SOP manual, or in a Quick Reference Handbook (QRH). The normal checklist is expected to be used routinely while the non-normal checklists are seldom needed on revenue flights. Most pilots are and should be aware of the criticality of the non-normal checklist.

The expanded checklist is usually found in an equipment or training manual. The expanded checklist is a supplement to either normal or non-normal checklists and is not used in flight. It often covers in considerable detail all items needed to fly the airplane safely and efficiently. In what may be merely an exercise in semantics, some do not even call it an expanded checklist but simply a detailed list of the items in normal and abnormal procedures. Expanded checklists are an excellent training or review of procedures tool.

Display of Checklists

The method used to display most normal checklists is to use a 'paper checklist'—a stiff piece of cardboard listing the specific items. It is held in the hand or clipped to one or both of the cockpit yokes. Modifications may have one of the checklists mounted to the instrument panel or printed on a placard, which is permanently attached to the yoke. Other checklist types include the 'scroll checklist', which consists of a narrow strip of paper that scrolls vertically between two reels; 'mechanical or electro-mechanical checklists' that have plastic slides which cover checklist items; and another type (also called mechanical) that has toggles that illuminate items to be checked and then are turned off as they are completed. There are also 'vocal checklists', that are audible checklists that the pilot activates with a rotary switch that enables selection of either normal or non-normal checklists, and computer-aided check lists that include a display and pointer. Finally, one of the latest types is a computerized checklist that includes a feedback loop indicating completion of the checklist item.

Undoubtedly the advanced electronic checklist, as in the B-777 and the A-340, is the checklist of the future. It can include both normal and abnormal checklists. Electronic checklists are shown on a CRT or flat panel in front of the pilots. They usually indicate that a given item has been completed and display the next item to be acted upon. They also do things like detecting and indicating unfinished items and automatically forwarding non-normal preparation items to the 'before landing checklist' so they will not have to be remembered during what may be a busy period.

Checklist Development

While nearly everyone agrees that checklists are important, unfortunately many present checklists are hard-to-read and have poor design or organization. Early checklists were often originated by fiercely independent and strong-minded people. Pilots were expected to adapt to the checklists they were given and they adapted. It is not surprising that checklists did not develop in a systematic fashion.

Checklists have expanded over the years, with items being added in response to everything from operational system changes and improvements to passenger concerns and considerations of potential liability. Accident investigators have discovered checklists that had out-of-sequence procedures, checklists that covered non-essential items, and have even found some emergency checklists that were buried in flight manuals and difficult to locate in an emergency.

Regulatory Requirements

A fundamental concept in checklist preparation in the US and in many other countries is that the regulatory authority must approve checklists. In the US, this means that the FAA's Principal Operating Inspector (POI) must approve the checklist for that specific airline. Many other countries have essentially the same process (Germany for example) and there are variations to the point that one cannot be accurate with generalities.

An admitted problem within the US FAA has been to achieve greater uniformity in opinions and regulation interpretations among its more than 2,900 POIs (Principal Operations Inspectors) and Aviation Safety Inspectors. In an effort to deal with this problem, the FAA publishes an Air Transportation Operations Inspector's Handbook (8400.10). It is a useful detailed document, which provides guidance for its Operations Inspectors. The FAA report: *Human Performance Considerations in the Use and Design of Aircraft Checklists* (1995) is particularly helpful in evaluating checklists.

Specific regulatory requirements for airlines in the US are provided in FAR 121.315— *Cockpit Check Procedure*—which reads:

1. Each certificate holder shall provide an approved cockpit check procedure for each type of aircraft.

2. The approved procedures must include each item necessary for flight-crew members to check for safety before starting engines, taking off, or landing, and in engine and system emergencies. The procedure must be designed so that a flight crewmember will not need to rely upon his memory for items to be checked. (It should be noted that immediate actions, which must be performed from memory, are required in some emergency actions the correct completion of which is in the checklist.)

3. The approved procedures must be readily usable in the cockpit of each aircraft and the flight crew shall follow them when operating the aircraft.

FAR 121.315 does not give any guidance for the 'type, concept, method, philosophy (except minimally) or presentation' of an airline's checklists. All of these factors are important considerations in checklist development. Even as straight-forward as this regulation appears, and despite the best efforts of its writers, this regulation, like many others, contains several words which can be interpreted in more than one way. For example, people can have very different opinions as to whether an item is 'necessary'. Also, there is nothing in the regulation that limits items to those for safety only. Neither is there a requirement that checklists be reasonably consistent between fleets on an individual airline. There are variations in the type, concept, method, philosophy, and presentation between airlines and even between fleets in the same airline.

Long-haul vs. Short-haul Checklists

A basic problem in checklist design for all airlines is that there is often a significant difference between the usage (frequency of performance) and the time available to perform the tasks of the short-haul and the long-haul pilot. A detailed checklist can be a major nuisance for the short-haul pilot, who may have to perform exactly the same task several times during a single day. The identical checklist may not be a nuisance at all and a good reminder for the long-haul pilot who may perform a given task only one time per trip, and then in some cases as few as three times per month. The ideal checklist for the long-haul pilot may be barely acceptable for the short-haul pilot. This is one of several cases where standardization across aircraft fleets may not be the best answer. It is difficult to depend upon compliance with a checklist that becomes a nuisance.

Checklist Standardization Within the Airlines and the Industry

Achieving standardization of checklists within the industry is ultimately not as important as standardization within the airline, and as Degani and Wiener point out, there are real problems even here. A major reason for differences between airlines is that individual airlines frequently have different checklist philosophies. One difference is that some airlines maintain that checklists should be as simple as possible and contain only critical items. This is essentially the position of the manufacturers. Other airlines maintain that the checklist should cover each item that should be performed during each operation and that it should include non-safety items it considers important.

Degani and Wiener cite an extreme example where one company has 50 items that pilots are required to check on the engine start before takeoff and another company that has only 13 items for the same airplane in the same phase (Degani and Wiener, 1990). These are not frivolous differences, but they do show that there are major differences in the automation and checklist philosophies of the two companies. We would only add that checklists should not be used as a substitute for adequate training or good judgement.

Another reason that there are checklist differences both between and within airlines is that checklists vary between manufacturers. Most airlines have airplanes made by more than one manufacturer in their fleets. The airlines and the FAA are inclined to accept the basic checklist suggested by the manufacturer, who should know the critical items for operation of the airplane it is selling, unless they have a very good reason to modify it. Unfortunately, legal liability issues can be a factor in this decision. Complete standardization across all fleets may well be impossible.

One of the worst examples of a problem associated with the lack of standardization is in the checklist of one aviation system manufacturer's checklist which utilizes the color green to indicate accomplished items. Another manufacturer of the same system uses the color green to indicate items that have not yet been accomplished (Degani and Wiener, 1990). It takes very little imagination to see the problems created for the cockpit crew if they are forced to use both types of checklists.

Achieving good and standard checklist design and practices among international airlines is a difficult area. While there are major cultural differences among ICAO's nation States, the same basic operational items must be covered whether the airplane is flown by Australian, Dutch, Brazilian, Egyptian, or the pilots of any other country. A recent ICAO Global Flight Safety and Human Factors Symposium, which was held in Auckland, New Zealand, was attended by participants from 54 sovereign States. Securing international uniformity in the philosophy, use, and content of checklists in as diverse a group as was represented at that Symposium is an extremely difficult, if not a virtually impossible, task. It is a daunting task for ICAO.

One of the many reasons it is important to achieve greater international conformity and standardization is that international interchanges of airplanes and crews (including wet leases[2]), sometimes with modifications, frequently occur. When this happens, flight crews can be forced to use airplanes and their procedures (including checklists) that differ from the cockpit instrumentation and procedures with which they are familiar.

A classic case of differences between airlines was the insistence of a high operating official of an international airline that the actuation switches on all of his airline's airplanes be wired in opposite directions from the way that they were manufactured. While the switches and the checklists for them were consistent within that airline, the actuation switches were wired directly opposite from the way they were originally manufactured and used on all other airlines. For the conventionally wired airplane, a switch that was 'on' was in the 'down' position; on this airline, an 'on' position required the switch position to be 'up'. This modification required special rewiring for each airplane. It was an expensive process. The unique wiring also created very real problems for pilots, who were trained in a conventionally wired airplane and were then forced to use an airplane in line operations with the special wiring. It also created problems for pilots who were involved in an interchange that included both types of aircraft.

The Checklist Problem

Checklist problems have been getting attention from the industry for several years. One of the leaders was the Japanese airline, All Nippon Airways (ANA), which following the accident of one of its B-727s in Tokyo Bay, began a

---

[2] Wet leases usually involve the hiring of a commercial transport from another carrier that includes a complete crew. Problems occur with on-the-spot temporary modifications of the pure wet lease process and with unforeseen substitution of equipment. This can result in a mixture of each airline's philosophy, crews, manuals, and procedures.

program to standardize its operations and, to the extent possible, its checklists. At the time its project started, ANA had six aircraft types and six types of checklists. Their checklists varied in headings, phraseology, size, color, and in specific items. The resulting hodgepodge was typical of the checklists used by airlines in all countries. A very hard working committee spent considerable time improving and standardizing all checklists in the ANA fleet to the extent it felt possible.

The industry has continued to have accidents and incidents involving the misuse of existing checklists. This has led to a concerted effort to re-think the whole question. This is a complex process. With most airlines, redoing checklists involves a variety of airplanes as well as basic CRM and operational procedure issues. Nearly all airlines have made, or are making, a special effort to redo their checklists in a consistent usable fashion.

Revitalization of Checklist Content, Use and Philosophy

Degani and Wiener began their study of the normal checklist by concentrating on the human factors of checklists per se. However, they soon discovered '...that this was only the outer shell of the problem'. The core of the problem emerged as 'the design concepts and the social uses surrounding the use of the checklist that have led some pilots to misuse it or not use it at all'. Air transport operations vary by company and by country. There are differences in checklist implementation.

Revitalization of checklist content, its use, and philosophy was further stimulated when, in slightly over one year, the misuse of checklists was considered by the NTSB to have been one of the probable causes of three air transport accidents. Two of these involved no flaps/no slats takeoffs. In the third, a B-737 ran off the runway at LaGuardia airport and continued into the adjacent waters following a misset rudder trim that was missed on the pre-takeoff checklist.

The desirability of the standardization of checklist philosophy remains a basic question. Unfortunately, neither US or Western European databases nor the current literature showed any systematic human factors effort to determine how checklists should be designed until the work of Degani and Wiener and the FAA's 1995 report on checklists. An exception is the work of Aer Lingus on checklists which has been described by Captain Neil Johnston (Johnston, 1995), but which was not available at the time of the Degani and Wiener study.

Checklist problems are not restricted to aviation. For example, the probable cause of the 1987 marine industry accident of the capsizing of the *Herald of Free Enterprise* was due to the omission of a pre-departure check item. Swain and Guttman have indicated that the nuclear industry has problems similar to those in aviation. A well-written book by Valerie Barnes discusses the development of procedures and their checklist in the nuclear power industry.

The Role of Checklists in Air Transport Crashes

Checklists, if properly performed, ensure that important routine or potentially critical situations do not result in improper operation of the items listed. The omission of a routine item that was on an omitted checklist and became critical is shown in air transport crashes in which takeoffs were attempted without takeoff flaps/slats. While there were usually additional causal factors, the resulting crashes may have been prevented with the careful use of checklists. Several reasons for such omissions are discussed in chapter 9 and in our references, particularly in the reports by Degani and Wiener and in the FAA report: *Human Performance Considerations in the Use and Design of Aircraft Checklists.*

Often cited examples of checklist problems are the August 1987 crash of a Northwest Airlines DC9-82 at Detroit Metropolitan Wayne Airport and the Delta Airlines B727-232 crash that occurred 31 August 1988 at the Dallas-Fort Worth International Airport in Texas (NTSB, 1988 and 1989). In these accidents, the flaps were not extended for takeoff contrary to operating procedures, the omission of the flap/slat extension was not detected, and the appropriate checklists were not performed. Captain Robert Sumwalt III gave a more complete list in an article, which was printed in US Airways' *Safety Hotline.* Captain Sumwalt's list, which is shown in Table 8.1, gives a list of major aircraft accidents to US air transports between December 1968 and August 1988 that were attributed to checklist misuse. Tragically, a great number of persons were almost certainly unnecessarily killed in these accidents.

Checklists are Not an Operational Panacea

It would be a mistake to assume that the proper use of checklists is some sort of an operational panacea for developing a safe and efficient operation. Crew coordination, quality control, the team concept, decision-making, and many other items are also involved. Checklists are simply one of several necessary and important factors. They are necessary in normal routine operations because the correct completion of a series of normal, but sometimes critical, required

Table 8.1   Major Aircraft Accidents Attributed to Checklist Misuse

Dec.	1968	ANC	Pan Am 707, Flaps not set for takeoff
Dec.	1974	NY	Northwest B727, Pitot heat left off
Jan.	1982	DCA	Air Florida B737, Engine Anti-ice left off
Jan.	1983	DTW	United DC8 Cargoliner, Stabilizer trim not set for takeoff
Aug.	1987	DTW	Northwest DC9-82, Flaps not set for takeoff
Aug.	1988	DFW	Delta B727, Flaps not set for takeoff

Source: US Airways' *Safety Hotline*

actions cannot routinely be depended upon without a standardized method that verifies their correct completion. Checklists can be considered the flight crew's safety net. Checklists are also necessary for both normal and non-normal or emergency conditions. Developing good checklists requires a good understanding of the airplane's operation and an understanding of the reasons for the checklists.

Today, there is a growing agreement that cockpit checklists should be simple, include a minimum number of items, be appropriate for the operation involved, and, to the extent possible, be standardized between fleets. As we have noted, the standardization issue is particularly difficult because there are often major differences between fleets of aircraft, even on the same airline.

Checklist Usage

The proper use of checklists is a relatively simple affair. A first requirement is that users of the checklist understand the reason for checklists and the reasons for the items that are on them. Good checklist design will always assure that the checklists produce a double-check on the proper actuation of critical items. Careful verification of checklist items is the key to the effective completion of any checklist. Routine use of checklists should be part of every operation. One captain advocates the regular use of checklists 'simply because most checklists are written in the blood of others'.

A human characteristic is the very real tendency for people to see only what they expect to see and hear only what they expect to hear. Pilots are particularly vulnerable during the routine occurrences that are a part of day-to-day operations, and during those rare occasions that they must operate with unusually high stress levels. Aviation professionals should be fully aware of this natural characteristic of all humans and therefore guard themselves against it. An effective way to guard against the very human characteristic to see what you expect to see and hear what you expect to hear is to meticulously follow good checklist procedures. The double check in the checklist procedure makes it an effective safety net.

An excellent review and description of checklists, their purpose and problems can be found in *Human Factors of Flight-Deck Checklists: The Normal Checklist*. In spite of the fact that the authors do not always agree with each suggestion in this thoughtful study, it is a recommended reference to this chapter. Other highly recommended papers are Neil Johnston's 'The Development and Use of a Generic Non-Normal Checklist with Applications in Ab Initio and Introductory AQP Programs', and, of course, the FAA's *Human Performance Considerations in the Use and Design of Aircraft Checklists*.

Maps, Charts, and Approach Plates

The maps, charts, and approach plates used in aviation are highly specialized documents that are required to provide a great deal of information in an obviously limited space. Each has a specific purpose. The industry has come a very long way from the night that Jack Knight borrowed a road map from the Omaha station manager so that he could continue the scheduled coast-to-coast airmail flight by flying the critical first night segment from Omaha to Chicago.

Maps

In the US, the principal maps used in aviation are the regional and sectional maps published by the Federal Aviation Administration. The FAA also publishes special maps that are used for long-range planning. FAA maps are designed to display geographical and other features that are of interest to pilots, but these maps are of little use to other kinds of users.

Maps are used primarily by pilots using visual flight rules (VFR), by helicopter pilots, and by other pilots flying relatively low level flights and using geographical (either natural or man-made) entities for navigation. Pilots flying jet airplanes spend very little time at altitudes low enough to discriminate other than the largest geographical features. Their principal navigational documents are the aeronautical charts that will be discussed in the next section.

It is unfortunate that flying at high altitudes in the jet era has taken away much of the romance that is described in the books of such authors as Wolfgang Langewiesche and Guy Murchie. We particularly recommend Guy Murchie's *Song of the Sky*, as a combination of the science and romance that can be found only by those who are fortunate enough to spend time in unpressurized airplanes at the lower altitudes.

Charts

Aviation charts are unique in that they are designed specifically to portray the information needed to fly from one radio facility to another using only radio navigation capability. The adoption of the Global Positioning System (GPS), which will be discussed in Chapter 22—The Air Transport Future—can change much of this, although there will continue to be a real need for the information on present charts for the foreseeable future.

In the US and in most of the world, a majority of aviation charts and approach plates are produced by Jeppesen Sanderson, Inc. Charts and approach plates also are produced by the Aeronautical Charting and Cartography Division of the US government National Oceanic Services and in other countries by some separate companies or organizations. Approach plates are highly specialized and are discussed in the next section. The principal charts used by pilots throughout the world are those portraying low altitude airways and those portraying high-altitude jet airways. Each has relevant intersection, waypoint, and navigational radio bearing and frequency information. Standard terminal arrival routes (STAR) charts and standard instrument departures (SID) charts are specialized charts that are published for selected busy airports. Some US high-density areas such as Minneapolis/St. Paul, Chicago, Boston, Atlanta, Miami, Dallas-Ft. Worth, Los Angeles, San Francisco, and Seattle also have local area charts which display navigational and terrain data that would make low altitude charts too congested. Special local area charts also have been created for high density areas in foreign countries.

Approach Plates

Much of the continuing concern regarding approach plates and the information displayed on them was reinforced by the results of a questionnaire distributed by a major US airline and reported at IATA's Twentieth Technical Conference (IATA, 1975). The results of that questionnaire were disturbing and clearly showed the need for improved presentation, increased emphasis on approach plate data, and the need for improved training involving approach plates.

Approach plates involve the most critical phase of flight. They are used routinely, often under very demanding lighting and operational conditions, and contain a great deal of data that are frequently changing. Revisions to many approach plates and aviation charts are published weekly although sometimes distributed by the operators less frequently. Research projects to improve the presentation of approach plate data to the pilots has been conducted by many groups, including the John A. Volpe National Transportation Systems Center, the FAA's Cartographic Standards and Flight Procedures Branches, the Society of Automotive Engineers G-10 Committee, the National Oceanic and Atmospheric Association, the Defense Mapping Agency, and Jeppesen Sanderson, Inc. As we have noted, the latter publishes most of the approach plates used throughout the world.

Virtually all approach plates have six general classes of information. One is the name, location, identification and frequency of the appropriate radio facility or facilities and a graphic depiction of minimum safe altitudes (MSAs) by sector. The second class of information is the radio frequency for the facility and the voice frequencies of approach control, tower, and ground control. The third is a plan view of the immediate area, which includes the airport and obstructions in the area and the fourth is a profile view of the instrument approach and the missed approach. The fifth class of information gives minimums data involving the specific approach diagramed. The sixth class of information is on the back of an approach plate and is a diagram of the airport giving airport elevations, runway numbers and dimensions, taxiway identifications, airport area obstructions, and other relevant airport detail. Approach plates are very carefully designed and constructed.

A major redesign of approach plates was announced in 1997 by Jeppesen Sanderson, Inc. The new approach plates should result in reduced workload for flight crews and a higher level of accuracy in their use of approach plate information. The new plates are the result of exhaustive studies by the organizations previously mentioned. The new format incorporates the lessons learned in a comprehensive human factors evaluation of approach plates, utilizes a standard pre-approach briefing sequence of briefing information, facilitates basic CRM techniques, and emphasizes day-to-day usability and legibility under all operating conditions. Details of the new approach chart formatting are given in Appendix F, which is a Jeppesen Briefing Bulletin that is highly recommended. It illustrates the kinds of information that are required in an all-weather operation and the complexity and importance of good approach plates.

The introduction of shaded color lines to those approach plates that have particular topographic problems[3] has been a very welcome improvement. Continual efforts are made to display important information in a user-friendly manner and to reduce unnecessary clutter. There can be considerable disagreement about what constitutes clutter because these plates must satisfy the needs of many users whose needs vary. An additional use of color has been in the use of color photographs of airport areas in special pages designed specifically for airport qualification. A very big advantage of this qualification method is that the pages are available for review at any time they might be needed. Some of the problems associated with the use of color in aviation are discussed in Chapter 4—The Physical Environment and the Physiology of Flight.

Today's airplanes increasingly use CRTs or flat panels to display approach data. Flat panels have several advantages over CRTs as a display medium, and as technology is improving, have become the cockpit display method of choice by manufacturers and users. However, a majority of users still depend on paper charts because they fly airplanes that do not have CRTs or flat panel. Additionally, paper charts have information that is not shown on CRTs or flat panels. They are also required for data verification. The problem of developing electronic displays that are compatible with present paper charts remains.

Some non-US airlines, notably British Airways, Lufthansa, KLM, and other members of the ATLAS group, print their own approach plate charts or have them specially printed. Many of the US and international airlines that use Jeppesen Sanderson, Inc. as the provider have special modifications made in the approach plates used by their pilots that reflect their own operating specifications and other customized data.

Information Management

A great deal of data and other information, including that in maps, charts, and approach plates, is available in the cockpit. The management of this information is a very real problem for pilots, airline management, manufacturers and regulators. Today, it is a growing operational problem for at least two reasons.

---

[3] Jeppesen criteria for a significant terrain feature is any spot elevation or obstacle more than 2,000 feet above the airport reference point within a circle of six nautical miles or above 6,000 feet within a circle of 25 nautical miles of the airport reference point.

One reason is that close to the maximum amount of information that can be assimilated under operating conditions is available in an uncoordinated fashion. The second reason, which is closely related to the first, is that any part or parts of that wealth of information ultimately may be required.

Information is available in paper documents, through voice communications with ATC or company personnel, and through information transmitted through the airplane instruments, especially in glass cockpit airplanes where the amount of information available has increased dramatically. Managing or prioritizing that plethora of information is not an easy task. While technologies in the new glass cockpits have reduced the number of instruments the flight crew must scan, technology has also provided the capability of producing even more information on these fewer devices. Human factors problems or opportunities abound.

Virtually all of the research being done on information management at this time involves work in glass cockpit airplanes and in developments, such as the electronic flight bag, that may be available in the future. The management of operational information involves more than just communication with the cockpit crew. An example of its scope in air transport operations is the agenda of a research program that is part of the Aviation Safety/Automation Program at the NASA Langley Research Center. The research program uses a working definition that defines the information environment as consisting of, 'the people, technology, and other attributes associated with the aircraft, airlines, aircraft manufacturers, Federal Aviation Administration (FAA), and ATC' (Ricks and Abbott et al., 1991). In this book we have a somewhat narrower view and are principally concerned with information management in the cockpit. Even with that narrower view, cockpit information management is involved with all of these entities.

Changes in the operational environment, especially changes that are planned in the future raise additional questions. New national airspace goals that result in the need for increased ATC capacity inevitably means more traffic and closer aircraft spacing. This almost certainly will result in a higher operational workload for the pilots and a need for greater coordination within company operations, with ATC, and with other users of the rest of the National Airspace System. While these are not new demands, their amount and sophistication have increased, and the increased demands can very well result in a cockpit information overload, especially if the changes result in significantly increased traffic separation responsibilities for the cockpit crew. This subject is discussed in connection with TCAS in Chapter 21—Current Safety Problems—and also in Chapter 22—The Air Transport Future.

Paper documentation is the principal medium used for much of the operational communication today. In addition to basic human factors logic, the operational problems, or opportunities in human factors, consist largely of first establishing useful and logical indexing and less difficult formatting. Others include reducing the amount of cross-referencing required, stowage of both used and seldom-used information, the quality and effectiveness of the technical writing, and the general typography used. This includes the utilization of optimum fonts, colors, weight of paper used, design, etc. Finally, the amount of information in the operational documentation must stay well within a person's capabilities.

The introduction of the 'glass cockpits' has changed the nature of the information problem by increasing the total amount of information available. A new problem is determining the proper selection of the electronic pages picked on the FMC. Among other things, it requires a good understanding of the particular FMC in use and its operational logic. This is not a simple process because FMCs vary and do not always follow straightforward logic.

The need for good indexing in FMC control and display units and for good presentation has given emphasis to the need for a priority system in the mass of information that is available. Much information can be of interest during low workload periods that has a limited operational priority during those periods of moderate to high workload that occur during many normal and almost all non-normal operations.

The Continuing Omnipresence of Information Management Problems

Some years ago, Billings and Cheaney noted that over 70% of the 28,000 reports submitted to ASRS over a five year period involved problems in the transfer of information (Billings and Cheaney, 1981). Much of this is an information management problem. Billings and Cheaney concluded that both human attributes and system factors contribute to information transfer problems and deficiencies. The human attributes they identified as contributing to the information transfer problems they examined were: 'distraction, forgetting, failure to monitor, and non-standard procedures and phraseology'. The system factors they identified were: 'nonavailability of traffic information, degraded information, ambiguous procedures, environmental factors, high workload, and less commonly, equipment failure'. Most of the problems they identified have human factors ramifications. It is hoped that increased data link will alleviate at least some of them.

There is no reason to think that information management problems will not increase in the future. Today, the voice channel is close to saturation during some critical phases of flight. As Billings and Cheaney demonstrated so clearly 16 years ago, information transfer problems and deficiencies are a major problem in the aviation system. There is little in the change from a three-person to a two-person crew that ameliorates those difficulties, although it is clear that further automation and increased reliability have helped. The use of computer-based data link from the ground to the airplane and its return will be aimed at reducing the present load on voice channels, but is almost certain to present problems of its own. The total information available will be substantially increased. Almost by definition, an increase in the amount of data available increases the importance of information management. If the increased data is not well managed, there is no way that it can increase the safety and efficiency of air transport operations.

Specific problems in information management are also discussed in Chapter 11—Automation, Chapter 13—Crew Resource Management (CRM) and the Team Approach, and in Chapter 20—The Worldwide Safety Challenge. It is an important and difficult area. Virtually none of the old problems will go away even with new technology. The introduction of 'glass cockpits' presents a challenge to experts in human factors for they must meet all the old problems and in addition, the new problems generated by the advanced technology.

References

Braune, Rolf J., Hofer, Elfie F., and Dresel, F. Michael (1991). 'Flight Deck Information Management', *Proceedings of the Sixth International Symposium on Aviation Psychology*, The Ohio State University, Columbus, Ohio.

Degani, Asaf (1992). *On the Typography of Flight-Deck Documentation*, NASA Contractor Report 177605, December 1992, Ames Research Center, Moffett Field, California.

Degani, Asaf and Wiener, Earl L. (1994). *On the Design of Flight-Deck Procedures*, NASA Contractor Report 177642, June, 1994, Ames Research Center, Moffett Field, California.

Degani, Asaf and Wiener, Earl L. (1993). 'Cockpit Checklists: Concepts, Design, and Use', *Human Factors*, Human Factors and Ergonomics Society, Inc., Santa Monica, California.

Degani, Asaf and Wiener, Earl L. (1991). 'Philosophy, Policies, and Procedures: The Three P's of Flight-Deck Operations', *Proceedings of the Sixth International Symposium on Aviation Psychology*, Columbus, Ohio.

Degani, Asaf and Wiener, Earl L. (1990). *Human Factors of Flight Deck Checklists—The Normal Checklist*, NASA Contractor Report 177549, Ames Research Center, Moffett Field, California.

Federal Aviation Administration (1995). *Human Performance Considerations in the Use and Design of Aircraft Checklists*, Federal Aviation Administration, Washington, D.C.

Hawkins, Frank H. (1993). *Human Factors in Flight*, Second Edition, edited by Orlady, Harry W., Ashgate Publishing Ltd., Aldershot, Hants, England.

Howard, Benjamin (1954). 'The Attainment of Greater Safety', presented at the 1st Annual ALPA Air Safety Forum, and later reprinted for presentation at the Aircraft Accident Prevention Course, University of Southern California, July 1957.

IATA (1975). 'Approach Plates', Twentieth Technical Conference, 10-15 November 1975, International Air Transport Association, Montreal, Quebec, Canada.

Johnston, Neil (1995). 'The Development and Use of a Generic Non-Normal Checklist with Applications in Ab Initio and Introductory AQP Programs', Aviation Psychology Research Group, Trinity College, Dublin, Ireland.

Lautman, Lester G and Gallimore, Peter L. (1987). 'Control of the Crew-Caused Accident', *Airliner,* April-June, 1987, Seattle, Washington.

Murchie, Guy (1954). *Song of the Sky*, The Riverside Press, Cambridge, Massachusetts, and Houghton Mifflin Company, Boston, Massachusetts.

Mykityshyn, Mark, Kuchar, James K, and Hansman, R. John (1991). 'Electronic Presentation of Instrument Approach Chart Information', *Proceedings of the Sixth International Symposium on Aviation Psychology*, The Ohio State University, Columbus, Ohio.

NTSB (1988 and 1989). *Aircraft Accident Reports NTSB/AAR-88/05 and NTSB/AAR-89/04*, National Transportation Safety Board, Washington, D.C.

Palmer, Everett and Degani, Asaf (1991). 'Electronic Checklists: Evaluation of Two Levels of Automation', *Proceedings of the Sixth International Symposium on Aviation Psychology*, The Ohio State University, Columbus, Ohio.

Ricks, Wendell R., Abbott, Kathy H., Jonson, Jon, Boucek, George, and Rogers, William H. (1991). 'Information Management For Commercial Aviation', *Proceeding of the Sixth International Symposium on Aviation Psychology*, The Ohio State University, Columbus, Ohio.

Rogers, William (1991). 'Information Management: Assessing the Demand for Information', *Proceedings of the Sixth International Symposium on Aviation Psychology*, The Ohio State University, Columbus, Ohio.

Ruffell Smith, H.P. (1979). *A Simulator Study of the Interaction of Pilot Workload With Errors, Vigilance, and Decisions*, NASA Technical Memorandum 78482, Ames Research Center, Moffett Field, California.

9 Man's Limitations, Human Errors, and Information Processing

Man's Limitations

In 1952, in an early attempt to promote design safety, Jerome Lederer, the grand old man of aviation safety, noted in a lecture given to the Royal Aeronautical Society that:

> ...the average man has only one head, two eyes, two hands, two feet, his response to demands cannot be guaranteed within plus or minus five per cent; his temperature cannot be allowed to vary more than a few degrees; his pump must operate at constant speed and pressure; his hydraulic system is accustomed to relatively stable conditions; his pressure containers, both hydraulic and pneumatic, have limited capacity; his controls are subject to fatigue, illness, carelessness, anger, inattention, glee, complacency and impatience. This mechanism was originally designed to operate in the Stone Age. It has not since been improved. The problem consists of permitting this ancient mechanism designed to function within narrow tolerances to control its destiny in a strange environment of very wide ranges in operating conditions.
>
> (Jerome Lederer, 1952)

Today, we are still facing the same problems that Jerry mentioned nearly 50 years ago. It is a fair tribute that the industry has made substantial progress in the intervening years, still operating with humans who have 'not since been improved'. While we have continued to operate in a strange environment, today we have an even wider range in operating conditions. However, the essential truth and the relevance of Jerry's comment should be remembered.

Anchard Zeller made this point and its consequences very clear when in 1966, only 31 years ago, he told attendees at a Flight Safety International Air Safety Seminar that:

> One basic truth which must always be remembered any time a human operator is used is that regardless of how well-selected, how well trained, or how optimally used, the human is subject to limitations. If these limitations are exceeded even on a momentary basis, the potential for an accident has been created. If this potential is repeated often enough, an accident will result.
>
> (Anchard Zellar, 1966)

Individual Differences

It should be remembered that while people have similar general limitations, people in a given country and people from different countries are different. They have at least two basic differences. One of them is anthropometric, and unfortunately, most of our anthropometric data have been based upon research that was conducted in the US and in Europe (frequently with a military population). The other basic difference depends upon items that are part of their different cultural backgrounds. There is considerable evidence that the characteristics of people from another nation may be quite different in many ways. For example, we know that in addition to physical differences, motivational factors can vary considerably as a function of nationality and cultural background. Even perception has been shown to be different in some cases. Beyond that, within the same national culture, different organizational and regulatory cultures can affect behavior. Under these conditions, one cannot rationally expect the same behavior from all individuals.

There are differences in performance, attitudes, and personalities; even among as highly selected a group, as are pilots. There are also differences among non-pilots, who are an important part of the aviation system. Usually, the factors we are discussing influence most people in the same direction. The primary variability is in terms of the degree, not the nature, of the effect. However, it should always be remembered that people are different and that these differences can be significant.

Generic Human Errors

Two thousand years ago Cicero told us that 'it is the nature of man to err', and experience in the intervening years strongly supports that observation. Rasmussen, Duncan, and Leplat (1987) caution us by suggesting that the immense variety of human errors that one observes may 'reflect complexity of the environment, rather than complexity in the psychological mechanisms involved'.

It is a basic fact that all people (including pilots) make errors. They make them normally, rarely, and they make them randomly. There is really very little disagreement regarding this point, although for a long time legend or tradition had it that air transport Captains never made any mistakes or errors. If they did, they clearly were being derelict in their duty. We now know better than that.

There are a great many definitions of error. From our perspective, one of the better definitions is that of Senders and Moray (1991) who define error as 'a human action that fails to meet an implicit or explicit standard'. They further state that: 'An error occurs when a planned series of actions fails to achieve its desired outcome and when this failure cannot be attributed to the intervention of some chance occurrence.' We are not as concerned with a specific definition as we are with recognition of the phenomena of human error and its commonality.

Man's Senses and Some Sources of Error

Most human experience starts with a form of physical energy being received by the senses. Therefore, it seems appropriate that we start our discussion of human error with some of the factors associated with the receiving of sensory stimuli.

Man's Receptors

All people are equipped to sense light, sound, smell, taste, movement, touch, heat, and cold. Many of these senses are required to enable an appropriate response to external events and to carry out required tasks. These senses are perceived through the eyes, ears, taste buds, the various touch organs, the vestibular or inner ear sense, which is used in determining body orientation, and the proprioceptive sense, which utilizes sensors in the joints and muscles to help us position our body members. Some senses are more sensitive than others are and most can be subject to degradation. All of them have limitations. Knowledge of physiology is required for any human factors expert working in this area. International regulations developed under the auspices of ICAO now recommend that a limited knowledge of physiology should be a part of licensing examinations for future pilots. Relevant areas are covered in more detail in Chapter 4—The Physical Environment and the Physiology of Flight.

Sensing is Different from Perception

It has been estimated that approximately 80% of all the information concerning the outside world is obtained visually. The percentage may even be higher than that in aviation for, in flying, vision is the most important of our senses. In this book, most of our discussion regarding the senses involves vision. In depth discussions can be found in the References at the end of this chapter.

One of the basic facts we know about vision is that sensing is different from perceiving. The sensors of the body convert input energy from the retina to the brain where the information is perceived, interpreted, and attaches meaning to the information. Perception is different than sensing because under a given stimulus the sensing is always the same while perception of that stimulus can vary considerably. Figure 9.1 at first looks like a series of unrelated shapes, but with a little coaching, or after some study, an individual can see something entirely different.

Attention, Set, and Motivation

While we can never be sure that all of the stimulus energy available to an individual is perceived, stimulation never falls upon a completely passive receiver. Prior learning and current motivation are both important determinants in the quality and usefulness of the information that is perceived from a given stimulus. A limitation in man's perceptual capability restricts the ability to perceive signals emanating from more than one sensory source at the same time.

There is considerable evidence that tells us that the brain can control not only the intensity of stimulation, but also can actually tune out some, but not all, unwanted stimuli. If two sets of stimuli are competing for our attention, the advantage generally falls to the stimuli of greatest size, intensity, most frequent repetition, and the most vivid contour, contrast, or color. However, ultimately the specific needs or interests of the individual overcome all of the other factors. The stimulation is the same but perception can vary considerably.

We have seen in Chapter 4—The Physical Environment and the Physiology of Flight that the brain can force fixation on what the brain considers important stimuli. This often leads to tunnel vision or fixation on one or two elements, ignoring others, which may also be important. This is a form of selective attention where the operator is concentrating so hard on one aspect of the task that he/she fails to notice other, and perhaps even more important elements. Under these conditions, the operator may sense the desired information but fail to perceive and act appropriately

A very famous experiment involving a cat and a mouse demonstrates the extent to which the effectiveness of distracting stimuli can be restrained if attention is focused on another object. A cat was implanted with electrodes to measure the amount of neural excitation going from the ear to higher centers in the brain. Then noises in the form of 'clicks' were introduced. When the cat was presented with a mouse inside a jar, its attention was devoted to the mouse.

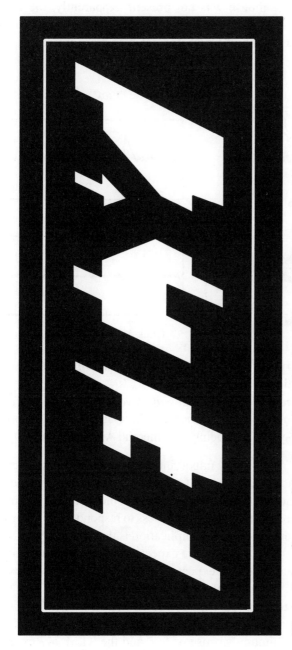

Figure 9.1   Sensory Stimulus that can Lead to More than One Perception
Source: A Review of Some Universal Psychological Characteristics Related to Human Error, page 4, Gabriel, 1975. Reproduced courtesy of The Boeing Commercial Airplane Company

Clicks introduced during this situation resulted in much less neural activity in the cat than when the mouse was not present. Apparently, the incoming stimulation of the clicks was inhibited by the cat's brain in order to minimize distraction from the mouse in which the cat was more interested. This is illustrated in Figure 9.2

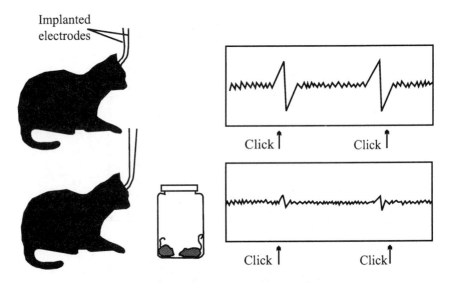

Figure 9.2   Modification of Sensory Stimulation Resulting from Attention Elsewhere

Source: *A Review of Some Universal Psychological Characteristics Related to Human Error*, page 30, Gabriel, 1975. Reproduced courtesy of The Boeing Commercial Airplane Company

The concept of 'set' is very important in operational aviation because it can influence behavior in many ways. One definition of set is 'the tendency to use a particular method or type of solution to a problem based upon previous experience or directions', another is 'a readiness to respond to the environment in a particular manner'. The general implication is that we perceive from and respond, in part at least, to what we are expecting. The folk wisdom inherent in the phrase 'you see what you expect to see, and hear what you expect to hear' illustrates the importance of the set phenomenon.

Many believe that the concept of set was an important factor in the collision on the ground of a KLM 747 and a Pan American 747 at Tenerife. Visibility was at a minimum because of fog, and the KLM 747 expected to hear a clearance to takeoff. Unfortunately, the takeoff clearance was partially misheard and then misinterpreted. The result was a collision with Pan Am, the world's worst air disaster, and 582 lives were lost. The effect of habits and the social environment can have considerable influence on the day-to-day manifestations of set perceptions.

When we are faced with indistinct or ambiguous cues, motivation becomes an important factor. It is entirely normal for a person to project into the stimulus situation the things he/she wants to see. Put another and more common way, the more we want to hear something, the more likely we are to hear it; the more we want to see something, the more likely we are to see it. This tendency is important in operational aviation, for example when marginal weather is encountered during landings. Pilots inevitably want to see approach lights and the runway so they can make the landing. Lights have been misidentified. The consequences of error can be disastrous if careful attention and reliance is not placed on hard instrument data that is not susceptible to sensory interpretation. As is so graphically illustrated in the Tenerife catastrophe, no hard instrument data are available to verify an ATC clearance. The system depends upon effective aural communication.

Motivation is a complex component of behavior and is particularly important because it is a factor that is central to behavior. Motivation arouses, directs, and integrates all behavioral aspects. Motivation incites the will and this leads to a requisite action whether expressed as purpose or motive. Lack of motivation can also stifle action.

A purpose can be defined as a goal and is usually a relatively simple process. Motives are more complex and can be analyzed by dividing them into two categories. The first category of motives is a 'drive' for an internal state such as hunger or thirst and which spurs us into action. The second category is a goal that also can be responsible for behavior and can at least temporarily terminate an initial drive. Motives involve leadership and morale. At a very basic level, motivation includes Maslow's famous hierarchy of needs, which is shown in Figure 9.3.

Peter F. Drucker has noted that: '...a want (or need) changes in the act of being satisfied.... As a want or need for either a drive or goal, approaches satiety, its capacity to reward and with it its power as an incentive diminishes fast. However, its capacity to deter, to create dissatisfaction, and to act as a disincentive rapidly increases' (Drucker, 1974). This is a particularly apt observation.

Some typical types of errors that can occur as a result of inappropriate attention, set, or motivation are shown in Figure 9.4. These types of errors are

Self-Actualization
(Maximizing Individual Potential)

Esteem
(Recognition, Respect, Self-esteem)

Social Needs
(Love, Belonging, Friendship, etc.)

Safety
(Security and Freedom from Physical Danger)

Physiological Needs
(Hunger, Thirst, Shelter, Sleep)

Figure 9.3   Maslow's Hierarchy of Needs
Source   Adapted from Maslow, A.H., 'A Theory of Human Motivation',
               Psychological Review, July, 1943

not limited to the examples given.

An additional complication in better understanding motives and behavior is that man is a complex animal. Motives do not always influence behavior in simple and direct ways. Frequently, several conflicting motives may be operating simultaneously with incompatible responses needed to satisfy each motive. The result of the incompatible responses is a conflict that must be resolved. Methods used to resolve these conflicts include: 'a delay in a decision or no decision at all; attempts to escape responsibility by finding something else to do; for just forgetting the situation; or by finding another way to compromise the conflict' (Gabriel, 1975). Conflicting motives can also cause other forms of behavior such as unstabilized behavior, behaving in a stereotyped manner (which essentially ignores the conflict), reacting in an emotional fashion, or becoming overly aggressive. Critical factors in determining the response to conflicts are the relative strength of the motivations, nearness of the attainment or exposure to the object or condition, and exposure to the conflict situation.

Several aviation psychologists believe that a very large percentage of

Attention
- Pilot so intent on tracking the flight director that he/she ignores sink rate, altimeter, airspeed, or raw ILS data.
- In weapon delivery, pilot so intent on tracking target that he/she flies aircraft into target.
- Pilot distracted by malfunction and forgets to maintain flight.
- Pilot does not acknowledge or correct too high a sink rate or too low an altitude and undershoots.

Set
- Pilot expecting to be at 10,000 feet and misreads 1,000 foot altimeter indication as 10,000 feet.
- Spoilers deployed instead of gear retracted on a go-around.
- Unintentionally continuing below minimums when inaccurate weather received.

Motivation and Conflict
- 'Cutting corners' to maintain schedule.
- Deviation from flight path to please passengers by providing view of geographic phenomena.
- Consciously continuing below minimums.
- Noise abatement approaches and takeoffs.
- Making repeated attempts to land when weather deteriorating.
- Reluctance of crew or the traffic controller to call captain's attention to an omission or error.

Figure 9.4   Typical Types of Error Occurring as a Result of Inappropriate Attention, Set, or Motivation

Source: *A Review of Some Universal Psychological Characteristics Related to Human Error*, page 40, Gabriel, 1975. Reproduced courtesy of The Boeing Commercial Airplane Company.

aviation accidents can be attributed to conflict factors. While many of the causes of human error can be eliminated by better human engineering of the cockpit, many of the factors associated with motivation and attention cannot. These sources of error need to be corrected by considering their existence in the design of regulations, procedures, and company rules. Finally, it is necessary to effectively stress the importance of routinely complying with established rules and procedures. The constructive use of peer group pressure is always helpful.

Accident Rates and Human Error in Aviation

For many years from 60 to 80 percent of all air transport accidents (and incidents) have been attributed to the performance of the flight crew. We now know that human errors are inevitable and that all people (even design engineers and pilots) make errors. They make them normally, inadvertently, comparatively rarely, and randomly. There is really very little disagreement regarding this point. There are at least two reasons for the pilot/accident record, which today most would agree is an oversimplification and at the very least misleading. One of the reasons is the tendency of society to fixate on a single cause as the reason for its catastrophes. The second reason is that the pilot or flight crew, is always the primary and the last person(s) available to cope with the consequences of the accident-enabling behavior of any parts of the air transport system.

One of a pilot's principal jobs is to recover from mistakes or failures of elements in the system, including his or her own. Gerald Bruggink (1975), talking to a group of pilots, told them very bluntly, and correctly: 'You are the last line of defense in a system operated by individuals susceptible to error.' The unfortunate fact is that errors are inevitable. The industry must improve in learning to minimize the number of errors that are created and then control the operational consequences of those errors that do occur. Problems associated with controlling these consequences are discussed in Chapters 19—Some Ramifications of Accident Analysis, and Chapter 20—The Worldwide Safety Challenge.

The Classification of Errors

Errors are classified so that we can better describe and understand them. There are many ways to do this. A common method is to categorize them first as errors of omission—where inaction or untimely action takes place and instead a specific action is required. Second, as errors of commission—where the wrong action is selected, the correct action is started but not completed, or where controls are manipulated incorrectly while trying to execute a correct action. Finally, errors can be categorized as errors of substitution—where an incorrect action is executed in place of a different action that would have been correct. An even simpler method of classification is to simply call errors either reversible errors or irreversible errors. An irreversible error is one that cannot be corrected.

Another classification, and one we think can be helpful in aviation safety, is to divide errors into random errors, systematic errors, or sporadic errors. Random errors are very difficult to predict, can be caused by a variety of conditions, and may or may not be related to the immediate task. The aviation system must learn to tolerate random errors for they are very difficult to eliminate through better training, better procedures, or better design. Systematic errors are predictable and, while they may not occur at every opportunity, are the products of less than optimum design, procedures, or training. The cure for systematic errors is to identify them first and then secondly, to correct the deficiency in design, procedures, or training. Sporadic errors occur during or after a routinely good performance and are difficult to predict or to prevent. Hawkins (1993) illustrates these three types of errors in an example of rifle shots at a target in Figure 9.5.

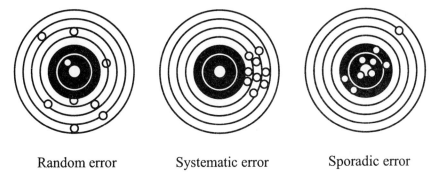

Random error   Systematic error   Sporadic error

Figure 9.5   **The Classification of Error as Random, Systematic, or Sporadic**

Source: *Human Factors in Flight*, Second Edition, page 47, Hawkins, and used with permission of Ashgate Publishing Company

Hawkins also relates this classification of errors to aviation with the following:

> If we relate these three error types to the flying situation, we could say that the pilot whose landing touch-down point varies without a recognizable pattern is committing random errors. The one who consistently undershoots is demonstrating a systematic error. While the one who normally lands the aircraft accurately, but then inexplicably makes a rare undershoot is experiencing a sporadic error. The same kind of error classification can, of course, be applied to the performance of other tasks, such as a stewardess making coffee or the mechanic carrying out maintenance work.   (Hawkins, 1993)

Classification in the NTSB Study

In an effort to better understand the human errors involved in air transport accidents, the NTSB classified each of the 302 errors that it identified in 37 major US air carrier accidents that occurred between 1978 through 1990. The Board's Safety Study found 8.162 errors per accident (302/37=8.162). The NTSB adopted a classification scheme that was used by Dr. H.P. Ruffell Smith in his classic study, *A Simulator Study of the Interaction of Pilot Workload with Errors, Vigilance, and Decisions*. This classification scheme, which illustrates the complexity of the error process, is shown in Figure 9.6. In addition to the classification described below, the NTSB study team also analyzed each of the 302 errors into an error of omission or commission.

While it is clear that there are many ways to classify errors and that most of them have further subdivisions, it is well to remember the admonition of Rasmussen, Duncan, and Leplat (1987) who remind us that, 'the seemingly immense variety of human errors observed may reflect complexity of the environment, rather than complexity in the psychological mechanisms involved'. World air transport operations are varied, complex, and dynamic. No single error taxonomy can satisfy all possible needs. In order to keep this problem in perspective, it is important to understand the limitations of the analysis.

Slips and Mistakes

Still another classification scheme, and one of the most helpful, divides errors into slips or mistakes. The operation of modern air transports in an increasingly complex environment is not a simple task. The inevitable errors that occur in any operation can be divided into either slips or mistakes. Slips have been called an incorrect and inadvertent action, while a mistake has been called an error of intention, understanding, or knowledge. For example, making a takeoff when there is a 20 knot tailwind (10 knots is the usual limitation), or making a takeoff when there is ice on the wings (a prohibited condition) is a mistake— a rule-based mistake. The plan is to takeoff, but the plan itself is bad. There are definite rules against virtually all mistakes. In contrast, slips are simply errors of execution. They consist of such things as taking off with the flaps up, landing with the gear up, or taking off or landing without a clearance. Checklists are designed to eliminate many inadvertent slips. Failure to use a checklist as stipulated is a mistake.

Primary and Secondary Errors

The first eight errors 'are considered primary errors; that is they are not dependent on making a prior error'. The ninth error type is considered a secondary error which is 'dependent on another crewmember previously or simultaneously making a primary error'.

1. *Aircraft Handling*: Failing to control the airplane to desired parameters.

2. *Communication*: Incorrect readback, hearback, failing to provide accurate information, providing incorrect information.

3. *Navigational:* Selecting wrong frequency for the required radio navigation station, selecting the wrong radial or heading, misreading charts.

4. *Procedural:* Failing to make required callouts, making inaccurate callouts, not conducting or completing required checklists or briefs, not following prescribed checklist procedures, failing to consult charts or obtain critical information.

5. *Resource Management:* Failing to assign task responsibilities or distribute tasks among crewmembers, failing to prioritize task accomplishment, overloading crewmembers, failing to transfer/assume control of the aircraft.

6. *Situational Awareness:* Controlling aircraft to wrong parameters.

7. *Systems Operation*: Mishandling engines or hydraulic, brake and fuel systems, misreading and miss-setting instruments, failing to use ice protection, disabling warning systems. (Note: the pneumatic system should also be included here.)

8. *Tactical Decisions:* Improper decision making, failing to change course of action in response to signal to do so, failing to heed warnings or alerts that suggest a change in course of action.

9. *Monitoring/challenging:* Failing to monitor and/or challenge faulty action or inaction (primary error) by another crewmember.

Figure 9.6   Error Classification used in NTSB Safety Study
Source: Adapted from NTSB Safety Study, January, 1994

The dividing line between these error categories is sometimes quite clear and at other times difficult to determine explicitly. In another explanation, James Reason has defined slips and mistakes in this fashion. 'Slips are defined as errors which result from some failure in the execution stage of an actions sequence, whereas mistakes have their origins in the planning phase' (Reason, 1987).

Mistakes are frequently due to inadequate training, although there can be other reasons, while slips are more often associated with stress, fatigue, complacency, high or even low workload, and with many other conditions that exist either on or off the job. Whether mistakes have their origins in the planning or pre-planning phase or whether the origin of slips is the result of some failure of an action sequence—the result is an error. The result of that error can be an outcome that was not intended. Effective fixes for mistakes can be quite different from effective fixes for slips. The potentially adverse consequences of either a slip or a mistake must be controlled, for either of them can be a major factor in an accident or in a serious incident.

Information Processing

Man receives information from the environment through his senses and then processes that information in his/her brain. In this process, a very basic question is whether or not a person can process more than one channel of information at a time, i.e., whether he/she is a single-channel or a multi-channel processor. Despite the fact that people can do things like drive a car and carry on a conversation or use a mobile telephone, or ride a bicycle and eat a sandwich at the same time, most contemporary evidence is that, at least for higher order mental processing, people have a single channel. This channel has a limited capacity.

Information processing is critical in the operation of today's aircraft. While humans have remarkable sensing capabilities, we also know that every individual's information processing capability has limitations. Because of these limitations, a problem in modern 'glass cockpits' is the amount of information available. It does very little good to provide more information than can be assimilated. It does very little good to provide information from displays without having an understanding of how effectively the information presented can be processed by the human who must use it. These are continuing problems for design engineers.

Less than optimum instrument and warning systems have resulted from a failure to take into account the capabilities and limitations of the human processing system. One of our most successful and prominent transports has over 400 bells, lights, whistles, and other aural or visual indicators. Each one conveys important information. However, it is sometimes difficult to identify the specific information each of the 400 represents, and potential combinations can come close to overloading even a well-trained crew. Problems such as these do not necessarily go away when the crew becomes more experienced.

Digital electronics in CRTs and flat panels are fixtures in 'glass cockpit' airplanes. Their flexibility and capacity have many advantages, but they have also exacerbated a very real human factor problem. While digital electronics and CRTs or flat panels increase the amount of information that is available, we have learned that man has only a limited capacity for processing information. Especially with the newer airplanes, a continuous flow of massive amounts of data can be easily obtained. Unfortunately, attention can be devoted only to a part of that information. There is an inevitable trade-off between speed and accuracy for those tasks that have not become essentially 'automatic'. There is an obvious need to provide some sort of filtering or priority system in those cases where multiple signals must be evaluated. We do know that when manual operation is required and speed is increased, errors increase, and when accuracy is required, time increases.

The mass of information that is available in a modern glass cockpit creates problems for design engineers, regulators, operators, and pilots. Initial and recurrent training as well as short- and long-term memory limitations are involved. Most of these considerations raise valid human factor problems. This is a highly complex subject and many of its human factor problems are dealt with in the discipline of psychology. For further discussion on information processing, we highly recommend Wickens' *Engineering Psychology and Human Performance,* which is listed as a reference for this chapter.

Short Term and Long Term Memory

Short term memory is often abbreviated as STM. A common example of STM in action is the use of telephone numbers. Nearly everyone has looked up a telephone number, been distracted, and found that it was necessary to look up the number again because of an entirely normal limitation of short-term memory.

Ordinarily STM is limited to a capacity of seven ± 2 items and, while it is not greatly influenced by the type of information, that information is rapidly forgotten if the information is not given sustained attention. If the information exceeds the capacity of the STM system, some of it is lost even if no interruption occurs. This is among the reasons that good cockpit procedure requires at least one of the pilots to use a scratch pad, or other memory device, for ATC clearances. If information can be recoded into larger conceptual units (called 'chunks') instead of the original 'bits', STM capacity can be significantly increased. As an example, WAM JAV is much easier to remember than 926-528 because WAM JAV can be processed as two 'chunks'.

With long term memory (LTM), both continuous attention and rehearsal are required for new information to be retained. When continuous attention and rehearsal occur, the information occupies the central processor of information in the brain and limits the processing of other information.

There is considerable evidence to indicate that something once learned or experienced is never lost from long-term storage. Therefore, the problems of long-term memory may well be primarily matters of retrieval rather than storage. Nearly everyone has had the experience of not being able to remember the name of a familiar friend. Words are often 'on the tip of the tongue' but simply not available when needed.

As a result of experience and training, the skilled operator learns that much of the incoming information received is either redundant or not important so it can be ignored. Professor James Reason has stated that the input selector or focuser of attention, is probably the single most important difference between the skilled and the unskilled or novice operator (Gabriel, 1975). The input selector (in the brain) which is guided by information stored in long-term memory, sifts and orders or reorders the incoming information to maximize the amount of vital information that flows through the brain's limited capacity system. The skilled operator, having filtered out useless or redundant information, then has time to anticipate further plans or actions which can further reduce load and make it manageable. With repeated attention and practice, an action or set of actions, which originally required central processing becomes more automatic. When this happens, for all practical purposes the automatic program has been placed in long-term storage, but is available on very short notice if needed.

Skill Requirements, Skill Recall, and Skill Degradation

Skills require either manual or mental coordination or both. Many of them are often slow and difficult to acquire, and some skills can be lost quickly if they are not practiced. Although the information involving the skill may never be entirely lost, some kinds of skills are relatively fragile things that readily degrade if not practiced. The degradation can be subtle and not noticed unless the skill is suddenly needed. This is one of the reasons for required proficiency training and checking every six months for captains and annually in the case of copilots. The FAA's Advanced Qualification Program (AQP) is reexamining the degradation of important skills to determine reasonable time limits for individual maneuvers that are specifically required and that may or may not be used in normal routine flying. The Air Line Pilots Association believes that while some present frequency requirements are excessive, or even unnecessary, any required skill should be part of a recurrent training and checking program and actually demonstrated at no longer than two-year intervals.

Fortunately, most skills are usually quickly relearned even after long periods of disuse. They have varying degradation or deterioration problems. Good examples are the skills required to swim or to ride a bicycle, which seem to decline at a much slower rate than many others and which return very quickly. As we shall see in Chapter 11—Automation, the degradation of manual flying skills that are not used because of the automatics is a matter of genuine concern. It is much easier to recall or to relearn a given skill than it is to learn it initially.

The need for higher mental involvement can be significantly reduced or even eliminated after considerable practice. This happens in basic instrument flying. Responses become a pattern and happen automatically without involving the central processor of the brain. After the requisite skill is acquired, a formerly difficult task requires considerably less mental effort and becomes easy. Unfortunately, highly acquired skills are seldom completely transferable. For example, expert proficiency in aircraft operation and in the interpretation of the now old-fashioned analogue instruments by no means ensures that an equal level of proficiency using the digitally actuated instruments and displays in the 'glass cockpits' of our latest transports can be easily obtained. This is in spite of the fact that the basic physical control movements required by the airplane are the same.

Man's Brain—More than Just a Human Computer

Man's brain, of course, functions as a computer for some areas, but the brain is much more than just a computer. However, while it does many simply marvelous things (such as using judgment and making complex decisions that the computer cannot), the brain computer processes data very slowly if compared to modern digital computers. For simple tasks, man's maximum processing rate is approximately two to three decisions per second, while in a modern computer the processing rate is much faster. Even in continuous tracking tasks, the human computer performs intermittently, not continuously. If the task becomes more complex, processing time in the brain increases. In the human brain, the time required for decision is increased with uncertainty or if there are multiple possible responses.

There are substantial advantages in using computers to perform (or direct) many of the routine functions that are required to operate a modern air transport safely and efficiently. However, the performance of the computer must be monitored at all times. Unfortunately, people are not very good monitors, especially of infrequent events. The important topic of monitoring the automatics (computers) is discussed in detail in Chapter 10—Workload.

The Routine Use of Established Procedures and Their Rare Deliberate Disregard

The deliberate disregard of established rules or procedures is fortunately rare in air transport operations. However, in some accidents there seems to have been an almost certain disregard of established rules or procedures. It should be recognized that the disregard of established rules or procedures is not limited to pilots. The questions and the discussion regarding this behavior are equally applicable to individuals who are engaged in any type of operations.

One reason for poor compliance is that sometimes the procedures themselves are poor. All procedures should be examined routinely. Poor procedures are not acceptable in any highly critical operation, and certainly not in air transport. Procedures, including checklists and required callouts, should be well developed and efficient. Lufthansa's Heino Caesar put it nicely when he said, procedures (including SOPs) have to be 'realistic, advantageous, easy to use, and reasonable.' We would add that they should be developed with good communication with the people that will use them.

Developing procedures that are 'realistic, advantageous, easy to use and reasonable' is not as easy as it may sound. Differences in equipment and differences in operating cultures, even if only between pilot domiciles within the same airline, can create genuine problems. Another problem is that there are very real differences between short-haul and long-haul operations. Procedures or checklists that are not a problem for the long-haul pilot can become a very real nuisance for the short-haul pilot who must use them frequently, sometimes six or more times a day. It is not surprising that pilots sometimes modify or ignore procedures that are not 'realistic, advantageous, easy to use, and reasonable' for their particular operation. The development of checklists is discussed more thoroughly in Chapter 8—Documentation, including Checklists and Information Management.

The problem of achieving a very high percentage of compliance with established procedures is not a simple one. For example, we have the problems mentioned in the preceding paragraph, and then there always are a few and fortunately very rare number of individuals who simply rebel.

There will always be considerable differences in motivation, character, life-style, personal goals, personality and ego strength among pilots or any other groups. Regardless of the care with which pilots are selected, monitored and tested, many of these attributes can and do change over the course of long careers. However, regardless of whether they are temporary or permanent changes, virtually all pilots want to be considered a 'good pilot' by their peers. In addition, and even more important, they do not want to be considered a 'poor pilot' by their peers. Pilots, like the members of many other groups, are particularly sensitive to peer group pressure.

It would be nice if all it took to achieve compliance with good and established procedures were authoritarian pronouncements, but unfortunately, it is not that easy. Good communications with the entire pilot group is essential. However, even good communications will not always be enough for the willful deviator. For example, some common reasons for willful deviations are:

- The pilot, or other individual, may think the established procedure is simply wrong.

- The pilot, or other individual, may think the established procedures are OK for the 'average' pilot or other individual, but that he/she is different.

- The pilot, or other individual, may think his/her procedure is either just as good or better than the one established.

- The pilot, or other individual, may think the procedure is not important or not necessary—or just not worth the bother—just this once, frequently, or always.

- Lastly, in some instances, the pilot, or other individual, does not really object to the established procedure but consciously or subconsciously just wants to defy 'them'—i.e., meaning management or authority in general.

In most of these cases, the pilot, or other individual, believes that safety is not jeopardized significantly by not following the procedure. Even if safety is slightly jeopardized, these individuals frequently think it is worth the risk because they are completely convinced that an accident cannot happen to them—at least not this time. Of course there can be other reasons for willful deviations and variations within them. With slight modification, the reasons listed above can be the underlying causes for rule breaking in many areas.

The willful deviations also have at least three common characteristics. In varying degrees:

1. Each one defies authority.

2. Each one may mask a degree of insecurity that is frequently displayed by overcompensating—often with an overly 'macho'image (or its feminine equivalent).

3. Each one reinforces individual egos.

The message that has to get through to flight crewmembers and their supervisors is that professionals follow established operating procedures that are 'realistic, advantageous, easy to use, and reasonable,' for sound reasons they know and understand. Procedures that are realistic in the environment in which they will be used are a requirement. The same will apply to any other person required to follow established procedures in an effective manner.

It is important to realize that deviations from established procedures are entirely normal reactions to the stresses of day-to-day living. Most egos are helped with reinforcement. Any solution to the difficult but very real problem of dealing with the 'willful violator' should recognize the needs expressed in this kind of behavior and, in addition, attempt to sublimate the hostility, and, in some cases, the insecurity implicitly suggested in the defiance of authority. While peer group pressure should not be considered a panacea, it does have elements that deal directly with this problem.

Keeping Flight Crew Errors in Perspective

The air transport industry is concerned with flight crew errors because a very high percentage of air transport accidents is associated with them. In an early burst of enthusiasm at the introduction of the latest new generation of airplanes, some design engineers and aviation safety experts believed that increased automation would control, or even eliminate, human error. Unfortunately, further automation, as exemplified in the latest transports, has done neither, although it may very well have changed the form or manifestation of many common errors (Wiener, 1983). Meanwhile the world-wide accident rate continues to show that from 60 to 80 percent of air transport accidents are crew related, and that a very high proportion of them are attributed to flight crew performance. The role that latent conditions within the aviation system can play in these accidents is now recognized. Latent conditions are discussed in Chapter 20—The Worldwide Safety Challenge.

At IATA's 20th Technical Conference, 10 November 1975, Dr. Richard Gabriel after considering the problems associated with human error, concluded:

> ...It is difficult to understand how man ever does anything without making an error. His perceptions are subject to illusions, he can only attend to one thing at a time, he is limited to the rate at which he can process information, he can't always directly control responses once they have become automated, his memory can play tricks on him, he is beset by conflicts in motivation, he gets tired; etc., etc., etc.

> And yet despite all of these limitations, man is highly reliable and can do things automatic systems cannot. One author, using automobile traffic data, has concluded that man makes literally millions of decisions without an accident resulting (Reason, 1974)....and this record is achieved in spite of drunks, pedestrians and the frustrations of poor roads, bad visibility, incomprehensible signs and other less-than-optimum conditions.

> Unfortunately, despite this impressive demonstration of reliability, estimates of the proportion of all accidents attributable to human error remain at approximately 80 percent. Many of these accidents can be avoided. Fifty percent of all propellant leaks in a missile program were caused by overtorquing nuts although the torque was specified in the procedures. Another study found that improperly written handbooks and manuals caused 33 percent more errors than did poorly designed indicators and controls....

> Safety campaigns ought to be effective. Our basic motivation for survival and self-interest should be powerful forces in promoting safety. But safety

propaganda has been found to be relatively ineffective in achieving improvements. One author (Kay, 1971) has suggested this is because we all tend to believe accidents happen to someone else, not to ourselves. He points out that admonitions lack direct reinforcement. The person reading a safety poster has probably not suffered the type of accident described, and probably believes himself {or herself} to be too skilled to make such an error.

The situation isn't hopeless. Where effort has been made to overcome dangerous conditions, the pattern of accidents has changed (Kay, 1971). Systematic attacks on accidents will pay dividends. Too often in the past, we have resorted to patches rather than solutions. Rather than developing a real understanding of the problem we have added a warning or automated the task which may have compounded the problems rather than solved them. Understanding of human characteristics has progressed sufficiently to be of real value in the design of jobs.... If adequate priority and emphasis is placed on reducing human error, there is no doubt that the goal can be achieved.   (Gabriel, 1975)

We should never forget that a lot of people do a lot of things in the air transport industry very well. As Gerry Bruggink has frequently pointed out, one reason that the safety record of air transport is so good is because of the performance of the operational participants, and certainly the performance of the pilots. Undeniably, some errors are attributed to carelessness, neglect, and even continued poor judgment. On occasions, performance is not always what might be expected of well-trained, well-paid, professionals. However, by no means are all pilot errors of that category. Benjamin Howard, a pioneer racing pilot, airline pilot and aircraft engineer, made a telling point regarding pilot errors when many years ago he told a group of aviation experts: 'I ... submit that we are evading our responsibility when we charge a crash to pilot error when the pilot is only guilty of doing what other pilots have already established as something to be expected of a qualified pilot' (Howard, 1954). These wise words are not restricted to professional pilots. We should always remember that making slips, mistakes, or errors of commission or omission are simply characteristics of all human beings.

References

Aviation Safety Commission (April, 1988). *Volume I Final Report and Recommendations* and *Volume II Staff background papers*, US Government Printing Office, Washington, D.C.

Braune, R.J. (1989). *The Common/Same Type Rating: Human Factors and Other Issues*, SAE Technical Paper Series 892229, presented at Aerospace Technology Conference and Exposition, 25-29 September 1989, Anaheim, California, Boeing Commercial Airplanes, Seattle, Washington.

Bruggink, Gerard M. (1975). 'The Last Line of Defense', presented at the Special LEC Meeting of ALPA Pilots in New Orleans, LA., 14 March 1975.

Chapanis, Alphonse (1959). *Research Techniques In Human Engineering*, The John Hopkins Press, Baltimore, Maryland.

Drucker, Peter F. (1974). *Management: Tasks, Responsibilities, and Practices*, Harper and Row Publishers, New York.

Gabriel, Richard F. (1975). *A Review of Some Universal Psychological Characteristics Related to Human Error*, presented to International Air Transport Association 20th Annual Meeting, Douglas Aircraft Company, Long Beach, California, now Boeing Commercial Airplane Company, Seattle, Washington.

Grieve, B. S. (1990). 'The Terrible Risk - It's Worth A Thought', presented at the 43rd International Flight Safety Foundation Symposium, Flight Safety Foundation, Arlington, Virginia.

Hawkins, Frank H. (1993). *Human Factors in Flight*, (Second Edition), ed. by Orlady, Harry W., Ashgate Publishing Ltd., Aldershot, England.

Hofer, E. F., Palen, L.A., Dresel, K.M. and Jones, W.P. (1991). *Flight Deck Information Management—Phase I*, FAA Contractor Report DT FA01-90-C-00055, Document D6-56305, Boeing Commercial Airplane Group, Seattle, Washington.

Hofer, E.F., Palen, L.A., Higman, K.N., Infield, S.E., and Possolo, A. (1992). *Flight Deck Information Management—Phase II*, FAA Contractor Report DTFA01-90-C-00055, Document D6-56305-1, Boeing Commercial Airplane Group, Seattle, Washington.

Hofer, E.F., Kimball, S.P., Pepitone, D.D.T., Higman, K.N., Infield, S. E., and Possolo, A. (1994). *Flight Deck Information Management—Phase III*, FAA Contractor Report DTFA01-90-C-00055, Report No.: FAA-RD-94-XXXXX, Federal Aviation Administration, Washington, D.C.

Howard, Benjamin (1954). 'The Attainment of Greater Safety', presented at the 1st Annual ALPA Air Safety Forum, and later reprinted for presentation at the Aircraft Accident Prevention Course, University of Southern California, July 1957.

ICAO (1993). *Human Factors Digest No. 10, Human Factors, Management and Organization*, International Civil Aviation Organization, Montreal, Canada.

Kay, H. (1971). 'Accidents: Some Facts and Theories', in *Psychology at Work*, ed., P. Warr, Penguin, Hammondsworth, England.

Lederer, Jerome (1952). Infusion of Safety Into Engineering Curricula, Royal Aeronautical Society, Brighton, England.

Lundberg, Bo (1966). The 'Allotment-of-Probability-Shares' - APS - Method, presented at the International Symposium on Civil Aviation Safety, Stockholm, Sweden.

Lysaght, Robert J., Hill, Susan G., Dick, A.O., Plamondon, Brian D., and Linton, Paul M., Wierwille, Walter W., Zaklad, Allen L., Bittner, Alvah C. Jr., Wherry, Robert J. (1989). *Operator Workload: Comprehensive Review and Evaluation of Operator Workload Methodologies*, Technical Report 851, United States Army Research Institute for the Behavioral and Social Sciences, Alexandria, Virginia.

Maurino, Daniel E., Reason, James, Johnston, Neil, and Lee, Rob E. (1995). *Beyond Aviation Human Factors,* Avebury Publishing Ltd., Aldershot, England.

Norman, Donald A. (1980). *Errors in Human Performance*, Center for Information Processing, La Jolla, California.

Norman, Donald A. (1988). *The Psychology of Everyday Things*, Basic Books, Inc., New York.

Perrow, Charles (1984). *Normal Accidents—Living With High Risk Technologies*, Basic Books, Inc., New York.

Prendal, Bjarne (1974). Management and Communication: Discipline and Motivation, presented at the 27th International Flight Safety Foundation Symposium, November 1974, Williamsburg, Virginia, Flight Safety Foundation, Arlington, Virginia.

Rasmussen, Jens, Duncan, Keith, and Leplat, Jacques (1987). *New Technology and Human Error*, John Wiley & Sons Ltd., Chichester, Great Britain.

Reason, James (1987). 'The Psychology of Mistakes', in *New Technology and Human Error*, ed. by Rasmussen, Jens, Duncan, Keith, and Leplat, Jacques, John Wiley & Sons, Ltd., Chichester, Great Britain.

Reason, James. (1974). *Man in Motion: The Psychology of Travel,* Walker & Co., New York.

Ruffell Smith, H.P. (1979). *A Simulator Study of the Interaction of Pilot Workload with Errors, Vigilance, and Decisions*, NASA Technical Memorandum 78482, Ames Research Center, Moffett Field, California.

Sears, Richard L. (1985). 'A New Look at Accident Contributors and the Implications of Operational and Training Procedures', Presented at the 38th International Flight Safety Foundation Symposium, November 1985, Boston. Flight Safety Foundation, Arlington, Virginia.

Senders, John W. and Moray, Neville P. (1991). *Human Error: Cause, Prediction, and Reduction*, Lawrence Erlbaum Associates, Hillsdale, New Jersey.

Stewart, Stanley (1989. *Emergency: Crisis on the Flight Deck*, Airlife Publishing Ltd., Shrewsbury, England.

Taylor, Donald H. (1987). 'The Hermeneutics of Accidents and Safety', in *New Technology and Human Error*, ed. by Rasmussen, Jens, Duncan, Keith, and Leplat, Jacques, John Wiley & Sons, Ltd., Chichester, Great Britain.

Wickens, Christopher D. (1992). *Engineering Psychology and Human Performance*, Harper Collins Publishers, New York.

Wiener, Earl L. (1983). 'The Human Pilot and the Computerized Cockpit', presented at Air Line Pilots Association Symposium 'Beyond Pilot Error', December 1983, Air Line Pilots Association, Herndon, Virginia.

10 Workload

Workload and Pilot Performance

Workload is an important term, and one of the most misunderstood in aviation psychology and in air transport. It has a great deal of influence on human performance. While any reasonably sophisticated pilot or other aviation professional should be generally familiar with workload, it becomes a very complex phenomenon if it is thoroughly studied. The following discussion is an overview of this important, difficult, and complex subject. One of the better books on all aspects of human performance, including workload, and one that we highly recommend for anyone interested in further study is the second edition of *Engineering Psychology and Human Performance* by Christopher D. Wickens of the University of Illinois. It is listed as a reference at the end of this chapter.

Workload is important to us because it is central to an understanding of pilot performance and therefore to the efficient operation of the aviation system. In this, or any other context, a major reason for the difficulty in dealing with workload either theoretically or practically is that there is no single, commonly accepted, definition of workload (Meister, 1985). It is not surprising that the workload problem is further complicated when, as often happens, the term is used without any definition at all.

Workload Examples

Lysaght et al (1989) used the following example of the common task of driving a car to illustrate how workload can vary with only minimal changes in performance and to demonstrate that workload is not the same as performance. Exactly the same principles apply in an airplane.

- Since you are an important personage, the State Police have closed the interstate to all other drivers. You are cruising down the highway at the speed limit on a nice, sunny day. Easy driving, right?

- You have just passed the state line. This second state doesn't think you are quite as important and now you have some traffic. Still not bad.

- You have been driving for a while; it is approaching the rush hour near a metropolitan area and the traffic is picking up.

- It is Friday afternoon and everyone wants to get home or out of town before the storm hits. Traffic is now much heavier than normal and slowing down.

- You left early this morning and didn't realize you hadn't stopped for lunch. You're tired and hungry.

- Traffic is now reduced to a crawl. You also forgot to get gas when you forgot lunch. You've got to get to an exit and find a gas station.

- While you are crawling along, the weather has turned. It is now raining.

- It has also gotten dark and visibility is not good. The highway is not well marked, and you must be careful not to miss your turnoff.

- Worse, the car in front does not have brake lights so you have to pay very close attention to this stop-and-go stuff. Eyes on car in front.

- A few miles are covered, but with the dark, the outside temperature has also dropped. It is no longer raining, it is freezing. Several cars are off the road. Still bumper to bumper and gas is getting very low.

- Your two-year old, who was sleeping in the back seat, wakes up. He is hungry, scared, and crying.

- It is not a lot of fun with all that is going on. In addition the engine sounds like it is missing and you know you are not yet quite out of gas. (You've turned the radio down and would like to turn the kid down.)

The report continues: 'You are about to "lose it" as anyone who has been in a similar situation can attest. Improbable, yes, but not impossible. Performance remained acceptable.' And while not stated explicitly in the report, clearly the margin of safety decreased significantly.

For a second example Lysaght et al. (1989) give a particularly graphic description of the importance of having appropriate tools and procedures for a given task. Their example very nicely illustrates the effect of mental workload, and the way it can affect performance. The task they used is:

- Recite the alphabet.

- Count from 1 to 26.

- Now do both, interleaving the alphabet with the counting, A-1, B-2, etc., saying the answers.

Most people have difficulty getting past about G-7 or H-8. This is a very difficult task. However, it becomes a relatively easy task if one is given a pencil and paper and told to write down the answers before saying them. The paper and pencil reduce the heavy burden on short-term memory. This task gives us a very good example of performance under a cognitive workload. The important point to remember is that the same task can be relatively easy or difficult depending on how we do it. Invariably, workload is very high, and there is a performance failure on the attempt without the pencil and paper. Performance of the task is not acceptable. In the second case when a pencil and paper are provided, workload is much lower, and the performance is completely acceptable. The difference is in the tools and procedure used for the second task.

Workload Generally

Workload involves several variables. Many of them are difficult to quantify. Workload can be expressed as an internal phenomenon that if too high, can cause the operator to experience difficulty, discomfort, or anxiety. Workload varies with training, procedures, experience, and sometimes with stress levels. In some cases, workload has been used interchangeably with stress. The two terms are not the same although workload increases with high stress levels.

Workload can also be expressed as an external phenomenon—as output. This is a narrow and common usage. Two layman definitions of workload taken from two different popular dictionaries are: 'The amount of work assigned to or done by a worker or unit of workers in a given time period,' and 'the amount of work carried by or assigned to a worker or position.' Neither is particularly helpful in the context of our discussion. NASA's Sandra G. Hart has defined workload as, '...a hypothetical construct that reflects the interaction between a specific individual and the demands imposed by a particular task. Workload represents the cost incurred by the human operator in achieving a particular level of performance.' This definition is much closer to our need to better understand the workload concept in air transport operation. Pilot characteristics that affect workload are the individual's capabilities, skills, prior experiences, and biases.

David Meister wrote about workload with characteristic clarity when he wrote: '...whatever W/L is, it is affected by tasks and individuals, involves comparison between demand and ability to satisfy that demand, and produces an effect on the human and the total system' (Meister, 1985). Few would disagree. There is no question that workload, however it is defined, is an important aviation human factors consideration.

Automation and Workload

Automation, which is discussed in chapter 11, influences the workload associated with the latest airplanes. A major concern with automatics is determining the optimum level of automation. Automation is usually designed to reduce workload, but it is now feared by some that automation can lower pilot workload below its optimum point so that boredom and complacency will set in. A valid research question is to determine the proper level, or levels, of automation so that the automation level being used is appropriate for the phase of flight or other condition being experienced. This has been called by Parasuraman and others as the issue of 'adaptive automation' (Parasuraman et al., 1993). It is an erroneous oversimplification to believe that any decrease in workload is automatically good, and that any increase in workload is automatically bad.

Another area of concern involving automation and workload involves the issue of how much 'trust' should be given to automation. This is a difficult issue, and the level can vary by the individual making the evaluation. The level of trust is a workload issue because workload is increased if the pilots consider the automation unreliable. We know that neither humans nor automation devices always function with 100% perfection. Pilots can become complacent and too trustful with very reliable automation. They also can become overly critical and distrustful of slightly fallible automation, even if the automation performs at levels equal to or better than they expect of manual performance. These concerns are very closely related to the need to develop a realistic philosophy and practice of the use of automation. Each of the concerns affects workload.

Factors Affecting Workload

Training and experience also affect workload. People who are well and appropriately trained find it much easier to perform a given task than do those who have not received adequate training. Those with relevant experience find it easier to perform a given task than do those for whom it is a new experience. Another factor can be simple ability. It is a common experience to see some people perform easily and well while others have difficulty with the same task. There are large differences in the ability to perform specific skills. It is not easy to determine whether these differences are due to natural ability, to experience, or to better training. In many cases, the differences can be due to a combination of all three.

Factors such as circadian dysrhythmia, sleep deficits, poor schedules, and excessive fatigue inevitably can lead to degraded performance. Obviously, they also can affect workload. Other factors that can affect workload are procedures, the internal and external environment, and the operational demands of the phase-of-flight. Still another factor is whether a specific condition with peak or minimum workload or an overall evaluation is being considered. Still other factors are the mental requirements of the automation or of the operation involved, and whether the operation is conducted in a low-density or high-density environment. Real-world workload is a very complex phenomenon with virtually all factors that affect it affected by the quality of training. Generalities that imply that they cover all cases have definite limitations.

The practical effect of items such as those in the previous paragraphs is that the actual workload experienced can and will vary depending on the individuals and the conditions in existence at the time of measurement. For example, assuming that there are no individual differences, the workload for a given piloting task can be quite different with a rested crew on a sunny afternoon than it would be at three a.m. on a dark and stormy night and at the conclusion of a long period of multi-stop operations. The normal flow of adrenaline, especially in the latter case, is an additional complication.

In an effort to minimize the effect of such considerations in a reasonable fashion, the FAA, which is responsible for safety, has developed an integrated process for the certification and then for the operation of any new airplanes, or of systems that modify already certificated airplanes. The rules and procedures promulgated by the FAA are a requirement for any air transport operation in the US. Many countries have similar provisions, and several have reciprocal agreements so that duplication of certification procedures is not required. (See Appendix O of this book.) The system seems to work well for the industry has a proven record of safe operations. However, as we can see when we look at Appendix D of Part 25 of the Federal Aviation Regulations, workload is not a simple subject.

Appendix D of FAR 25

The Federal Aviation Regulations (FARs) covers workload in Part 25 which specifies, as described in its title, Airworthiness Standards: Air Transport Category. Paragraph 1523 (Minimum Flight Crew) of Part 25 states that the minimum flight crew must be established so that it is sufficient for safe operation considering, among other things, 'the workload on individual crewmembers.' It specifies the items that must be considered in making determinations in Appendix D of Part 25.

Revised paragraph 25.1523 rules became effective in April of 1965 and made obsolete the 80,000 lb. rule, which required the use of a flight engineer for airplanes weighing more than 80,000 pounds.[1] After April of 1965, crew complement was determined by cockpit workload not the gross weight of the airplane. Cockpit workload must be determined according to the considerations specified in the new Appendix D of Part 25. Three decades later, these provisions are essentially unchanged although their implementation and administration has been considerably refined.

The seven workload considerations that Appendix D lists and that must be considered are:

1. Flight Path Control
2. Collision Avoidance
3. Navigation
4. Communications
5. Operation and Monitoring of Aircraft Engine and Systems
6. FMS Operations and Monitoring
7. Command Decisions

Appendix D also lists the following ten workload factors that have to be part of the listed considerations. The ten workload factors are:

1. Access and Operation of Controls
2. Access and Conspicuity of Instruments/Displays
3. Number and Complexity of Procedures
4. Degree/Duration of Mental and Physical Effort
5. Extent of Monitoring Required
6. Crew Member Unavailability

---

[1] An expanded discussion of the 80,000 lb. Rule occurs in Chapter 3—A Brief History of Human Factors and its Development in Aviation.

7. Degree of Automation
8. Communication
9. Emergencies
10. Incapacitation

There has been little contention among the parties involved (pilots, manufacturers, airlines, or researchers), regarding these workload functions and factors. Nevertheless, there has not always been agreement as to how they should be measured. There has also been considerable contention regarding the minimum crew required to operate safely with airplanes that are becoming more complex and are operated in an environment whose complexity is increasing. However, because of engineering advances and economics, a two-person crew has become the de facto crew complement standard in current transport airplanes.

Selected References of the President's Task Force

The importance of workload received additional impetus with the *Report of the President's Task Force on Aircraft Crew Complement* (McLucas, Drinkwater, and Leaf, 1981). The Task Force was established to determine if the jet transports that would be introduced over the next decade could be flown safely by a flight crew of two members. It put considerable emphasis on workload. The *Report* stated that while the FAA had utilized present state-of-the-art methods in evaluating cockpit workload in present transports, the Task Force believed that the workload process could be improved. It recommended among other things:

- That certification procedures relating to crew workload evaluation should be improved and strengthened,

- That the traditional task/time-line analysis method be supplemented by improved subjective evaluation methods applied by qualified pilots in demonstrating compliance with FAA crew complement criteria,

- That the evaluation of such areas as crew procedures, workload evaluation, and training requirements be supplemented by using qualified line pilots to augment FAA certification teams, and

- That the MEL (Minimum Equipment List) be prepared, and related tests that examine combinations of failures be conducted during the crew complement certification process and during the development of the air carriers' operating specifications.

The FAA agreed with the recommendations of the Task Force. Minimum crew certification procedures were modified and since that time have remained essentially unchanged. Throughout the industry, there is general agreement with the President's Task Force that there is a need for more valid and reliable workload assessment techniques, and for improved assessment of cockpit workload. As Barnes and Adams (1997) and others have pointed out, this is especially true with two-person crews and the installation of advanced avionics systems.

Workload Evaluation Today

Spurred by the *Report of the President's Task Force*, a great deal of effort was spent in an effort to evaluate and update the then present state of workload knowledge. State-of-the-art efforts were described in *The Assessment of Crew Workload Measurement Methods, Techniques, and Procedures* (Corwin et al., 1989). This is a report that summarizes work conducted as a part of a FAA/ Air Force sponsored contract. State-of-the-art methods were also evaluated in a report, *Operator Workload: Comprehensive Review and Evaluation of Operator Workload Methodologies* (Lysaght et al., 1989). This report was sponsored by the US Army Research Institute for the Behavioral and Social Sciences and analyzes the contemporary scientific literature on operator workload. Finally, several studies on workload were published by the Human Performance Group at NASA/Ames Research Center.

Methods of Determining Workload

Workload researchers have long searched for a straightforward, easy, and reliable method of determining workload. It was hoped that a physiological trait or a measurable performance characteristic could be discovered. A very big advantage of a physiological trait or a performance characteristic, if an appropriate one could be discovered, is that it could then be measured. No individual subjective judgment by either the subject or the researcher would be required.

Unfortunately, there is too much individual variability and too little reliable direct correlation between the physical traits being measured and workload. Eyeblink rates, pupil diameter, eye movements, critical flicker frequency (CFF), blood pressure, respiration rates or content, electroencephalograms (EEGs), electromyography (EMG), body fluid analysis, galvanic skin response (GSR), etc. have all proved unreliable. Heart rate was thought to hold some promise but it is so confounded with stress, arousal and physical exertion that it too is unreliable. One of the heart measures that comes closer than most to providing accurate answers is the interbeat interval of the heart as measured by Dr. Allan Roscoe of Great Britain.

Performance measurements of wheel (aileron) inputs, elevator control inputs, or rudder control inputs are all too variable to be reliable indicators of workload. Lysaght et al. said it well when they stated that: 'Every technique reviewed has been shown to be sensitive to workload and almost every technique has been shown to have failures.' None of them, including Roscoe's heart beat interval measurement, is reliable enough to be used in air transport certification.

One of the many difficulties in selecting a single measure is that task demands vary. A measure that gives reliable results in determining workload for one task may not be at all satisfactory for another. Not surprisingly, subjective workload measures have proven to be the most useful. They can be relatively easy to use and if taken at appropriate times do not interfere with the task being measured. While subjective ratings do not represent the inherent properties of a task, they do give the apparent interaction between an operator, a task, and the environment. This is precisely the information needed in most air transport considerations of workload, including certification. As the President's Task Force noted:

> There is, after all, inherent validity in the basic process by which pilots decide that an aircraft, its systems, and its crew complement are acceptable from the point of view of system safety and efficiency.
>
> (McLucas, Drinkwater, and Leaf, 1981)

The importance of an appropriate pilot sample cannot be overestimated. For example, the use of only test pilots for the evaluation of transport aircraft cannot help giving a biased result. Pilots with minimal training give completely different answers than pilots who are familiar with the aircraft and who have been well trained for the maneuver or scenario used in the evaluation.

Task/Time-Line Analysis (TLA)

As was stated in the *Report of the President's Task Force*, a traditional method of measuring workload was to use Task/Time-Line Analysis (TLA). TLA represented state-of-the-art of workload evaluation at that time of the *Report* but was not completely satisfactory.

TLA computes the ratio of time required to the estimated time available. This must be done for each operating procedure and for every action of the flight crew. Measurements are taken for vision, manual motions left, right, and both, and for the required auditory, cognitive, and verbal actions. A refinement (used by Boeing) added eye point-of-regard and dwell time in order to make estimates of the mental workload that is involved. However, many believe that even with this refinement TLA cannot adequately account for the non-observable or mental workload of the flight crews, nor can it measure the effect of monotony.

A major criticism of using TLA in other than in design and in an advisory or supplemental role is that it is very difficult to estimate either the time required to accomplish a given task or to estimate the time available. For example, two of the FAA's senior test and certification pilots (Jim Richmond and the late Berk Greene), wrote: 'The trend over the last decade is that the time required for both normal and abnormal control of the airplane and system's demands is almost ridiculously low. That doesn't mean that time-line analysis is not a good tool for design, but it is beginning to lose its usefulness as a workload certification tool' (Greene and Richmond, 1993).

Another criticism is that TLA takes a serial approach in calculating tasks when it is known that pilots can and do conduct multiple tasks in parallel. Not recognizing that some tasks are, or can be, conducted in parallel can lead to an overestimation of workload. Fortunately, this is an error on the side of safety and is acceptable, particularly if designing or evaluating a new airplane, system or device.

At best, TLA is a complex, somewhat esoteric, method of workload analysis that requires special skills for the analyst. Most observers agree that it is of real value in the design process for it enables the designer to make relatively conservative estimates of what the crew's task-demands will be so that the actual workload of the crew probably will be acceptable.

Modified Cooper/Harper Scale

The original Cooper-Harper Scale was created in 1969 to develop a handling-qualities scale for new aircraft or modifications (mostly military) so that they could be evaluated by the test pilots in a consistent fashion. It is a subjective scale that uses a branching technique with ten categories of aircraft flyability ranging from 1 (excellent) to 10 (representing major deficiencies).

The Modified Cooper-Harper Scale was developed in 1983 for use in those cases where the tasks were not primarily motor or psychomotor in order to better evaluate cognitive functions such as perception, monitoring, evaluation, communications, and problem solving. All of these are an important part of air transport operations. Modifications from the original Cooper-Harper scale (and there are many) included changing the rating scale, asking the pilots to rate mental workload rather than controllability, and emphasizing difficulties rather than deficiencies. A general conclusion reached by many was that the Modified Cooper-Harper Scale could be used in experiments where overall mental workload is to be assessed, but that it may not be as effective in a situation requiring an absolute diagnosis of a subsystem. The Modified Cooper-Harper Scale, as used by the developers Wierwille and Casali, is shown in Figure 10.1.

Subjective Workload Assessment Technique (SWAT)

The SWAT has been used by the US Air Force and the US Army. Concern has been raised because in one helicopter study SWAT could not discriminate control configurations in a pilot-copilot situation, which had been clear in a single-pilot operation. A more basic concern has been that the SWAT uses only three levels—high, medium, and low—in the three major scales it uses to determine workload, instead of the generally accepted five or seven point scales. The objection to the three level scale is that if pilots are given only three choices, they tend to select the middle category.

The three scales SWAT uses to determine workload are (1) time load, which reflects the amount of time available in planning, executing, and monitoring a task; (2) mental load, which estimates how much conscious mental effort and planning were used to perform a task; and (3) psychological stress load, which estimates the amount of risk, confusion, frustration, and anxiety that were associated with task performance. It then uses a conjoint analysis to convert individual ratings to an overall workload score based on the importance the pilot-raters place on each of the three major dimensions.

While there is general agreement that the SWAT system has produced both validity and reliability, a practical objection for its use, in other than the military environment, is its initial requirement for pilot evaluators to make a card sort of 27 different ratings before the program can be started. If done manually, this sorting takes from 20 to 60 minutes. The sorting procedure can be a distraction from the essential research being done, particularly if a transport operation is being simulated. The efficacy of the SWAT system has not been

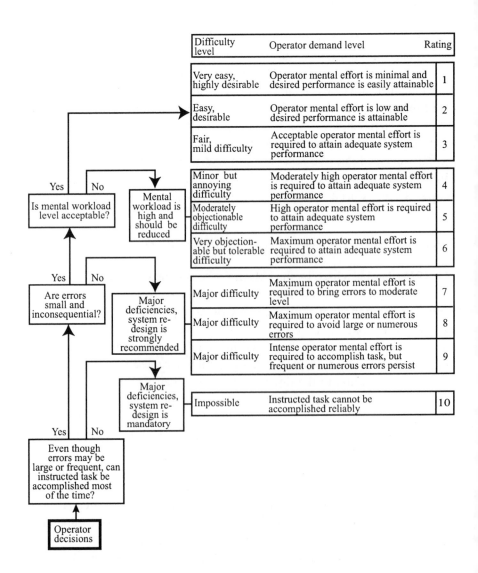

Figure 10.1 The Modified Cooper-Harper Scale

Source: *Operator Workload: Comprehensive Review and Evaluation of Operator Workload Methodologies*, page 87, Lysaght et al., 1989

demonstrated in the air transport environment.

NASA-Task Load Index (NASA-TLX)

The NASA-TLX (Figure 10.2) grew out of the NASA-Bipolar Scale. The NASA Bipolar Scale used nine scales representing workload plus an overall workload scale. The NASA-Bipolar Scale, as well as the later NASA-TLX, was concerned with both the multi-dimensional nature of workload (resulting in multiple workload dimensions), and the individual nature of the dimensions of workload as it was assessed by individual operators. The latter led to the development of individual weighting procedures.

The ten scales in the NASA-Bipolar Scale represent overall workload, task difficulty, time pressure, performance, mental/sensory effort, physical effort, frustration level, stress level, fatigue, and activity type. The NASA-TLX, which is a refined and shorter procedure, also uses a multi-dimensional rating scale but uses only six workload dimensions—mental demand, physical demand, temporal demand, performance, effort, and frustration. A very good description and discussion of these and the other methods for determining

Title	Endpoints	Description
Mental demand	Very low/ Very high	How mentally demanding was the task.
Physical demand	Very low/ Very high	How physically demanding was the task.
Temporal demand	Very low/ Very high	How hurried or rushed was the pace of the task.
Performance	Perfect/ failure	How successful were you in accomplishing what you were asked to do.
Effort	Very low/ Very high	How hard did you have to work to accomplish your level of performance.
Frustration	Very low/ Very high	How insecure, discouraged, irritated, and annoyed were you.

Figure 10.2 The NASA-TLX Rating Scale Description

Source: *Operator Workload: Comprehensive Review and Evaluation of Operator Workload Methodologies*, page 94, Lysaght et al., 1989

workload can be found in Lysaght et al., and in Corwin et al.—the main references for this section. These two references also give illustrations of each of the workload scales.

Bedford Scale

The Bedford Scale was created at the Royal Aircraft Establishment in Bedford, England, and is a modification of the Cooper-Harper Scale. It too uses a decision tree scale and uses a three-rank structure to ask pilots whether it was possible to complete the task, whether the workload was tolerable, or whether the workload was satisfactory without reduction. Its rating scale end points range from 'workload insignificant' to 'task abandoned'. An interesting new concept is that it asks pilots to use their estimated spare capacity to measure levels of workload. A very big advantage is that the Bedford Scale is relatively easy to use and pilots seem to like it. The Bedford Scale is shown in Figure 10.3.

Dynamic Workload Scale

The Dynamic Workload Scale (Figure 10.4) is still another multi-dimensional subjective rating scale. It was developed by Airbus Industrie and has been used to certify Airbus transports. The Dynamic Workload Scale uses a five-point scale for the pilots used in the evaluation and an overlapping seven-point scale for observers. The pilots and the observer are asked to include reserve capacity, interruptions, and stress in determining the workload they encounter. The technique developed to use the Dynamic Workload Scale requires the observer to make a rating whenever the workload has changed since the last rating, or when five minutes have passed. In the latter case, the observer gives an electronically signaled cue to the pilots telling them they should make a rating.

Comparative Evaluations

If a comparative concept of evaluation is used, some of the objections and the lack of a universally accepted scale are minimized. Here the workload of the new or derivative aircraft can be compared to a baseline aircraft, which has been previously certificated and is efficiently operating in the commercial airline environment. One objection to the comparative concept of evaluation has been that individual variations among the raters can make results questionable. An answer to that objection, supporting the importance of a

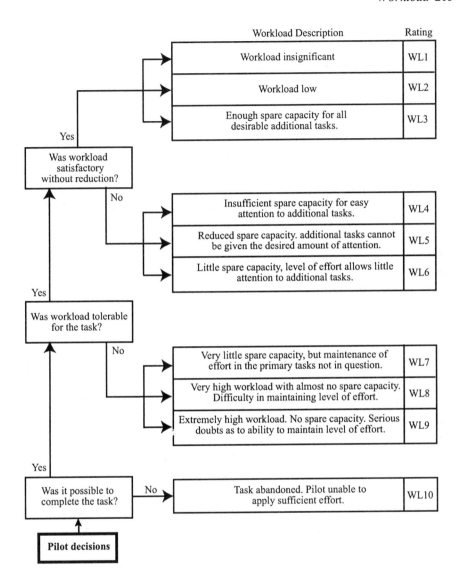

Workload Description	Rating
Workload insignificant	WL1
Workload low	WL2
Enough spare capacity for all desirable additional tasks.	WL3
Insufficient spare capacity for easy attention to additional tasks.	WL4
Reduced spare capacity. additional tasks cannot be given the desired amount of attention.	WL5
Little spare capacity, level of effort allows little attention to additional tasks.	WL6
Very little spare capacity, but maintenance of effort in the primary tasks not in question.	WL7
Very high workload with almost no spare capacity. Difficulty in maintaining level of effort.	WL8
Extremely high workload. No spare capacity. Serious doubts as to ability to maintain level of effort.	WL9
Task abandoned. Pilot unable to apply sufficient effort.	WL10

Was workload satisfactory without reduction?

Was workload tolerable for the task?

Was it possible to complete the task?

Pilot decisions

Figure 10.3 The Bedford Scale developed by Roscoe and Ellis at the Royal Aircraft Establishment in Bedford, England

Source: *Operator Workload: Comprehensive Review and Evaluation of Operator Workload Methodologies*, page 89, Lysaght et al., 1989

Start assessment	Subject-pilot scale	Observer-rater scale	Description of the scales
Is workload light?	yes, it is. A	2 very light 3 light	Low levels of workload so that all tasks are accomplished promptly.
Is workload moderate?	yes, it is. B	4 very acceptable 5 fair & acceptable	Moderate levels of workload with some task interruptions but significant reserve capacity.
Is workload heavy?	yes, it is. C	6 just acceptable	High levels of workload so that workload relief is desirable. Frequent task interruptions or significant levels of mental effort.
Is workload extreme?	yes, it is. D	7 not acceptable continuously	Extreme levels of workload unacceptable on a continuous basis. The possibility of error omission could become high. Continuous task interruptions or extreme levels mental effort or stress.
Is workload supreme?	yes, it is. E	8 not acceptable instantaneously	Supreme levels of workload unacceptable on an instantaneous basis because of physical or mental impairment.

Figure 10.4 Dynamic Workload Scale

Source: *Airbus Industrie Workload and Vigilance,* page 480. Collection of Airbus Training papers, courtesy of Airbus Industrie

representative sample of pilots, is that in the real world the airplane or system must be used by a wide variety of individuals. This is one reason for assuring that a representative sample of pilots is used. Another answer to the same objection is that there are also statistical methods for determining the extent of inter-rater reliability so that its effect can be considered. The number and make-up of the raters used is obviously a very important question.

Pilot Subjective Evaluation (PSE)

The PSE is a combination of a subjective rating scale and a comparative scale. Boeing developed it in conjunction with the FAA for use in the certification of the Boeing 767 and it has since been used for other airplanes. PSE includes both a seven-point rating scale and an integral questionnaire. A particularly interesting aspect of the PSE is that it uses a previously certificated reference airplane, which is chosen by the pilot, in order to make a comparative evaluation of workload. The pilots rated whether the operation of the 767 was more, the same, or less demanding in terms of mental effort, physical difficulty, and time required than it was in the reference airplane. Any areas of greater workload identified areas that provided an opportunity or even a need for design improvement. At the end of each session, pilots were interviewed regarding areas that caused greater workload. The PSE is shown below in Figure 10.5.

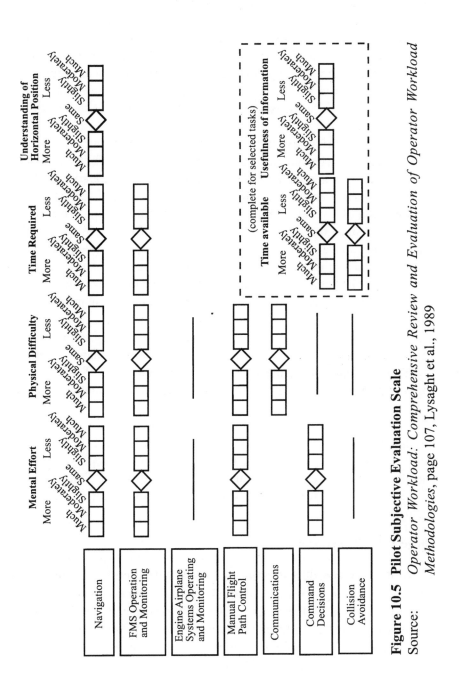

Figure 10.5 Pilot Subjective Evaluation Scale

Source: *Operator Workload: Comprehensive Review and Evaluation of Operator Workload Methodologies*, page 107, Lysaght et al., 1989

Modified Pilot Subjective Evaluation (MPSE)

The MPSE is a custom modification of the PSE and has been used as part of a total evaluation process that was developed by Orlady and Barnes in support of after-market aircraft supplemental type certifications that require crew workload evaluation. It, like the PSE, uses a seven-point rating scale and an integral questionnaire. It uses an expanded interview at the end of each evaluation session.

If used, the MPSE can and should be modified according to the needs of the operation being examined. One of its strengths is its flexibility. The MPSE has three main target sections. The first is Flight Identification and Pilot Information and is essentially self-explanatory. The second section includes an evaluation of normal operations, including departures and arrivals. Data for arrival includes the airport being used, the time of arrival, flight conditions at and during arrival, ATC conditions, the type of approach being used, and the AP/FD modes that were used, if any. (FMS modes would obviously be of interest if a FMS was available.) Pilots are then asked to evaluate the Appendix D workload considerations. The third section covers irregular and emergency operations. In addition, it includes alerting indications and procedures and whether the irregularity was planned or unplanned. The MPSE form for normal arrival is shown in Figure 10.6. The form for additional, irregular, or emergency operations is shown in Figure 10.7. The number and place of the N/As (not applicables) show in the arrival form will vary depending upon the equipment or system being tested.

The MPSE evaluation forms and all comments, either good or bad, are discussed in an expanded interview following each flight. The expanded interview has proved to be particularly valuable. A sophisticated and knowledgeable interviewer is a requirement. It is an advantage if the interviewer has also been an observer in the simulator or flight exercise that is being evaluated.

Other Considerations

As we have previously indicated, there is a wide variation in the physical dimensions, training, experience, and cultural background of pilots in the world aviation system. All of these differences are important considerations, and all of them can affect the workload that is experienced. Anthropometric analyses are an important part of the design process. Other considerations, such as training, experience, and cultural background, can be equally important. Each

Procedures: arrival

Conditions:

Arrival workload functions

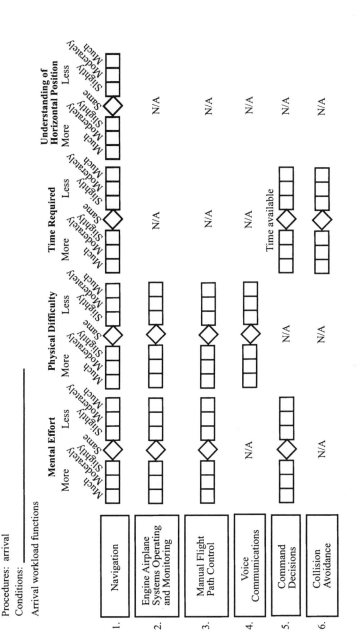

Figure 10.6 The MPSE Form for Normal Arrivals

Source: The MPSE methodology and its related forms are copyrighted by H.W. Orlady and R.B. Barnes and used herein with permission

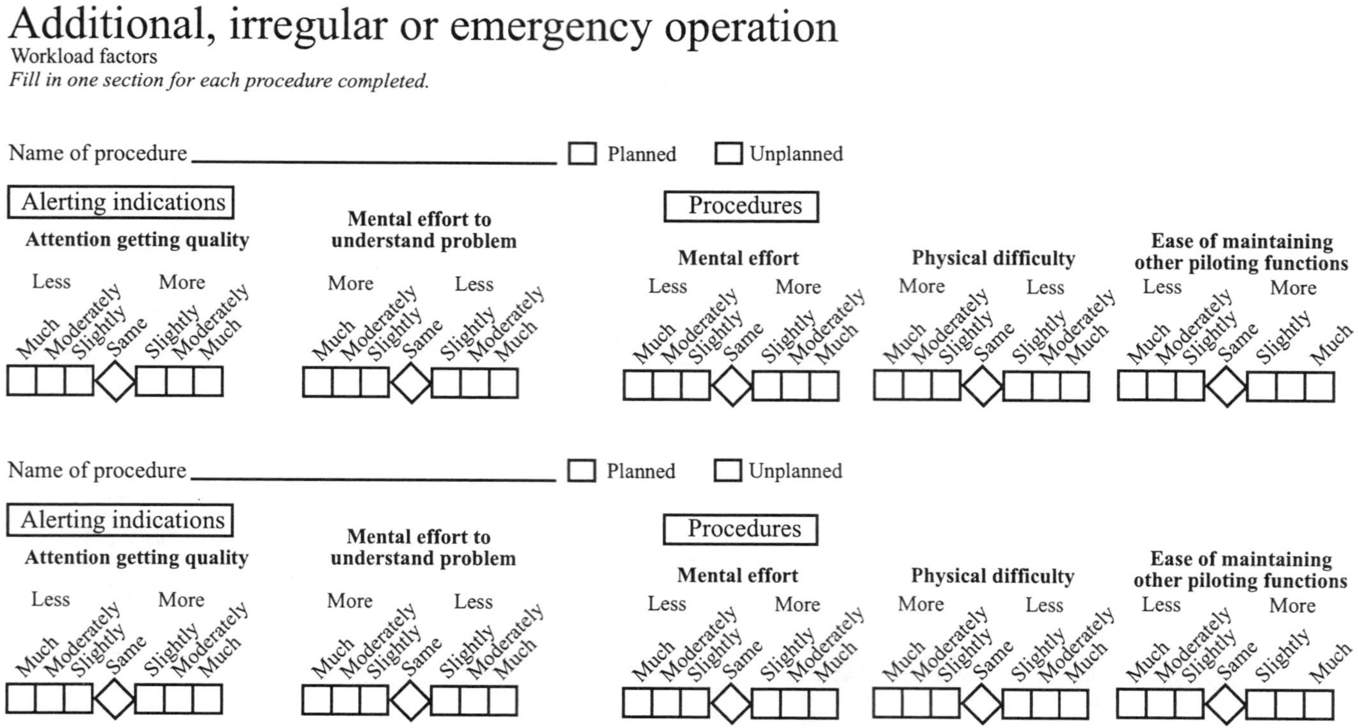

Additional, irregular or emergency operation

Workload factors

Fill in one section for each procedure completed.

Name of procedure ____________ ☐ Planned ☐ Unplanned

Alerting indications

Attention getting quality: Less — Much, Moderately, Slightly, Same, Slightly, Moderately, Much — More

Mental effort to understand problem: More — Much, Moderately, Slightly, Same, Slightly, Moderately, Much — Less

Procedures

Mental effort: Less — Much, Moderately, Slightly, Same, Slightly, Moderately, Much — More

Physical difficulty: More — Much, Moderately, Slightly, Same, Slightly, Moderately, Much — Less

Ease of maintaining other piloting functions: Less — Much, Moderately, Slightly, Same, Slightly, Much — More

Name of procedure ____________ ☐ Planned ☐ Unplanned

Alerting indications

Attention getting quality: Less — Much, Moderately, Slightly, Same, Slightly, Moderately, Much — More

Mental effort to understand problem: More — Much, Moderately, Slightly, Same, Slightly, Moderately, Much — Less

Procedures

Mental effort: Less — Much, Moderately, Slightly, Same, Slightly, Moderately, Much — More

Physical difficulty: More — Much, Moderately, Slightly, Same, Slightly, Moderately, Much — Less

Ease of maintaining other piloting functions: Less — Much, Moderately, Slightly, Same, Slightly, Much — More

Figure 10.7 The MPSE Form for Additional, Irregular, or Emergency Operations

Source: The MPSE methodology and its related forms are copyrighted by H.W. Orlady and R.B. Barnes and used herein with permission

of them can affect workload.

A basic difference in workload assessment measures is that while most of them seek to find a valid and reliable tool to determine the level of workload, only the PSE and MPSE measure comparative workload. The rationale for the comparative approach is that any airplane, or even system, whose workload is equal to or less than the workload in a proven and successful airplane or system, is obviously acceptable. It avoids the difficult question of determining an acceptable level of workload, even if workload could be determined in a satisfactory fashion, without having an agreed upon baseline.

It would be nice for engineering designers if workload researchers could develop a measure that could accurately predict the workload that would be created by a given system or by an airplane in advance of a real-world test. It would also be nice if researchers could develop a non-subjective measure that could simply be measured in an actual test scenario. However, the researchers have been unable to do so to date for some of the reasons we have discussed. Actual workload assessments when the system, systems, or airplane are being used in real-life conditions are most accurately made with a subjective methodology. Sandra G. Hart said it nicely in an AGARDOGRAPH[2] edited by Alan Roscoe when she said:

> Subjective ratings may come closest to tapping the essence of workload and provide the most generally applicable and sensitive measure. This is because they provide a direct indication of the impact of flight-related activities on pilots that integrates the effects of many workload contributors. (Hart, 1984)

---

[2] An AGARDOGRAPH is a paper or report prepared for AGARD (see Glossary) by a committee selected for a specific purpose.

References

Barnes, Robert B. and Adams, Charles F. (1997). 'Minimum Crew Certification: Human Factors Issues and Approaches', in *Proceedings of the Ninth International Symposium on Aviation Psychology*, The Ohio State University, Columbus, Ohio.

Corwin, William H., Sandry-Garza, Diane L., Biferno, Michael H., Boucek, George P. Jr., Logan, Aileen, L., Jonson, Jon E., Metalis, Sam A. (1989). *Assessment of Crew Workload Measurement Methods, Techniques and Procedures, Volume I - Process, Methods and Results and Volume II - Guidelines for the Use of Workload Assessment Techniques in Aircraft Certification*, WRDC-LTLR-89-7006, Federal Aviation Administration, Washington, D.C.

Greene, Berk and Richmond, Jim (1993). 'Human Factors in Workload Certification', presented at SAE Aerotech '93, Federal Aviation Administration, Seattle, Washington.

Lysaght, Robert J., Hill, Susan G., Dick, A.O., Plamondon, Brian D., and Linton, Paul M., Wierwille, Walter W., Zaklad, Allen L., Bittner, Alvah C. Jr., Wherry, Robert J. (1989). *Operator Workload: Comprehensive Review and Evaluation of Operator Workload Methodologies*, Technical Report 851, United States Army Research Institute for the Behavioral and Social Sciences, Alexandria, Virginia.

McLucas, J.L., Drinkwater, F.J. III, Leaf, H.W. (1981). *Report of the Task Force on Aircraft Crew Complement*, US Government Printing Office, Washington, D.C.

Meister, David (1985). *Behavioral Analysis and Measurement Methods*, John Wiley & Sons, Inc., New York.

Meister, David (1989). *Conceptual Aspects of Human Factors*, The Johns Hopkins University Press, Baltimore, Maryland.

Parasuraman, J., Molloy, R., and Singh, I.L. (1993). 'Performance Consequences of Automation-induced "Complacency"', *International Journal of Aviation Psychology*, 3(1), 1-24., Lawrence Erlbaum Associates, Mahwah, New Jersey.

Weimer, Jon (1995). *Research Techniques in Human Engineering*, Prentice Hall, Inc., Englewood Cliffs, New Jersey.

Wickens, Christopher D. (1992). *Engineering Psychology and Human Performance*, Harper Collins Publishers, New York.

11 Automation

— Lest We Forget—

Man is not as good as a black box for certain specific things, however he is more flexible and reliable. He is easily maintained and can be manufactured by relatively unskilled labour.

(Wing Commander H.P. Ruffell Smith R.A.F., 1949)

Introduction

Automation of an airplane's operating functions to assist the pilot has been a part of aviation since even before the Wright Brothers proved that powered flight was possible. Probably the first example of aircraft automation was the gyroscopic device of Sir Hiram Maxim,[1] who patented it in 1891 in London. This was 12 years before the Wright Brother's successful flight at Kitty Hawk. Maxim's device provided stability augmentation to the fore and aft elevators of a highly unstable experimental airplane that was built and tested on two very ingenious fixed wood and iron tracks which controlled the airplane. While Maxim's airplane never flew because it could not be controlled, he was among the several scientists of that era working very hard to develop an airplane which could be flown under its own power and which could be controlled in the air (Moolman, 1980; and Billings, 1997).

The Growth of Automation

The Wright Brothers, who achieved the first powered flight in 1903, began working on an autopilot in 1905 and applied for a patent for an automatic

---

[1] Hiram Maxim was a Maine Yankee, born in 1840. He invented a machine gun toward the end of the 1880s that the US War and Navy Departments found impractical. In 1881, the British War Office showed considerable interest and Maxim settled in England and became a British subject. In 1884 he perfected the Maxim machine gun which fired 600 rounds per minute and made him famous. He was knighted by Queen Victoria. Maxim had been interested in aeronautics since boyhood, but was unable to solve the problems of aerial steering and control. His 1894 aeronautical machine weighed an incredible 8000 pounds but did confirm that a machine could be lifted off the ground with its own power.

stabilizing device in 1908. They received patents for both in 1913 and were awarded the prestigious Collier Award on 5 February 1914 for these accomplishments. That same year 21 year-old Lawrence Sperry invented his famous gyroscopic stabilizer which has formed the basis for all future stabilizing systems including the complex inertial navigation systems (INS) that in 1969 guided the first men to the Moon (Crouch 1989). As F. Howard's *Wilbur and Orville* stated: 'In the evolution of flight vehicles, an increasing use of automation in certain aircraft offsets direct pilot motor responses and contributes greatly to reductions in workload requirements of pilots' (Mohler, 1997).

Barometric altitude adjustments to fuel-air ratios and controllable-pitch propellers were introduced in some of the early attempts at increased power control automation. Some years later these were followed by propeller auto-synchronizers that matched the propeller speeds in all engines thereby minimizing vibration and the annoying beat frequency noise. These automated devices became very important in the early four-engine piston airplanes such as DC-4s and Constellations and added considerably to passenger comfort and the marketability of aircraft using them.

The autofeathering devices, which were required in order to achieve acceptable single-engine performance in some of the twin-engine airplanes built or modified just after World War II, were a mixed blessing. The purpose of these devices was to minimize drag and controllability in case of an engine failure after takeoff. Unfortunately, several autofeathering devices were involved in accidents where fully functional engines were inadvertently shut down by this early automation. Autofeathering devices generally were not accepted because of their unreliability.

Autopilots

Autopilots, primitive by today's standards, were available for the early DC-3s, and many airlines used them. One of the holdouts was Eastern Airlines, whose pilots complained that, 'if it wasn't in Captain Eddie's SPAD,[2] he won't buy it'. In those days, 'Captain Eddie' was Eddie Rickenbacker, World War I hero, and the President and CEO of Eastern Air Lines. Regarding autopilots, Captain Eddie's response is said to have been: 'I pay those guys to fly, so let them fly. I'll be damned if I'll pay them just to sit there.'

---

[2] The SPAD was an early World War I fighter that was manufactured by the *Société Pour l'Aviation et ses Dérivés*.

Autopilots in those days were rough and far from 'user friendly'. They were used sparingly. Pilots have always tried to fly smoothly and with maximum comfort for their passengers and in the early days, it was usually just easier to fly manually. This was true because of the quality of the autopilots and the quality and characteristics of the four-quadrant low frequency radio ranges that pilots depended upon for navigation. As autopilots were improved and very high frequency (VHF) navigation became available, the autopilots made it much easier to fly smoothly enroute with precision. Manual flying, especially on long cruise legs, was a very fatiguing task. Big advances in autopilots occurred in the 1950s. Improvements in autopilot design have continued and the routine use of autopilots among transport pilots is common today. The latest autopilots are certified for use in all phases of flight except takeoff.

Autothrottles

Autothrottles, which controlled thrust to jet engines by controlling fuel flow, were initially quite rough and sometimes caused large power changes and disagreeable 'power hunting'. The resulting noise and vibrations were annoying to both pilots and passengers, and therefore, autothrottles were rarely used. Later more sophisticated systems were developed which overcame the early objections. These later autothrottles became commonly used because they were smooth and significantly reduced workload for the pilots. The latest development in autothrottles has been that of full-authority digital engine controllers (FADECs). They further improve the precision with which jet powerplants can be controlled. Many automatically prevent overboosting and temperature exceedances. Today the latest jet aircraft incorporate autothrust systems that are used to automatically determine engine parameters and to set engine power for nearly all flight phases, including takeoff.

Other Automated Systems

Another automated system in wide use today is the anti-skid (anti-lock) braking system. This is a pressure modulating system that gives the airplane maximum braking by automatically releasing any braking wheel just before it locks-up and skids. If a braking wheel locks up, the wheel's traction capability is significantly reduced.

In a further refinement, many of the latest airplanes also have an automatic braking system that provides maximum braking for landing or for an aborted takeoff. This system allows the pilot to select different deceleration rates for automatic braking on landing. For example, on a Boeing B-767, five different levels may be chosen. During normal landings, modulated braking, according to the selection made by the pilot, is automatically applied after landing gear wheel spin-up is achieved with ground contact. In addition, the pilot may arm the system to provide maximum braking for rejected takeoffs (RTOs). This RTO feature, when selected, operates automatically under a number of conditions such as when both throttles are retarded to idle above 80 knots (Boeing B-767), or when ground spoilers are deployed if the airspeed is greater than 72 knots (Airbus A-320).

In order to reduce cockpit workload, aircraft systems and subsystems have been simplified and gradually further automated. Electrical, hydraulic, pneumatic, and fuel systems, which formerly were the province of the flight engineer, have been largely automated to make two-person crew operations feasible. Seat belt and no smoking signs have been activated automatically, the automatic load shedding of electrical loads to simplify problems associated with a generator failure is provided, and air conditioning packs have been automatically deactivated following an engine failure on takeoff. These and other automatic devices have all reduced cockpit workload. They have not made these airplanes less complex.

The continued use and increasing level of automation in transport aircraft operations has had many ramifications. For example, significant improvements in radio communications supplanted the need for an airborne radio operator. In much the same fashion, automated navigation systems such as INS and DME have eased the workload on domestic flights, and they, along with Doppler and LORAN, have replaced the navigator that formerly was required for long trans-oceanic flights. Recent additions have been FMS, GPS, and incremental elements of FANS. (See the Glossary for a description of these acronyms. They are discussed further in later sections of this chapter and in Chapter 22—The Air Transport Future.) Automation has made practicable the use of a two-person crew for the complex transport aircraft that have become the new standard for today.

The Role of Automation and the Pilot

There has been considerable debate over whether or not today's automation has become so good that it has significantly changed the role of the pilot. While a great deal of this debate is probably semantics, we believe that the role of the pilot has not changed. The pilot's role is still to fly the airplane from point A to point B safely and efficiently. It was that in the day of the DC-3, and it is the role of the pilot today.

Increased automation is now, and traditionally has been, simply a tool to help the pilot fulfill his traditional role. Automation has not changed the pilot's role and we do not expect it to do so in the future. However, there is no question that the pilot's task required to perform his role has changed and changed considerably. The change has been gradual and evolutionary. Flying a DC-3 was different than flying a DC-6 or DC-7, and flying those airplanes was different than flying B-707s or DC-8s. It is not surprising that flying a B-707, B-727, or a DC-8 is also different than flying one of the latest airplanes. Flying in today's environment is different than was flying in the early years. Many of the changes in the latest airplanes are due to increased automation.

Today's cockpit instrumentation has changed from analogue to digital. The latest airplanes now have CRTs and flat panels to show cockpit information. They have a considerably greater level of engine and subsystem automation, can display additional information, and they have considerably more sophisticated autopilots. The latest airplanes are faster, larger, and are increasingly complex. In addition, they, as well as all other airplanes, are now flown in a considerably more complex environment.

Because the old analogue displays had several advantages over strictly digital presentation, digital displays have, in some cases, been modified to retain those advantages. One big advantage of the older analogue displays was that they usually consisted of a round dial and a pointer or pointers and presented a total picture that was not present when only a digital number was displayed. One has only to consider the information displayed on the old-fashioned (analogue) round-dial wrist watch with the information one can get from a digital wrist watch to see the difference that is portrayed in the total temporal picture.

The Five Sub-tasks of the Pilot

The five sub-tasks of the air transport pilot are:

1. to operate, manage, and to monitor the engines and airplane control systems;

2. to avoid inadvertent encounters with either unfriendly terrain or with objects on the ground;

3. to navigate efficiently to the destination airport;

4. to ensure comfort to the passengers and crew (by operating and monitoring such systems as pressurization and temperature control);

5. to communicate with company operations and with air traffic control.

These subtasks occur in all airplanes. Automation in transport aviation is concerned with automating parts or all of the pilot's sub-tasks.

Theoretically, and in most cases actually, by making these sub-tasks easier and thereby reducing workload, automation enables the pilot to accomplish his/her tasks with increased safety and efficiency. Today's automation is evolving from simply taking tasks away from pilots to both making these tasks easier to perform or manage and by providing better total operational information. In some cases, automation also accomplishes the pilot's operational tasks better than the pilots can manually perform them routinely and consistently.

Control, Information, and Management Automation

Boeing's Delmar Fadden has divided modern automation into 'control automation', which assists the pilot in airplane control tasks or substitutes his/her manual manipulation with the automatics, and 'information automation', which includes all of the display and avionics that are devoted to navigation and environmental surveillance and to the digital communications that are a growing part of both air traffic control and airline operations. Charles Billings has added a third category which he has called 'management automation' (Billings, 1997). Management automation includes all of those things that aid the pilot in the management of the mission, i.e., in flying from point A to point B safely and efficiently. Management automation guides the airplane, performs the necessary flight functions, and furnishes the pilot with information involving both the state of the airplane and of progress toward the mission's goals.

The difference between 'control automation' and 'management automation' has become an important distinction. In control automation, the pilot physically selects the particular automation system that will accomplish the task or sub-task that is required. These things are done individually and separately. In management automation, the pilot simply selects the overall goal and the automation selects, and then in an integrated fashion, performs the tasks required to achieve the selected goal. Another way of thinking of the difference is that control automation involves inner loop control, while management automation involves the outer loop.

Automation and Safety

A major question raised during the introduction of the latest generation of increasingly automated airplanes was whether or not increased automation actually increases safety. Some have stated that the present level of automation actually decreased safety, or at least that there was a strong possibility that it would. This is an important question, for there is no doubt that safety is the preeminent consideration in all air transport operations.

The Safety Record with Increased Automation

Air transports' hull loss accident rate during the jet era is shown in Figure 11.1. It clearly shows the improvement in the safety record established by later and increasingly automated airplanes. While the entire improvement in the safety record for later airplanes may not be entirely due to automation, the improvement in the record is obvious. Equally obvious is the fact that the improvement has not changed the traditional role of human factors in the designated causes of air transport accidents. Whether or not the flight crew is the inheritor or the instigator of air transport accidents (Reason (1997), the flight crew is still directly involved in from 60 to 80 percent of all major accidents.

Airbus Industrie provided a graph (Figure 11.2) at Airbus Industrie's Fourth Flight Safety Conference that shows essentially the same information but in a slightly different format. The automated aircraft that it considered were the McDonnell Douglas MD-80, MD-11, MD-90, later versions of the Boeing B737, B757, B767, Airbus A310, and the A300-600. They all had better records than previous generations, and it appears 'that the fly-by-wire aircraft will be as good and possibly better than the third generation automated aircraft' (Birch, 1997). This prediction proved true, however, the benefits seem to get smaller with each aircraft generation. This suggests that in order to make substantial improvements in the safety record it will be necessary to more effectively use a total systems approach, to take advantage of operational human factors advancements, and to implement good CRM programs. It will also be necessary to take better advantage of the lessons that can be learned through increased operational feedback, and the lessons that can be learned

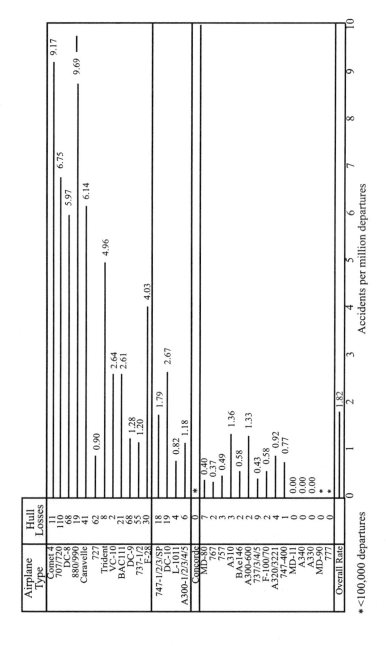

Airplane Type	Hull Losses	Accidents per million departures
Comet 4	11	9.17
707/720	110	6.75
DC-8	68	5.97
880/990	19	9.69
Caravelle	41	6.14
727	62	0.90
Trident	8	4.96
VC-10	2	2.64
BAC111	21	2.61
DC-9	68	1.28
737-1/2	55	1.20
F-28	30	4.03
747-1/2/3/SP	18	1.79
DC-10	19	2.67
L-1011	4	0.82
A300-1/2/3/4/5	6	1.18
Concorde	0	*
MD-80	7	0.40
767	2	0.37
757	3	0.49
A310	3	1.36
BAe146	2	0.58
A300-600	2	1.33
737/3/4/5	9	0.43
F-100/70	2	0.58
A320/3221	4	0.92
747-400	1	0.77
MD-11	0	0.00
A340	0	0.00
A330	0	0.00
MD-90	0	**
777	0	
Overall Rate		1.82

Accidents per million departures

* <100,000 departures

Figure 11.1 Hull Accident Rates: Worldwide Commercial Jet Fleet 1959-1996

Source: *Statistical Summary of Commercial Jet Aircraft Accidents—Worldwide Operations— 1959-1996*, page 15, Boeing Commercial Airplane Company, 1996

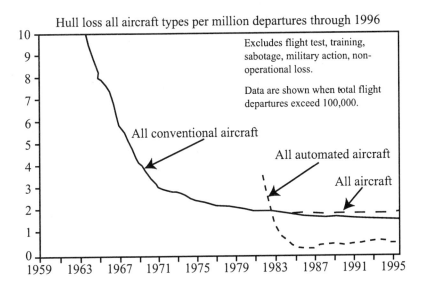

Hull loss all aircraft types per million departures through 1996

Figure 11.2 **Hull Losses in Automated vs. Conventional Aircraft 1959-1995**

Source: Courtesy of Airbus Industrie

through increasingly refined and individually sensitive training methods.

While the safety record strongly suggests that the argument against automation because of an adverse effect on air safety has little merit, it also makes it clear that automation has not reduced the problem of human error. That fundamental problem remains and, in addition, two basic automation questions that are still open and receiving considerable attention are (1) defining the optimum task allocation of automation in the two-person crew in increasingly automated airplanes, and (2) determining the ultimate extent of automation in air transport operations.

Are we Really Seeing New Problems?

In a discussion of advanced technology accidents, incidents, and human errors, John Lauber, then a Member of the US NTSB, posed a question that is at the heart of any discussion of automation when he asked:

> Are these examples of automation-caused problems, or are they simply human performance accidents which just happen to involve automatic systems, but which are otherwise not fundamentally different from any other human performance accident [or incident]?

In much more general terms, David Meister raised the same question when he wrote:

> It is possible that most behavioral problems do not change, but appear in a new guise only when one concentrates on the...hardware/software interface details in which they are wrapped.

The answers to the questions raised are still open and very important to any one interested in air safety and the human factors involved in air transport operations. The truth many very well require more than a simple 'yes' or 'no' answer.

The Role of the Human

In its 'National Plan to Enhance Aviation Safety through Human Factors Improvements', one of the ATA's Human Factors Task Force's assumptions was that: '.... Humans will continue to manage and direct the NAS (National Aviation System) through the year 2010...' (Air Transport Association, 1989). We, and many others, believe that the Task Force's hypothesis is overly conservative and that we are much farther from a fully automated system, including the pilotless airplane, than the year 2010. It is hard to imagine today's public accepting a pilotless airplane even assuming that the aviation system has made this technologically possible.

You should note that the year 2010 was simply an assumption stated in the ATA report. It is not a prediction of conditions beyond 2010. We have no reluctance to state that we believe that humans will perform critical functions in the National Aviation System for many years past the year 2010 and that human factors, including the optimum use of automation, will remain subjects of major concern for years well past 2010. In considering all of the interface problems involving the flight deck, the report of another group, the FAA Human Factors Team, noted that 'there is general agreement that the flight crew is and will remain ultimately responsible for the safety of the airplane they are operating.'

Nearly two decades ago, Wiener and Curry noted that, 'Any task can be automated.' They then asked, 'whether it should be...?' (Wiener and Curry, 1980). There is little doubt that we are close to being able to completely automate the airplane and to have a pilotless airplane, impracticable as such a plan would be. For both basic safety and socio-political reasons, we believe a pilotless airplane will be neither a feasible nor an acceptable possibility in the foreseeable future.

Fundamental Automation Problems

The growth of automation has continued in the latest airplanes that have been developed in the US and in other countries. One of the factors that has exacerbated many of the problems we have today was noted in the Report of the FAA Human Factors Task Force. It found that: 'Each airplane manufacturer has a different philosophy regarding the implementation and use of automation.' This is not surprising for the latest airplanes represent new levels of automation and at least slight differences in automation philosophy.

The competition among manufacturers is fierce. From the government's or the public's point of view, progress would be impossible if the industry was standardized to the point that innovations were automatically restricted in the interests of standardization. The human factors and purely technical questions that are raised may well have more than one good answer. Complex issues are involved. There are not simple, black and white answers to many of the basic human factors and technical questions.

Existing Automation Philosophies

Orlady and Barnes have noted that the FAA has summarized three manufacturers' philosophies on automation as follows:

Airbus —

- Automation must not reduce overall aircraft reliability; it should enhance aircraft and systems safety, efficiency, and economy.

- Automation must not lead the aircraft out of the safe flight envelope and it should maintain the aircraft within the normal flight envelope.

- Automation should allow the operator to use the safe flight envelope to its full extent, should this be necessary due to extraordinary circumstances.

- Within the normal flight envelope, the automation must not work against operator inputs, except when absolutely necessary for safety.

Boeing —

- The pilot is the final authority for the operation of the airplane.

- Both crewmembers are ultimately responsible for the safe conduct of the flight.

- Flight crew tasks, in order of priority, are safety, passenger comfort, and efficiency.

- Design for crew operations based on pilot's past training and operational experience.

- Design systems to be error tolerant.

- The hierarchy of design alternatives is simplicity, redundancy, and automation.

- Apply automation as a tool to aid, not replace, the pilot.

- Address fundamental human strengths, limitations, and individual differences — for both normal and non-normal operations.

- Use new technologies and functional capabilities only when they result in clear and distinct operational or efficiency advantages and there is no adverse effect to the human-machine interface.

McDonnell-Douglas —

- Uses technology to assist the pilot naturally, while giving the pilot the final authority to override the computer and use skill and experience.

(Orlady and Barnes, 1997)

Automated Systems and Pilot Workload

To date, reducing pilot workload at any time and at any level has been considered desirable. As we have noted, that concept now is being seriously questioned. The question raised is not a simple one. While it may be true that we went through a period of almost having automation for the sake of automation, that is not true today. Major concerns regarding increased automation are the negative effect it may have on the flight crew's ability to monitor effectively, the on-the-job challenge that would remain in operating a fully automated airplane, and the effect of increased automation on overall safety.

In today's airplanes, many normal subsystem functions formerly performed by the flight crew have been automated. The handling of faults in these subsystems is often designed to be largely automatic (Billings, 1991). Billings quotes the chief of MD-11 operations: 'One of our fundamental strategies has been: if you know what you want the pilot to do, don't tell him, do it' (Billings, 1991 and 1997).

A human factors question regarding automatics is whether a given system should be shut off or changed without any prior pilot notification. We should note that if a given system can be shut off or changed without any prior pilot notification and without any adverse ramifications, it would probably reduce pilot workload. However, we believe that it is fundamental that the automatic operation is accompanied by appropriate feedback and that the pilot is immediately notified of the action. As Charles Billings has stated, and we completely agree, it is essential that the human operator must both monitor the automatic system and at all times be properly informed. There is a modification or expansion to this concept. In this case, systems would be allowed to change status automatically, including being shut down, without prior notification to the pilots. However, after receiving notification of the system or status change, the automatics would permit the pilots to return the systems to their original state.

The Flight Management System (FMS)

The Flight Management System (FMS) is the central component of advanced technology aircraft. The FMS is capable of four dimensional area navigation, integrating latitude, longitude, altitude and time parameters while optimizing aircraft performance to allow the pilot to select specific flight profiles such as most economical, fastest, company-preferred, etc. It is easy to become confused as some of the acronyms and terminology used to describe the different components of flight management systems can vary both by manufacturer as well as across different aircraft of the same manufacturer. For example, Airbus refers to the FMS as FMGS which stands for Flight Management and Guidance System.

The MD-80 and the Boeing 757-767 marked the first systematic effort to integrate a variety of automatic devices into a seamless process for use in line operations. While previously long-range aircraft with inertial navigation systems (INS) capability were able to program flight paths with INS, the FMS was the first system designed to be the primary means of navigation for all flights under all conditions. Some of the pilot's tasks are relieved by FMSs. However, pilots now have additional knowledge requirements, and most experts agree that the cognitive load of the pilots has been increased.

The FMS integrates a number of different systems to provide automatic lateral and vertical navigation as well as performance optimization. The components included in the FMS can vary by aircraft type, by manufacturer, as well as by operator specifications. For example, the FMS on a Boeing B737-300 (through B737-800) is comprised of, but not limited to four component systems:

- Autopilot/Flight Director Systems (AFDS),

- Autothrottle (A/T), and

- Inertial Reference Systems (IRSs) and, on later aircraft, Global Positioning Satellites (GPSs) and satellite communications (SATCOM),

- Flight Management Computer/Control Display Units (FMC/CDU).

On the Airbus A-320, the FMGS (or FMS) consists of the following units:

- Flight Management and Guidance Computers (FMGCs),

- Multifunction Control Display Units (MCDUs),

- Flight Control Unit (FCU),

- Flight Augmentation Computers (FACs),

- Throttles, and

- EFIS CRTs.

The flight management computers (FMCs, or in Airbus terminology, FMCGs) represent a major step in automation development and will be the focus of this discussion. The FMCs combine three types of information:

- flight plan and performance information entered by the pilot,

- information received from supporting systems (ie., IRSs, navigational aids),

- information from within its own database.

Interface with the FMC is obtained through a control and display unit (CDU) usually installed on the center cockpit pedestal with one being supplied for each pilot. The CDU is the hardware used by the pilot to program and use the flight management computer (FMC). The CDU used in the MD-80, which is typical, is shown in Figure 11.3.

The FMC is a complex system. The FMC contains a large number of pages, each of which contains up to 14 lines of alphanumeric information. It is a major source of the additional cognitive load placed upon pilots and a major reason that the pilot training load has increased, not decreased.

FMC internal software includes a complex operational program which is the navigation and performance database that is specific for the airplane and the operator. This updatable software package stores a great deal of information. The operational program can contain over 1,400 software modules. The navigation database covers all the areas where the airplane is normally flown. It is tailored to specific airline customers and can contain over 32,500 navigation points and airway route structure data. Typically, up to 40 additional waypoints can be entered into the database by the pilots.

The FMC performance database provides much of the information that is found in airplane manuals in predecessor airplanes. It provides speed and altitude guidance in all flight regimes and is used by the FMC to provide detailed projections for waypoints along the entire flight. The performance database also provides accurate airplane aerodynamic data (such as drag) that can vary with engine model and gross weight. The performance database also provides maximum altitude and maximum and minimum speeds for all weights and aircraft conditions.

While much has already been done to simplify and clarify FMC/CDU operation, the complexity of the mode and display architecture creates substantial operational issues. It is difficult, for example, to enter alphanumeric data routinely without making an occasional error. Inputting an east longitude when one intended to enter a west longitude can create significant operational problems. This example actually occurred at one airline when the pilots were initializing the system on the ground during part of their normal cockpit setup. They received several error messages indicating that the 'box' did not like the inputted position entry. With some persistence and creativity, they were able to override these messages and force the incorrect entry.

Performance and monitoring standards must be very high. Considerable

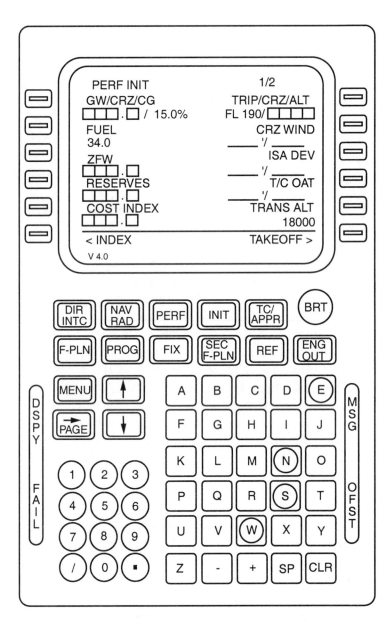

Figure 11.3 Honeywell Flight Management System Control and Display Unit (CDU)

Source: Reproduced by permission of Douglas Aircraft Corporation, now Boeing Commercial Airplane Company

effort is being made to simplify these types of problems and to make the CDU entries more error resistant. Later models are a definite improvement, and as Billings explained:

> Standard or frequently used routes are stored in the navigation database and may be recalled by number. SIDS and STARs[3] are also in the database, if a change is required by ATC, only the name of the procedure need be entered. Changing the arrival runway automatically changes the route of flight. Appropriate navigation radio frequencies are autotuned as required. Perhaps most important, newer FMSs interact directly with navigation displays; pilots are shown the effect of a change of flight plan in graphic form. They can thus verify that an alternative flight plan is reasonable (although not necessarily what was requested by ATC) before putting it into effect.   (Billings, 1997)

Approach and runway changes are frequently necessary and seem inevitable in our present ATC system. Some level of reprogramming is required whenever changes in the approach or runway being used are made by ATC if the crew wishes to make full use of the flight managment systems. This reprogramming frequently takes excessive pilot time, especially with the early FMCs. The reprogramming task itself is usually not that difficult; some pilots engage in it simply because they know the system has the capability and because they know how to do it. However, punching CDU buttons during the approach phase is a distraction for both the PNF and the PF, who should verify the programming before it is executed. This is time during the approach that is needed to keep a good outside-the-cockpit look for other airplanes.

To meet this problem many pilots made it a rule not to reprogram below 10,000 feet, a provision that has been adopted by many airlines as a SOP. This solution may result in the approach being manually flown but permits the human resources available to be used for higher priority tasks. The cost of the solution is the loss of the benefits of the FMS automation in terminal areas. Approach and runway change was as engaging or distracting problem in older airplanes. Pilots have found that the alphanumeric entry of data in the FMC is more time and attention consuming than simply turning the selector knobs in predecessor aircraft.

Much of our abbreviated discussion of FMSs is admittedly based on pages 107 to 117 of *Aviation Automation: The Search for a Human-Centered Approach*. In this highly recommended text, Dr. Charles Billings fully discusses

---

[3] SIDs (Standard Instrument Departures) and STARs (Standard Arrival Routes). These routes are promulgated by the air traffic control system.

FMSs and their use. It is an excellent source for additional information.

The Proliferation of Warnings and Alerts

One of the consequences of increasingly complex and sophisticated airplanes and the reduction of the standard cockpit crew from three to two persons has been an increase in the sophistication of the alarm and warning systems. These warnings can be important elements in achieving effective system awareness. The increased sophistication of alarm systems and the introduction of new alarm systems such as Enhanced Ground Proximity Warning Systems (EGPWS), Traffic Alert and Collision-Avoidance Systems (TCAS), and Wind Shear Avoidance Systems (WSAS), and now the Integrated Hazard Avoidance Systems (IHAS), have not eliminated the very real problem of false, multiple, or confusing alarm events. These systems are described in Chapter 21—Current Safety Problems—and Chapter 22—The Air Transport Future. Researchers, designers, and operational experts are working diligently and with some success to eliminate, or at least minimize, the adverse operational consequences that can accompany these problems.

In today's high technology airplanes, alarms can be displayed by either one or a combination of cues that include visual cues such as lights, icons, or text; auditory cues such as bells, whistles, sirens or artificial speech; or tactile cues such as stick shakers or stick pushers.[4] This proliferation of warnings is controlled to some extent by the inhibition of certain warnings at times in the flight when the distraction caused by the warning could jeopardize safety of the flight or, at the very least, be simply an unnecessary annoyance. Examples are the inhibition of the landing gear warning horn at speeds above 210 knots and, in one of our latest transports, the inhibition of the fire-warning bell and associated Master Warning Lights during takeoffs. Further examples are given in the priority system in IHASs.

Because 'the number of combinations of multiple alarms grows as a factorial of the independent alarms (that is, three alarms allow for six combinations, four alarms allow for 24 combinations, and so on)', the problem of priority is a very real one for designers of our latest transports (Gilson et al., 1996). It is important that this problem be kept in perspective. To date,

---

[4] Actually, the stick pusher is more than just a cue for it is a positive stall-prevention system that forces the stick forward commanding the aircraft to rotate from climb to a shallow dive. The stick shaker does not push the control stick forward but only shakes it.

multiple failures seem to have been rare and well controlled in transport operations, and the occasions when there are more than two such failures are extremely rare. However, all who are involved in air transport operations should be aware of the potential problem, especially those involved in training.

Gilson et al., believe that, 'When multiple failures do occur, a successful outcome will depend more on problem-solving skills than on any set of rules.' They stated the problem nicely when they wrote:

> A balance is needed between the undisputed capabilities of the computer and the unique problem-solving and pattern recognition skills of the human operator.
> (Gilson et al., 1996)

Pilot/Computer Interface

Many researchers believe that a very real problem in today's transports is what is often called the pilot/computer interface. A series of recent accidents involving Airbus transports—Habsheim (1988), Bangalore (1990), Strasbourg (1992, Nagoya (1994), and Toulouse (1994), and the serious incidents that occurred with an A310 at Moscow (1991) and with an A310-300 at Paris, Orly (1994) have caused concern. ASRS data makes it clear that the interface problems, whether they are basic design problems or a combination of design problems and training, are not limited to the Airbus.

When considering the incidents given in the preceding paragraph, one should remember that items like mother tongue, the social environment, the '4 Ps'[5] and the quality of training should be considered. These are all important in determining the role and demands of the pilot/computer interface. Consideration of this problem involves automation in all modern transports.

A major positive factor in dealing with the error dilemma is the development of 'human-centered automation', a concept championed by Charles Billings and discussed in later paragraphs. The main precepts of human-centered automation have been adopted by US and European manufacturers, operators, pilot associations, and regulators. The concept of human-centered automation is already making significant progress toward reaching the ultimate, and probably illusory, goal of 'zero' accidents.

---

[5] The '4 Ps' are Philosophy, Policies, Procedures, and Practices. They were discussed in Chapter 6—The Social Environment.

The Proliferation of Control Modes

A problem that is causing considerable uneasiness involves the proliferation of control modes in the newest systems, particularly when the airplane is under the control of the autopilot and the pilot crew is monitoring the performance of the autopilot and the airplane. Unfortunately, the training that the pilots received in this area has often been less than optimum. Training devices, especially some of the early part-task trainers, have been inadequate in providing realistic training. As an example, many of them provided only single-path scenarios, e.g., a single button is the only correct response in the trainer, when it is quite clear that several responses can achieve the desired outcome. All of the responses are available and most of them are used in the actual airplane.

A training factor in the real world is the reality that if all of the available modes are covered in training, then the pilots are responsible for demonstrating knowledge and competence in all modes during the checkride. This inevitably increases training times. One airline, as an example, specifies that pilots are responsible for only seven modes on a checkride, and these are the only modes taught during training. The official checking potential is limited to these seven modes. While these modes do not represent all that can be used in normal aircraft operation, they do represent a good foundation for the pilot to start with and upon which to continue building.

As an example of this problem in another airplane, pilots can choose from at least five different levels of automation to change altitude (Sarter and Woods, 1995). The late Berk Greene and Jim Richmond raised a very provocative question when they asked, 'do we need all these modes?' (Greene and Richmond, 1993). Particular problems have arisen because the common autopilot modes of 'vertical speed', 'flight path', and 'flight level change' all have similar functions, yet have different constraint limits and affect aircraft behavior in different ways when limits are reached. The 'open descent' mode in the A-320 and its coupling with the flight director has caused concern as a potential factor in the landing accident at Bangalore. An expedient solution by airlines operating the A-320 has been to prohibit the 'open descent' mode below a safe transition altitude on final approach (Billings, 1991). Obviously, a basic human factor problem involving this problem was not recognized during the design or certification stage.

Autopilot Mode Confusion

The increase in automation, and the pilot monitoring of the automated systems that is implied, has human factor ramifications and will continue to have them. Most observers would agree that further automation would have an adverse effect on the effectiveness of pilot monitoring, some of this because humans are not good monitors of rare events. There is considerable concern that the automatics are so reliable that monitoring them has become a very boring task and that the effectiveness of monitoring is inevitably degraded. The further challenge this gives designers is discussed later in this chapter.

The automation problem is exacerbated by the logic that is automatically built into some automation. For example, if one of our latest transports is using the vertical speed mode of the autopilot system, the aircraft will not capture the desired or selected altitude and if on the selected altitude, is allowed to fly away from it. In another instance, if an operational constraint is reached, the only annunciation to the pilots can be a fairly innocuous change in the mode status readout. This is inadequate feedback and not always effective. In some cases pilots have wanted to remove just part of the automation and utilize the remaining features but are unable to do so because 'all or nothing' are the only options. For example, pilots may want to change their navigation or speed manually but maintain altitude-hold. In some of today's airplanes, this is not possible. This is probably a computer programming problem undoubtedly caused because the computer programmer did not know some of the intricacies of flying airplanes in the real world. It is one reason that it is highly desirable to have pilots involved in the design stage.

When studying air transport automation, two university researchers, David Woods and Nadine Sarter, found a great deal of what they called 'mode confusion'. They noted, for example, that there are numerous vertical modes, and that it is difficult at times for pilots to understand how the modes are functioning, especially at the margins of the aircraft's performance envelope. Whether this confusion can be solved with better design, better displays, or better training (or a combination of them) is one of many issues now being explored. A Massachusetts Institute of Technology (MIT) study found that 74% of 184 mode awareness incidents involved vertical navigation, while only 26% involved horizontal navigation (Hughes, 1995). Table 11.1 lists the many autoflight modes found in the Airbus A-320.

Table 11.1   FMS and Autoflight Modes in the Airbus A320

Autothrust modes	Vertical modes	Lateral modes
TOGA	SRS	RWY
FLX 42	CLB	NAV
MCT	DES	HDG/TRK
CLB	OPEN CLB	LOC*
IDLE	OPEN DES	LOC/AP NAV
THR	EXPEDITE	LAND
SPD/MACH	ALT	ROLLOUT
ALPHA FLOOR	V/S-FPA	
TOGA LK	G/S-FINAL	
	FLARE	

Source: Reprinted with permission from *Human Factors*, Vol. 37, No. 1, 1995. Copyright 1995 by the Human Factors and Ergonomics Society. All rights reserved

Complacency or Overreliance on Automation

An interesting and very real problem involved with the almost complete automation present in today's transports is pilot complacency and overreliance upon automation. This pilot response occurs in normal operations and also is reflected in the pilot's reliance on the system to automatically make the correct response during abnormal operations. If highly reliable automation is routinely used (as it is in both short- and long-haul flights), there is an understandable tendency for flight crews to rely upon the automation to the point that the normal checks that are inherent in good manual operations are sometimes disregarded. Inattention and boredom, particularly on long flights, seem virtually inevitable. Make-work provisions to keep the pilots occupied, and therefore at least theoretically alert, are unacceptable to all parties. A major reason is that they simply do not work. 'Busy work' does not necessarily raise awareness of the system to be monitored.

The classic case used to illustrate overreliance of the automatics is the overrun of a SAS DC10-30 at John F. Kennedy airport. The ceiling was 300 feet broken and the visibility 1 1/2 miles. The calculated approach speed was 154 knots and the threshold speed was 149 knots. During the approach, the airplane, which was set up for an automatic approach, was reasonably normal until the airplane descended through 800 feet. Then the airspeed increased and was still about 50 knots above threshold speed when the aircraft passed the threshold. The airplane landed more than 4,000 feet down the runway. Even the theoretical stopping distance of the DC-10 on a dry runway was longer than the runway length remaining for this crew. This runway was wet; the resulting overrun was inevitable.

The NTSB determined that: 'the probable cause of this accident was the flight crew's (a) disregard for prescribed procedures for monitoring and controlling of airspeed during the final stages of the approach, (b) decision to continue the landing rather than to execute a missed approach, and (c) overreliance on the autothrottle speed control system....'

The accident is of particular interest to us because it happened more than a decade ago (28 February 1984), it happened to an airplane that is not as fully automated as today's glass cockpit airplanes, and it was the first time that pilot 'overreliance on the automatics' was given as a probable cause. While there have been no further accidents with explicitly stated similar causes, there have been several disturbing incidents involving overreliance on the automatics. There is consensual agreement among operational experts that pilot complacency with automatics is a valid concern with today's airplanes.

Despite all of its problems, automation is generally accepted as the way of the future and most of the controversy regarding automation is on how best to use it. Dr. Billings has noted a point that should always be remembered. It is that, while: 'Individual human capabilities have not changed very much in the short history of aviation, ... human operators, considered collectively, have changed a great deal, in the course of learning to design and understand how to operate the advanced technology that characterizes aviation' (Billings, 1991). This is a normal evolutionary process. In the future, human operators (in this case pilots) will continue to change. Since the beginning of aviation, successful pilots have been forced to and been able to change and to adapt. As we move into the 21st century, successful pilots will also change as they continue to adapt to and learn to operate the advanced technology that characterizes aviation.

Automation, Training, and Manual Skills

One of the great myths associated with increased automation is that automation has (or will) reduce training requirements. This is simply not true. Automation has created training requirements that add to the previous requirements. The skills and knowledge needed to take full advantage of increased automation must be added to the training curriculum. Sarter and Woods (1995) state flatly that: '...new automation produces new training requirements.'

There is no question that increased automation is a welcome tool and that it is used by pilots to make flight safer and more efficient. However, using today's automation efficiently does not mean that today's pilots do not need all the old skills and knowledge. Today pilots need more. They need all of the old skills plus the new skills required by the automation. One of the lessons learned in more than a decade of operation with increasingly automated airplanes is that a high level of manual pilot skills is still required. Manual skills must be a part of any recurrent or transition training and checking program in addition to the emphasis given to the proper use of the automatics.

There are many reasons for skill deterioration, including such items as scheduling or even motivation, but in most manual skills deterioration cases, the problem is simply a lack of practice. In one case, a major international airline has reported that its long range fleet schedules permit only one and one/half takeoffs and landings per month for each pilot. This calculation assumes that all takeoffs and landings are shared equally among those pilots who are responsible for having and maintaining takeoff and landing skills. Another airline has reported that because its landings are at the end of a long and fatiguing flight, its pilots sometimes just let the automatics make the approach and landing simply because it is easier than doing it manually. Additionally, it provides a more efficient way for them to monitor inside and outside of the cockpit and to manage overall workload, particularly in adverse weather conditions and unfamiliar airports.

Many carriers have been forced to schedule the pilots of highly automated aircraft that are used exclusively for long range trips for simulator training every ninety days simply to maintain the pilot's takeoff and landing currency. One US carrier has even reduced their recurrent training timetable for these pilots from one year to nine months. The maintenance of manual skill proficiency in highly automated airplanes is a very real problem. Other training considerations are discussed in Chapter 16—Selection and Training.

Automation and the 'Error' Dilemma

An obvious dilemma for the designer of a transport airplane or any of its systems is that both humans and automated machines and their components make errors. Neither is completely reliable at the levels required for a continued reduction in air transport accidents and incidents.

It is a mistake to consider errors as a separate issue for automation, although it is certainly possible that automation itself makes some kinds of errors more probable. We know that automation is not perfect. We also know that inevitably all humans make inadvertent, rare and random errors. Today, air transport operations have increased to the point that the aviation system cannot tolerate even inadvertent, rare, and random operational errors without at the very least controlling their operational consequences. (See Chapter 20— The Worldwide Safety Challenge—for further discussion of this major industry problem.)

On very infrequent occasions the increasingly automated aircraft systems and the computers that control them either make mistakes, or the airplane encounters a situation that was not foreseen by the designers of the automated systems. The only resource available to correct or obviate the consequences of these inevitable mistakes, or to deal with an unforeseen situation, is the flight crew or the air traffic controller. In each case a human operator is the ultimate back up. In an interesting and provocative comment, Lisanne Bainbridge noted that, '...the increased interest in human factors among engineers reflects the irony that the more advanced a control system is, so the more crucial may be the contribution of the human operator' (Bainbridge, 1987).

With these basic ground rules, a major design objective must be to make each airplane or system 'error resistant' to the extent possible—to make it very difficult for a human to make an error with the system, or if automation is involved, to make it difficult for the automation to make a mistake. However, even with an error resistant system, errors will still occur. Therefore, a second design goal must be to make all systems, especially the critical systems, 'error tolerant', i.e., to ensure that inadvertent errors are detected and alleviated, or at least diminished, to the point that they are not critical. Simplification, in addition to its other advantages, can be a great tool to make systems error resistant for either humans or machines. Simplification also makes it easier to detect inadvertent errors and easier to take appropriate remedial action.

The safety problem that is caused because pilots or controllers make rare errors, is minimized by having two well-trained pilots monitor the behavior of the automatics, monitor the operational performance of each other, and by having the automatics monitor the performance of the pilots. This process is enhanced by the automated warning systems that are discussed in the following paragraphs.

Automated Warning Systems

Automated warning systems provide a significant contribution to the increased help that automation has given the pilot. While all of them are not new, they are prime examples of machines (or automation) monitoring the human. Today, modern aircraft have three types of automated warning systems: 1) those that ensure that the aircraft is configured for the phase of flight it is in or about to enter; 2) those that monitor the aircraft's systems; and 3) those that involve environmental threats to flight safety.

Automated warning systems monitor the execution of the pilot's sub-tasks. The first type involves the configuration of the aircraft and is used to ensure that the aircraft is configured for the phase of flight it is in or is about to enter. These, as well as the second type of warnings, have been used for many years. Examples are landing gear warnings if the gear is not down when the throttles (or power levers) are closed for landing, and flap position warnings if the throttles are advanced for takeoffs if the flaps are not in position for takeoff. A second set of warning systems is concerned with the airplane's systems and includes hydraulic system state, engine fuel or oil conditions, electrical systems operation, and pressurization system state. Basically, the second type provides information including system malfunctions and seldom involves or is caused by a human error.

The third set of warning systems involves environmental threats to the safety of flight and includes Ground Proximity Warning Systems (GPWSs) and the newly developed Enhanced Ground Proximity Warning Systems (EGPWS), Traffic Alert and Avoidance Systems (TCASs), and Wind Shear Avoidance Systems (WSASs). Wind Shear Avoidance Systems can operate both on the ground for controllers and in the air for pilots. Minimum Safe Altitude Warnings (MSAWs) probably should also be included in this classification. They are entirely in the province of controllers and enable them to warn the flight crew that they are flying below minimum flight altitudes. MSAWs are found in the US because of frequently irregular terrain in the US and because of software compatibility with current US ATC ARTS III approach radar.

Whenever EGPWSs, MSAWs, TCASs or the ATC Center controller's Conflict Alert System is triggered, an error of some sort has been made somewhere in the system. The controller's Conflict Alert System utilizes aircraft Mode C transponders and warns of any time standard separation of altitude or distance is not maintained for any reason. Both pilots and controllers sometimes irreverently call it the 'snitch patch'.

An effective device, used mainly in Europe without the punitive aspects of the 'snitch patch', is the controller's Short Term Conflict Alert (STCA). This is a software-device that alerts the controller whenever a loss of separation is imminent. Because the parameters of STCA can be set to go off well before there is an actual loss of separation, it can have a high nuisance warning rate. Conversely, if its parameters are set close to the actual loss of separation, there is very little time for recovery of the situation. STCA is a tool to help controllers prevent loss of separation as opposed to TCAS which is an anti-collision device for the pilots. Further discussion is beyond the purview of this book. Readers interested in additional information regarding controller operations are encouraged to consult *Air Traffic Control: Human Performance Factors*. This book, co-authored by Dr. Anne Isaac and Bert Ruitenberg,[6] is presently in press and will be available in May of 1999.

Automatic warnings of any kind can be entirely visual (as in a steady or flashing red light), can be a combination of visual and aural signals, or can include tactile warnings. The form and intensity of these warnings is a human factors problem. It is of some interest that GPWS, TCAS, and WSAS were all mandated by congressional action, an indication of a very high level of congressional interest in aviation safety.

There is little doubt that, in spite of some early 'shakedown' problems, automated warnings have significantly increased the safety of air transport operations. Details regarding various types of automated warnings have been discussed in previous chapters, e.g., the configuration warnings and warnings about environmental threats in Chapter 8—Documentation, including Checklists and Information Management. Automation warnings are also discussed in Chapter 21—Current Safety Problems. System warnings were discussed in Chapter 10—Workload.

Human-centered Automation

Dr. Charles Billings, former Chief Scientist at the NASA/Ames Research Center, and now Professor Emeritus at The Ohio State University, has examined, precisely defined, promoted, and vigorously articulated the concept

---

[6] Dr. Anne Isaac is with EUROCONTROL, Belgium. Bert Ruitenberg has been representing the International Federation of Air Traffic Controllers Association (IFATCA) since 1992; he works as an Air Traffic Controller at Schiphol Airport, Amsterdam, Netherlands.

of human-centered automation. It is not known who first used the phrase, human-centered automation for as Dr. Billings has noted, 'Sheridan, Norman, Rouse, Cooley, and many others wrote for many years about human-centered or user-centered technology' (Billings, 1997). The human-centered concept is important and is now generally recognized and accepted by·all elements in the air transport system.

Human-centered automation is an important, useful and descriptive phrase. As we have noted, its theory and philosophy is supported in virtually all circles. *Aviation Automation: The Search for a Human-Centered Approach,* which was authored by Billings, is must reading for anyone interested in fully exploring automation in transport aviation. Much of the following discussion is based upon that book.

The principles of human-centered automation have been identified by Dr. Billings as follows:

Premise:
>The pilot bears the responsibility for safety of flight.
>Controllers bear the responsibility for traffic separation and safe traffic flow.

Axiom:
>Pilots must remain in command of their flights.
>Controllers must remain in command of air traffic.

Corollaries:
>The pilot and controller must be actively involved.
>Both human operators must be adequately informed.
>The operators must be able to monitor the automation assisting them.
>The automated systems must therefore be predictable.
>The automated systems must also monitor the human operators.
>Each intelligent element of the system must know the intent of other intelligent system elements.

These concepts are fundamental, straightforward, and critical. We have expanded, and in some cases slightly rewritten, the corollaries from the perspective of a pilot. In contrast to this book, which is primarily concerned with flight operations, the Billings text cited above considers all aspects of automation in aircraft operations. For the reader who is interested in air traffic control and the role of the air traffic controller, we highly recommend not only Dr. Billings' book, but as we have previously indicated, also *Air Traffic Control: Human Performance Factors* by Dr. Anne Isaac and Bert Ruitenberg.

The Corollaries Expanded

To command effectively, the human operator must be involved.

This is a central principle. It matters little whether the pilot completes his/her task by controlling the aircraft directly or simply manages the other human or machine resources that are being used. The human operator must be directly involved and be fully aware of that involvement. Human factors experts have known for years that very highly skilled personnel (e.g. pilots or controllers) do not do a good job of monitoring for events that have a very low probability of occurrence. Those who maximize the use of technologically possible automation often miss this very critical point.

In order to be involved, the human operator must be informed.

The information available has to include all of the data that is necessary to keep the pilot actively involved in the operation. This must include the information required to keep the pilot fully informed regarding the state, progress, and intention of the system or systems. Otherwise the human operator cannot hope to be meaningfully involved.

A very important human factors issue is determining the form, time, and manner in which information is presented. Data is not information. As Dr. Billings has stated: 'It becomes information only when it is appropriately transformed and presented in a way that is meaningful to a person who needs it in a given context.' Data that is clear and meaningful when examined on a desk in a quiet well-lighted room during daylight hours, may not be as clear and meaningful in the cockpit on a dark and stormy night during stressful flight conditions.

The pilot must be able to monitor the automated system.

This is closely related to the preceding corollary. Automated systems are fallible and pilots are the last line of defense capable of controlling, or in some cases preventing, a system failure. A system failure can lead to a catastrophe if the failure is in a flight-critical automated system. The economic and moral stakes are too high to permit an operating philosophy that does not enable the pilot to monitor automated systems effectively. In most cases it is simply not possible to monitor a system effectively without knowing how the system is planning to accomplish its task (Billings, 1991). Adequate training is a crucial element.

Automated systems must be predictable.

There is no way that an automated system can be monitored effectively if it is not predictable. There is no way that departures from the normal behavior of an automated system can be promptly detected if the pilots are unaware of the expected behavior of that system. Training must include the normal operation of each automated unit as well as its behavior during any failure modes.

For some time we have had fully automated systems such as the yaw damper, which automatically counters the tendency of swept wing aircraft (all modern jets) to create adverse yaw during banked turns and also dampens the effects of turbulence. Another automatic system is the pitch trim compensator, which counters the tendency of jet aircraft to pitch down at high speeds. These are predictable systems and normally they work automatically. If they do not work or work improperly, the pilots are informed with a warning light and by the performance of the airplane. The flying task for the pilots is increased if these automatic systems do not work. Latest airplanes also include automatic aerodynamic trimming in order to keep the airplane's center of gravity within its operational control limits. On very long flights fuel burn can cause major changes in the airplane's center of gravity.

The automated systems must also be able to monitor the human operator and the human must be able to monitor the automatics.

This corollary emphasizes two very real problems. First, humans are fallible and are not good monitors. Secondly, even the highly capable computers available today can fail partially or completely and cannot anticipate all of the circumstances that might be encountered in a line operation. Therefore, the performance of the computers must be monitored and so must the performance of the human monitor. Automated warning systems have provided a very constructive role. For example, when they are used properly, the Ground Proximity Warning Systems (GPWSs), which are part of any modern airliner and required in the US and several other countries, and Minimum Safe Altitude Warning Systems (MSAWS), which are a part of the US air traffic control system, have virtually eliminated controlled flight into terrain (CFIT) accidents. As we have noted, the new version of GPWSs, called Enhanced GPWS (EGPWS), provides better warnings and more warning time than the original versions. While some human factors experts believe there are design flaws in EGPWS, it represents a further improvement in a very good initial design.

'Envelope protection' (or limitation), which provides protection in critical aircraft speed and configuration conditions by making it physically impossible to exceed certain operating parameters, is a design feature in a very well-known European-built transport. These transports have been certified by the US FAA and are flown by US airlines. US pilots and manufacturers have generally opposed the absolute performance limitation found in these aircraft in order to give the pilots final control and permit them to try and exceed the performance limits under extreme conditions. Many pilots who fly these airplanes with full envelope protection like them very much. In fact, one international pilot flying these aircraft, says, 'the only pilots that don't like them are the ones that have never flown them'.

In an obvious and perhaps evolutionary compromise, one way that US manufacturers have modified the original envelope protection concept is by designing new transports that provide similar envelope protection with the added provision that only by exerting significant extra force on the flight controls can design envelope limits be exceeded. There is far from international agreement on whether or not it is a better and safer philosophy to permit the pilot to try and exceed the envelope limitations in a true emergency.

Each element of the system must have knowledge of the others intent.

A very basic principle in cross-monitoring, which must be effective in achieving maximum safety, is that it can only be effective if the monitor knows what the system (human or machine) is trying to accomplish. This is implied in the second, and third, and fourth corollaries. It also is implied in the monitoring of air traffic control by pilots in flight and is a problem for both pilots and controllers in present TCAS. This principle has significant training implications for it requires knowledge of each automatic system so that pilots flying one of the new airplanes will never have to ask: 'What is it doing now? Why did it do that?' or 'What is it going to do next?' The principle also requires good communication between the pilot flying (PF) and pilot-not flying (PNF) for it is virtually impossible to be sure of intent without effective communication.

Human-centered Design and Automation in the Future

It seems clear that air transports of the future will make human-centered principles a key part of design and automation, certainly in the cockpit, and hopefully in other parts of the airplane and in other parts of transport operations. In the meantime, because present airplanes will be in active use for many years, it is reasonably certain that present state-of-the-art design concepts will continue to be used. It is very difficult and very expensive to make design changes in a certificated airplane.

A factor that is often not appreciated is that the most human factors work must be done in the five years before the 'design freeze' that is required in the manufacture of any new transport. The 'design freeze' occurs approximately three years before certification and delivery of the first airplane to an airline. Then, after the new airplane is placed in service, researchers and its users critique it. It is virtually impossible to make major changes in the airplane at this stage. Any problems discovered during early operations that result in major recommendations are evaluated carefully. Unless a major safety item is involved, recommended changes become a part of the next family of airplanes. Only if the problem discovered is life threatening, can major changes be made after the 'design freeze'.

References

Abbott, K, Slotte, S.M., and Stimson, D.K. (1996). *Federal Aviation Administration Human Factors Team Report on: The Interfaces Between Flightcrews and Modern Flight Deck Systems*, US. Department of Transportation, Federal Aviation Administration, Washington, D.C.

Air Transport Association (1989). 'National Plan to Enhance Aviation Safety through Human Factors Improvements', Air Transportation of America, New York.

Bainbridge, Lisanne (1987). 'Ironies of Automation', chapter in *New Technology and Human Error*, ed. by Rasmussen, Jens, Duncan, Keith, and Leplat, Jacques, John Wiley & Sons Ltd., Bury St. Edmunds, Suffolk, England.

Billings, Charles E. (1989). 'Toward A Human-Centered Aircraft Automation Philosophy', *Proceedings of Fifth International Symposium on Aviation Psychology*, The Ohio State University, Columbus, Ohio.

Billings, C.E. (1991). *Human-Centered Automation: A Concept and Guidelines*, Technical Memorandum 103885, NASA-Ames Research Center, Moffett Field, California.

Billings, C.E. (1997). *Aviation Automation: The Search for a Human-Centered Approach*, Lawrence Erlbaum Associates, Mahwah, New Jersey.

Birch, Stuart (1997). 'Safety', *Aerospace Engineering*, August 1997, Society of Automotive Engineers, Warrendale, Pennsylvania.

Crouch, Tom D. (1989). *The Bishop's Boys: A Life of Wilbur and Orville Wright*, W.W. Norton & Company, Inc., New York.

Degani, Asaf, Chappell, Sherry L., and Hayes, Michael (1991). 'Who or What Saved the Day? A Comparison of Traditional and "Glass" Cockpits', *Proceedings of the Sixth International Symposium on Aviation Psychology*, 29 April-2 May 1991, The Ohio State University, Columbus, Ohio.

Federal Aviation Administration (1996). *Federal Aviation Administration Human Factors Team Report on: The Interfaces Between Flightcrews and Modern Flight Deck Systems*, Federal Aviation Administration, Department of Transportation, Washington, D.C.

Funk, Ken, Lyall, Beth, and Riley, Vic (1995). 'Flight Deck Automation Problems', *Proceedings of the Eighth International Symposium on Aviation Psychology*, Columbus, Ohio.

Gilson, Richard D., Deaton, John E., and Mouloua, Mustapha (1996). 'Coping with Complex Alarms', *Ergonomics in Design*, October 1996, Human Factors and Ergonomics Society, Santa Monica, California.

Greene, Berk and Richmond, Jim (1993). 'Human Factors in Workload Certification', presented at SAE Aerotech 1993, Federal Aviation Administration, Seattle, Washington.

Hughes, David (1995). 'Incidents Reveal Mode Confusion', *Aviation Week and Space Technology*, 30 January 1995, McGraw-Hill, New York.

Isaac, Anne and Ruitenberg, Bert (1999). *Air Traffic Control: Human Performance Factors*, Ashgate Publishing, Aldershot, United Kingdom.

Meister, D. (1989). *Conceptual Aspects of Human Factors*, John Hopkins University Press, Baltimore, Maryland.

Mohler Stanley R. (1997). 'Wright Brothers', *Air Line Pilot*, May 1997, Air Line Pilots Association, Herndon, Virginia.

Moolman, Valerie (1980). *The Road to Kitty Hawk*, Time-Life Books, Inc., Chicago, Illinois.

National Transportation Safety Board (1984). *Aircraft Accident Report, Scandinavian Airlines System, Flight 901, McDonnell Douglas DC-10-30, John F. Kennedy International Airport, Jamaica, New York, February 28, 1984*, NTSB/AAR-84-15, National Transportation Safety Board, Washington, D.C.

Orlady, Harry W. (1988). 'Training for Advanced Technology Aircraft', Unpublished paper, Aviation Safety Reporting System, Moffett Field, California.

Orlady, Harry W. and Barnes, Robert B. (1997). *A Methodology for Evaluating the Operational Suitability of Air Transport Flight Deck System Enhancements*, Society of Automotive Engineers, Warrendale, Pennsylvania.

Rasmussen, Jens, Duncan, Keith, and Leplat, Jacques (1987). *New Technology and Human Error*, John Wiley & Sons Ltd., Chichester, Great Britain.

Reason, James (1997). 'Reducing the Impact of Human Error in the World-Wide Aviation System', keynote address at the Ninth International Symposium on Aviation Psychology, The Ohio State University Aviation Psychology Laboratory, Columbus, Ohio.

Sarter, Nadine B. and Woods, David D. (1995). 'How in the World Did We Ever Get Into That Mode? Mode Error and Awareness in Supervisory Control', *Human Factors*, March 1995, Human Factors and Ergonomics Society, Santa Monica, California.

Wiener, Earl L and Curry, Renwick E. (1980). *Flight Deck Automation: Promises and Problems*, NASA Contract Report 81206, NASA-Ames Research Center, Moffett Field, California.

12 Situation Awareness and Operating in Today's Environment

Situation Awareness

Situation awareness represents an important, though by no means, a new concept. Slightly more than a decade ago, the US Air Force called it the single most important factor in improving mission effectiveness (Final Report: Intraflight Command, Control, and Communication Symposium, 1986). Situation awareness is a phrase that is used meaningfully by many people in many contexts. We believe that situation awareness includes virtually all of the elements involved in the operation of safe and efficient air transportation.

Situation awareness covers several areas, often combining areas that are frequently considered separately. Total situation awareness is important in air transport for as Harvard professor, Lee Bolman, tells us, 'pilots always operate on their theory of the situation. If they see their situation improperly or do not see changes in what can be a very dynamic total situation, they have poor situation awareness' (Bolman, 1979).

The Problem of Definition

Situation awareness has no commonly accepted definition. This is in spite of the fact that it has become an almost ubiquitous phrase in the aviation industry. Sarter and Woods, who have noted the lack of a commonly accepted definition, perhaps 'tongue in cheek' and almost heretically, even suggest that: 'It is not even clear whether situation awareness really denotes a distinct phenomenon or only illustrates the tendency of applied cognitive science to coin new terminology in the face of ill-understood issues' (Sarter and Woods, 1991).

One reason for the definition difficulties is that the same term is used in several contexts. These range from references to military combat to references in air transport operations or to references in running a municipal government (Sarter and Woods, 1991). The mission of a fighter pilot, with concern about such things as anti-aircraft radar, opposing fighter pilots, military targets, etc. is vastly different than the mission of an air transport pilot. In these paragraphs, as in this book, we are concerned with air transport aviation. Our perspective is from the viewpoint of the informed and concerned air transport pilot. Regal et al. told us several years ago that at the highest level situation awareness 'simply means that the pilot has an integrated understanding of the factors that will contribute to the safe flying of the aircraft under normal or non-normal conditions' (Regal, Rogers, and Boucek, 1988). The 'bottom line' of that definition is that the flight crew must know what is going on and what in the foreseeable future will be going on. The Regal et al. definition ensures that due consideration is given to the complexity of an air transport operation. It is a good definition.

Douglas Schwartz at the 1989 Fifth International Symposium on Aviation Psychology gave another good definition when he stated:

> Situation Awareness is the accurate perception of the factors and conditions that affect an aircraft and its flight crew during a defined period of time. In simplest terms, it is knowing what is going on around you—a concept embraced to the need to "think ahead of the aircraft".   (Schwartz, 1989)

A scholarly definition given by Mica Endsley is: 'Situation awareness is the perception of the elements in the environment within a volume of time and space, the comprehension of their meaning and the projection of their status in the near future.' She notes that 'it is necessary to determine exactly what the elements in the definition are' (Endsley, 1993). In a paper entitled 'Towards a New Paradigm for Automation: Designing for Situation Awareness', she stated that situation awareness is 'a person's mental mode of the state of a dynamic system (and that it) is central to effective decision making and control and is one of the most challenging portions of many operator jobs.'

What is Situation Awareness in Transport Aviation?

Ultimately, situation awareness covers at least five areas. Seldom are all areas involved in a single incident or a single bit of research. One area gives status information—the physical state or condition of the airplane, e.g., the amount of flaps or spoilers, the amount of power being drawn from the engines, the fuel state, or the position of the landing gear or flaps. A second area is the position of the airplane in respect to the flight plan, to any natural or man made obstructions, or to any other airplanes of interest from an avoidance point of view. A third area is the total external environment, including the present and future weather and details of the aviation infrastructure.[1] The fourth area involves: the time the airplane will meet its next navigational fix; the time it will reach its destination; the time available for holding or diversion; the time limit for the fuel available; the time before the weather will change, etc. The fifth area is the state of the other members of the operating team (the cockpit and cabin crew), the passengers, and even the cargo that might be aboard.

Some researchers do not like to include the fourth area—the time dimension—as being a separate entity in situation awareness. They say 'time' is important in virtually all other parts of situation awareness and can fit nicely into Endsley's Levels 2 and 3. (Endsley's taxonomy includes three levels of situaiton awareness and are discussed later in this chapter.) However, we prefer considering 'time' as a separate area because of its importance in certain circumstances ('time' is particularly important in 'place' information) and because it can be very important in other cases. Actually, the differences in opinion may well be primarily semantic and we are not particularly concerned. The important point is to be sure that 'time' is considered. 'Time' is important in air transport operations and its importance will increase as it becomes an additional factor in advanced ATC concepts.

Present day airplanes are complex machines. For example, the first B-747s had over 400 assorted bells, whistles, visual, auditory and tactical alarms, each of which presented situation awareness information to the pilot that was considered important. It is a far from easy task to be sure that such information is effectively available when it is needed in the midst of other emergency, caution, and advisory signals. Designers and operational experts have spent a great deal of effort to reduce these numbers in subsequent airplanes. Some of this effort is seen in efforts to prioritize warning information.

Aircraft Status Information

The physical state of the airplane is usually shown by cockpit dials, gauges, or instruments. The position of the flaps is indicated by a cockpit gauge, which is usually shown on an analogue gauge in older airplanes and which may be shown on a CRT or flat panel in analogue or digital fashion in high technology airplanes. Selected flap position is confirmed by checking the position of the

---

[1] The aviation infrastructure includes airports (runways, taxiways, and ramps), navigation radio, communication radio, fueling facilities, the ATC system, etc.

flap position handle. Leading edge flaps, slats, and trailing edge flaps change the shape of the wing, making it possible to utilize an airfoil that is very efficient at cruise speeds. By adding carefully designed slats and flaps, the airfoil changes so that it is now efficient at takeoff, initial climb, approach, and landing speeds.

Spoilers disturb the airflow over the wing, increase drag and decrease lift. Their position is an important situation awareness consideration. Spoilers must be used properly, for partial spoilers significantly reduce climb capability and full spoilers can make an airplane unflyable. Full spoilers are used only after landing in order to get the full weight of the airplane on the wheels and increase the effectiveness of braking. The use of spoilers in the air with flaps extended is prohibited with some airplanes, and in those where use of partial spoilers is approved, it is very important that the flight crew is aware of the position of the spoilers at all times. Failure to retract the spoilers when a full performance climb was indicated is thought by some to have been a factor in the American Airlines B-757 crash at Cali, Columbia.

Flap position is extremely important. Modern airplanes simply cannot make normal takeoffs without leading edge flaps, slats and trailing edge flaps in a takeoff position. Flaps and slats are also required for a normal landing. The failure to have takeoff flaps has been the major cause of such airline accidents as Northwest Airlines (NWA) at Detroit, 16 August 1987 and Delta Air Lines (DAL) at the Dallas-Ft. Worth airport on 31 August 1988. In each case the NTSB found that the leading edge and trailing edge flaps were in the fully retracted position when the airplanes took off. In each case, proper use of the taxi and pre-takeoff checklists would have assured proper positioning of the flaps for takeoff. In each case, failure of the automatic takeoff warning system for unknown reasons was a contributing factor. The failure of the flight crew to be aware of the configuration of the airplane made the accidents virtually inevitable.

There have been recent accidents where, for reasons that are hard to understand, the flight crew seemed to be unaware of the power being used, or of the flight's profile. Both should be clearly indicated by the cockpit instruments. An example is the Airbus 320 accident at Bangalore, India where the Air India pilots, with apparently idle power, allowed the airspeed to decrease to 25 knots below the target speed late in the descent. One could not fly a DC-3, or any other airplane, that way. It seems to be another case where a perfectly good airplane was allowed to crash under completely normal conditions. While there is not always complete agreement as to the causes of these accidents, there seems little doubt that there was a substantial lack of situation awareness during a critical phase of flight.

Another and a continuing aircraft status situation awareness problem in advanced technology airplanes involves the mode of the automatic flight control system being used and the consequences of being in a particular mode at a particular time. There are many real-world examples of confusion about the particular mode that was in operation and also examples of the confusion regarding the consequences of being in a mode that became inappropriate as the flight progressed to another altitude or phase-of-flight. Pilots have committed erroneous actions because an action was executed that was incompatible with their situation. Many of these problems are apparently due to a lack of understanding of the system (Sarter and Woods, 1995). Wiener (1989) has suggested that many of them are actually the result of what he calls 'clumsy automation'. There is little question that in several cases, inadequate training is involved.

Several papers have been written regarding mode awareness. It is perhaps the largest single problem that pilots have in present advanced technology airplanes. Mode awareness is being given special attention in a program that is being run at NASA's Ames Research Center (Degani, Shafto, and Kirlik, 1995a, and Degani and Kirlik, 1995b). In addition, the Boeing Commercial Airplane Company is in the midst of a specific mode awareness program as this is being written. There should be substantial improvements in the pilot's mode awareness in airplanes of the future.

Place Information

Place information is of major importance in transport operation. Most air transport flights are made via IFR (instrument flight rules), and this means that the pilot's navigation is conducted by information secured from cockpit navigational radio and the cockpit instruments. Place information can be verified by visual contact with terrain and with other airplanes under VFR conditions, and also by attention to maps, charts, the flight plan, and the cockpit instruments. Properly used cockpit instruments, including navigational radio, are critical during all phases of flight. They are particularly critical during IFR, night, or in any restricted visibility operations.

CFIT (controlled flight into terrain) accidents have been the largest single category of air transport accidents. Most of them occur because the flight crew was not aware of their exact location. There can be many reasons that the crews had faulty place information The development of Ground Proximity Warning Systems (GPWS), Minimum Safe Altitude Warning Systems

(MSAWS), and now Enhanced Ground Proximity Warning Systems (EGPWS) each represents a significant advance in the long quest to minimize the operational consequences of errors of less than optimum operations that have led to many CFIT accidents. These accidents are discussed at greater length in Chapter 21—Current Safety Problems. Underlying reasons for some of these unwanted incidents are discussed in Chapter 9—Man's Limitations, Human Errors, and Information Processing.

The moving map display in glass cockpit airplanes is an outstanding feature that both reduces pilot workload and gives pilots additional and more accurate place information. When selected, a moving map gives the pilots a continuous 'bird's eye' (sometimes called a 'God's eye') view of the airplane and its position in the outside world. This display shows the aircraft position directly. Formerly, it was necessary to mentally integrate data from a chart and several cockpit instruments in order to get the same information. Associated computers even perform valuable inferential tasks (such as calculating wind speed and direction) that can be shown on the display. Pilots are given accurate real-world information in a display that requires virtually no additional mental integration.

Map displays can be used during all phases of flight and can portray a great deal of place and other information. For example, they can display TCAS information showing other aircraft's relative position and altitude. Radar weather information can also be displayed and is adjusted automatically for the range chosen by the pilot. Another option available is to display all airports and navaids, again within the specified range.

During cruise the displays show the desired track, give graphic advanced warnings for interceptions or track changes, (which may or may not be on an airway), will show proposed route changes or vectors, and, of course, show the exact position of the airplane. Another nicety is a displayed arc that shows airplane climb or descent performance potentials and limitations in the present configuration of the airplane. This can be critical information in respect to ATC clearances.

During an ILS approach, moving map displays show the position of the airplane relative to the procedure, again graphically give advance warning for required turn-ons and altitude changes, and display the position of the airplane during holding patterns. When combined with information on the PFD (primary flight display), these displays give the exact position of the airplane on the localizer and glide slope. These features increase accuracy and reduce pilot workload, for they eliminate interpretive steps that are otherwise required. Pilots have called the moving map display the single most helpful feature of the new glass cockpits. Managing the plethora of information available is an important skill.

Total Operating Environment, including the Weather

The operating environment of the aviation infrastructure includes the facilities available, the existing traffic density, and the weather. Flying in the US eastern corridor is quite different than flying in one of the less populated southwestern states. Flying in some foreign countries can be quite different than either. Increased traffic density results in an increase in the number of traffic delays and increases the hazard of a catastrophic mid-air collision. Flight crew workload inevitably increases with increased traffic density. Pilots must be concerned with internal cockpit indications and the external view through their windscreens. Both winter and summer weather varies considerably with the geographical area. Such meteorological phenomena as windshear, ice, snow, and daytime and nocturnal thunderstorms have to be experienced to be appreciated.

Weather is an important part of the operating environment. Existing and forecast weather in either winter or summer are important considerations for any operation. They can determine the feasibility of the operation itself, the potential holding time, and the alternates that may be required. The predicted movement of cold fronts and warm fronts and the existence of such meteorological events as a 'Chesapeake Bay Low' are important. Another is the existence or forecast of an east wind with dew points and temperatures close to each other at stations in the great plains with its almost certainty of ground fog as the air mass cools adiabatically with the steady rise in terrain. Other geographical areas can have their own meteorological problems.

These, and a number of other meteorological phenomena that vary throughout the world, are all considerations for pilots and dispatchers. They are a meaningful part of situation awareness. Meteorological phenomena of interest include such things as predicted and unexpected wind shears, tornadoes, volcanic ash, hurricanes, thunderstorms, freezing rain, and hail. In many areas, the quality of weather and traffic situation awareness varies with the relevant topical knowledge of the operators. Even the simple question often asked of a pilot, 'How's the ride?', involves situation awareness.

The proximity of other aircraft is another important situation awareness area. Minimizing the risk inherent in crowded skies is helped by TCAS, its modifications and the still developing operating rules for TCAS use. In spite of the fact that mid-air collisions are extraordinarily rare in transport aviation, the threat of a mid-air collision is rated high by a great many pilots and by many other aviation safety experts.

Only in air transport could the equivalent of a 'Queen Mary' be destroyed by a collision with the aerial equivalent of a rowboat, e.g., in the 31 August 1986 collision between an Aeromexico DC-9 and a small single-engine aircraft while the DC-9 under complete ATC Control was approaching Los Angeles. What is even worse is the potential loss of life that could be caused by the mid-air collision of two fully-loaded large transports. Inevitably, in addition to a major loss of life, such an occurrence would create a great deal of emotional, adverse, and critical publicity at both national and international levels.

Mid-air collisions are a greater aviation safety threat in general aviation. In the years 1983 through 1985, 237 mid-air collisions occurred in the US. Most of them occurred in general aviation and in the traffic pattern at uncontrolled airports.

The Temporal Element and Time

The concept of situational awareness also has a temporal dimension. It pervades most of the other situational awareness areas. A pilot automatically thinks ahead of the aircraft if he/she has a good understanding of situation awareness for his/her aircraft operation. Both the present and future status of the aircraft and its crew are involved.

At present, time involves three kinds of situation awareness and a fourth will be added to it the future. They are important on every flight. The first area is related to fuel. Airplanes have only a limited amount of fuel and when it is expended the airplane can no longer fly. Two extreme cases of fuel exhaustion were the United Airlines DC-8 accident at Portland, Oregon in 1978 and the Avianca accident at Cove Neck, New York in 1990. Although the cause of these accidents varied considerably, in both cases the airplanes ran out of fuel and crashed before reaching their destination airport. It has been said that a tank of fuel is nothing more than a tank of time. When that time is up, you are out of fuel and out of flying.

Lack of sufficient fuel has caused serious problems short of crashes such as landing without planned reserve fuel and, in at least two recent cases, the forced landing at an intermediate airport because fuel on board was not checked properly before takeoff. Fuel is an important aspect of situation awareness.

Time is also important in the aircraft's progress in relation to its flight plan, including its time over planned navigational fixes. If there are time variations from the flight, they should be noted as soon as possible and the reason for the discrepancy identified. This can be a critical element, especially on long flights. Flight time is also important in its conformance with the flight's schedule times and in relation to the clock hours that the flight is scheduled and in which it operates.

Finally, time is an important consideration in weather forecasts. It is of particular importance to operational planning personnel, who must make worldwide plans for schedules that can be flown. The time associated with weather forecast is important for flight crews. They are always concerned with minimum weather for takeoffs and landings, and with enroute winds and other enroute weather phenomena.

The time element that will become increasingly important in the future is associated with a revised and improved ATC system. Aircraft on selected routes are already receiving clearances with specific time constraints over navigational and approach fixes, for example, entering Russian airspace from the east. Airspeed must be managed to achieve specified times, and if the aircraft is unable to meet those times, the clearance will be voided. This capability to manage time constraints can be particularly important in long-range and supersonic flying. The flight management system software in many modern aircraft already has this capability although it is not used by all operators.

Status of Other Team Members and the Passengers

The status of other team members and passengers is an area that is seldom considered in situation awareness studies, perhaps because it is very seldom an operational problem. However, the status of other team members, both in the cockpit and in the cabin, are of interest to crew members if that status will degrade operational performance for any reason.

Incidents of subtle or obvious incapacitation (which are discussed in Chapter 15—Fitness to Fly), or instances of unusual stress or fatigue (which are discussed in Chapter 14—Fatigue and Stress), are obvious concerns that do degrade performance. Individuals affected by any of these conditions cannot be depended upon to perform in their otherwise normal fashion. While limited crew performance degradation can usually be compensated for with awareness of the remaining crewmember or members, situational awareness may be reduced for the cockpit crew if performance degradation problems are not considered.

The general health and well-being of passengers is beyond the responsibilities of the flight or cabin crew, however, they are directly concerned if any of the passengers become a threat to the airplane, the other passengers, or themselves, and also if one or more of them require sophisticated medical attention.

Three Levels of Situation Awareness

In a highly successful effort to better analyze situation awareness incidents, Endsley developed a 3-level taxonomy (Endsley, 1995). The three levels are:

- Level 1 - Failure to correctly perceive the situation.
- Level 2 - Failure to comprehend the situation.
- Level 3 - Failure to comprehend the situation into the future.

Using the taxonomy, ASRS reviewed 113 situation awareness incidents that were reported to it and found 169 situation awareness errors. Of the 113, 80.2% were classified as Level 1 errors, 16.9% as Level 2 errors, and only 2.9% were classified as Level 3 errors. These findings strongly suggest that Level 1 errors (failure to correctly perceive the situation) need the most attention. There can be several reasons for these errors. One is that the information may be known but temporarily forgotten. This can be because of a shortcoming in the system design, a failure in the communication process or because of inadequate training. The data may have been available but difficult to detect or perceive; or it may have been clearly available but simply missed. The data may have been missed because of distractions, complacency, the narrowing of perception, because of concentration in other areas, or because of a too high task load. Situation awareness is not a simple process, but neither is the safe and efficient operation of an air transport a simple process.

Situation Awareness and Automation

It is postulated by some that automation can be at such a high level that the pilots are gradually eased out-of-the-loop because their role has been reduced to that of simply monitoring the automatics. If that happens, the concern is that pilots can lose a meaningful sense of system or situation awareness to the point that they lose their ability to recognize and correct the automatics and/or their ability to take over the airplane if that is required. Of equal concern is the loss of skill which may be needed on-the-spot but which has degraded because it has not been sufficiently exercised for long periods of time because of automation and the practice and philosophy of using manual skills. These problems are discussed in greater length in Chapters 11—Automation, and 16—Selection and Training.

Summary of Situation Awareness

Throughout the industry, there is general recognition that good situation awareness increases safety, reduces workload, enhances pilot performance, expands the range of pilot operations, and improves decision making (Regal, Rogers, and Boucek, 1988). Achieving and maintaining a high level of overall situation awareness, however it is defined and however it is measured, is a product of a good operating philosophy, good training, good SOPs, and good crew coordination. As we will discuss in Chapter 13—Crew Resource Management (CRM) and the Team Approach—principles play a significant role. All aspects of situation awareness are important and must be taken seriously by professional pilots.

Economics and operational efficiency have made two–person crews the new standard for future airplanes. We have learned that one person simply cannot routinely fly today's airplanes alone in the environment in which these airplanes must be operated. This has led to the team approach in the operation of the latest airplanes and in an expanded role for the copilot. Recognition of the universality and inevitability of human error, including recognition that even a captain can make an inadvertent error and that the operational concept of the inevitable error must be controlled, reinforce these concepts. They are an inherent part of situation awareness.

References

Andre, Anthony and Degani, Asaf (1996). 'Do You Know What Mode You're In? An Analysis of Mode Error in Everyday Things', *Proceedings of the 2nd Conference on Automation Technology and Human Performance*, University of Central Florida, Daytona Beach, Florida.

Armstrong, Donald (1991). 'Enhancing Information Transfer: The Aircraft Perspective', presented at an AIAA/NASA/FAA/HFS Conference, 15-17 January 1991, entitled 'The National Plan: Challenges to Aviation Human Factors', Federal Aviation Administration, Long Beach, California.

Bolman, Lee (1979). 'Aviation Accidents and the "Theory of the Situation"', *Resource Management on the Flight Deck*, in Proceedings of a NASA/Industry Workshop Held at San Francisco, California, 26-28 June 1979, Ames Research Center, Moffett Field, California.

Degani Asaf, Shafto, Michael, and Kirlik, Alex (1995). 'Mode Usage in Automated Cockpits: Some Initial Observations, *Proceedings of International Federation of Automatic Control (IFAC)*, 27-29 June 1995, Boston, Massachusetts.

Degani, Asaf and Kirlik, Alex (1995). 'Mode in Human-Automation Interaction: Initial Observations about a Modeling Approach', *Proceedings of IEEE International Conference on Systems, Man, and Cybernetics (SMC)*, 22-25 October 1995, Vancouver, Canada.

Edwards, Elwyn (1988). 'Introductory Overview, The SHEL Model', *Human Factors in Aviation*, Academic Press, Inc., San Diego, California.

Endsley, Mica R. (1993). 'A Survey of Situation Awareness Requirements in Air-to-Air Combat Fighters', *The International Journal of Aviation Psychology*, Vol. 3, No. 12, 1993, Lawrence Erlbaum Associates, Hillsdale, New Jersey.

Endsley, Mica R. (1995). 'Situation Awareness: Where Are We Heading?', *Proceedings of the Eighth International Symposium on Aviation Psychology*, The Ohio State University, Columbus, Ohio.

Gilson, Richard D., Deaton, John E., and Mouloua, Mustapha (1996). 'Coping with Complex Alarms', *Ergonomics in Design*, October, 1996, Human Factors and Ergonomics Society, Santa Monica, California.

Greene, Berk and Richmond, Jim (1993). 'Human Factors in Workload Certification', paper given at SAE Aerotech 93, Federal Aviation Administration, Seattle, Washington.

Hawkins, Frank H. (1987). *Human Factors in Flight*, Gower Publishing Group, Ltd., Aldershot, England.

Hawkins, Frank H. (1993). *Human Factors in Flight*, Second Edition, edited by Orlady, Harry W., Ashgate Publishing Ltd., Aldershot, England.

Jones, Debra G. and Endsley, Mica R. (1995). 'Investigation of Situation Awareness Errors', *Proceedings of the Eighth International Symposium on Aviation Psychology*, The Ohio State University, Columbus, Ohio.

Orasanu, Judith (1995). Situation Awareness: Its Role in Flight Crew Decision Making, *Proceeding of the Eighth International Symposium on Aviation Psychology*, The Ohio State University, Columbus, Ohio.

Regal, David M., Rogers, William H., and Boucek, George P. Jr. (1988). *Situational Awareness in the Commercial Flight Deck: Definition, Measurement, and Enhancement*, SAE Technical Paper Series 881508, given at Aerospace Technology Conference and Exposition, Society of Automotive Engineers, Warrendale, Pennsylvania.

Ruffel Smith, H.P. (1979). *A Simulator Study of the Interaction of Pilot Workload With Errors, Vigilance, and Decisions*, NASA Technical Memorandum 78482, Ames Research Center, Moffett Field, California.

Sarter, Nadine B. and Woods, David D. (1991). 'Situation Awareness: A Critical But Ill-Defined Phenomenon', in *The International Journal of Aviation Psychology*, Vol. 1, No. 1, 1991, Lawrence Erlbaum Associates, Hillsdale, New Jersey.

Sarter, Nadine, B. and Woods, David D. (1995). 'How in the World Did We Ever Get into That Mode? Mode Error and Awareness in Supervisory Control', *Human Factors*, Human Factors and Ergonomics Society, Santa Monica, California.

Schwartz, Douglas (1989). 'Training for Situational Awareness', *Proceedings of the Fifth International Symposium on Aviation Psychology*, The Ohio State University, Columbus, Ohio.

Wiener, Earl L. (1989). *Human Factors of Advanced Technology ('Glass Cockpit') Transport Aircraft*, NASA Contractors Report 177528, National Aeronautics and Space Administration, Ames Research Center, Moffett Field, California.

13 Crew Resource Management (CRM) and the Team Approach

The Team Approach and Multi-Crew Operation

Perhaps the greatest change in aircraft transport operations in the past several decades has been recognition that safety and efficiency require a 'team effort' and additionally, that the team involves more than just the flight deck crew. It was not always thus. In August of 1987, and while he was still the US FAA Administrator, Allan McArtor, very succinctly stated the importance of the cockpit team concept when he told a group of airline executives: 'Individuals don't crash, flight crews do.' Regulators had not always spoken that way and neither had most operational experts.

Historically, a 'single pilot' (the captain) was considered the prime and really only important individual involved in an air transport flight. This stereotype began with the white-scarfed, goggled pilot and personified individuals who had such personality traits as independence, good judgement, resourcefulness, calmness under stress, machismo, and especially in the early days, bravery.[1] Individualism and individual performance were stressed. As late as 1932, and after significant improvements in the safety record, one of 50 airline pilots was still killed each year in aircraft accidents (Lederer, 1962).

As airplanes grew larger and the operations grew more complex, a copilot was added to the flight crew. Those first copilots, and all copilots for a considerable time, were considered redundant pilots. Their function was simply to provide an operational backup in the extremely rare condition that the captain for any reason became incapacitated, and to provide support and reduce workload for the captain if they were asked by that captain to do so. Initially, many captains did not particularly like the idea. For several years many copilots did little more than make out flight plans for the captain to approve and sign. Their main job in flight was to handle the radio communications.

---

[1] See Appendix J which details an incident in the air mail days and a letter by Captain Larry Letson describing the details of the flight and the crash of a B-247 in 1934.

Why CRM?

Until the mid-1980s most of the human factor interfaces in air transport operations were concerned with the physical pilot-cockpit instrumentation relationships. While research and investigations had been productive in those areas, even with the restricted concept of limiting human factor interfaces to the physical pilot cockpit instrumentation interrelationship,[2] it became obvious that the air transport safety problem involved more than just this physical relationship. As accidents and incidents were evaluated, it became clear that the technical ability of the crew was very seldom the sole cause of accidents and incidents and that frequently there was considerably less than optimum communication within the cockpit. There were also crew interface problems that included inadequate leadership, poor cockpit management, poor followership, and less than optimum group decision making. All of these subjects are discussed later in this chapter.

As we have seen in Chapter 2—The Industry and its Safety Record, from 60 to 80 percent of hull loss accidents in commercial air transport have been attributed to the flight crew for almost as long as records have been kept. Something different had to be done. Despite improvements in the overall safety record, neither industry nor regulatory efforts had been able to change the disheartening and unsatisfactory relationship between accidents and the operational behavior of the cockpit crew. As a result of much investigation and much soul searching by virtually all aspects of the industry, the Cockpit Resource Management (CRM) concept developed, has grown, and been modified and refined. Today Cockpit Resource Management has expanded and is now called Crew Resource Management. It is almost universally accepted.

CRM History

Research that was begun in the mid-seventies by Billings, Lauber, and Foushee at the NASA Ames Research Center provided a major stimulus to the CRM movement. The research was aimed at addressing some of the more perplexing problems underlying the so-called 'pilot error' accidents and incidents that had plagued the industry for many years. As we have previously indicated,

---

[2] The 'basic T', which we was discussed in Chapter 3—A Brief History of Human Factors and its Development in Aviation—and which is a central part of cockpit instrumentation today, was a product of this period.

the early observations indicated that many of the operational problems had nothing to do with 'stick and rudder' skills, but instead seemed related to other areas, such as decision-making, crew coordination, command, leadership, and communications skills—both within and outside of the cockpit. Of particular interest was the fact that pilot training programs scarcely touched upon the areas that were creating the problems.

From its inception, there have been definitional problems as to what a good CRM program should include. This problem was well stated in the report of one of the Working Groups at a NASA/MAC Workshop held in San Francisco May 6-8, 1986. Participants at this, NASA's Second Workshop on CRM, included NASA's researchers, operational experts, and interested academicians from the US and abroad. Among other things the report of Working Group 3-B noted:

> ...that this working group, if not the entire Workshop, had a definitional problem with the term Cockpit Resource Management (CRM). Strictly interpreted, the term was too restrictive, liberally interpreted, it became all things to all people and its borders became blurred. Working Group 3-B elected to refer to that body of knowledge normally labeled CRM as 'it'. 'It' is not just training in a particular concept or to a particular objective, 'it' is an operational style, a way of life that exists before, during, and after the operation of the aircraft. 'It' incorporates comprehensive orientation to, and training in, multiple aspects of human performance and resource management.   (Butler and Reynard, 1986)

In the early years of CRM, cockpit management focused on interfaces within the cockpit crew. Several airlines were even more narrow in their focus and initially created CRM courses for captains only, missing the basic point that flying a transport aircraft was a team operation and that the entire team should be involved in any CRM training program. These airlines designed courses only for captains as though the content of the courses was a secret and should be kept from the other crewmembers.

In the broader and better courses, copilots had to learn that it was important to be a good team member for the same reasons that it was important for the captain to be a good team leader. The concept that all resources of any type should be considered and used is exemplified in John Lauber's 1984 definition of CRM when he called it 'the effective utilization of all available resources—hardware, software, and liveware—to achieve safe, efficient flight operation' (Lauber, 1984).

The importance of 'team' and the team concept as it applies to air transport was emphasized and expanded by Harvard's J. Richard Hackman. He had studied team operations in many contexts for many years. He told a NASA/ MAC Workshop on Cockpit Resource Management Training that:

> ...the overall task of successfully completing a flying mission is always a team task....cockpit crews always operate in an *organizational context*, and the transactions between the crew and representatives of that context (e.g., organizational managers or air traffic controllers) are consequential for any crew's performance. (Hackman, 1986)

Unfortunately however, even in the late seventies and the early eighties the team concept was given little emphasis in normal operations or in training, checking, or regulatory activities. Most pilots, the operators, pilot unions, the FAA, and the public did not consider team effort an important part of air transport operations. This in spite of the fact that a growing number of thoughtful pilots knew better and utilized many now defined CRM principles on their own.

In some circles, the CRM definition problem still exists although it continues to be refined and clarified. One of the clarifications involves an expanded label for CRM. It is now called 'Crew' Resource Management to ensure that the cabin crew is included, and not just 'Cockpit' Resource Management. The new terminology is now accepted in most operating circles throughout the world.

Today, the US FAA has become a strong supporter of CRM. The FAA's Advisory Circular 120-51B and its references as well as ICAO Human Factors Digests, especially Digest 2, are must reading for anyone interested in developing a deeper understanding than is covered in this overview. A further indication of the importance of CRM in the FAA's overall thinking is that CRM training has been made a requirement for FAA approval of all Advanced Qualification Programs (AQPs). AQP is a key element in the FAA's reassessment of the airline training process. The AQP is discussed in greater detail in Chapter 16—Selection and Training.

The Essentials of CRM

Because there have been definitional problems with the term CRM, there has not always been agreement on what a CRM program should entail. One reason is that if the airlines are from different countries, with major differences in cultures and operating conditions, there frequently are major differences between these airlines. There can even be differences between airlines in the same country. CRM programs that are very good programs for airline 'A' may be inappropriate for airline 'B'. What has been a successful program for one carrier may be quite inappropriate for another. We also now know that a one-shot CRM course is not appropriate for any airline, and can even be counter-productive. As Professor Hackman has stated, CRM as a concept must be embedded in the organizational culture.

Practical and Philosophical Requirements

In spite of the differences between airlines, there are some basic conceptual and philosophical elements that can serve as the starting point for any CRM program. These assumptions are important because they form the foundation on which a good program can be built. A successful CRM program requires active participation and support from all relevant parties. These include top management, training and checking personnel, line pilots and pilot association representatives. Dispatchers, flight attendants, and their representatives also should be included. Ideally, they should be involved from the beginning of the process.

If an assessment of the level of knowledge, attitudes, or perceptions of the airline's pilot group toward CRM is made before the program is begun, later problems can often be avoided. It is important that CRM programs get started off on the right foot, that all participants recognize them as being practical and useful, and that the programs are recognized as a thoughtful and sometimes new approach to a very real problem. A fair number of airlines have spent much time describing and discussing what their program will not entail, for example, that it will not diminish the authority or responsibility of the captain. This too often reported canard is not true and creates negative and unproductive time.

Basic practical and philosophical requirements for a good CRM program include recognition that:

- The main purpose of any CRM program is to increase flight safety and the effectiveness of flight crews.

- Flight crews must be considered as an operating team, not as a collection of technically competent individuals. The training should focus on the functioning of teams not individuals. The crew must be considered as the unit of training.

- CRM training exercises should include all crewmembers. The exercises should be done with full line-qualified crews and the crewmembers should function in the same roles that they normally perform in flight.

- CRM training should both instruct crewmembers how to behave in ways that foster crew effectiveness and reinforce such behavior.

- Crewmembers should practice the skills necessary to be both effective team leaders and effective team members. Both are important and CRM training should provide opportunities to perfect them.

- CRM training should include effective team behaviors routinely achieved during both normal and non-normal operations. Every simulated trip should be operated with all crewmembers utilizing and exhibiting good CRM behavior. All simulated trips should be operated in the normal and expected way.

- CRM can and should be blended into all forms of aircrew training. CRM programs must not consist solely of stand-alone courses.

- CRM instructors and evaluators will undoubtedly need special training. They should be familiar with CRM principles and comfortable discussing them. In addition, their behavior and performance on the line should reflect an understanding and support for CRM concepts.

Introduction and Implementation of a Good CRM Program

Introduction and implementation of a good CRM program in an airline where these are new concepts will never be easy. There is general agreement that CRM training, as recommended in FAA 120-51B, should include at least three distinct phases:

- An awareness phase where CRM issues are defined and discussed.

- A practice and feedback phase.

- A continual reinforcement phase where the routine use of CRM principles are ensured on a long-term basis by their utilization in all normal and non-normal training and checking activities.

A good CRM program is not self-sustaining but must be continually nourished and supported by all levels of operational and safety management, by training personnel, and by pilot representatives.

The awareness phase is particularly important because it sets the stage for what may be a major change in corporate and regulatory operational philosophy. Again, the importance of early involvement of all relevant parties in the program design phase cannot be overstated. The awareness phase should include instructional presentations that inform trainees of the reasons for CRM training and the ways that CRM factors that were poorly performed have contributed to accidents and incidents in the past. It can be particularly helpful to use incidents or accidents from one's own carrier if these are used sensitively and deidentified wherever possible. Emphasis should be on the event and its cause or causes. It should not be on the individuals involved. This first phase should also cover the role of interpersonal and group factors in the maintenance of crew coordination and should introduce, what will be for many, a new terminology and perhaps a new way of thinking.

The practice and feedback phase (usually in the simulator and in line-checking and line operation) is necessary because CRM skills cannot be learned by classroom instructions only. Continued reinforcement is necessary because no matter how effectively the CRM classroom in the awareness phase is conducted, single exposures are insufficient. It is unrealistic to expect that a short training program can change overnight attitudes and operational philosophies that have developed over a crewmember's lifetime, and that previously have been continually reinforced by official company and regulatory checking.

The last phase of CRM is probably the most neglected. For the CRM principles to take hold they must be integrated into all aspects of training and reinforced at all operational levels. Furthermore, the practices, procedures, policies, and general philosophy of the carrier must be supportive of sound operating principles. These principles have been called the 'Four Ps' and were discussed in Chapter 6—The Social Environment. Such an emphasis on CRM does not imply a diminished importance of manual flying skills as is sometimes suggested by critics. Rather, CRM expands and amplifies the skills needed to successfully fly today's aircraft in today's environment. It must be an inseparable part of the organization's culture, something that is taken for granted as 'this is the way we fly our airplanes'.

CRM Subject Domains

There has been much discussion in the industry as to what material should be in a CRM program. As often happens with new concepts, there was, and still is, considerable variety in the way that some of the original premises of CRM are used. The end result is that, in some cases, CRM becomes a term with little specific meaning. The FAA and ICAO attempted to address this problem by specifying the requirements for a CRM program. As Table 13.1 illustrates, the ICAO Human Factors Digest on CRM and the FAA Advisory Circular on CRM—AC 120-51B take slightly different approaches, although the subject domains they recommend are directly related. The FAA suggests two main groupings, while ICAO is more detailed and suggests six topical areas. One possible reason for these differences is that ICAO is concerned with a much broader and diverse group. ICAO must consider all airlines in all countries, while the FAA is concerned only with US airlines. Another reason is simply that CRM is still a developing concept. While the ICAO domains are more detailed, it is important to recognize that there are no substantive differences between the two. An expanded list of the ICAO and FAA domains and their subtopics is given in Appendix G.

Table 13.1   CRM Subject Domains

ICAO	FAA
• Communications	• Communications Processes and Decision Behavior
• Situational Awareness	
• Problem-Solving/Decision-Making/Judgement	• Team Building and Maintenance
• Leadership/Followership	
• Stress Management	
• Critique	

Source: *ICAO Human Factors Digest No. 2*, FAA *Advisory Circular 120-51B*

Communication processes include both inter- and intra-cockpit communication. The captain must create an atmosphere in the cockpit that facilitates good communication. 'Assertiveness', which is encouraged in CRM programs, has created some problems because a few atypical and overly aggressive first officers have felt that assertion meant more that simply being sure that their views are given appropriate consideration. In spite of the fact that there is no suggestion or implication in any CRM program that subordinate crewmembers are considered to be co-captains, a perceived erosion of the captain's fundamental authority has created real problems in the minds of some. A common dictionary definition that assertion can be 'a declaration stated positively but with no support or attempt at proof' has not helped. This important point that copilots are not co-captains is one of the CRM tenets that must be reinforced frequently, even at 'mature' CRM carriers.

Team decision-making is another area that has created several problems. Decision-making is a complex process that involves more than just communication. It will be discussed in the following paragraphs.

Decision-Making in the Cockpit

From the flight crew's standpoint, the process of making decisions actually starts during dispatch, when the captain prepares or approves the paperwork for the flight. The process continues in acceptance of the airplane, and is only finished when the airplane is docked at its gate after landing and all of the paperwork associated with the flight is completed.

Ultimately, all decisions have three characteristics: (1) they involve a choice among alternatives, (2) the specific nature of the problem must be accurately assessed, and (3) they involve risk assessment. The risk assessment may or may not be explicit but is inherent in all decision making. Several of these factors may merit group discussion with the relevant parties. A complication is that most of the decisions are routine and so similar to previous decisions that the response becomes, while guarded, almost automatic.

Flight crews, especially captains, make many decisions during a flight. To discuss each decision would make the process tedious and boring. A very real problem is to decide which decisions are not just routine. These should be discussed. Other crewmembers should not hesitate to ask the reasons or rationale for decisions that they do not understand or that they do not consider routine.

As we have noted, in some cases such as in the decision to abort a takeoff or to continue the flight, there may not be time enough for either discussion or explanation with the other crewmembers and certainly with a dispatcher. This then is a situation that is entirely dependent upon a captain's decision although a captain with sound CRM skills would have briefed the contingencies and

and duty allocations of a rejected takeoff, as an example, and would also make sure that the entire crew has a chance to debrief and learn from such incidents. In other cases, such as determination of an alternate airport, the best method of handling an actual or impending system problem, for most cabin problems, and for the best method and prioritizing of rare but occasional multiple problems, there can be time for ample discussion.

The oft-quoted CRM definition of making best use of the resources available in the *time* available can have some interesting ramifications for long-haul operations. One captain of a long range flight even noted that potentially he has enough time to sleep on the problem if, for example, the problem involves operation at a destination which is still eight hours away. The resources available include the crewmembers (including cabin crew, and if appropriate, other crewmembers who might be travelling in the back as passengers), dispatch, mechanics, engineering, central operations, air traffic controller, etc. The captain, and indeed the full crew, should not hesitate to use whomever might be helpful.

A Rule-based Operation

It is the nature of an air transport operation to be a rule-based operation[3] to the extent possible, always with the understanding that the rules can never cover all situations that might be encountered. Rules are never a substitute for good judgement. Frequently, decisions are simply to decide whether or not the situation is covered by an existing rule, and if so by what rule.

The strong tendency to emphasize rules is because in most cases the operation is fairly fixed. When new situations arise or are anticipated, or when the airline flies into a new airport or area, it is perfectly reasonable for operating management to decide how it wants its operation conducted. When flight operations management promulgates the operating rules it believes will be necessary, the operation becomes simpler for all concerned. Beyond that, it is also acknowledged that it is practically impossible to closely supervise day-to-day flight operations, even if that were desirable, and therefore it helps everyone to have good rules. The rules should contain sufficient operational flexibility to facilitate safe and efficient flight operations.

It has become an industry practice to promulgate a new rule to deal with the stipulated causes of accidents or incidents. This satisfies the perceived needs of the operators, the regulators, and the investigators. In most cases, it

---

[3] Rule-based operations are one of three types of cognitive behaviors analyzed by Jens Rasmussen. The other two are skill-based and knowledge-based.

is much easier to promulgate a new rule than it is to ensure that all pilots understand the ramifications that might have been involved in the accident or incident and to ensure that the next crew involved would react properly under the stress of similar conditions. The promulgation of a new rule answers an obvious problem. An additional reason for the preponderance of rules is that the practice meets the perceived needs of the FAA, the NTSB, the traveling public, and congressional oversight committees.

An inevitable result of this rule proliferation has been the growth and expansion of Flight Operation Manuals (FOMs). The main purpose of a FOM has always been to state rules to govern the flight operation of the airline, including rules to govern an ever-expanding operational universe. As the industry grew and as accidents or incidents occurred, Flight Operations Manuals gradually grew in size. New rules were added to cover actual unfortunate occurrences and hopefully to cover any such occurrences that were foreseen. This is both a reactive and a proactive process and increases an already strong tendency to make an air transport operation a rule-based operation. However, even the best rules often need interpretation. This requires that decisions are made regarding them and conditions arise that are not covered by the rules. Frequently, flight crews are forced to interpret rules under less than optimum conditions.

Ordinarily, the decision process is fairly straightforward and easy, especially if there is little time pressure. Pilots are required to first recognize that a problem exists, second to define the problem accurately, third, to determine if a rule already exists that covers the situation, and fourth to evaluate the risks associated with feasible alternatives. The ultimate choice, or decision, can enter a very gray area considering such things as the nature of the problem, the condition of the crew (fatigue, etc.), actual runway condition, the aircraft's condition, meteorological visibility, time required and available, and the skill of the captain and the other crew members. These are the decisions that can be discussed with other members of the team. The overriding consideration in air transport is always safety. Therefore, crew decisions tend to be reasonably cautious (O'Hare, 1993).

Team Building and Maintenance of the Team Concept

The importance of good team building and the importance of maintaining the team concept should be part of every CRM training program. Their importance has been stressed throughout this chapter. Team building can be best facilitated by reinforcement during CRM training (where the team concept should be used to solve problems), by example as displayed by check pilots or other supervisory personnel flying on the line, and by obvious approval and commendation during required company or government check exercises. Both leadership and followership skills are required. Both can be taught and reinforced.

Maintenance of the Team

Once the team is created by the operational circumstances that have put the members together, it is the job of the captain to ensure that the operation utilizes the team concept and that it is maintained. The team concept can be maintained in several ways. Initially, it is maintained by the captain's general attitude and behavior. Other crewmembers should be fully aware of the team concept. The team concept is also maintained by making sure that all crewmembers are kept in the operating loop and by both sharing relevant experiences and making sure that the experiences of the other crewmembers are shared. Captains should refer to and think of the trip they are flying as 'our trip', not 'my trip'. Other crewmembers should refer to and think of the trip as 'our trip' and not the 'captain's trip'. Maintaining the team concept requires cooperation, understanding, and appropriate participation by all members of the crew.

Leadership

We have already seen that most flight operations should be a team activity. The publication of and training for Standard Operating Procedures (SOPs) that recognize team operations is a good start. A flight operations organization that ensures that SOPs are regularly and routinely followed is an obvious requirement. Equally important are the development of leadership and followership among all flight crewmembers.

A first requirement to be a good leader is to have the respect of the people with whom you work. These are the other members of the flight crew—pilots and flight attendants, the existing peer group, the immediate managers, and other supporting members of the operating team. To have their respect you must know and respect your job and recognize the importance of their jobs and respect them.

A good leader develops a good team. To develop a good team, the leader needs to delegate responsibilities when appropriate, and if a decision is required to listen first of all, then evaluate, and then decide. While final decisions are always the captain's, it is important to be sure that the opinions of other members of the team are considered if sufficient operational time permits, and it is not simply a routine decision. Others who are doing their best to solve a joint problem should get an explanation of the reasons the leader made the final decision in the way it was made. This includes those rare occasions where a decision was required on the spot.

If the leader was wrong, he/she should admit it. There are few single instances that establish a true leader better than one in which he/she freely admits they made a mistake. It keeps the entire team context in perspective and demonstrates that all members of the team are truly involved and have a common goal. Well before the inauguration of CRM programs, a senior operations official, whose managerial duties allowed him to keep only minimally qualified in line flying operations, had a standard introduction to his copilots. He always told his new (to him) copilot that on this trip the copilot's main job was to keep him (the senior operations official) out of trouble. Another manager's introduction was to suggest to his copilot that he was flying that day with one of the weaker pilots in the system as the captain manager did not get out to fly on the line as much as he would have liked. The point from both of these examples is not to suggest technical proficiency problems with management, but much more importantly to illustrate different ways that captains disavow perfection and solicit feedback.

It should be recognized that under some conditions the captain, while remaining overall leader of the flight crew, might not be the leader in a specific circumstance in which another member of the team has greater knowledge. An easy example to imagine is one where a subordinate crewmember has a great deal of experience with a particular foreign airport as compared to the captain. Soliciting inputs and utilizing this experience from other crewmembers can be a confidence-building situation for all concerned, yet does not mean that the captain has relinquished his command. More precisely, the captain is making good use of the resources in the cockpit. It might also be mentioned that this type of leadership behavior by the captain is radically different from that of the stereotypic autocratic captain.

Followership

Followership is an important concept in any team operation, for any team must have both a leader and followers. A dangerous combination can be the pairing together of two autocratic captains, neither of whom knows how to be a follower. The task of any leader is made much easier if he/she has good followers. A good team always has good followers.

The term 'followership' is a new term, so new that it cannot be found in many dictionaries. The term covers the ability to be a good team player and the ability to effectively help a recognized leader. Some have said that it is not translatable into some languages. This is one of several reasons that the concept of followership is not well understood in all countries. The best leaders also have the ability to act as good followers. Good leaders understand the system. They recognize that they operate under a strict seniority system in a dynamic and often changing industry. Positions frequently change. Captains can become followers, and followers can become captains. As an example, some US airlines have permitted their pilots to move to a flight engineer position after reaching the FAA-mandated age sixty retirement. Many of these pilots had been captains for a long period, and by most reviews, made the transition to a followership position very successfully. It is important for everyone to know and understand the system.

Workload Management and Stress Management

The management of cockpit workload is an important part of CRM. No longer can today's air transport operation in any way be considered a 'one-man-band' operation. It is a responsibility of the captain to ensure that workload is allocated among cockpit participants so that no crewmember is overloaded. The foundation of this distribution is good standard operating procedures (SOPs). The captain in non-normal situations may modify SOPs in order to keep the workload of the individual well within his/her personal limits. The bottom line is that the distribution of cockpit workload should not allow the effectiveness of the cockpit team to be diminished. Cockpit workload and the elements that create it are discussed in Chapter 10—Workload.

Stress management is an important part of CRM because it affects the operational effectiveness of the team. Cockpit workload and cockpit stress should be controlled to the point that the effectiveness of the cockpit crewmembers is maintained. Under no circumstances should either excessive workload or excessive stress jeopardize the safety of the flight. There are times, unfortunately, when this statement is more easily made than it is followed.

Stress can be either created by the operational circumstances and/or by domestic or other conditions that arise outside of the flight itself. These subjects are more fully discussed in Chapter 14—Fatigue and Stress. However, from a

CRM perspective the main point is to recognize that stress, for any reason, must be controlled. If stress ever reaches the level at which it may affect the performance of any crewmember, the operation must be modified to the point that the ultimate safety of the flight is not jeopardized.

Critical Background and Supporting Roles

Because a CRM program can involve a major change and reorientation in a company's operating philosophy, it is particularly important that such a program be actively endorsed at all levels of corporate management. Too often, CRM programs that have been promoted and finally approved have received only initial support at the top levels of management. That is simply not good enough. Active and visible support at the top levels is an absolute requirement if a traditional operational philosophy is to be effectively changed. It is important for all levels of management to recognize that this new program may represent a major change in organizational operating culture. To be truly effective, the new program must be visibly supported at the top and intermediate management levels.

Active and visible support by employee organizations (primarily the unions involved) is equally important if day-to-day line operation is to be affected. This area is sometimes neglected. Some carriers have had strong support from employee organizations because representatives were included in the initial planning of the program and they were kept as an important part of the operating loop as the CRM program progressed. Their input was solicited and used. Here they have a common goal. If the employee organization is neglected, the result can be a program that may look great on paper but which has little effect on day-to-day operations because it cannot be implemented effectively.

Support by the operating divisions of the regulatory authorities is another vital link. The FAA plays this role in the US. Individuals like the Principal Operations Inspector and the agency's Flight Inspectors also may need special training. Because in the US the FAA has to approve any changes or modifications in each company's training program and operating rules, it is very difficult to effect changes in operating philosophy and practice without FAA support and approval.

The Flight Instructor, the Check Pilot, and the FAA Inspector

Flight instructors, check pilots, and FAA inspectors are key people in transitioning to a CRM operation. It is essential that flight instructors, check pilots, and FAA inspectors know, understand, and support the reasons that are inherent in the new operating culture. This involves considerably more than just knowing the principles of CRM.

In many cases, flight instructors, check pilots, and FAA inspectors are a product of the 'old school'. They have been successful, and rewarded, because they performed well under the old system. It is not surprising that many of them look with a jaundiced eye at any changes in a system that has served them well in the past. With at least some of them, the changes may be perceived as a threat to their position. Careful training and selection of flight instructors and check pilots is an important first step in the development of any good CRM program. This training is a critical problem for both the companies and the FAA inspectors.

The 'good' flight instructors, 'good' check pilots, and 'good' FAA inspectors can be genuine role models. They can be major and effective supporters of a CRM program. Few people have the power to keep a good CRM program from being effective as do flight instructors, check pilots, and FAA flight examiners.

The Rare 'Boomerang'

A specific problem that has been identified by Professor Robert Helmreich is the boomerang CRM pilot, i.e., the pilot who probably had negative attitudes toward CRM prior to any CRM training and who develops specific negative attitudes toward CRM after going through training. Although 'boomerangs' are admittedly rare, the small population of CRM boomerangs is not restricted to the line pilot. A boomerang can also be a flight instructor, a check pilot, or a FAA inspector. It is important that they be identified early for boomerangs can do damage to a good program that goes well beyond their small numbers. As Helmreich, Merritt, and Wilhelm have later noted, 'boomerangs' are called other names as well, including 'cowboys' or 'drongos'.[4] While we prefer to call them 'boomerangs', the important point is that they should be identified and dealt with.

---

[4] A drongo is a small Australian bird that flies around and defecates on the heads of unsuspecting passersby.

The Role of the Captain

In the early days, there was an informal but rigid understanding that the captain was always right. His decisions were never questioned by a subordinate crewmember. The captain was so preeminent in all operational matters that he/she was seldom questioned by anyone.

An oft retold example involving the captain's decision-making was the case of a DC-3 crew flying a schedule that consisted solely of a northbound flight from Miami to New York in the morning and then a return southbound trip from New York back to Miami. This route is just east of and parallel to the Appalachian Mountains. During the summer it is characterized by a long row of building and eventually towering cumulus and cumulo-nimbus clouds, the centers of which should be avoided if at all possible. This was a serious problem in the days of unpressurized airplanes and their limited altitude capability. Vicious thunderstorms and extreme turbulence develop from these clouds nearly every afternoon. There was no way for a DC-3 to climb faster than the build-ups.

It so happens that a fledgling copilot was scheduled to fly these DC-3 trips with a very senior captain. The eager young copilot was anxious to learn as much as he could from his veteran captain. However, he was forced to admit to himself that he just could not figure out which way to detour when they had a large and actively building cumulus cloud directly ahead of them. If he was flying he never knew which way to detour. Ultimately, the captain simply told him which way to turn. Finally, as they were reaching the end of the month they were scheduled together, the copilot finally blurted out his question. The captain looked at him rather quizzically and said, 'It's easy son, if you are northbound just detour to the West and if you're southbound detour to the East.' This really confused the copilot for he had not heard of any meteorological principles that were as easy and straightforward. He finally confessed to the captain that he simply didn't understand. The captain's reply was logical and laconic. He simply replied, 'It may not be scientific, but it sure keeps the sun out of your eyes.'

Actually, this is a good example of the way that rules (in this case, to stay out of building cumulus clouds and thunderstorms) are frequently applied and decisions are frequently made. Obviously it does not represent optimum performance. As Gary Klein and Judith Orasanu have written later, experts seldom follow the classical decision making models even when making good decisions. In classical decision making, alternatives are examined and disregarded until the best solution can be discovered. Here, the experts instead simply look at the situation, short cut the alternative examinations, and use a solution that has worked well on previous similar occasions (Klein, 1985 and Orasanu 1993). There are obvious limitations to this sort of decision-making. Because in the example of deciding the way to detour to avoid cumulus build-ups there was plenty of time for discussion available before and certainly after the incident, it is clear that the performance of the laconic captain did not represent good CRM behavior.

The concept of the captain's basic responsibility is written in the FARs and in operations manuals. It has been continually reinforced at all levels, even when it involves purely judgement decisions. For example, the Flight Operations Manual of a prominent airline contained the following statement: 'Nothing in this manual should be considered a substitute for good judgement.' While the judgement of all crewmembers is important, the captain's judgement is preeminent. Everyone, the pilots, the company and the FAA, recognized that the captain's judgement, which might be later questioned but seldom was, was the judgement that counted. Unfortunately, in the early days of CRM when team operation was stressed for the first time, some saw CRM as erosion of captain's authority. Nothing could be further from the truth.

The New Zealand Air Line Pilots Association (NZALPA) gave a good example of the new and changed cockpit philosophy that is part of the operations of progressive airlines throughout the world. The occasion was the Royal Commission inquiry into the 1979 crash of Air New Zealand's Flight 901 at Mt. Erebus in Antarctica. In the midst of the inquiry a representative of NZALPA explained the policy of Air New Zealand and the pilots' association by stating:

> ...it is the inherent responsibility of every crew member, if he be unsure, unhappy or whatever, to question the pilot in command as to the nature of his concern. Indeed, it would not be going too far to say that, if a pilot in command were to create an atmosphere whereby one of his crew members would be hesitant to comment on any action then *he would be failing in his duty as pilot in command.*
> (Vette, 1983, italics supplied)

Obviously, this cockpit philosophy does not degrade a captain's authority. He or she must make the final decision after considering the opinions of other crewmembers whenever operational time permits.

The Captain's Dilemma

Virtually all of today's senior captains have been a successful part of an operations culture in which the captain was not only king, but also one whose behavior was very seldom questioned. A prominent US airline had a flight operations manual containing a sample dispatch release form where the name, used in the example for the captain, was 'I. M. King'. This example illustrates what was the prevailing operational culture in many carriers. The copilot was considered an apprentice and was definitely subordinate in all respects. Historically, the captain was routinely checked for technical proficiency, but rarely for anything to do with managing his cockpit team and getting the best performance from it.

The dilemma facing some of these captains was that they were being pressed to change an operating philosophy that they had used successfully for years. When these senior captains were being tested and interviewed for initial employment, there were few, if any, behavioral measurements used in the process. The selection people were almost exclusively interested in the technical aspects of flying. A number of experienced and successful captains felt that CRM, as they perceived it, was a definite threat to their operations and their careers. Virtually all of them had been copilots under the pre-CRM system, learned to adjust to a variety of captain's personalities, styles, and in some cases, operational quirks, and were successful in that they were in turn promoted to captain. They expected to continue to be operationally successful in the organizational culture with which they were familiar. For some, it has not been easy to change. It should be mentioned, however, that for other captains, CRM simply formalizes the way that they had been operating their cockpits for years.

The Role of the Copilot

Today, in most cases flight operations are different. One of the major changes in the operation of transport airplanes has been an upgrade of the copilot's skills and knowledge. This has been an evolutionary change and has inevitably resulted in an increase in training for copilots, an increase in the copilot's operational importance, and an accompanying increase in training time and costs. An upgrading of the copilot's skills and knowledge has been required because:

- The importance of the effective monitoring of all manual or automatic operational behavior is recognized, even if the captain is the pilot-flying.

- There is no way that one can be expected to monitor the operational behavior of another without having a clear understanding of what the individual (or automation) is trying to do.

- If a long-range operation is involved, rest provisions have to be provided for all crewmembers (including the captain). The air transport system has to be assured that competent and fully-qualified crewmembers are on the flight deck at all times. Today, copilots in augmented long-range crews are required to have full captain ratings in the airplanes they fly and demonstrate their competency in FAA required proficiency checks.

The Copilot's Dilemma

The record shows several accidents or incidents in which copilots had detected serious problems in the performance of the captain, but in which the copilots were unwilling or unable to effectively alter the course of events. Monitoring of the Captain (acting as PF) by the copilot (acting as PNF) had been neither a sufficiently practical nor an effective method for achieving redundancy or double-checking. In some cases, the problem was compounded because inexperienced copilots, may also have been oblivious to the dangers their captains were unaware of or ignoring. However, these cases are relatively rare. An early ASRS study (NASA CR 166433) showed that the captains detected 33% of the anomalies reported when he or she was the pilot flying and 35% when he or she was the pilot not flying. The copilots' percentages were 15% and 18%. The copilots' poor monitoring performance was not because he/she did not understand the operation.

In some cases, the reason was simply, the reluctance of the subordinate crewmember (the copilot) to interfere with the operational behavior of the captain, or the failure of the captain to reexamine his/her operational behavior if it was questioned. Another possible reason for these operational anomalies was fear of the potential consequences of the copilot's temerity in violating the traditional concept that whatever the captain did was always approved, and was always right. Another possibility is that the copilot was reluctant simply because he/she was unsure of or concerned with the captain's reaction to his/her interjection. If the copilot was still on probation with the company, this was another meaningful factor. On some carriers, the copilot while on probation faced the possibility of being summarily discharged after a charge of poor performance by his/her captain. There could be no appeal after a captain's complaint. Fortunately, most airlines now have a different system.

The issues discussed above are CRM issues. To create a good CRM climate, it should be the goal of all concerned to create a cockpit atmosphere where shared decision making is not considered an adverse reflection on the captain's competency. Instead shared decision making should be considered an example of good crew coordination, of good leadership by the captain and an example of good followership by the remaining participants. A further discussion of decision making is beyond the scope of this book but can be found in Orasanu's chapter on decision making in Wiener, Kanki, and Helmreich's *Cockpit Resource Management.*

Cockpit Crew/Cabin Crew Interface

Cockpit crew/cabin crew interface is a difficult and complex problem for pilots, cabin attendants, companies, and the FAA. It has become even more important with two-person cockpit crews and with the large number of flight attendants that are needed on each airplane to serve an increasing number of passengers of various ages, physical condition, and language capability. Cockpit crew/ cabin crew interface is an important part of any good CRM program.

Cabin crew performance has been receiving increased attention in accident reports. From an accident prevention standpoint, it has become clear that cockpit crew/cabin crew coordination and communication have not always been satisfactory. There are several reasons and the industry's traditional quest for 'zero' accidents, 'zero' incidents makes it clear that more attention must be given to the cabin crew/cockpit crew interface.

This was graphically illustrated in the Air Ontario accident at Dryden, Ontario on March 28, 1989 when an Air Ontario F-28 crashed on takeoff due to an accumulation of ice on the wings. One of the flight attendants saw the accumulation of wet snow building on the wings but feared that the pilots would not be appreciative of that sort of operational information from a flight attendant and so did not inform them. Other factors were involved, but a central problem was cockpit crew/cabin crew coordination and communication (Moshansky, 1992).

While in the real world cockpit crew and the cabin crew work together, however imperfectly, and in the larger sense even have common goals, the two crews are the products of separate cultures that must interface effectively. Chute and Wiener said it well when they noted that: 'There are historical, organizational, environmental, psychosocial, and regulatory factors that have led to misunderstandings, problematic attitudes, and suboptimal interactions between the cockpit and cabin crews' (Chute and Wiener, 1995). These problems include CRM issues. The history of the cockpit crew/cabin crew interface and some of their inherent problems are discussed in greater depth in Chapter 17—The Challenging Role of the Flight Attendant.

Environmental Factors

Several task requirements and environmental factors interfere with better interface between the cockpit and the cabin. The first starts before the flight when the flight crew's duties require them to report in dispatch to either make a flight plan or approve a flight plan produced on a computer, study the weather, and plan or approve the required fuel load. A preliminary check-in with the scheduling crew desk is an important but usually simply perfunctory procedure.

Cabin crews have different problems, usually check in to a different crew desk, and then go to their own briefing room. Their crew desk is virtually always separate and can even be on a different floor. Both crews are busy during this pre-flight time and the pre-flight briefing of the cabin crew by the captain is frequently perfunctory at best. Latest aircraft have only two pilots but can have as many as 19 or 20 cabin attendants so that anything approaching a personal interface with each of them during this busy time is virtually impossible.

The cabin-cockpit door is another environmental factor that militates against an ongoing interchange between the two crews. It provides a physical barrier that isolates each crew from the other. Regulations in the US require that the cabin-cockpit door be kept locked in flight. The existence of this door makes it virtually impossible for either crew to know what the other is doing. The physical isolation problem is exacerbated with double deck airplanes and a passenger section that is divided into sections and holds up to 400 people.

None of these conditions make a meaningful interface easier. They make a good CRM training program and a CRM-based operation that includes flight attendants difficult and even more important.

Other Factors

A major additional factor that militated against cockpit crew/cabin crew interface was the elimination of the flight engineer and, therefore, elimination of the three-person crew. Traditionally, the flight engineer had been the cockpit's intermediary with the cabin. The flight engineer could fix faulty movie projectors, recalcitrant ovens, help with unruly passengers, determine if a gray mist coming from the ventilation system was simply fog or the more serious smoke emanating from one of the air conditioning packs, and provide a myriad of other functions when the cabin crew needed additional help to solve a particular problem.

A good deal of this natural cockpit crew/cabin crew interface was the product of the physical location of the flight engineer's seat. It was located behind the copilot, faced sideways and could be swiveled forward as needed. The flight engineer performed a perfect filter or buffer from the cabin crew for the pilots if they were busy. From the pilot's point of view this was a valuable, if unspecified flight engineer function. The pilots and especially the captain, who was also responsible for safety in the cabin, could be kept informed of unusual activities in the cabin without inappropriate distractions from their principal job of flying the airplane.

Another important additional factor is the problem and period of the 'sterile cockpit'. As a result of several accidents and incidents in which operationally unnecessary cabin crew/cockpit crew interface seemed implicated, the FAA promulgated FAR 121.542. It stipulates that no flight crew shall perform 'any duties during a critical phase of flight except those duties required for the safe operation of the aircraft.' By FAR definition, critical phases of flight include all ground operations involving taxi, takeoff, and landing, and all other flight operations conducted below 10,000 feet, except cruise flight.

The 'except cruise flight' clause was recognition that some commuter short-haul flights never get above 10,000 feet. Without that escape clause, the regulation would have made virtually all cabin crew/cockpit crew communication prohibited on those flights because the regulation further stipulates that: 'Non-essential communications between the cabin and cockpit crews is prohibited.' This later clause has created a great deal of interpretative confusion among cabin crews, cockpit crews, all airlines, and the FAA itself.

Whether or not a given communication is essential or whether or not it is about a valid safety consideration is a very gray area. It is particularly difficult for a non-pilot with no instrumentation available to determine correctly if an airplane is above or below 10,000 feet. It requires a decision that must be made on the spot and frequently with very little time to consider. This problem involves both cockpit crews and cabin crews. It is not limited to CRM training. The sterile cockpit concept is discussed in greater detail in Chapter 17—The Challenging Role of the Flight Attendant.

Conclusions

In the foreword to *Cockpit Resource Management*, a book which was written in 1993, John Lauber referred to CRM as 'an unfolding but already proven human factors success story' (Lauber, 1993). This was, and still is, certainly true. Good cockpit management was not a new concept. Forward thinking and progressive captains have utilized its principles for a great many years. The British, and some others, frequently called at least parts of it airmanship.

However, the CRM movement marks the first time that these principles were formally acknowledged and taught and the first time that the performance of the flight crew, as a team, was considered a regular part of training, checking, and line operations. While CRM is not a panacea, today it, or something very close to it, is considered an essential part of airline operation. CRM is becoming a regulatory requirement in the US and in some other countries. Parts of the world have been reluctant to adopt the specific words or the acronym CRM, but there is virtually no quarrel anywhere with its concepts.

An important step, and a very basic difficulty for the airlines and for the FAA, has been to determine a means of evaluating the effectiveness of CRM skills. Determination of the effectiveness of CRM programs and objective evaluation of the CRM skills of the individuals (and teams) that participate have been given a very high priority within the regulatory agencies and airlines of a great many countries.

The process toward evaluation of CRM behaviors and skills has been evolutionary and controversial, at least in some areas and to some pilots. From the perspective of the FAA, if CRM skills are really that important, then the carrier should ensure that its pilots have them, and it becomes the FAA's role to ensure that they are evaluated. However, the development of valid and reliable measures has been difficult. Professor Robert Helmreich and his team from NASA/UT have been instrumental in some of the early research and measurement development. In fact, his 'behavioral markers' are given as an appendix in the FAA's CRM Advisory Circular and are an excellent starting point.

It is not surprising that progress in CRM has been slow and uneven in an industry that is basically conservative. CRM involves a major official philosophical change and this is true both in the United States and in the rest of the world. Implementation of a CRM program is not an easy task. In 1993, Bill Taggart, of the NASA/University of Texas project and who has been involved in CRM from its beginning, wrote an article 'How to Kill Off a Good CRM Program' that contains a number of items to consider if one is initiating or reexamining a CRM program.

Unfortunately, some of Taggart's top ten ways to kill off a good CRM program have been demonstrated too many times. The ten ways are shown in Appendix H and indicate traps that can catch the unwary. One should never forget that it is not easy to implement or to maintain an effective CRM program. Even as the airline matures in its CRM attitudes and as CRM behaviors become behavioral norms, it will be necessary to revisit the fundamentals.

While a good CRM program should ultimately result in increased safety and efficiency in the airline's operation, a good CRM program takes additional training and that can cost big money. It should not be relegated to 'this year's safety theme', thereby giving participants the false impression that it is of transient interest to the corporation. The program should be productive, and visibly so, or it will probably fail.

Each program must meet the needs of its particular airline and consider the operating philosophy and the social environment in which it will be used. Airlines vary both within a given country, across different cultures, and across national boundaries. Some airlines will be able to embed the CRM principles more easily than others. The main objective must be to keep the concepts active and true to their original goals of safety.

An outstanding book on CRM, and one that we heartily recommend, is *Cockpit Resource Management*, edited by Earl Wiener, Barbara Kanki, and Robert Helmreich. This book includes writings on the various aspects of CRM by 23 authors who are acknowledged experts in their field, and who represent US and non-US academia, regulators, airline management, and military transport. Government research organizations such as the US NASA are also represented. Other major and important references are ICAO's Human Factors Digest 2, and the FAA's Advisory Circular 120-51B and its revisions. All are major references for this chapter. Airline operations are not a simple process. Neither is a very dynamic air transport industry in a rapidly changing and expanding world a simple industry. There are a great many relevant variables.

References

American Heritage Dictionary of the English Language (1978). Houghton Mifflin Company, Boston, Massachusetts.

Butler, Roy and Reynard, W.D., (1996). 'Integration into the total training curriculum', from *Proceedings of the NASA/MAC Workshop*, held in San Francisco, 6-8 May 1986, NASA Conference Publication 2455, edited by Harry W. Orlady and F. Clayton Foushee, NASA/Ames Research Center, Moffett Field, California.

Chute, Rebecca D. and Wiener, Earl L (1995). 'Cockpit-cabin communication: a tale of two cultures', in *The International Journal of Aviation Psychology*, 5(3). 257-276, Lawrence Erlbaum Associates, Inc. Mahwah, New Jersey.

Chute, Rebecca D. and Wiener, Earl L. (1996). 'Cockpit-Cabin Communication: II. Shall We Tell The Pilots'. In *The International Journal of Aviation Psychology*, 6(3), 211-231, Lawrence Erlbaum Associates, Inc., Mahwah, New Jersey.

Driskell, James E. and Adams, Richard J. (1992). *Crew Resource Management: An Introductory Handbook*, Final Report DOT/FAA/RD-92/26 and DOT/VNTSC FAA-92-8, Department of Transportation, Washington, D.C.

Federal Aviation Administration (1993). *Advisory Circular—Crew Resource Management Training*, AC No. 120-51B, Federal Aviation Administration, Washington, D.C.

Hackman, J. Richard (1986). 'Group Level Issues in the Design and Training of Cockpit Crews', *Cockpit Resource Management Training, Proceedings of the NASA/MAC Workshop*, ed. by Harry W. Orlady and H. Clayton Foushee, NASA Conference Publication 2455, 6-8 May 1986, Ames Research Center, Moffett Field, California.

Helmreich, Robert L. and Wilhelm, John A. (1989). 'When Training Boomerangs: Negative Outcomes Associated with Cockpit Resource Management Programs', in *Proceedings of Fifth International Symposium on Aviation Psychology*, The Ohio State University, Columbus, Ohio.

Helmreich, Robert L., Merritt, Ashleigh C., and Wilhelm, John A. (1999). 'The Evolution of Crew Resource Management Training in Commercial Aviation', *The International Journal of Aviation Psychology*, Vol. 9, Number 1, 1999, Lawrence Erlbaum Associates, Publishers, Mahwah, New Jersey.

Human Factors Digest No. 2 (1989). 'Flight Crew Training: Cockpit Resource Management (CRM) and Line-Oriented Flight Training (LOFT)', International Civil Aviation Organization, Montreal, Canada.

Klein, Gary A., Calderwood, Roberta, and Clinton-Cirocco, Anne (1985). 'Rapid Decision Making on the Fire Ground', Klein Associates, Yellow Springs, Ohio.

Lauber, John K. (1984). 'Resource management in the cockpit', *Air Line Pilot*: 53, 20-23, Airline Pilots Association, Herndon, Virginia.

Lauber, John K. (1993). In Foreword to *Cockpit Resource Management*, edited by Wiener, Earl L., Kanki, Barbara G., and Helmreich, Robert L., Academic Press, Inc., Harcourt Brace Jovanovich, Publishers, San Diego, California.

Lederer, Jerome (1962). *Perspectives in Air Safety*, Daniel Guggenheim Medal Award Lecture presented at 1962 ASME Aviation and Space Conference, Washington, D.C.

Moshansky, V.P. (1992). *Commission of inquiry into the Air Ontario crash at Dryden, Ontario*, Catalog No. CP32-55/1-1992E), Minister of Supply and Services, Toronto, Canada.

National Transportation Safety Board (1988). *Hazardous materials incident report: Inflight fire, McDonnell Douglas DC-9-83, N569AA, Nashville Metropolitan Airport, Nashville, Tennessee, February 3, 1988* (NTSB/HZM-88/02). Washington, D.C.

National Transportation Safety Board (1972). *Aircraft Accident Report: Allegheny Airlines, Inc. Allison Prop Jet Convair 340/440, N5832, New Haven, Connecticut, June 7, 1971* (NTSB-AAR 72-20), Washington, D.C.

National Transportation Safety Board (1979). *Aircraft Accident Report: United Airlines, Inc, McDonnell-Douglas DC-8-61, N8082U, Portland Oregon, December 28, 1978* (NTSB AAR-79-7), Washington, D.C.

National Transportation Safety Board (1982). *Aircraft Accident Report: Air Florida, Inc. Boeing 7737-222, N62AF Collision with 14th Street Bridge, Near Washington National Airport, Washington, D.C., January 13, 1982,* (NTSB AAR-82-8), Washington, D.C.

National Transportation Safety Board (1984). *Aircraft Accident Report: Eastern Air Lines Inc., Lockheed L-1011, N334EA, Miami International Airport, Miami, Florida, May 5, 1993* (NTSB AAR 85-01), Washington, DC.

National Transportation Safety Board (1994). *National Transportation Safety Board Safety Study,* (NTSB/SS-94/01), Washington, D.C.

O'Hare, David, (1993). Book review of *Human Factors for Pilots,* in *The International Journal of Aviation Psychology,* 3(1), 83-85, Lawrence Erlbaum Associates, Inc., Hillsdale, New Jersey.

Orasanu, Judith (1993). 'Decision making in the cockpit', in *Cockpit Resource Management,* edited by Wiener, Earl L., Kanki, Barbara G., and Helmreich, Robert L., Academic Press, Inc., Harcourt Brace Jovanovich, Publishers, San Diego, California.

Regal, D, Rogers, W., and Boucek, G. (1988). *Situational awareness in the commercial flight deck: Definition, measurement, and enhancement,* SAE Technical Paper Series 881508, Warrendale, Pennsylvania.

Ruffell-Smith, H.P. (1979). *A Simulator Study of the Interaction of Pilot Workload with Errors, Vigilance, and Workload,* NASA Technical Memorandum 78482, NASA/Ames Research Center, Moffett Field, California.

Sarter, Nadine B. and Woods, David D. (1991). 'Situation Awareness: A Critical But Ill-Defined Phenomenon', in *The International Journal of Aviation Psychology,* 1(1), 45-57, Lawrence Erlbaum Associates, Inc., Hillsdale, New Jersey.

Taggart, William R. (1993). 'How to Kill Off a Good CRM Program', *The CRM Advocate,* October, 1993, Resource Options, Inc., Charlotte, North Carolina.

Vette, Gordon and Macdonald, John (1983). *Impact Erebus,* Hodder & Stoughton, Auckland, New Zealand.

Wiener, E. L., Kanki, Barbara G., and Helmreich, Robert L. (1993). *Cockpit Resource Management,* edited Wiener, E. L., Kanki, Barbara G., and Helmreich, Robert L., Academic Press, Inc., Harcourt Brace Jovanovich, Publishers, San Diego, California.

14 Fatigue and Stress

Fatigue and Tiredness

Many books and articles have been written about fatigue. It is often used interchangeably with tiredness, and both words are used in a variety of contexts. Two dictionary definitions of fatigue given that illustrate general perceptions of fatigue are: 'Physical or mental weariness or exhaustion resulting from exertion,' and 'The decreased capacity or complete inability of an organism, organ, or part to function normally because of excessive stimulation or prolonged exertion.' This dictionary[1] also uses fatigue as a synonym for 'tiredness' In another definition, fatigue was described as '...an unpleasant experience (that) has entered into the life of everyone.' Dr. Kenneth Bergin, in his book, *Aviation Medicine*, describes fatigue as 'a progressive decline in man's ability to carry out his appointed task, which may become apparent through deterioration in the quality of work, lack of enthusiasm, inaccuracy, lassitude, ennui, disinterestedness, a falling back in achievement or some other more definable symptoms.' Finally, 'since fatigue is taken to be an experience, it is an expression of the whole person' (Bartley and Chute, 1947).

Common sense and appropriate research has shown that fatigue in its broadest sense is not localized but is general and not specific to a given body member. The occasions for the production of fatigue are endless. In most cases situations that can cause fatigue are things like long hours, sleep cycle interruptions, overwork, circadian disrhythmia, or in some cases simple boredom. Fatigue is not a new or unique phenomenon.

Bartley and Chute state that while boredom may form a part of the fatigue picture, the reverse is not possible. That is because boredom is caused by environmental events, while fatigue is a product of the total situation. They also note that boredom is much more transient than fatigue and may be alleviated simply by escaping the situation that is causing the boredom. That is not true with fatigue. While an individual may be both bored and fatigued, fatigue is a much broader term. Variations of the problem, but with different underlying causes, may explain why in the literature one finds references to

---

[1] The American Heritage Dictionary of the English Language.

'mental fatigue', 'nervous fatigue', 'muscle fatigue', 'combat fatigue', 'operational fatigue', etc.

The Basic Problem

Fatigue and tiredness have been called ill-defined concepts that are part of everyone's life. As Graeber has written, 'Tiredness is experienced by all of us almost every day of our lives. It varies in intensity but is a familiar, sometimes welcome feeling which is rarely given serious thought' (Graeber, 1988). Unfortunately, fatigue is still a general concept and does not have a specific meaning or a good definition. While even today fatigue cannot be measured directly, there is no question that it exists and fatigue can certainly be felt (Davenport and Jensen, 1989). Fatigue can be a serious problem in aviation if it becomes excessive.

One of the first organized studies of the fatigue phenomenon in aviation was performed in England in the famous Cambridge studies. They were conducted in the 1940s and concluded that with fatigued pilots:

1. The timing of motor responses suffered more and more as fatigue developed;

2. Subjects became increasingly willing to accept lower standards of accuracy and performance;

3. They shifted from following the six primary flying instruments to making more automatic reactions; and,

4. They increasingly forgot to check instruments out of their immediate range of vision.

Since that time a great deal of additional research, not all of it in aviation, has been undertaken. It is universally agreed that fatigue can adversely affect performance, that it is a very complex problem, and that it is an unavoidable consequence of operations that continue throughout all 24 hours in a day. While to many, it seems almost as if the researchers are 'reinventing the wheel', there is genuine concern that the fatigue problem in aviation be understood and handled right. The airlines' exemplary safety record, the relatively small number of accidents, particularly when compared to exposure, and the difficulty of showing a correlation between fatigue and the accident record has not made the problem easier. The record may well be a tribute to man's resourcefulness, adaptability, and the drive to survive.

Fatigue in Air Transport Operations

Fatigue, or at least tiredness, has been a problem since the beginning of commercial aviation. Both pilots and the operators long have recognized its existence. Fatigue was a problem for Jack Knight on the first scheduled night flight on February 23, 1921. After flying his part of that first scheduled night flight, Knight arrived at Omaha at 2 a.m. He was exhausted and looking forward to his normal scheduled rest, but because there was no other pilot available in Omaha, he agreed to continue the flight to Chicago. He was very tired but flew on to make aviation history.

In another example of fatigue in the industry, captains in the 1940s for at least some airlines were required to make comprehensive written evaluation reports on their copilots at the conclusion of a joint scheduled tour of duty. One of the questions asked was simply 'Does he (the copilot) have the stamina required of an airline pilot?' It was an important question and there were good reasons for concern. The basic question was did the copilot deal appropriately with the fatigue that was inevitable in their operations.

Fatigue was first officially cited as the 'probable cause' of an accident in the NTSB Report (NTSB/AAR-94/04) on American International Airways Flight 808 which crashed at Guantanamo Bay on 18 August 1993. The report stated that among the 'probable causes of this accident were the impaired judgment, decision-making, and flying abilities of the captain and flight crew due to the effects of fatigue...' All three crewmembers had suffered a cumulative sleep loss over the preceding 65 hours, had circadian disruptions, and had prolonged periods of wakefulness prior to the accident. All of these support the conclusion that fatigue was very much involved in that tragic event. Jim Danaher, chief of the NTSB's Operational Factors Division has stated that if the aircraft had not crashed: 'The Company had intended to ferry the airplane back to Atlanta after the airplane was offloaded at Guantanamo Bay. This additional segment would have resulted in a total duty time of 24 hours and 12 hours of flight time' (Duke, 1997).

One reason that fatigue had not been given as a 'probable cause' in an accident is that fatigue was simply not considered as a possible 'probable cause' in accidents until very recently. No one really knew very much about fatigue. For example, in the late 1940s, an Alitalia flight became hopelessly lost, ran out of fuel, and crashed in Connecticut farmland. The flight crew had left Rome, landed at London, returned to Rome, and were enroute to New York when the flight crashed over 24 hours from their original departure from Rome. The 'probable cause' given was fuel depletion. There was no mention of fatigue.

It would be nonsense to suggest that fatigue can be totally eliminated in air transport operations. Aviation by its nature is conducive to fatigue, and insidious as fatigue may be, operational professionals must accept the fact that it is a part of air transport. Pilots should expect to encounter fatigue during their career, and they should be aware of the problems it can create. If the level of fatigue is excessive, the operation must be modified.

Unfortunately, there is no magic bullet or quick fix to deal with the fatigue that is often a part of either long- or short-haul operations. Today, there is no simple solution that will address all individuals, all operational demands, and all of the technology that is currently involved in the aviation industry. Air transport requires 24-hour operations. The challenge to the industry is to incorporate present relevant scientific and physiological knowledge into areas that will maintain and increase present safety margins, and, at the same time, maintain a viable air transport operation.

Circadian Rhythm and the Time of Day

Circadian rhythms, sleep irregularities, and the operating time of day can create performance problems that are similar, however they have quite different causes.

Circadian (*circa-dies*) rhythm is the most common of the body rhythms. The word circadian comes from two Latin words, 'circa' meaning around and 'dia' meaning day. The circadian cycle is related to the earth's rotation time of 24 hours. The 24-hour time is only approximate for there are individual variations. Scientists have found that circadian rhythm varies from 24 to 27 hours in isolated humans. In the real world the circadian cycle is regulated by entraining agents called *zeitgebers*. Daily light and darkness are the most powerful *zeitgebers*, while others such as meals and social or physical activity also have considerable influence.

Sleep, which is a requirement for all humans, is also cyclic and is directly related to our normal circadian rhythm of light and darkness, as anyone who has tried to spontaneously sleep in the daytime can attest. Circadian rhythms obviously affect flight activities. For example, in the contiguous US there are four time zones. A pilot who lives on the East Coast and flies to the West Coast will have his/her normal circadian cycle disrupted by three hours. All of the relevant *zeitgebers* —light and darkness, meals, and any social or physical activity—will be displaced. If the flight continues on to Honolulu, an additional three time zones are involved. Other long-range flights can involve as many as 12 to 16 time zone changes and the basic problem is exacerbated.

Performance is also affected by time of day. This is one of the major problems of the night shift. Individuals work at a time they would normally be sleeping. In transport aviation, which is a round-the-clock industry, this is a problem for individuals in all parts of the system. Performance degradation and selected remedies that to some extent alleviate it also will be discussed in the next chapter—Fitness to Fly.

Sleep Loss, and Circadian Disrhythmia

Sleep is an active complex physiological state that is vital to survival. Sleep loss creates sleepiness that is more than a minimal nuisance that can be overcome with will power. Unfortunately, sleepiness can potentially degrade most aspects of human capability. Without question it can degrade those aspects that are important in flight operations.

We know that sleep is an absolute requirement for all humans. It is an elemental physiological need. Obtaining even one hour less than that required by the individual can affect waking levels of sleepiness. Sleep has been defined by two researchers in the field, Carskadon and Dement, as 'a reversible behavioral state of perceptual disengagement from and unresponsiveness to the environment'. While human physiological capabilities and limitations remain central to the safe operation of transport operation, disruptions to a normal sleep cycle are inevitable. Therefore it is imperative that all parts of transport operations learn as much as they can about sleep and sleepiness, and about measures that can be taken to minimize the sleep problems that are inherent in air transport operations. Unfortunately, there is no single easy solution to the sleep problems that arise from some operational demands.

The two principal sources of fatigue in aviation are sleep loss and circadian disruption. Both are related although one does not have to have a circadian disruption to develop a sleep loss. Sleep loss results in sleep debt. Sleep debt can be cumulative and can result in uncontrolled sleep episodes (i.e. microsleeps lasting only seconds or extended episodes of sleep which can last for minutes). Uncontrolled sleep episodes can occur when driving a car, flying an airplane, operating machinery, or just standing. Uncontrolled sleep episodes are obviously incompatible with flying an airplane or with other activities that put people at risk

While a complete discussion of sleep loss and fatigue is beyond the purview of this book, all aviation professionals should know that sleep loss can be acute and cumulative and leads to sleep debt. Sleep loss definitely can occur and adversely affects waking performance, vigilance, and mood. Even though minimal increased sleepiness can often be individually overcome, it can potentially degrade most aspects of the concept of human capability. Sleep loss is a concept that should be known to all pilots. Sleep loss, if not replaced, accumulates into a sleep debt. Sleep debts are very real and can only be replaced by sleep. It usually takes at least two to three consecutive nights of normal sleep to stabilize an individual's sleep pattern and restore optimal performance and alertness after a long westbound flight. A long eastbound flight is more difficult and it can take even longer to return to normalcy. With some individuals and some schedules, additional time may be required.

The Circadian Clock

All operational personnel should be aware of the circadian clock. Every human has a circadian clock in the brain that regulates such physiological and behavioral functions as the sleep/wake cycle, body temperature, hormones, performance, mood, digestion, motor activity, and many other functions on an approximately 24 hour basis.

When the circadian clock in our brain is moved to a new work/rest or sleep/wake cycle (e. g., simply moved to a new time zone), it does not adjust immediately. One well-known effect is jet lag. Unfortunately, it can take from several days to more than a week to completely adjust to major changes. Complete adjustment is not possible in some scheduled air transport operations. It becomes very important to utilize countermeasures that can minimize and control fatigue problems when they arise.

A complicating factor is that there is considerable variation among individuals in the need for sleep and the effectiveness or time needed for circadian adjustment of all of the physiological factors affected. For example, most individuals need about eight hours of sleep per night, but some need as little as six hours while others may need as many as ten to feel wide awake, alert, and able to function at their peak level.

Factors that affect circadian disrhythmia and the resultant fatigue are age, individual sleep requirements, experience, overall health, and exercise patterns on both layovers and at home. Personal strategies are important. Preventive strategies, which can be used in the periods before duty and on layovers, include obtaining an appropriate quantity of maximal quality sleep prior to duty, maintaining good sleep habits, scheduling sleep periods intelligently during layovers, napping, exercising, and maintaining balanced nutrition. As pilots have known for years, westward flights are easier than eastward flights. It is easier to lengthen the circadian day as happens on westward flights than it is to shorten it as must be done if the flight is eastward.

Commuting

Another factor that must be considered is commuting to and from work assignments. Many pilots have decided for personal, economic, and other reasons, to live and reside outside their domicile. Another situation that results in a commute for the pilot occurs during a merger or when pilot domiciles and bases are realigned. Making the personal decision to commute to the job puts additional responsibility on the commuting pilot so that he or she can ensure that they report for duty in a rested and alert condition. It also puts additional responsibilities on companies to be sensitive to this issue. There is now general recognition that scheduled commuting should be considered as duty time as far as fatigue is concerned.

For many, commuting creates some unique problems. For an extreme example, one pilot with a schedule out of Guam lived in Miami. To commute from Miami took two days travel, which required flying via Detroit and Tokyo. Going home required flying through Honolulu, Los Angeles, and Minneapolis. Because he crossed the international date line, he would arrive at home the same day he left Guam, although it did not feel the same, and the total travel time required was still two days. More normal commutes are for a pilot flying out of Minneapolis but living in Albuquerque, the pilot flying out of Chicago but living in northern Maine or New Orleans, or the pilot who flies out of Pittsburgh but lives in North Carolina. For these pilots, the commuting flights are a part of their regular schedule. It should be noted that flight attendants also commute. One airline, for example, has a significant number of Hong Kong-based flight attendants who commute from Honolulu and other locations on the US mainland.

The Air Transport Association (ATA) says that nearly half of the pilots who work for the largest US carriers have substantial commutes that average 922 miles each way. This creates a different way of life and a unique culture. Some say that it is dangerous, but are countered by others who point out that in the long history of air transport operation, no accidents have been attributed to the fact that the crew commuted, perhaps to a considerable extent because most flight crew members commute responsibly.

There are an infinite number of variations in the commuter scheme. The FAA maintains that it cannot legislate what people do with their time off, much less dictate the place they live. Because the problem is also rife with collective bargaining ramifications, companies are reluctant to tamper with travel privileges that many employees (not only pilots) consider legitimate perquisites that attract people to the industry.

Operational Countermeasures

Operational countermeasures that can be used during flight include physical activity, strategic caffeine use, and operationally feasible social conversations with other cockpit crewmembers and the flight attendants. Countermeasures are important, but unfortunately, no individual countermeasures work every time for every individual. Individuals and conditions vary. What helps in one case may provide only minimal help in another.

An effective countermeasure that is sanctioned by some foreign airlines is the use of planned in-flight naps. Swissair, Air New Zealand, Lufthansa, and other international long-haul carriers are already practicing prescheduled cockpit napping (Aviation Week and Space Technology, 1996). Other carriers that sanction scheduled naps enroute are British Airways and KLM. The trend in the use of in-flight naps is supported by sleep research and is growing among those airlines that have more regulatory flexibility than US airlines. In the US, thanks to the NASA-Ames Fatigue Countermeasures Program, which is discussed later, planned in-flight naps are under serious consideration by the FAA. While planned in-flight naps are not officially sanctioned in the US as this chapter is being written, approval is expected in the future.

Scheduled Enroute Naps

Enroute naps have been a difficult and controversial subject for a very long time. That method of dealing with the very real problem of operational fatigue, mainly caused by interruptions of the normal sleep cycle, has been used by the working pilot more or less surreptitiously since DC-3 days. A not completely uncommon scheduling irregularity problem that was alleviated by preplanned but unsanctioned scheduled enroute naps, was informally related by a Canadian copilot, who explained to one of the book's authors that:

- The previous winter he was holding a regular line of flying and was not on reserve, when he got a phone call from the crew desk at about 8 p.m. He was told the company was entirely out of available reserves and was asked if he could be out at the airport for a trip leaving for London at midnight. The trip had just come up, and the lack of coverage resulted from a last minute request flight operations had received from the sales department.

- At this point he had spent a normal uneventful scheduled day off and was planning going to bed at a normal time. The crew desk said they had just received notice of this last minute flight from operations and were in desperate straits to get the flight covered. The copilot agreed to fly the trip, did not have enough time for even a short pre-flight nap, and left home about ten p.m. in order to get out to the dispatch office by eleven for the midnight departure.

- When he arrived at the airport, and incidentally for the first time, met the captain for this trip, he found that the captain was not on reserve either and had had almost identical experiences at home and with the crew desk. The North Atlantic and London had typical winter weather and it was bad.

- The captain said he planned to make the approach at London's Heathrow airport and that he would take about an hour's nap prior to the descent into Heathrow. He also said that if the copilot thought it would help him he could plan on taking an earlier nap after they were all squared away at cruise.

- The trip was flown that way, each pilot took his scheduled nap, a good approach was flown to minimums at Heathrow, and all in all it was a very good and safe operation.

In order to understand this very real operational issue, one has only to imagine being an airline or regulatory official or a passenger for that flight. Under those circumstances, would he have been most comfortable during the approach to minimums at Heathrow if that trip was flown by a crew that followed the regulations exactly and took no naps, or by the crew which took two scheduled enroute naps?

Economic factors force airlines to fly throughout a 24-hour day. Understandably, many passengers like schedules that enable them to conclude a normal working day and then fly on to another city or to fly on long distance flights to arrive in time for a productive working day at a distant destination. Sleeping or relaxing on the airplane is exactly what these passengers want. A flight schedule that is ideal for the passengers can completely disrupt normal sleep for the flight or cabin crews. In addition, it is expensive to keep airplanes on the ground and companies like to keep their airplanes in the air. The inevitable result is that flights are not scheduled for optimum working conditions for the flight or cabin crew

Because of all of these factors, individuals and airlines sometimes are forced to compensate for known fatigue situations. There are limited options available. Pilots' ability and willingness to adapt to sometimes very difficult conditions has made it possible for the air transport industry to continue to have its really excellent safety record

Reserve Pilots and Long-haul Pilots

Fatigue is an even greater problem for long-haul pilots and for reserve or standby crews. Reserve pilots are required to be available to fly trips on very short notice anytime they are called to do so. The amount of notice required varies with different carriers and is usually a contractually negotiated item. Reserve pilots are called out for a variety of reasons including the sudden illness of a regular scheduled crewmember, up-line flight cancellations, diversions, equipment substitutions, and the addition of extra sections. It can be very difficult or virtually impossible for a reserve pilot to plan for all possible trips.

It is clear that both the operators and the pilots have a joint responsibility to minimize operational fatigue. The problem is exacerbated by such things as duty schedules that are poorly designed from a fatigue standpoint, unanticipated operational delays, and by pilots who report for duty without proper rest. While short-haul trips can require long duty days, early report times, night flying, or in some cases adverse time zone changes (Foushee et al., 1986), long-haul trips frequently include lengthy transmeridian flights as well, and thereby impose further circadian desynchronization problems. The operational significance of this factor has not been clearly identified; however, its likely impact on performance is underscored by the consistently higher accident rate for long-haul versus short-haul commercial sectors (Graeber, 1988 and Caesar, 1987).

Heino Caesar also points out that the long-haul pilot is faced with several additional adverse factors. They include the handling difficulties of large, heavy multi-engined aircraft (especially when abandoning take-offs and performing critical landings), and such additional long range stresses as different climates, jet lag, unfavorable arousal levels because of boredom, night flying, and lack of sleep. Other adverse factors are serious fuel problems, sometimes badly equipped remote airports, unfamiliar weather phenomena, seldom flown routes, language and communication problems, and decreasing experience in the manual handling of the airplane due to increased automation and minimal opportunities to practice. Comparing long-haul and short-haul safety is not a simple problem.

The air transport industry is a global industry and requires 24-hour activities to meet operational demands. Shift work, night work, irregular or unpredictable work schedules, and time zone changes are simply a part of the business. Over the years, human physiological requirements for sleep have not changed despite the evolutionary and technical advancements made in the industry with its 24-hour-a-day operational demands.

The NASA-Ames Fatigue Countermeasures Program

In 1980 the US Congress asked NASA to examine a possible 'safety problem of uncertain magnitude, due to transmeridian flying and a potential problem due to fatigue in association with various factors found in air transport operations'. The problem was given to the NASA-Ames Research Center and its response started with a NASA/DOT Workshop on Pilot Fatigue and Circadian Desynchronosis that was held 26-28 August, 1980 in San Francisco. The Workshop participants concluded that 'there is a safety problem, of uncertain magnitude, due to transmeridian flying and a potential problem due to fatigue in association with various factors found in air transport operation'.

The Workshop was followed by an extensive NASA-Ames program called the Fatigue/Jet Lag Program that in 1991 evolved into the Fatigue Countermeasures Program. The goals in both programs were the same, namely, to:

- determine the extent of fatigue, sleep loss, and circadian disruption in flight operations;

- determine the impact of these factors on flight crew performance;

- develop and evaluate countermeasures to mitigate the adverse effects of these factors and to maximize flight crew performance and alertness.

The 16-year old NASA Program is one of the few that has seriously looked at the fatigue problem in air transportation. It has studied cockpit fatigue by integrating controlled laboratory studies with full-mission high-fidelity simulation studies and, most important of all, with real-world studies in both domestic and international flight operations. Each of these research approaches has strengths and weaknesses from both a scientific and an operational standpoint. For example, field studies conducted during line operations obviously can more accurately reflect real-world conditions as compared to the other types. However, field studies are inherently difficult to conduct because the studies must not interfere with regular operational procedures, or more importantly, with flight safety. In addition, it is virtually impossible to control all potential contributory factors in any field study. Laboratory studies provide a controlled environment and can provide precise values for specific outcomes. However, it is frequently difficult to generalize their findings to real-life operational reality.

Full-mission, high-fidelity flight simulation studies have many advantages for they can manipulate trip scenarios and measure a wide range of flight variables. However, they are very expensive and still must face the problem of operational reality. An important advantage of the NASA program is that it integrates all three kinds of studies enabling it to take advantage of the unique strength that each method offers.

The NASA program has looked at the fatigue associated with a variety of operations ranging from short-haul air transport operations (all flights were less than eight hours) to long-haul flights in an international pattern with all flights scheduled for more than eight hours. The field studies of short-haul air transport operations involved 74 pilots from two airlines before, during, and after three- and four-day scheduled trips. The study examined the extent of sleep loss, circadian disruption, pilot performance and fatigue in these typical short-haul operations. The long-haul studies looked at such things as duty requirements, pilot performance, local time, and the ways that the circadian system affected the timing, quantity, and quality of sleep. This study used 29 international pilots who were flying B-747 aircraft on one of four international trip patterns (Rosekind et al., 1994).

An earlier study by Foushee et al. examined the performance of 20 volunteer twin-jet transport crews in a full-mission simulation that included most aspects of an actual line operation. The crews flew the simulation before and after an acknowledged fatiguing duty sequence of trips. The abstract of this study revealed that:

> ...not surprisingly, Post-Duty crews were significantly more fatigued than the Pre-Duty crews. However, a somewhat counter-intuitive pattern of results emerged on the crew performance measures. In general, the performance of Post-Duty crews was significantly better than the performance of Pre-Duty

crews....

Further analyses suggested that the primary cause of this pattern of results is the fact that crewmembers usually have more operating experience together at the end of a trip, and that this recent operating experience serves to facilitate crew coordination, which can be an effective countermeasure to the fatigue present at or near the end of a duty cycle....   (Foushee et al., 1986)

One of the results of the NASA-Ames Fatigue Countermeasures Program has been the development of a NASA/FAA education module and which is entitled 'Alertness Management in Flight Operations'. It was created to return worthwhile and useful results to the operational community. The objectives of the Module are 1) to explain the current state of knowledge about the physiological mechanisms that underlie fatigue, 2) to clarify misconceptions, and 3) to give fatigue countermeasure recommendations. The Module was developed as a one-hour live presentation that provides an opportunity for interaction and a discussion focused on application of the principles discussed. It is complemented by a NASA/FAA Technical Memorandum that provides all of the slides from the presentation (National Transportation Safety Board and NASA Ames Research Center, 1995).

Much of the material in this chapter is taken from reports of the NASA Fatigue/Jet Lag Program and its successor the NASA Fatigue Countermeasures Program. It is an international program and has received support and collaboration from the US Federal Aviation Administration (FAA), the National Transportation Safety Board (NTSB), air carriers, pilot unions, and the military. International studies of this program involved worldwide collaborations with research and flight operations groups from the United Kingdom, Germany, Japan and other countries.

Keeping Tiredness and Fatigue in Perspective

Theoretically, governmental regulations are supposed to create conditions which control undue operational fatigue. However, the regulations have been far from an unqualified success. The working conditions for many pilots are more influenced by their current contracts (Working Agreements and Side Letters of Agreement) that are the result of collective bargaining than they are influenced by regulations. However, there are a number of commuter and other airlines who have not been able to reach agreements that are satisfactory in this area. For these operators, their limiting conditions are those stipulated in sometimes inadequate federal regulations.

Recently in the US and in other parts of the world, there has been a movement to revise national regulations in regard to the working conditions of flight crews. Stimulated by the NASA program, there is now a great deal of interest in considering 'duty time' in addition to 'flight time' and 'rest time'. Not surprisingly, airline organizations and pilot associations do not see eye-to-eye in these areas. Appendix I shows one of the ways that a major US airline (United Airlines) addresses this difficult and controversial area in their flight operations manual. Operational flexibility, the quality of working conditions, and ultimately big money is directly involved. If a collective bargaining problem is to be determined with regulations, the details of that regulation are of real interest to both sides. One side or the other, or even both, may benefit. One can only be sure that the issues will be complex and that administration will not be easy. With a good solution, the result can be in the best interest of both parties.

The 1986 NASA Technical Memorandum 88322—*Crew Factors in Flight Operations: III. The Operational Significance of Exposure to Short-Haul Air Transport Operations* by Foushee, Lauber, Baetge, and Acomb helps keep the question of fatigue in perspective. As we have previously discussed, the study revealed that: '...not surprisingly, Post-Duty crews were significantly more fatigued than Pre-Duty crews. However, a somewhat counter-intuitive pattern of results emerged on the crew performance measures. In general, the performance of Post-Duty crews (on identical full-mission simulations) was significantly better than the performance of Pre-Duty crews.'

In this study, other factors seemed more important than fatigue at the level of fatigue measured. 'Further analyses suggested that the primary cause of this pattern of results is the fact that crewmembers have more operating experience together by the end of a trip. This recent operating experience serves to facilitate crew coordination, which can be an effective countermeasure to the fatigue present at or near the end of a duty cycle'.[2]

Our final word on fatigue in air transport operations is that tiredness and

---

[2] This assertion is supported by the NTSB Safety Study—*A Review of FlightCrew-Involved Major Accidents of U.S. Air Carriers, 1978 Through 1990* NTSB/SS-94/01). The report of the study states that: 'The Safety Board's understanding of industry practices related to air carrier crew scheduling suggests that the percentages of accident crews who were on their first flight or first day together are greater than would be expected.'

fatigue are an inherent part of air transport operations. They are very complex issues, and issues which professional pilots should understand. Tiredness and fatigue are certainly not limited to fatigue in long-range transmeridian flights. We highly recommend the latest reports from the NASA-Ames' Fatigue Countermeasures Program, the NASA-Ames/FAA Education and Training program, and Graeber's chapter, Aircrew Fatigue and Circadian Rhythmicity, in Wiener and Nagel's *Human Factors in Aviation* for anyone wishing to know more about this very real problem.

Stress

In general, one may say that whenever man is forced to act upon conflicting forces, stress is created. Stress is a complex phenomena and can be either acute (where it is being caused by a situation) or chronic (where it is not of a temporary nature and is being caused by an ongoing situation or life event). It is manifested in an internal state that is brought about by the pressures that life, or in some cases a particular situation, can bring to a specific event or events.

Like fatigue , stress is a real part of an air transport operation, and it is a phenomenon whose direct application to air transport operation is very difficult to determine. Two of our most perceptive aviation psychologists, Robert Helmreich[3] and H. Clayton Foushee,[4] have noted that the question: "'How

---

[3] Robert Helmreich is professor of psychology at the University of Texas and Director of the NASA/University of Texas/FAA Aerospace Crew Performance Project. He was chair of an FAA working group to develop the *National Plan for Aviation Human Factors*, has conducted a great deal of aviation research for NASA and the FAA, and is recognized world-wide as an outstanding aviation psychologist and aviation human factors authority. He continues to be one of our most active and innovative research scientists, has received many honors, and is a fellow of the American Psychological Association and the American Psychological Society.

[4] H. Clayton Foushee is a former Principal Scientist of the Crew Research and Space Human Factors Branch at NASA-Ames Research Center and former Chief Scientific and Technical Advisor for the FAA. Dr. Foushee headed a joint NASA/FAA effort (with some Department of Defense assistance) to implement the comprehensive National Aviation Human Factors Plan for the US—a plan that still forms the government's coordinated effort in aviation human factors. Since leaving the FAA, Dr. Foushee has held several prominent positions at Northwest Airlines, most recently as Vice-President of Regulatory Affairs.

does stress induced by fatigue, emergencies, and personal experiences influence the way teams communicate and operate?", is among the important and theoretically interesting questions regarding flight crew group processes that we know very little about.'

Our major interest is in the effect of stress on performance, and we should note that all stress is not bad. Each of us has seen instances where a seemingly small or moderate amount of stress or stimulation has resulted in a feeling of well being and high levels of performance. In fact, very few instances of outstanding performance occur in the absence of some kind of stress.

High levels of stress are associated with unpleasant psychological and physiological symptoms such as fear, anxiety, sweating, and fatigue. Lower levels of stress can result in a feeling of well being and a feeling of being completely on top of the situation. It should be noted that the level of stress perceived varies between individuals and can vary over time with each individual. For example, a stressful task that is completed successfully results in less stress if it is attempted another time.

Arousal, while it should not be used interchangeably for stress, is sometimes used as an indicator of the amount of stress existing in a given situation. Figure 14.1 shows the relationship between arousal and performance.

This figure shows that at low levels of arousal, as might exist in the cruise phase of a long night flight, stress levels are low and performance for some tasks may be degraded. This has been at least a theoretical concern in long-range flights that are largely flown with the automatics. As might be expected, arousal is increased during descent, approach, and landing, and performance can be expected to improve. However, the expected improvement in performance is often mitigated for many pilots by circadian disrhythmia and tiredness or fatigue. Fortunately, even these phases of flight, including the landing in some airplanes, can be made with the automatics. In these cases sometimes any increase in arousal is minimal. This factor has been discussed in Chapter 11—Automation. Figure 14.1 also shows that if arousal (stress levels) become too high, performance will decrease. One reason for the decreased performance can be that there is a tendency to narrow or restrict the focus of attention to what is perceived as the primary demand and miss other important information.

At the low levels of arousal the nervous system is not fully functioning, and the processing of any sensory information is slow. Moderate levels of arousal produce interest in external events and in performing tasks. The level of performance increases. Unfortunately, at high levels of arousal, the

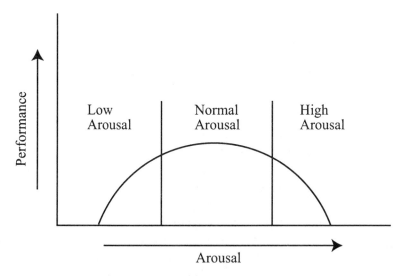

Figure 14.1 The Relationship Between Arousal and Performance
Source: Adapted from *Human Factors for Pilots*, page 67, Green, 1991,
            and used with permission from Ashgate Publishing Company

sympathetic nervous system, over which we have no control, produces extra
adrenaline that provides the body with increased heart and respiration rates,
increases blood pressure, and increases blood flow to the muscles in order to
cope with the new stress. This response is sometimes known as the 'fight or
flight' response. Historically, when man was a hunter, it was a very useful
reaction. Today, however, this sudden physical release of energy is seldom
what is needed, and in most cases is detrimental to desired behavior. Even
when the source of the stressful experience has passed, the 'fight or flight'
reaction continues and continues to be inappropriate. We now know that the
stress reaction can occur without the actual occurrence of an event but simply
as a response to the anticipation of the perceived demand or threat.

Environmental Stress

On the flight deck, environmental stresses include excessive noise, heat,
vibration, and low humidity. Helicopters are particularly vulnerable to noise
and vibration stress.

Heat

The ambient temperature preferred by most people in ordinary clothing is around 70°F. (21°C.), although this varies slightly by national habits and customary climatic exposure. Temperatures outside the range of from 60° to 90° (15°-32°C) usually cause discomfort and lower resistance to other stresses. Both are occasionally found in normal flight operation.

Noise

Noise has been used to help maintain arousal levels during periods of boredom and fatigue and may even be used to mask other distracting sounds. However, excessive noise, in addition to being uncomfortable, can disrupt the performance of a task, cause annoyance and irritability, and lead to cardiovascular and other physiological reactions. Some of the sounds used for cockpit warnings can produce these unwanted responses. Recent research has attempted to design sounds that attract attention without causing a startle response. Green et al. have noted that excessive noise has an effect similar to excessive heat in narrowing perception and causing attention to become focused and restricted (Green et al., 1991).

Vibration

Working in a vibrating environment for any length of time affects both visual and psychomotor performance. Depending upon the frequency and amplitude of the vibration, it can cause chest and abdominal pain, interfere with breathing, cause backache, headaches, eyestrain, pains in the throat, result in disturbances of speech, and cause muscle tension. In an attempt to provide the pilot with a vibration-free working area, designers try first of all to minimize or eliminate any vibration. If that is not possible, they attempt to isolate him/her from the source of the vibration, and minimize the vibration with a carefully designed seat. Most helicopter pilots believe that the designers still have a long way to go. Turbulence also can be a form of vibration.

Low Humidity

Because the outside air temperature at jet cruising altitudes can easily reach minus 60° F. (-51.6°C.) and has almost no water vapor, the relative humidity within the aircraft can fall to around 5 per cent when it should be closer to something between 40 and 60 percent for optimal comfort. We have already discussed some of the effects of exposure to very low levels of humidity such as a drying of the mucous membranes of the eyes, nose, and throat.

The low humidity problem is greater on long haul flights where some researchers have noted other effects such as water retention and the manufacture of less urine. Because of these potential effects, Green et al. (1991) suggest that it may be better to drink only the amount of fluid needed for comfort (avoiding large quantities of caffeine), and 'dealing with the skin or lips with moisturizing agents or water sprays'. More traditional and conventional advice is given by the Aerospace Medical Association's Passenger Health Subcommittee which states that: 'The common recommendation of increased fluid intake during air travel is all that is needed to control the mild side effects of low humidity.' Many passengers and crew have found that simply increasing the intake of drinking water has been very helpful. It should be remembered that lack of humidity problems for the passengers may be exacerbated for cockpit and cabin crews because of their continued exposure to a very low humidity environment.

Occupational and Domestic Stress

Occupational and domestic stress situations are reasonably common. One is associated with the job, the other with just daily living. Both occupational stress and domestic stress are sometimes chronic and can be very hard to quantify. If not well controlled, either can adversely affect on-the-job behavior. Both the details of the particular stress situation and the individual's reaction to it can vary tremendously. For example, some pilots have maintained that only when they were out flying on their job could they be free of the very real problems they had at home (personal conversation with W.P. Monan, a very experienced pilot and flight manager). We do know that pressure in any sphere is a potential cause of stress. It should be considered on an individual basis because of varying reactions to the stress. The effect of virtually all stress is dependent upon the kind and amount of stress and the individual's reaction and coping techniques.

The Job (Occupational Stress)

Stress that is caused by requirements of the job can vary considerably with the individual. Frequently it is time related, depending upon cockpit position, domicile, reserve or scheduled status, immediate supervision, the competence of the individual, the economic stability of the airline, and in somewhat rare cases, the job itself. Pilots react to job requirements quite differently. Many of them, for example, find that any type of required checks or required training are very stressful because of the consequences of failure. Others find these situations merely a temporary challenge. A well-respected flight manager upon retirement after 35 years of successful performance confessed, 'I have been going out to the Training Center (for checks and training) for 35 years—probably at least 90 times—and I hated every minute of it'. These were stressful times for him.

Domestic Stress and Home Relationships

Virtually any changes in a person's domestic situation such as divorce, marital separation and even marriage can be a source of stress. Both pilots and managers should be aware of this potential problem and of the effect it can have on operational behavior and performance. Events such as the death of a spouse or child, divorce, a partner either leaving or returning to employment, etc. can all lead to increased levels of stress depending, to a great extent, on the conditions in place and the success of the individual's coping strategies.

For some time efforts have been made to see if there were acute situational factors that might precipitate an accident. Admittedly, because such factors are by nature short-lived, they are difficult to identify in this context. A veteran aviation psychologist—Robert A. Alkov—has written that 'pilot error' should be regarded as a result of precipitating factors rather than a cause of accidents' (Alkov, 1977).

In an intriguing study, two psychiatrists—Thomas H. Holmes and Richard Rae—found that many diseases were caused by changes in the life events of their patients. This choice of restricting the sample to the patients of Doctors Holmes and Rae may have created at least a partially biased sample. The results of the study may or may not have been affected by the bias.

Changes in the life events studied could be either desirable or undesirable, the important aspect seemed to be the change. These life events were things that nearly everyone experiences at some time in their life. They included events such as:

- the death of a spouse or other family members

- a change in job

- divorce

- moving and relocating to another area

- the birth of a child

- economic events such as either the origination or the termination of a mortgage

- promotions or demotions

- vacations

- retirement

- pregnancy (of oneself or of any female relative or companion)

- a son or daughter leaving home or returning home to live

The life events that seemed important clustered in a 24-month period and the health changes followed the life crises in about one year. An interesting finding was that the effect of the life changes was cumulative. For example, on an arbitrarily created scale, which listed the death of a spouse at the top of the scale with 100 points, the higher the accumulated point total, the greater was the expectancy of a serious illness.

Significant life changes in aviation are changes in residence, family separations, changes in working conditions, furloughs, mergers, sleeping and eating habits, social activities, and personal habits in general. Their total effect may tax the ability of the pilot to cope, even though he or she has adapted to the particular stress. If the additional stresses that are brought on by life crises in his personal life are added to the stresses that are a part of the pilot's job, it may create a severe burden. These stresses may have little influence on performance until they add up to a psychological burden that is unbearable for the individual.

In order to put these intriguing findings in an aviation operational context, Alkov looked at 737 Class A flight or flight-related mishaps that occurred to the US Navy during the calendar years 1979-82. One of the most interesting findings of this study is that there are substantial differences in stress coping, and that: 'The aircraft mishap due to aviator error might be seen as a symptom of inadequate stress behavior' (Alkov, 1984). It is important to realize that this is another area in which individuals are different and effective preventive treatment varies with the individual. There are, of course, many other possible reasons for each so-called 'pilot error'.

Serious Accidents, Illness, or Bereavement

Special categories of stress are auto accidents, airplane accidents, or home accidents that happen to individuals or to family members. All can be a source of emotional stress. Recovery may or may not take a considerable amount of time. The same is true of serious illnesses that affect family members or close friends. Individual reaction to misfortunes of this nature varies widely.

Stress related bereavements usually occur with the death of a family member due to an accident or to an illness that is either long and difficult or, conversely, very short in duration. Closely related is a major adverse change in the health of a family member. A classic case is that of a military pilot who made two gear-up landings in six weeks following the day that he took his wife to the hospital for the birth of their child which was stillborn (Green et al., 1991).

Conclusions on Fatigue and Stress

Both fatigue and stress are difficult because of their nature and because both are a part of normal transport operations. It would perhaps be nice if pilots never got tired, never had to fly when they are tired, always lived in domicile, or if a reserve pilot was never called out on short notice for an all-night trip. It also would be nice if pilots or their families never became ill, had domestic problems, accidents or bereavements. Unfortunately, that is not the real world. Professional pilots are forced to adapt to sometimes very difficult but legitimate needs of the service.

Fatigue is not a problem for pilots alone, but also can affect other operational personnel in the air or on the ground. It is axiomatic that operational management must always remember that the safety of a trip can never be jeopardized with an excessively fatigued cockpit crew. Operational management must also be concerned with other members of the crew. Deciding at what point a crew could reasonably be expected to become excessively fatigued is not an easy task. There are obviously legitimate commercial pressures for management to complete an operation and to maximize flight and duty times. The NASA Fatigue Counter-Measures Program is designed to furnish guidelines for both pilots and management in this very difficult area.

Stress is an equally difficult concept to deal with because it is so varied. Some level of stress is an inevitable and even desirable part of daily life, because a certain amount of stress is a part of day-to-day flight operations, and because individual variations, particularly for domestic or personal stress, make any sort of meaningful measurement virtually impossible. This by no means suggests that the concept of stress can be ignored. Excessive stress can result in significantly degraded performance. Domestic or personal stress is one of the most difficult areas. In most cases domestic and personal stress must be handled with empathy and understanding between the pilot and the first level of operational management.

While both fatigue and stress could very well have been included in the following Chapter 15—Fitness to Fly, we believe both fatigue and stress are important subjects in their own right and deserve separate treatment. Clearly, if one suffers from excessive fatigue or from either chronic or acute stress to the extent that performance is significantly degraded, that person is not fit to fly until the situation is corrected.

References

Aircraft Accident Report (1993). *Uncontrolled Collision With Terrain, American International Airways Flight 808*, NTSB/AAR-94/04, National Transportation Safety Board, Washington, D.C.

Alkov, Robert A., (1977). 'Life Changes and Pilot Error Accidents', presented at Air Line Pilots Association Human Factors Symposium, Air Line Pilots Association, Herndon, Virginia.

Alkov, Robert A. (1984). 'Aviator Stress Overload', *Approach*, May 1984, Naval Safety Center, Norfolk, Virginia.

Bartley, S. Howard, and Chute, Eloise (1947). *Fatigue and Impairment in Man*, McGraw-Hill Book Company, Inc., New York.

Beaty, David, (1969). *The Human Factor in Aircraft Accidents*, Martin Secker & Warburg Limited, London.

Caesar, Heino (1987). 'Safety Statistics and Their Operational Consequences', Flight Safety Foundation's 40th International Air Safety Seminar in Tokyo, Flight Safety Foundation, Arlington, Virginia.

Clark, Ron E. Nielsen, Ronald A., and Wood, Rawson L (1991). 'The Interactive Effects of Cockpit Resource Management, Domestic Stress and Information Processing in Commercial Aviation', *Proceedings of the Sixth International Symposium on Aviation Psychology*, The Ohio State University, Columbus, Ohio.

Duke, Tom (1997). 'Battling Fatigue: The Challenge is to Manage It', *Air Line Pilot*, Air Line Pilots Association, Herndon, Virginia.

Foushee, H. Clayton, Lauber, John K., Baetge, Michael M., and Acomb, Dorothy B. (1986). *Crew Factors in Flight Operations: III. The Operational Significance of Exposure to Short-Haul Air Transport Operations*, NASA Technical Memorandum 88322, Ames Research Center, Moffett Field, California.

Gabriel, Richard F. (1975). 'A Review of Some Universal Psychological Characteristics Related to Human Error', presented to International Air Transport Association 20th Annual Technical Meeting, Istanbul, Turkey, Douglas Paper 6401, Douglas Aircraft Co. (now Boeing), Long Beach, California.

Gander, Philippa H., Graeber, R. Curtis, Foushee, H. Clayton, Lauber, John K., and Connell, Linda J. (1994). *Crew Factors in Flight Operations II: Psychophysiological Responses to Short-Haul Air Transport Operations*, NASA Technical Memorandum 108856, Ames Research Center, Moffett Field, California.

Gander, P.H., Nguyen, D., Rosekind, M.R., and Connell, L.J (1993). 'Age, Circadian Rhythms, and Sleep Loss in Flight Crews', *Aviation, Space, and Environmental Medicine*, Aerospace Medical Association, Alexandria, Virginia.

Graeber, R. Curtis (1988). 'Aircrew Fatigue and Circadian Rhythmicity, in *Human Factors in Aviation*', ed. by Wiener, Earl L. and Nagel, David C., Academic Press, Inc., San Diego, California.

Green, Roger G., Muir, Helen, James, Melanie, Gradwell, David, and Green, Roger L., (1991). *Human Factors for Pilots*, Avebury Technical Academic Publishing Group, Aldershot, England.

Hawkins, Frank H. (1993). *Human Factors in Flight*, Second Edition, ed. by Orlady, Harry W., Ashgate Publishing Co., Ltd., Aldershot, England.

Rosekind, Mark R., Gander, Philippa, H., Dinges, David F., (1991). *Alertness Management in Flight Operations: Strategic Napping*, SAE Technical Paper Series 912138, Society of Automotive Engineers, Warrendale, Pennsylvania.

Rosekind, Mark R., Gander, Philippa H., Miller, Donna L., Gregory, Kevin B., Smith, Roy M., Weldon, Keri J., Co, Elizabeth L., McNally, Karen L., and Lebacqz, J. Victor, (1994, June). 'Fatigue in Operational Settings: Examples from the Aviation Environment', *Human Factors* 36(2), Santa Monica, California.

Rosekind, Mark R., Gander, Philippa H., Connell, Linda J., and Co, Elizabeth L. (1994, December). *Crew Factors in Flight Operations X: Alertness Management in Flight Operations*, NASA Technical Memorandum in press. DOT/FAA/RD-93-18, Ames Research Center, Moffett Field, California.

Rosekind, Mark. R., Wegmann, Hans M. (1995). *Principles and Guidelines for Duty and Rest Scheduling in Commercial Aviation*, in press as a NASA Technical Publication, Ames Research Center, Moffett Field, California.

15 Fitness to Fly

The General Premise

The aviation system traditionally expects pilots to be rested, alert, and in good physical condition (as is proven by the possession of an airman's medical certificate and the use of their own good judgment). In no way should pilots, or other of the system's professionals, be involved with any sort of substance abuse (alcohol or drugs contraindicated for flying, whether the drugs are purchased over-the-counter or obtained as prescription drugs). Most of this is viewed as an individual responsibility, is straightforward, and is agreed to by all reasonable people.

Today 'Fitness to Fly' has a broader connotation. An area that is not well understood, or at least not effectively recognized, is that accountability for fitness to fly does not end with the individual. Both the operators and regulators have areas in which they have important roles

Physical Well-being

Physical well-being for pilots, especially at the home residence, or at the beginning of a trip, is a fairly easy concept. The requirements to be well rested and alert and to have a current airman's medical certificate are individually reassuring. The system requiring them seems to work well. Accidents have virtually never been attributed to the physical condition of the flight crew.

Professional pilots have a large stake in maintaining their ability to pass the required FAA physical because it is a requirement for their job. All pilots are required to hold both a current pilot certificate and medical certificate. The ramifications for the medical examination are quite simple and dramatic. If a pilot cannot pass the physical, he/she is not legal to fly. Airline captains, who are required to hold a first-class medical, must pass a physical examination every six months. As a part of that examination, they must have an initial electrocardiogram (ECG, also referred to as EKG) at age 35 and then an ECG yearly, beginning at age 40. The importance of the medical certificate is a basic reason that most pilots have moderate and sensible diets, some sort of an exercise program—either formal or informal, keep their weight within fairly reasonable limits, take their health seriously, and can be considered as a group to be health conscious. The medical standards used for the three classes of medical certificates issued by the FAA are listed in Appendix P.

Diet

A great deal of the information in the following discussion of diet and exercise is based upon information from The American Medical Association's *Encyclopedia of Medicine* and the University of California's *The Wellness Encyclopedia* (American Medical Association, 1989, and Health Letter Association). Those interested in more detailed information about diet, exercise, and general are urged to consult these or other reliable texts.

Normally, we eat three meals a day. This is a commonplace fact that is generally accepted and mostly just taken for granted. However, the specifics of what we eat, and how much we eat, is also important. Concern with our diet involves more than just concern about nutritional deficiencies. We all know that most people eat too much, especially too much fat, and consume more sugar than is good for them. Other areas should also concern us.

A Balanced Diet

Today, we are bombarded with a plethora of dietary advice. Fortunately, good dietary requirements are reasonably simple. The basic food groups of milk, meat, fruits and vegetables, and breads and cereals provide the essential elements of a good diet. It includes proteins, carbohydrates, fats, fiber, vitamins, minerals, and water. A good balanced diet requires all of them.

There are three basic rules for maintaining a healthy diet. The first rule is to eat a wide variety of foods in order to take advantage of the different nutritional contributions each food makes. We have an almost infinite variety of things to eat. The second rule is that fruits, vegetables, grains, and legumes should make up more than half of the calories we consume. Fruits, vegetables, grains, and legumes are high in complex carbohydrates and fiber, low in fat, and free of cholesterol. The remaining calories in our diet should come from low-fat dairy products, lean meats, poultry, and fish. The third basic rule in a good diet is to maintain a balance between caloric intake and caloric expenditure. If we eat more food than we need, the balance is stored as fat and we gain weight. Most authorities believe that good weight control minimizes the chances of developing diabetes, arteriosclerosis, and hypertension. Exercise does help, but calories are the critical ingredient for weight control. Calories do count!

Cholesterol

Cholesterol is a waxy, fat-like substance that is manufactured in the liver, is carried in our blood and found in all tissues. It is an important constituent of body cells but too much of it is bad. Lipoproteins, which are also manufactured in the liver, are a significant part of the body's cholesterol package. There is overwhelming evidence that a high level of cholesterol in the blood significantly increases the risk of arteriosclerosis, coronary heart disease, or stroke. This is particularly true if the blood contains a high proportion of low-density cholesterol (LDL, or low-density lipoprotein), as opposed to high-density cholesterol (HDL, or high-density lipoprotein).

Strange as it may seem, high-density cholesterol (HDL) is a good kind of cholesterol. As the cholesterol package is carried in the blood, HDL not only carries less cholesterol than LDL but also seems to carry cholesterol back to the liver where it can be reprocessed or excreted. The level of blood cholesterol is lowered by the consumption of soluble fibers, monounsaturated or polyunsaturated fats, fatty fish, and by aerobic exercise. The level of blood cholesterol is raised by eating foods high in saturated fat, by eating foods such as eggs or organ meats that are high in cholesterol, by having excess weight, or by smoking (Health Letter Associates, 1991).

Proteins, Carbohydrates, and Fats

Proteins are needed for growth and for the repair of cells. A main function of proteins is to provide the amino acids that are required for a healthy life. We can get proteins from both meat and vegetables. For vegetarians, it is important to combine vegetables with whole wheat bread, rice, and other cereals in order to get all of the amino acids a healthy balanced diet requires and that most people get from meat, chicken, or fish. In general, animal proteins are more nutritious than vegetable proteins because they provide more essential amino acids.

Carbohydrates come from two food groups—the sugars and the starches. They furnish the main mechanism required for the chemical processes called metabolism that take place within the body's cells. Carbohydrates, especially the unrefined or unprocessed carbohydrates, should make up at least half of a normal diet. Unrefined carbohydrates have more desirable fiber than refined carbohydrates such as sugar and white flour.

Fats are a structural component of the cells and provide energy for metabolism. However, we don't need a lot of it. Fats should constitute less than 30 percent of total caloric input, but unfortunately often constitute considerably more. The excess of low-density cholesterol, which is associated with heart disease, seems primarily due to eating too much saturated fat. Most of the saturated fats we eat are found in meats and dairy products. Unsaturated fats, either monounsaturated fats, which are found in olive oil and avocados, or polyunsaturated fats, which are found in fish and vegetable oils, are better for us.

Fiber and Water

Fiber is the indigestible structural material found in plants and is highly desirable. A low fiber diet, which is almost always high in refined carbohydrates and fat often leads to constipation, diverticulosis (usually in the form of small sacs in the wall of the lower part of the colon which can become infected), and other disorders. Low fiber diets also encourage such unwanted effects as obesity and heart disease. A high fiber diet (including plenty of fruit, raw vegetables, grains, and cereals) provides desirable bulk without excess calories and minimizes the occurrence of the complications associated with a low fiber diet.

Water constitutes about 60 percent of our total body and is an absolute requirement for well being. The amount desired is often underestimated. Water is required to maintain metabolism, to maintain normal bowel function, and to maintain a desirable volume of blood. Many foods contain a great deal of water. Because they can be exposed to long periods of low humidity, flight crews are urged to drink relatively large amounts of water. The daily consumption of eight glasses of water is a reasonable goal.

Vitamins and Minerals

The question of required vitamins and minerals is somewhat controversial, especially among lay individuals. While a very tiny amount of 13 vitamins and 22 minerals are absolutely necessary for well-being, an inherent part of today's living is exposure to a very high-powered advertising campaign for the sale of supplemental vitamins and minerals. Most physicians and dietitians state unequivocally that required vitamins and minerals are provided in most normal diets. Many excess vitamins are eliminated in the urine. In fact, a physician friend maintains that Americans have the richest urine in the world.

Vitamin supplements are usually not required, although vitamin supplements, if taken moderately, do no harm. They should not significantly exceed the RDAs (recommended dietary allowances which are published by the Food and Nutritional Board of the National Research Council or the older and simpler US RDAs). The latter are the official standards found on food labels and are the US recommended standards. The vitamins A, D, E, and K can be dangerous if taken in excess, so caution is advised. If an individual believes he/she would be helped with supplemental vitamins or minerals, it is well to consult a knowledgeable physician or dietitian.

The minerals—calcium, zinc, magnesium, phosphorus, and iron—are all needed in appropriate amounts. Most are included in a normal diet, although recently there has been considerable interest in giving additional calcium, particularly to women after menopause, to prevent, or at least minimize, the onset of osteoporosis. While a very small amount of sodium chloride (table salt) is required to maintain fluid balance, adequate amounts of sodium chloride are found in a normal diet without the additional use of the salt shaker. Most Americans consume more salt than their bodies require. This is undesirable because excess salt often leads to hypertension.

All of these considerations also apply to food taken while on duty. Unfortunately, frequently there is often considerably less flexibility in the choice of foods available but the basic rules are the same. Special considerations involving food poisoning while in-flight are covered in later paragraphs dealing with in-flight incapacitations.

Exercise

Exercise and diet are two very important considerations in a professional pilot's career. Both are crucial factors in maintaining physical well being regardless of age. Exercise has a surprisingly large number of benefits. It can improve cardiovascular fitness and muscular endurance. An additional benefit is an increase of energy. Exercise can also significantly reduce the risk of coronary heart disease. It can help lower blood pressure and provide minimal help in reducing weight. A psychological advantage of exercise is that it gives an increase in self-esteem and improves an individual's overall sense of well being. Exercise is good for virtually everyone at every age. Exercise, like diet, is a very large subject and books are written solely about each of them.

The Four Types of Exercise

There are four general types of exercise. The first, and the most important, is 'aerobic'—from the Greek meaning 'with air'. Aerobic exercises such as brisk walking, jogging, swimming, or cycling force the body to continuously take in additional oxygen to meet the body's increased oxygen demands. Aerobic exercises increase cardiovascular and respiratory fitness. A second important general type of exercise is 'isometric' in which one group of muscles exerts pressure against a group of muscles or an immovable object such as a wall and in which essentially there is no movement. Isometric exercises can be very effective in increasing muscle strength but do not aid in exercising the cardiovascular system or increasing endurance. Isometric exercises are sometimes called anaerobic—also from the Greek—and meaning 'without air'.

A third type of exercise is 'isotonic'. In isotonic exercise there is movement while muscle tension remains more or less constant. The body works against external weights or against its own weight. Isotonic exercises increase muscle strength, size, and endurance. Weight lifting or calisthenics, with repetitious movements and little or no equipment, are typical isotonic exercises. The fourth type of exercise is 'isokinetic'. Isokinetic exercise involves both isometric and isotonic exercises. Isokinetic exercises combine strength training with some aerobic exercise and usually require special equipment.

Training Levels

In order to achieve a good training effect, the American College of Sports Medicine suggests performing aerobic exercise sessions of 15 to 60 minutes a day for a minimum of three (preferably five) days a week. Beyond that the College also believes that you should exercise at what it calls its training heart rate. This is sometimes referred to as the target heart rate. The easiest way to determine your training heart rate is to subtract your age from 220 (which is your maximum heart rate), then take 60 percent and 80 percent of that number. The results are the upper and lower end of your target heart rate zone. When you exercise, your heart rate should be somewhere between those numbers.

While there is not always complete agreement on the amount or type of exercise needed, there is general agreement that aerobic exercises are the most beneficial type. When aerobic exercises are regularly performed they help keep elevated blood pressure at normal levels, reduce the risk of heart disease, help control weight, and even raise the percentage level of HDL—the 'good' type of cholesterol. Recreational long-distance walking, running or jogging are ideal aerobic exercises for pilots because they require very little special provisions or equipment other than good walking or jogging shoes. Brisk walking or jogging can be conducted in virtually all locations, and has almost no seasonal limitations. Researchers have found that brisk walking can be one of the better aerobic exercises and it is easy to do. The importance of getting good footwear for any of these activities cannot be overemphasized.

Contrary to much popular opinion, regular effective exercise need not be markedly strenuous and age seems to be no barrier to its benefits. Starting in the 1980s, researchers became interested in moderate exercise and found that it may offer many of the same benefits as vigorous aerobic exercise. A study at the University of Minnesota found that men who routinely engaged in activities such as gardening, strolling, household chores, or bowling had stronger hearts and a lower risk of dying from a heart attack than their sedentary counterparts (Health Letter Associates, 1991). Particularly good news from this study—which included almost thirteen thousand middle-aged men at risk for heart disease—suggests that benefits from moderate activity seemed to stabilize at about an hour's physical activity and that greater amounts did not result in markedly better health. This is not to suggest that a more vigorous program is not more beneficial, but it does very strongly suggest that a small amount of exercise, if it is done regularly, is much better than no exercise at all.

While Not Flying

Most pilots live fairly conventional lives off the job. They have families and children. They are concerned with routine family matters, parent-teacher associations, soccer leagues, etc. in much the same way as their neighbors. The inevitable conflicts with their flight schedule and the adjustments that are required are considered simply part of the job.

Difficulties sometimes arise over fairly conventional happenings. A common one is a newborn infant or sick child who cries all night making it virtually impossible for anyone else in the family to get a good night's sleep. On occasion the pilot has a demanding flight the next day. There is little question that he/she cannot be normally well-rested for their flight after receiving virtually no sleep. The problem can become more difficult if the individual who is sick is the spouse. In most cases this means that the pilot must also function as a nurse for the sick spouse and as a caretaker for the children. None of these responsibilities abrogates the responsibilities he/she has as a professional pilot. This includes a responsibility, which is shared with management, not to fly a given trip in extreme cases.

Another occasional problem occurs when the pilot is due to fly, feels just a little 'under the weather', but still feels good enough to fly. It is very difficult to determine if he/she is experiencing just minor temporary and transient discomfort or if the discomfort is the beginning of a more serious illness. In the case of minor temporary symptoms, there is probably no question that the professional pilot should fly the trip. A complication can arise if the pilot takes an over-the-counter medication to alleviate the minor temporary symptoms. Many over-the-counter medications do a fine job of relieving some minor symptoms and help considerably. However, some of these over-the-counter drugs are illegal to take while flying. The problem with these over-the-counter drugs is the side effects they may create. The side effects may be relatively inconsequential for individuals on the ground and those who do not operate moving equipment, but are contraindicated for pilots. This is one of the many times that pilots need a well-qualified aeromedical physician for advice.

If the pilot is experiencing symptoms of a more serious illness, attempting to fly the trip can well result in severely degraded performance and the strong possibility of being unable to complete the schedule. The situation becomes more complicated if the pilot needs to be replaced at a location where reserve pilots are not available at that location.

No one—pilots or flight attendants—relishes the thought of becoming ill on a layover. When this does occur, it usually involves gastrointestinal distress due to different types of food and/or the quality of either food or water. The personal priority for each of them is to return home and recover. However, the pilot or flight attendant can face a dilemma in that while he/she should not fly if not physically fit, the flight may be delayed or even canceled because the company may have to fly out a replacement if one is not available at the particular station. The correct choice of not flying unless one feels well is an obvious although sometimes difficult one.

It is virtually impossible to suggest anything other than general guidelines for situations such as those delineated in the preceding paragraphs. Obviously, a pilot should not fly when he or she is not fit. To fly when it is contraindicated by one's physical condition is basically unsafe and can result in additional physical problems that would not otherwise have arisen. Not only can this be a very gray area with no clear lines of demarcation, but also there are several obvious conflicting demands. These demands can arise from basic safety issues (which are preeminent), the company's needs, the individual's physiological needs, in some cases peer group pressures, and in some cases a definite financial penalty for the individual involved. As we have noted before, this is one of the many cases where there is a great need for discussion with an aeromedical physician in whom the pilot has confidence and trust.

Substance Abuse

In the air transport industry there are three general categories of substance abuse—illegal drugs, alcohol, and tobacco. Their identification and treatment varies considerably.

Illegal Drugs

The US FAA requires random testing of any employee who performs a safety-sensitive function for evidence of marijuana, cocaine, opiates, phencyclidine (PCP), and amphetamines. The testing is done during designated test periods, which includes specific pre-employment testing and the periodic testing of any individual required to undergo a medical examination under Part 67. This testing is required unless the employer has an approved anti-drug program that is over one year old and that provides for unannounced testing based on a random selection of employees. Employees who must be tested include each person who performs a safety-sensitive function directly or by contract. These include the following employee groups:

- Pilots,
- Flight attendants,
- Flight instructors,
- Aircraft dispatchers,
- Aircraft maintenance technicians,
- Ground security coordinators,
- Security screening personnel,
- Air traffic controllers.

Regardless of the FARs, aviation has no place for individuals who place themselves under the influence of illicit drugs. Any use of such drugs is incompatible with air safety. Even the so-called 'soft drugs' can affect performance, mood, and health. Fortunately, illicit drugs in operational aviation are a virtually non-existent problem. However, if an isolated case of chemical dependence is discovered, specialist help is absolutely required. With such help it is possible that a pilot or other aviation professional can be returned to flying or to regular duty if the recommended treatment is successful. Such treatment is inevitably a slow process.

Alcohol

Alcohol is a socially accepted drug in most parts of the world. It is used to promote joviality and relaxation in spite of the fact that it is a drug that is both powerful and complex. The social use of alcohol is a fact of life in most cultures. While it may be simply a cyclic phenomenon, there is some evidence that the use of strong spirits is decreasing and is being replaced by an increased use of wine or beer. It does not change the incompatibility of alcohol and aviation.

Virtually all airlines have clearly written policies regarding the use of alcohol by aircrew members or dispatch personnel. Most airlines absolutely prohibit the drinking of any type of alcohol (including beer or wine) for at least 12 hours before any sort of flight activity, including the standing by associated with reserve duty. The consumption of any alcohol while wearing any part of a uniform that can be associated with the airline, even if the individual is not associated with flight duty, is strictly prohibited.

Alcohol Abuse

The excessive use of alcohol and its abuse is not always easily recognized or defined. A World Health Organization definition refers to alcohol abuse as being present when the excessive use of alcohol repeatedly damages a person's physical, mental, or social life (Green et al., 1991). However, for our purposes, that definition is too loose, for any significant level of alcohol has no place during aviation operations. Despite sometimes conflicting testimony regarding its attributes, 'alcohol is always a cerebral nervous system depressant'. Flying and alcohol simply do not mix.

A high social level of alcohol intake can be damaging to the individual even before reaching the disease level. Damaging levels have been defined as those that will cause some physical damage to liver, heart, blood cells, or other organs. They are surprisingly low in 50 percent of the population. For women, the levels are probably a third lower than for men. Danger signs of early problems with alcohol include regularly drinking alone, gulping the first drink, having to increase intake to feel good, memory loss regarding events of the night before, morning shakes, feelings of guilt about drinking, anger if criticized, and any adverse effects on family, work, or other social life (Green et al., 1991).

No occupational group is exempt from the illness of alcoholism. Aircrews frequently have a higher risk than others because they are exposed to many of the situations associated with the development of alcoholism as a disease. These include such factors as isolationism, high income, boredom on layovers, easy access to cheap alcohol, being part of a 'drinking culture', a perceived need to conform and be gregarious, the frequent use of alcohol to unwind and as an aid to sleep. One lesson learned early on is that alcohol, even if completely legal, does not really help sleep, but contributes to the poor quality and quantity of sleep that is often obtained on trip nights. There is a very real need to develop other 'unwinding' techniques.

It is now generally recognized that alcoholism is a treatable disease. The first signs usually appear to personal friends or working colleagues. Both are reluctant to identify the problem because in many cases the individuals seem to be reasonably 'sober citizens', and have responsible jobs as doctors, managers, pilots, and the like. A diagnosis of alcoholism can have very serious social and professional consequences, in spite of the fact that it is recognized as a treatable disease.

Prevention Programs

Alcoholism is a serious problem in most societies and more common than it is often perceived. Suspicion of the problem in a friend, colleague, or family member requires prompt, frank, and a positive approach with the knowledge that help is available. In the US, each airline (certificate holder) is required by FAR Part 121, Appendix J to establish an alcohol misuse program. The US ALPA, and other pilot representing groups, have developed very successful treatment programs that are available to their members. The ALPA program has been supported by the Federal Air Surgeon since its inception. Fortunately, treatment, while rigorous, can be effective, and pilots can return to flying if the treatment is successful. Total future abstinence is the only realistic goal for an alcoholic as it is the only sure end point. There is very little hope of a return to controlled drinking for a pilot who has had a serious alcohol problem.

Tobacco

Despite its practice in Western countries for more than 400 years, and even longer in other parts of the world, tobacco smoking has only relatively recently been recognized as a major health hazard. An early exception was England's King James I (1566-1625) who tried without success to banish use of the 'sot weed' from his realm.

Most of the recent research regarding the health hazards has involved cigarette smoking. Data regarding pipe or cigar smoking is less conclusive. Because pipe or cigar smokers usually do not inhale, they seem to have less risk of developing lung cancer than do cigarette smokers. There is no question that their risk is greater than the non-smoking population. Pipe and cigar smokers do have a higher risk of developing cancers of the oral cavity and upper respiratory tract, while the risk to snuff users is mainly associated with a higher risk of cancer of the oral cavity, specifically of the nasopharynx.

Smoking, like moderate drinking, is socially acceptable in many cultures. It is becoming unacceptable in some cultures because the evidence that smoking results in long term heart and lung problems is convincing. An additional problem is the second-hand and deleterious effect on the non-smoker who happens to breathe in exhaled air from a smoker. These individuals are considered 'passive smokers', and they are directly affected if they are exposed to smokers.

In the US smoking is prohibited in the passenger compartments of transport flights. A major reason is the known health hazard from passive or second-hand smoke for passengers who inhale the exhaled smoke from other passengers who smoke. The provision of a separate smoking section does not protect the non-smoking passenger because smoke remains in the recirculated cabin air. Second-hand smoke is considered a serious hazard in the United States. The FAA does not prohibit smoking in the pilot's compartment because of concern during the withdrawal period, although many airlines prohibit smoking by anyone—crewmembers or passengers—on their airplanes.

The objections to smoking are becoming well known as a result of an active campaign to prohibit all public smoking in the US and in many other countries. Increasingly convincing research shows that smoking significantly increases the likelihood of heart and pulmonary disease. Lung cancer is perhaps the best known harmful effect of smoking. It is particularly difficult to treat lung cancer successfully. Presently lung cancer causes about eight percent of all male deaths in the US and about four percent of the deaths of all women (Clayman, 1989).

Other important respiratory diseases associated with smoking are chronic bronchitis, emphysema, and combinations of the two. Typical symptoms are a shortness of breath, increasing breathlessness, and coughing up of sputum. Emphysema and chronic bronchitis cause tens of thousands of premature deaths in the United States each year. Smoking also increases the risk of mouth cancer, lip cancer, and throat cancer. *The Wellness Encyclopedia* dramatically states: 'There is simply no room for debate: smoking promotes heart disease and cancer, and is the major cause of premature, preventable deaths in the United States.'

IATA recently published the results of a survey which states that two/thirds of the business passengers flying world airlines now favor a total smoking ban on all international flights and that another eleven percent favored banning smoking on most flights (*San Jose Mercury News*, 1996). This IATA study illustrates the growing feeling in an important air travel class that smoking should not be a part of transport aviation. In the US, and in some other countries, several airlines have extended the domestic smoking ban to international flights. Smoking is prohibited also on flights on most, but not all, foreign airlines and the trend to ban smoking on all flights is continuing on a worldwide basis.

An additional incentive for airlines to ban smoking is that the tars that are an inevitable part of the air exhaled by smokers collect on air control or exhaust valves and create additional cleaning as well as other maintenance problems on air driven instruments and outflow valves.

The Benefits and Challenges of Not Smoking

One of the most compelling reasons for giving up smoking is that almost immediately the benefits of giving up smoking start to accrue. Cardiac benefits begin virtually with the cessation of smoking and the increased risk of lung cancer and other malignancies begin to decrease steadily. It has been said that one can do nothing for individual health—not even dieting and exercise—that pays dividends as quickly as giving up smoking.

Unfortunately, it is not easy to give up a well-developed smoking habit. Nicotine has been called 'one of the most addictive of all drugs' (Dr. Stanley Mohler, personal communication). M.A. Russell in his review of the subjective and behavioral effects of nicotine has noted: 'Nicotine...displays all the classical hallmarks of an addictive drug.'

Many smokers suffer withdrawal symptoms when they stop smoking. The heavy smoker's withdrawal symptoms can include increased 'tension, depression, irritability, difficulty in concentration, decreased heart rate, a rise in blood pressure, electroencephalographic changes, decreased attentiveness/ arousal, and an impaired performance in tracking and reaction time. These symptoms can occur as early as one hour after the person stops smoking' (Sommese and Patterson, 1996).

A further ramification is that while 'the cohesiveness of the entire flight crew could be affected by an irritable aircraft commander who is being forced to abstain from smoking, lack of crew coordination due to irritability, poor judgment, smoke allergy, or decreased attentiveness by an (other) essential crew member could be detrimental to the flight' (Sommese and Patterson, 1996). The same authors point out that the literature has not considered the nonsmoking crewmember's performance, mood, and health risk when exposed to secondary smoke on the flight deck. This is a valid consideration and an area of equal concern to anyone involved in air safety. Another complication is that it is an obvious mistake to assume that the only heavy smoker in the cockpit is the captain.

To keep the whole problem in perspective, Dr. Donald Hudson, ALPA's Aeromedical Advisor after reading the Sommese and Patterson article, points out 'that the benefits of not smoking far outweigh the minor symptoms that a person experiences from withdrawal from nicotine'. The Airline Pilots Committee of 1976, the Public Citizens Health Research Group, and the Aviation Consumer Action Group have petitioned the FAA to prohibit smoking by flight crewmembers for eight hours before, and during commercial flight operations. However, the FAA has concluded that the submitted data were insufficient and has rejected the petition. Individual airlines have set their own policies, many of them prohibiting smoking in the cockpit altogether.

Fortunately, at least in the US, smoking in the cockpit is becoming virtually a non-existent problem because of social, political, and organizational pressures. A small and isolated but very real dilemma is being faced by both pilots and cabin crewmembers who have no interest in giving up smoking but who cannot smoke at all during their sometimes very long trips. Some of them have resorted to using nicotine patches during the trip to keep them going until they arrive and can puff on the actual cigarette in a designated smoking area. This use of the nicotine patch is obviously a stopgap measure and can create additional health problems.

Other Problems With the Job

Two other problems associated with the job of being an airline pilot that are sometimes not given adequate consideration are obvious and subtle incapacitations and the maintenance of manual flying skills if flying with highly automated airplanes.

Incapacitation on the Job

Incapacitations are a difficult problem to deal with because they are rare and because the general subject is not a pleasant one. However, with the total number of departures and the total number of crewmembers involved, it is not surprising that incapacitation incidents occur. Incapacitation is a problem that should be squarely faced. (Appendix K—The Incapacitation Story, includes examples of actual pilot incapacitations.)

    The incapacitation of a pilot is one of the ten factors that must be considered in the certification of any new airplane or system (FAR 25.1523, Appendix D). Possession of a first class medical every six months is no assurance of freedom from this extremely rare event. The good news is that in a virtually all cases an easy and very effective way to control the potentially adverse consequences of an inflight incapacitation is simply to follow well-designed SOPs, use the 'Two Communication Rule',[1] and follow the four steps that should be taken in order in any incapacitation incident. The four steps are to (1) maintain control of the airplane; (2) take care of the incapacitated crewmember; (3) reorganize the cockpit and land the airplane; and (4) plan actions to be taken after landing.

Obvious Incapacitation

A great many of the maladies to which the general public is subject also happen to pilots. On fairly rare occasions, these occur in flight. The severity of these maladies ranges from fairly inconsequential discomforts to heart attacks causing death.

    'Obvious' incapacitations, as opposed to 'subtle' incapacitations, are

---

[1] The 'Two Communication Rule' states that one should have a high index of suspicion of an incapacitation at any time a cockpit crewmember does not respond appropriately to two verbal communications, or at any time a crewmember does not respond appropriately to any verbal communication associated with a significant deviation from either an SOP or a standard flight profile.

immediately apparent to the remaining crewmembers. They can occur suddenly; can be prolonged; and can result in a complete loss of function.

Perhaps the most prevalent and obvious type of incapacitation involves food poisoning, which can be caused by food eaten before or even in-flight on long flights. A very high percentage of reported incidents happen to passengers. Fortunately, there is some warning to the affected individual. If a crewmember is affected, crew duties can be adjusted to meet operational requirements. Other crewmembers should be immediately informed if an incident of food poisoning or other gastrointestinal upset is suspected. This is no time for secrets.

Airlines are sensitive to this problem and spend a great deal of money and effort ensuring that all food served on their airplanes is safe, although they have no control over food that is consumed before boarding. Virtually all airlines periodically warn their crews about this hazard. In most cases, passenger food and cockpit crew food is prepared separately and crews eat separate meals. Problems in food poisoning are rare, but must be considered because of the potential consequences. Fortunately, established policies and the good sense of the flight crews seem to work well and minimize incidents of this nature.

In spite of all precautions, incidents of crew food poisoning do occur to flight crew, as one did in mid-March of 1984. In this case a foreign airline, and one of the most meticulous in food preparation, found that more than 75 crewmembers who were scheduled to fly long-range trips over a four to five day period were stricken with suspected Salmonella poisoning. A majority of those affected were flight attendants, although some cockpit crewmembers were also stricken. Most of the attacks occurred on layovers because Salmonella symptoms usually do not occur for up to 24 to 48 hours after the contaminated food is eaten. In some cases the crews were unable to staff the flight they were scheduled to fly back to their home base.

The problem of potential food poisoning is taken very seriously by the airlines and by public health officials. Airline kitchens have a particular problem because they have to prepare meals in bulk, which are then stored, refrigerated and reheated. The culprit in the case referred to was suspected to be an aspic glaze on some hors d'oeuvres prepared in the airline's own flight kitchen.

Flight crews should be aware of the problem of potential food poisoning and be particularly sensitive to it when eating in foreign cities where the food is not always similar to the food that they normally eat. Minimal precautions such as the use of bottled water and the avoidance of unpeeled fruit or vegetables that are leafy and uncooked are recommended. Individual digestive systems can be disturbed by the unfamiliar food, fatigue, and the circadian dysrhythmia that often accompany long flights. Native populations develop a natural immunity to local bacteria that cannot be expected to develop in even frequent travelers.

Much the most dramatic 'obvious' incapacitation is a death in the cockpit. Fortunately, a death in the cockpit happens on only very rare occasions. Also fortunately, death in the cockpit very seldom creates a severe operational problem for a well-trained crew. Certification rules require that the airplane can be flown safely with one of the pilots completely incapacitated.

Subtle Incapacitation

'Subtle' incapacitations occur more frequently than the 'obvious' type. They are frequently unreported, partial in nature and usually transient—lasting for from a few seconds to several minutes. They are insidious because the affected pilot may look well and continue to operate but have only a partially functioning brain. The pilot may not be aware of his/her problem, nor capable of rationally evaluating it. 'Subtle' incapacitations are more difficult than 'obvious' incapacitations because there frequently is no warning except that the individual does not appear to function normally.

Jerome Lederer has identified a category of 'subtle' incapacitations he calls cognitive incapacitations. A cognitive incapacitation can create a difficult problem, especially if the cognitively incapacitated pilot is the pilot flying. The problem is how to handle the pilot flying when he/she is, by Lederer's definition, 'mentally disoriented, mentally incapacitated, or obstinate, while still physically able and vocally responsive'. Several airlines have found this incapacitation category quite helpful in analyzing and learning from some of their own incidents where, for example, an unstabilized approach resulted when the pilot flying became overly focussed or inappropriately focussed during the approach. Pilots sometimes refer to this condition as 'tunnel-vision' or say that the individual was 'out to lunch'. Often factors in the pilot's life outside of the cockpit are significant contributors.

Training for Inflight Incapacitations

Training for either obvious or subtle incapacitations that occur inflight is straightforward and relatively simple. It starts with a discussion in ground school that should include discussion of the higher than expected prevalence of in-flight incapacitations. From then on it is simply a matter of using the

Two Communication Rule to detect the incapacitation, and then following the four steps that are required to control the operational consequences of the incapacitation in the simulator and on the line. The regular routine use of SOPs and standard flight profiles are clearly requirements for the system to work.

Pilot Aging

Pilot aging is a particularly complex, difficult, and controversial issue. At least six factors are specifically involved, and none of them are simple. The six factors are the:

1. physiology of aging;
2. basic issue of safety;
3. issues involved in collective bargaining, including long-term pilot union goals;
4. considerations for the operator; including economic and long-term personnel policy goals;
5. philosophical questions raised in the concept of federal licensing and;
6. political considerations involved in public and congressional perception of these issues.

US FARs state that US carriers cannot utilize any pilots whose chronological age is more than 60 years, although such people can fly in air carrier operation as a second officer or flight engineer or as the first pilot in an air taxi operation. Other countries have different chronological age cutoffs. This is a very controversial issue with no general agreement on a specific number. In some countries the chronological age has been extended because of a critical pilot shortage.

Determination of a chronological age cutoff for pilots is sometimes considered primarily a physiological problem which also has a considerable cognitive testing or evaluation component. It is very difficult to make this determination based upon the scientific evidence, and there is very little consensus among aeromedical or aging experts on what chronological age should be established. There is obvious agreement that pilots should not be able to fly forever, but there is no agreement on what the cutoff age should be. In the US, many believe that it should be extended to the normal US retirement age of 65.

Determining the limits of a working career is also recognized by many as a legitimate collective bargaining problem. In the US, both of the parties involved—the management of the airlines and the unions representing the pilots, for quite different reasons support the age 60 regulation that is currently in effect.

In the US, the age 60 cutoff for airline pilots was established as a federal aviation regulation while Elwood Quesada, a retired Air Force general was the FAA Administrator. It was established just before the beginning of the jet era and while nobody really knew, several aeronautical experts expressed serious doubts that pilots approaching age 60 would be able to handle the demands of the new jet airplanes. At this time, the industry was so young that there had been virtually no age 60 pilots. Because the older and more senior pilots were the ones that would be trained first, the initial jets would all be flown by pilots approaching age 60 unless an age cutoff was established. This was also a period of pilot recession among the airlines, and there were many pilots on furlough. Obviously, if the older and more senior pilots were forced to retire, their jobs would become available for the more junior pilots, and those on furlough would be rehired. If a chronological age cutoff was established by a federal regulation and defended on the basis of safety, neither the union nor management would be faced with this very controversial issue.

This book does not attempt to deal with all of the issues that are involved. Anything approaching an acceptable treatment requires that all factors of this controversial question be considered. Such treatment is beyond the scope of this book.

References

Aviation Week & Space Technology (4 November 1996). McGraw and Hill, Inc., New York.

Bartley, S. Howard, and Chute, M.A. (1947). *Fatigue and Impairment in Man*. McGraw-Hill Book Company, New York.

Caesar, Heino (1987). 'Safety Statistics and Their Operational Consequences', Flight Safety Foundation's 40th International Air Safety Seminar, Tokyo, Arlington, Virginia.

Clayman, Charles B. (1989). *The American Medical Association Encyclopedia of Medicine*, Edited by Charles B. Clayman, M.D., Random House, New York.

Davenport, J.K. and Jensen, T.G. (1989). 'Fatigue Factor on Two-Man Crew', *Flight Safety Digest, September 1989*, Flight Safety Foundation, Arlington, Virginia.

Foushee, H. Clayton, Lauber, John L., Baetge, Michael M., and Acomb, Dorthea B. (1986). *Crew Factors in Flight Operations: III. The Operational Significance of Exposure To Short-Haul Air Transport Operations*, NASA Technical Memorandum 88322, NASA Ames Research Center, Moffett Field, California.

Graeber, R. Curtis (1988). 'Aircrew Fatigue and Circadian Rhythmicity', chapter in *Human Factors in Aviation*, edited by Earl L. Wiener and David C. Nagel, Academic Press, Inc., San Diego, California.

Green, Roger G., Muir, Helen, James, Melanie, Gradwell, David, and Green, Roger L. (1991). *Human Factors for Pilots*, Avebury Technical, Gower House, Aldershot, England.

Health Letter Associates (1991). *The Wellness Encyclopedia*, from the editors of the University of California, *Berkeley Wellness Letter*, Houghton Mifflin Company, Boston, Massachusetts.

National Transportation Safety Board (1994). *Aircraft Accident Report, Uncontrolled Collision With Terrain, American International Airways Flight 808, Douglas DC-8-61 N814CK, US Naval Air Station, Guantanamo Bay, Cuba, August 18, 1993*, National Technical Information Service, Springfield, Virginia.

National Transportation Safety Board and NASA Ames Research Center (1995). *Fatigue Symposium Proceedings*, National Transportation Board, Washington, D.C.

Robertson, Fiona A. (1996). 'Go Beyond Technical Correspondence', *Aviation Week & Space Technology,* 18 November 1996, McGraw Hill, New York.

Rosekind, Mark R., Gander, Phillipa H., Miller, Donna L., Gregory, Kevin B., Smith, Roy M., Weldon, Keri J., Co, Elizabeth L., McNalley, Karen L., and Lebacqz, J. Victor, (1994). 'Fatigue in Operational Settings: Examples from the Aviation Environment', *Human Factors*, Human Factors and Ergonomics Society, Santa Monica, California.

San Jose Mercury News (1996). Article published in the 26 October 1996 issue, San Jose, California.

Sommese, Teresa, and Patterson, John C. (1996). 'When a Person Stops Smoking', Article in the *Air Line Pilot*, February 1996, Air Line Pilots Association, Arlington, Virginia.

Taylor, Laurie (1988). *Air Travel - How Safe Is It?*, Blackwell Scientific Publications Ltd., Oxford, England.

16 Selection and Training

The Evolutionary Process of Selection

Early Pilot Testing

The first aptitude testing for pilots started for Army aviators with World War
I. Harvard's Professor Ross McFarland described some of these early attempts
in his *Human Factors in Air Transportation* in these terms:

> Although the tests used in the various allied countries were somewhat similar,
> there were variations in methods and interpretations. In Italy, for example, it
> was concluded that a good airplane pilot is one whose psychomotor activity is
> precise and well coordinated, whose perception is quick, and whose attention is
> constant and well distributed. The French stress the importance of emotional
> behavior, using vasomotor tests and variability in reaction times as a reflection
> of emotional instability. In England, psychological tests were used chiefly for
> determining the effects of altitude flying and for studying "staleness" in pilots
> rather than for giving routine examinations to candidates. Most of the tests
> were concerned primarily with such physiological data as pulse rate, blood pres-
> sure, and vital capacity, but certain psychological implications were involved.
> Volition or persistence, for example, was judged in terms of the candidate's
> ability to maintain a column of mercury by blowing into a manometer.
>
> (McFarland, 1953)

A major reason for the interest in preliminary testing was the high failure rate in training that all nations were experiencing. Training was expensive
both in terms of trainees, instructors, and equipment. The high failure rate in
flight training cost a lot of money and expended scarce resources. A high
percentage of the large number of accidents experienced was attributed to
pilot error. If the attrition rate could be reduced by reducing the number of
training failures and the number of accidents that were attributed to human
(pilot) error, and by reducing the loss of scarce airplanes caused by those
accidents, potential cost savings could be achieved. It seemed quite prudent
to spend considerable effort on pilot selection. It was hoped that a thorough
appraisal of psychological fitness at the time of selection could detect those

who would probably fail in training. It was claimed by many of these early aviation psychologists that established techniques of psychological testing could identify the potential failures.

While the early psychologists were at least partially successful, they faced at least two problems. One was that there was very little agreement, even among themselves, on the characteristics of the ideal pilot. The second problem was that the reliability of the available tests for the purpose of initial pilot selection had been overstated. There was a great deal of difference in the selection approaches used by different countries.

However, in spite of the fact that the testing was far from perfect, early testing in the military forces achieved a significant reduction in the number of training failures. Unfortunately and quite significantly, it did not achieve a comparable reduction in the percentage of accidents that were attributed to human error in either airplane training accidents or in accidents that occurred later in operational flying.

In the intervening years between World War I and World War II, attempts were made by the military (and the airlines) to improve the selection process for pilots and to take advantage of the selection work being done by the military. Considerable emphasis was given to the successful pre-testing that was used by the FAA's newly instituted and university oriented Civil Pilot Training Program (CPTP). This program began in 1938 and furnished a great many pilots to the airlines, armed services, and civilian instructor schools during its slightly more than three years of existence. The CPTP was terminated with the entrance of the US in World War II when all government-sponsored aviation training was done by the military. During World War II, Col. J. C. Flanagan developed the 'Stanine' (Standard 9) composite test battery which was used extensively by the armed services in selecting cadets for aviation. The 'Stanine' test was later used by many airlines for pilot selection.

Selection of Airline Pilots After World War II

The cessation of World War II had at least two major effects on air transport. The first was that additional, and much needed, airplanes suddenly became available, and the second was that a large school of experienced former Air Force, Navy, or Marine pilots wanted airline jobs. Flying as a pilot for an airline was: a job that permitted them to continue to fly; a job that involved desired skills that these pilots already possessed; a job that was relatively highly paid; and a job that occurred in a period when other good jobs were reasonably scarce.

As the airlines grew, the employment selection of all employees became an important responsibility of expanded personnel departments. The selection process varied widely among different airlines. In many airlines, the selection of all employees, which obviously included pilots, became a province of the personnel professionals. This change, from pilot selection being exclusively a responsibility of the flight department to a responsibility of the personnel department, was regarded as less than an optimum change among pilots and their supervisors.

The purpose of the selection process was to select individuals who could not only pass the initial training course, but who would also be good long-term career line pilots. There was a failure during this early period to conduct follow-up studies of the good and poor line performers. This deficiency was recognized, and in an attempt to deal with the problem of identifying good long-term performers, the airlines tried to develop some sort of a rating scheme. However, the operating and personnel departments could not agree on the basic requirements of a good pilot and good employee. Frequently, the line operating pilots disagreed with both. A major airline of that era tried to solve this problem by giving a series of psychological tests to a group of what it considered to be its superior pilots. There was so much variation in the group tested that the selectors finally gave up on this method of selection.

Generally speaking, airline supervisory or check pilots of this period, were not supporters of the psychological tests used during selection. Many of them were old airmail pilots. Several saw the psychologists as a threat and several of them still believed that the only test needed was to see whether or not the individual could manually fly the airplane well. This belief was not true then and is certainly not true today. It takes more than just 'stick and rudder skills' to be a good professional pilot.

During this period a mistake made by many airlines was the assumption that a good military pilot automatically would be a good airline pilot and a good employee. In spite of the fact that this idea was not true, the myth persisted at some airlines until comparatively recently. However, that statement by no means suggests that no ex-military pilots became good airline pilots and good employees. It is an important point for 1) many ex-military pilots became very good airline pilots and employees, 2) some ex-military pilots failed, and 3) many very good airline pilots had no military background. Today, in the US approximately one-half of all airline pilots have a civilian background, the others are ex-military.

Pilot Selection Today

The air transport industry has long recognized that few employment selection programs benefit more from improved selection procedures than do programs for pilot selection. Training courses for pilots are expensive and the costs of failure are high, not only in lost training time and revenue, but also in terms of the increased risk of operational failures and accidents.

While there are still wide variations in selection procedures among airlines, as there always have been, selection procedures have been refined and improved. Today, selection is based upon a broad concept of initial qualifications. The established operating community generally accepts rigorous pre-employment selection, but details are still controversial. A variety of written tests and interviews are used by US airlines. Typical minimum suggested requirements include a four-year college degree, an Air Transport Pilot (ATP) certificate, and the successful completion of the flight engineer written examination.

The college degree requirement is not typical throughout the world, although it does serve as an initial selection stage for most US airlines. Requirements for flying hours as well as requirements for the type of flying vary widely. The requirements also vary widely with the supply and demand of the market. As this chapter is being written, there is a large supply of available pilots in the US. This situation is not true throughout the world and can greatly influence the minimum flying qualifications required by an air carrier.

While requirements are usually less stringent for regional and commuter airlines, educational attainments as well as educational performance are considered important. The pilot certificates required for initial hire can be as minimal as a commercial license and instrument rating. Again, the demand of the marketplace and the available pilot supply play an important role. During early 1996 new hire flight times averaged 5,052 hours at major airlines, 4,375 hours at lesser trunk airlines, and 3,270 at all other carriers. About five percent of all new cockpit crew employed by the majors had less than 2,000 hours, with the lowest-time hired pilots having just 644 hours (Proctor, April 1996). While many training experts believe that total flying time has limited value, most concede that flying time does represent a certain accomplishment and most airlines require at least 1,000 to 1,200 hours. Among the minimum requirements of a major regional airline that is closely associated with a large US carrier are a total fixed wing time of 1,500 hours and multi-engined fixed wing time in excess of 300 hours.

Both regulators and operators now recognize that it is important to select individuals who can work well as team members, and recognize those who, if they do not have the required skills and knowledge, have the ability to acquire them. A great deal of effort has been focused on ensuring technical expertise, but we now know that attitudes associated with crew coordination and personality factors are also important. Obviously, they are important in pilot selection (Chidester et al., 1991).

One criticism of the selection policies of the major airlines is that they 'select people out', judging that a given person did not meet established standards, instead of 'selecting the proper people in'. We believe that this is not a valid criticism and that the criticism is mainly one of disagreement about hiring standards. While pilot hiring conditions vary throughout the world, in the US personnel people normally select aspiring airline pilots from among a very large group of applicants. One major airline had over 2,000 applicants for a pilot job (personal communication, 1996). We would have no quarrel with the belief that the personnel people may not be considering all the important characteristics of today's pilot and that selection policies could be improved. However, there is little question that the personnel people try very hard to select the most qualified of their candidates. There is also little question that the topic of pilot selection will always be a controversial one.

With all that, a healthy skepticism should still be maintained in any review of today's pilot selection. An example of such skepticism is that held by Dr. Diane Damos, who after a brief but critical examination of pilot selection batteries, observed that both their content and low predictive validity have remained remarkably stable over 50 years (Damos, 1995). Her skepticism is reinforced in a later article, the abstract of which notes that: 'First, the vast majority of pilot selection batteries predict training performance rather than operational performance; second, the batteries have low correlation between the predictions and the criterion' (Damos, 1996).

The frustration that is inherent in the selection process is not new and is well illustrated in the wry comment made several years ago by an interested and active pilot representative. While discussing the problem of pilot selection, he said, 'even if we could fire all pilots and start over again, I don't think we could end up with a better group or a better record'. While his opinion may be overstated, it does reflect the cynicism of many with the selection process and skepticism that new techniques will provide significantly better and safer employees.

Evolutionary Change, Training Needs, and Training Costs

Evolutionary change and technological innovation have been a part of air transport from its beginnings. Change always has been an important part of the industry. A basic principle is that successful pilots and successful administrators in air transport operations must be able to adapt to change. Most of the changes that have occurred in flight operations and in training over the years have been evolutionary rather than revolutionary, although sometimes the distinction between evolutionary change and revolutionary change becomes largely a matter of definition. Today, a significant change in airline flight operations is recognition of the importance of the behavioral aspects of human factors in flight operations. Recognition of, and the effective practice of, this new element have become a virtual requirement for airline survival.

It is a mistake to overgeneralize regarding training and training needs. Specific pilot training practices vary by world geographical area, by country, and by airline. It is a mistake not to recognize that there are meaningful differences in the skills and knowledge of individuals, and to assume that pilot training in each airline, or group of airlines, is, or should be, the same.

Changes in both the airplanes, and in the environment in which they operate, have required major changes in training. Both regulatory bodies, such as the FAA, and the airline industry have had to make training adjustments— the regulatory bodies to change the training requirements and to change the way they train their inspectors, and the airlines in the way they train their pilots and in the way they implement the changing requirements in their flight operations. Some pilots, some regulators, and some operators have found it difficult to adapt to these changes.

A complication, and still an inexorable fact, is that training is very expensive. Training is expensive, not only for pilots, but also for cabin attendants, mechanics, and other members of the operating team. In spite of the fact that flight simulators are considerably less expensive to purchase and to operate than the airplanes that were once used for training, simulators have by no means eliminated the real world economics involved in airline training.

There is no question that safety and efficiency are the industry's primary goals. Operators are forced to consider training costs in the same way that they consider other expenses, in spite of the fact that training deserves a higher priority than does almost any other kind of costs. There is often considerable disagreement with the kind and amount of training that is desirable. Frequently this creates a real dilemma for the operators. Invariably, there is considerable difficulty in justifying training that is beyond the minimum level required by regulations. In order to keep a reasonably 'level playing field' for training costs in a highly competitive industry, it is difficult for airlines to go much beyond the training required by regulation, even if an individual airline would like to do so.

Glass Cockpit Airplanes

It has been reported that the three comments most often heard in the new glass cockpits and highly automated airplanes are: 'Why did it do that?' 'What is it doing now?' and 'What is it going to do next?' Whether or not the basic problem is initial design, inadequate training, or a characteristic of the increased complexity of these airplanes is probably a moot question. However, a fundamental truth is that once an airplane is purchased, whether it is a good or marginally designed airplane, it becomes the responsibility of the airline that purchased it to fly this new piece of equipment safety and hopefully efficiently in its day-to-day operations. This can only be done with training that ensures that the pilots who fly the airplanes know them well and do not encounter situations that become a frequent nuisance or create an operational hazard.

The initial transition to a 'glass cockpit' can be a monumental step for pilots who only have flown analogue instrument cockpits (also colloquially known as 'steam-gauge' and 'rope-start' aircraft). Today, many airlines give early general training ranging from a two-day to a four-day indoctrination session to pilots who are transitioning from the older analogue airplanes to the new glass cockpits with their digital instrumentation. The operators and most participants have found this time well spent both in terms of the familiarization process as well as in diminishing anxiety levels about the new technology.

It is unfortunate that the introduction of glass cockpit airplanes was accompanied with the hope that advanced technology airplanes would not only increase safety and efficiency but also would decrease training costs. It was believed by many that the increased automation in these airplanes would reduce the amount of training required.

The record of these new airplanes proves that they did increase safety and efficiency. However, the hope that advanced technology airplanes would not require more training than their predecessors has not been fulfilled. Because advanced technology airplanes require all the old skills and knowledge plus the skills and knowledge required to operate the automatic systems safely and efficiently, training costs have not been reduced. Pilots, who fly these airplanes, almost universally like them but want more training and more systems knowledge. This is not because the airplanes cannot be flown with almost complete faith in the automatics, but because a great many pilots have learned the hard way that any system can malfunction if given sufficient time and opportunity to do so, or if programmed inappropriately by the operator.

Greene and Richmond have written about the difficulty of certifying systems that virtually never fail (Greene and Richmond, 1993). Their concern is certifying a system or component without observing a malfunction and its consequences. With some of today's airplanes, these malfunctions and failures are so rare that they may not ever actually occur during either the design or certification phases.

Ironically, a problem associated with flying airplanes that have automatic systems that virtually never fail, is an almost irresistible human tendency to depend at all times upon such systems. This human characteristic to depend upon reliable systems is a problem that must be considered in an air transport operation that makes 22,000 takeoffs and landings each day in the US and makes many more takeoffs and landings each day throughout the world. With these numbers, even rare events occur.

Therefore, the industry and individual pilots have to be concerned with rare events. The only protection available against these rare events—the adverse consequences of the highly unlikely failure of a remarkably reliable system—depends upon the working pilots. It depends upon effective operational monitoring and the effective routine double-checking of automatic systems. Unfortunately, humans are notoriously poor monitors of rare events and have difficulty with routine, and seemingly unnecessary, double-checking. Yet, human monitoring and the performance of routine double-checking tasks provide the industry's best and really the only protection against the adverse operational consequences of these rare events.

Most pilots have learned that it takes considerable training to use the automatics well because the automatics are complex. Substantial systems knowledge is required. Dr. Charles Billings' Second Corollary of 'Human Centered Automation' is that: 'The human operator must be able to monitor the automated system.' We completely agree. In order to monitor the system, the human operator must know what the automatic system is and should be doing, and recognize its failure if the automatic system partially malfunctions or fails. In order to do this, the human operator has to understand the system and that takes a high level of training. The pilot needs to know more than just which button to push to start or to change a given automatic mode.

Today's Training

Training is important and sensitive because it furnishes the primary interface between aircraft manufacturers, the companies that buy transport airplanes and the pilots who must fly them. Training also is the interface with the environment in which these airplanes must be operated. That environment has become increasingly complex. The airplanes themselves have become both more sophisticated and more expensive (Buck, 1995). The interfaces—between the manufacturers, the airline companies, the pilots, and the environment—as well as the pilot training associated with them—are critical factors in the safe and efficient operation of air transport (Orlady, 1993).

Airline pilot training programs have undergone important and largely evolutionary changes in the past two decades. Traditionally, airline pilot training had been concerned only with developing and maintaining technical skills. Today, meaningful behavioral skills are included and are considered critical.

Virtually all organizations involved in training realize that flying an air transport airplane in today's environment is a team task. This was not always true. As we have already noted, former FAA Administrator Allan McArtor put it pragmatically and succinctly in August of 1987 when he told a group of airline executives: 'Individual pilots don't crash, flight crews do.'

The role and responsibility of the copilot has been substantially increased and the importance of effective monitoring of the pilot-flying (PF) by the pilot-not-flying (PNF) is now stressed at all levels, especially when the copilot is the PNF. Equally important changes in present thinking are clear determination of PF and PNF duties and the almost universal recognition that any crewmember (even a captain) can make a reasonably rare and inadvertent mistake. Sound operation requires that such mistakes be promptly identified and that their operational consequences are minimized. This has both design and training implications. It represents a change toward an error management orientation in both training and operations.

Crew Resource Management and Training

Crew Resource Management, which was discussed at length in Chapter 13, has been defined as 'the effective utilization of all available resources—hardware, software, and liveware—to achieve safe, efficient flight operation' (Lauber, 1984). CRM is a philosophy of operation and a body of knowledge that is an essential part of all flight operations. It embodies such concepts as the 'the team approach', the expanded or enlarged role of the copilot, and the 'crew concept', in addition to the effective utilization of all other resources available to the crew. CRM principles should embody the way that the airline wants its flight operations conducted. These principles obviously must be an integral part of the entire training curricula.

Four Basic Types of Training

The airlines have four basic types of training. The first training is for new hires. It may typically last two weeks and indoctrinates the new hire on company, procedures, practices, and policies before they begin a specific aircraft transition course. The second type of training is transition or conversion training on a specific aircraft. Its objective is to teach pilots how to fly a different airplane safely and efficiently in the company's line operations. A third type of training is usually called upgrade training. Here the purpose is to prepare the pilot for a change in position, such as from copilot to captain. In some cases upgrading and conversion training are combined.

Finally, the fourth type of traditional airline pilot training is recurrent training. It affects all pilots and in the US is mandated by Federal Aviation Regulations at specified times—traditionally every 6 months for captains and annually for copilots. Its purpose is to ensure that the pilot has maintained his or her proficiency and the skills and knowledge required to fly as an airline pilot on a specific type of equipment. A secondary purpose is to reinforce proficiency in the equipment involved and make sure that the trainee has, or will acquire, the latest operational knowledge regarding that equipment. In recent years, in line with a gradual upgrading of the copilot's responsibilities, differences between captain and copilot checks have been considerably lessened. In many other countries, checks have always been close to identical for all crewmembers. Reexamination of recurrent training requirements is a part, and sometimes controversial part, of the FAA's Advanced Qualification Program (AQP). AQP is further discussed in later sections of this chapter.

Feedback Loops and Transfer of Training

Very little learning or effective training ever takes place without knowledge of the results. Knowledge of the results is called feedback. Feedback reinforces good responses and allows the individual to correct or eliminate those responses that are not effective. Errors increase with ineffective training.

Initially, feedback is obtained through the visual or auditory channels. These channels take up most of the information processing capacity of the single channel, limited capacity of the learner. Later, as learning progresses, some of the feedback can be obtained through sense receptors in the muscles, tendons, and joints (proprioception). One of the best examples of this is seen in the skilled typist who can type rapidly while reading from an appropriate draft. The skilled typist receives proprioceptive feedback from the fingers while reading from the draft.

Feedback is an important ingredient in other activities. When feedback is delayed, the limitations of basic memory play a larger role in ultimate behavior and the chance of error is increased. Prompt effective feedback should always be provided if it is at all possible. Effective feedback increases accuracy and can be a powerful motivator of good performance.

Several laws of learning are easier to understand if we remember that the brain acts as an information processing system. If a response is learned to a specific set of stimuli, and then additional learning takes place, the original learning can help or hinder the learning of the new material. If identical or very similar stimuli are intended to elicit essentially the same response, almost total transfer of training takes place. Very little new learning is required. This is an example of a positive transfer of training

In contrast, if new and sometimes difficult responses are to be made to old familiar stimuli, significant interference can result. Under these conditions, one can be sure that the new learning will take longer and more errors will result. The originally well-learned (but now inappropriate) response may occur, particularly in times of stress. This is a very common problem and is called negative transfer.

Negative transfer is particularly troublesome if the old and inappropriate response has become an automatic reaction to what actually are identical or very similar stimuli. A good example can be an engine fire where the fire indication is identical or nearly identical in two different airplanes but where a different procedure is employed in each airplane. Under these conditions, it is highly desirable that the procedures for both airplanes are made identical. If they are not, procedural errors become almost inevitable if the same pilots fly both airplane types. Small differences in procedures are more troublesome than large differences. Negative transfer is a particularly important consideration in transition training.

Ab Initio Training

A fifth type of training, although it is not a part of traditional airline pilot training in the US, is ab initio[1] training. Several European carriers already have successful programs, and some of them have had successful programs for several decades. Lufthansa,[2] Aer Lingus, British Airways, Japan Airlines and SAS have been pioneers in ab initio development. Other non-US airlines have also used the concept, with considerable success. With the desire of most nation States to have their airlines flown, to the maximum extent possible, by indigenous pilots, ab initio training is bound to increase.

There are several advantages to ab initio training. While there are variations in programs, the good ab initio programs include training in airline culture, aviation human factors, and multi-crew operations, in addition to standard instruction in the fundamental skills of an aviator. The traditional belief that total flying hours are a reliable indication of a pilot's level of skill and competency has been long dispelled. Proponents of ab initio training believe that pilots trained from the start for an airline career are more likely to succeed in that role and serve their airline longer and more productively than are pilots that adapted to that role later in their flying careers (Odegard, 1994).

The Aer Lingus Ab Initio Program

Airline ab initio programs can be meticulous and extensive. For example, Ireland's Aer Lingus begins with a careful initial selection process. Chosen individuals are then given a 14-month full-time ab initio training course that includes an augmented commercial pilot license and instrument rating ground school. They are also given 200 carefully supervised flying hours. On completing the initial program and first entering the airline, these students then are given a six-week generic training course which is used to provide a 'bridge' between the ab initio training they have received and the airline operations. The 'bridge' training consists of three weeks of ground-school and twelve four-hour simulator sessions using a simulator operating in a 'generic twin-jet' mode. Following this, the fledgling airline pilots go directly to Aer Lingus

---

[1] Ab initio is a Latin phrase meaning 'in the beginning'. In this context it involves an airline hiring persons for a pilot job, who have no, or very little, piloting experience, and then teaching them the flying and other desired skills required by the airline.

[2] Lufthansa Airlines started a serious and organized ab initio training program in the mid-1950s when it found that there were few available pilots in Germany. It developed its very successful ab initio training program in 1956 at its *Verkehhrsfliegerschule* (transport pilot school). To take full advantage of good flying weather all year, much of its flying now is conducted at its contract facility in Goodyear, Arizona.

specific type transition training. On Aer Lingus/Aer Lingus Commuter airlines, this training can range from training for the Saab 340 to training for all variants of the Boeing 737, depending upon the needs of Aer Lingus at the time of the students' graduation.

A key to the success of this program was the development of a standardized system of airplane operations that could be used for the entire Aer Lingus fleet. This included the development of a generic normal and non-normal initial checklist suitable for airplanes that ranged up to airplanes such as the Airbus 320 and the B-747. The use of these generic checklists was an important part of the ab initio program. Students learn Aer Lingus procedures from day one, learn them thoroughly, and learn to use them routinely.

Not surprisingly, the Aer Lingus ab initio program has been highly successful both operationally and economically. Captain Neil Johnston, in a highly recommended paper, has noted that 'type-specific instructors indicate that these trainees have markedly improved cockpit management skills in comparison to their predecessors' (Johnston, 1992). The fact that an eight-session simulator type transition training is routinely achieved successfully by 200-hour ab initio pilots is testimony to the effectiveness of the program. Equally impressive is the fact that Aer Lingus, which has a very good safety record, has had virtually no failures in their program. Almost one hundred percent of the individuals they selected and who have completed the course are now successful Aer Lingus pilots.

Ab Initio Programs in the US

There is a high probability that the seemingly inexhaustible supply of experienced potential airline pilots in the US will not last forever. That fact, coupled with the present rosy forecasts of the industry's future and declining estimates of the number of military pilots, have prompted several US universities and some airlines to become interested in ab initio programs.

Two successful ab initio programs in the US are a specialized four-year program at Embry Riddle University and the four-year ab initio pilot training program called Spectrum. Spectrum was developed jointly by the University of North Dakota (UND) and by Northwest Airlines. Some other universities such as Western Michigan[3] have similar programs, but most university programs, including Embry Riddle's, are not tied to a specific airline unless they

---

[3] Western Michigan University has just announced plans to acquire an advanced 737-400 flight simulator for its ab initio airline pilot training program. At present it has contracts to train cadets for Aer Lingus, British Airways, and Emirates Airlines.

have a contract with an airline.

Under the Spectrum program, Northwest Airlines agreed to select students in the University of North Dakota Aerospace Spectrum Program to fill pilot positions. Participating students were chosen as they entered their sophomore year and the school, and the airline then groomed them for a career at Northwest throughout the period they were at the university. The program included a rigorous screening process that was overseen by the airline in order to ensure that students are meeting the required airline standards. Presently, the program is not being implemented because of the abundance of qualified pilots already available.

The University of North Dakota also furnishes pilots for several regional airlines and has developed special courses for China airlines. They may or may not include ab initio training. (Odegard, 1994). Other US universities, such as Western Michigan University, are organizing training programs for specific foreign airlines. All of them anticipate a growing US airline need for ab initio pilots in the future.

Several other universities, who are members of the University Aviation Association (UAA), have ab initio programs whose contents vary widely. Many of them are not necessarily aimed at producing prospective airline pilots. The UAA membership includes 55 institutional members that furnish associate, baccalaureate and graduate programs in aviation. Its 1994 *Collegiate Aviation Guide* contains a listing for institutions in 46 states, the District of Columbia, Puerto Rico, and Canada. The *Guide* has more than 280 entries of institutions that have aviation offerings ranging from academic completion certificates and associate degrees to doctoral programs.

Line-Oriented Flight Training (LOFT)

As the name implies, LOFT is a form of training that utilizes full-mission simulation. LOFT training does not include any formal checking, and there is little doubt that removal of formal checking or evaluation considerably enhances the training potential of the session. A full LOFT session begins as a simulated line trip—with dispatch, load planning and any other relevant paper work included. Specific problems are introduced in the simulator as the flight continues with no 'simulator freezes' or other interruptions.

Unfortunately, there has been some confusion between LOFT and other types of full-mission simulation. Captain Tom Nunn, a LOFT pioneer, made the distinction between the two very clear at a NASA/Industry Workshop held by NASA's Ames Research Center in January 1981 (Lauber and Foushee, 1981). Among other things he stated:

> LOFT is *not* full-mission simulation. LOFT *utilizes* full-mission simulation to create a real-world environment but full-mission simulation has many uses beyond original LOFT concepts. Traditional LOFT is entirely a training concept. Full-mission may be used as a vehicle for experimental evaluations and other purposes. The primary thrust of LOFT is not specific procedure training and (it) is certainly not intended for flight checking. A proper distinction between any type of full-mission and LOFT must be maintained.

Modifications and improvements in the implementation of LOFT programs should be and are continuing to be made. A modified form of LOFT is line-oriented simulation, LOS, which involves only segmented parts of a LOFT scenario. It still maintains the LOFT concept and has no evaluation. LOS is frequently used for training in special situations and to avoid long unproductive simulated periods at cruise. As discussed in later paragraphs of this section, the shortened segments of LOS can be used for checking and evaluation but then they are separated and referred to as line-oriented evaluation (LOE). It is important that the LOE sessions used for required checking and evaluation are kept entirely separate from the LOS or LOFT training sessions.

LOFT has a great deal of conceptual support within the industry, but there is not always agreement on some of the details that should be included in a LOFT program. The 1981 report of a NASA/industry workshop on line-oriented flight training still is required reading for any airline interested in implementing LOFT (Lauber and Foushee, 1981), in spite of the fact that there have been changes and improvements in LOFT programs since that time.

LOFT Briefings and Debriefings

LOFT sessions are videotaped at most carriers so that they can be played back for the crew at debriefing. Using the videotape as a tool, a skilled instructor can involve the entire flight crew in an effective debrief and discussion of flight crew performance more easily than is possible without the realistic feedback provided by the video. The videotaped feedback enhances the potential of the training exercises because it gives a positive, convincing, and unmistakable record of the crew's performance. The entire crew becomes aware of the ultimate consequences (good or bad) of their actions.

In many countries, including the US, videotapes of the LOFT session are erased at the conclusion of the training sessions to be sure that performance during a LOFT will have no later checking or disciplinary ramifications. Under these conditions, it is much easier for pilots to view a LOFT session as entirely a learning situation.

However, in other countries there is a belief that routine erasure at the conclusion of a LOFT exercise can diminish the training potential of that session—that lessons learned can be reinforced and that videos of specific LOFT sessions should be given to the pilots involved and even shared with other pilots. One of the obvious potential problems with saving a videotaped session is adverse litigation to the pilots or the carrier. This is particularly true in a litigious society such as the US. For example, one US carrier was subpoenaed (unsuccessfully this time) by an attorney for the training tapes of a particular pilot involved in a crash. Such liability issues are further discussed in Chapter 18—Non-punitive Incident Reporting.

One of the areas in which improvement still is needed is in the quality of LOFT briefings, both initial and postflight, and specifically in getting more briefing leadership from the flight crew. Pre-flight and post-flight debriefings by the flight crew are an important part of LOFT. As flight crews become more used to discussing their performance, the quality of their debriefing increases. This is, of course, a LOFT objective.

A real and practical problem, especially for short-haul crews flying many segments in a single day, is that much of the recommended briefing for each leg becomes a nuisance, and almost boringly redundant in the real world on the line. Because giving the same full briefings does not meet the practical needs of their trips, pilots on the line consequently often do not give them except on the first segment of the day. This inconsistency can create a training problem for the pilots and for the LOFT instructors. The pilots are asked to fly the simulator in the same manner as they fly on the line, and the instructors are asked to reinforce the importance of all pre-flight and post-flight briefings. On the line, the extent of the briefings the Captain gives the flight attendants on short-haul flights usually depends on whether or not it is the first leg of the trip and whether or not the flight attendants are a new part of the crew. With a continuing flight attendant crew, repetitive briefings on each leg of short-haul trips become both redundant and a nuisance to everyone concerned.

Generic LOFT Problems

There seems little doubt that the air transport industry has entered a new era. Among other things, the new era has involved changes in general operating philosophy and changes in training. New operating philosophies must be taught and LOFT, which has been very well received, is a part of the new training. To deal effectively with these changes the Air Transport Association has formed a LOFT/Instructor Focus Group which is preparing a consensus paper to provide guidance to training instructors in facilitation of LOFT and related topics, and in the non-LOFT evaluation of crew performance. Because LOFT principles involve new concepts, instructor training is a very important part of LOFT development. These principles involve a major change in the way that flight crews and flight instructors historically have been taught and in the way they have operated for sometimes a great many years. Two NASA Technical Memoranda discuss these proposals and this area. They are references for this section and are worth reviewing (Dismukes et al., 1997, and McDonnell et al., 1997).

The Problem of Evaluation in a Non-jeopardy LOFT

Any flight operations training program has two basic problems for both the regulators and the carriers they supervise. The first is to ensure that the flight crew is capable of handling all of the problems it might reasonably be expected to encounter. It is not easy to develop appropriate LOFT scenarios that meet all of these conditions.

The second challenge, which is an entirely separate problem, is the management and disposition of a crewmember whose performance during the LOFT is not satisfactory. The purpose of the LOFT program is training and not checking and it is essential that the pilots in the LOFT program view it as such. Although the distinction can become very gray, when LOFT is integrated with the 'training to proficiency' concept, which is also becoming an accepted training philosophy, a LOFT exercise should not include any type of formal check or evaluation.

There is little question that a more productive training situation is created if checking is not a part of the LOFT process and if the pilot views the training as a 'no-jeopardy' situation. Many airlines simply require more training if performance has not been satisfactory during the LOFT. In most cases the additional training is done in a line-oriented simulation (LOS) that emphasizes problem areas. This is obviously a form of evaluation, but it does not involve formal checking. Then, in order to meet the moral and regulatory requirements of the real world and still retain the advantages of LOFT training, LOS sessions are usually followed the next day by a line-oriented evaluation (LOE).

At one airline a pilot not performing satisfactorily receives extra simulator training and is then also assigned to 'special tracking'. Instead of returning for recurrent training in one year, the pilot will return in six months. Although one training department manager has advocated that this tracking should be viewed positively as simply 'special handling', it is not usually viewed as such by the pilot needing to return. The dilemma, of course, is that this additional training is a very tangible consequence for the pilot that resulted from a theoretically non-evaluative situation. The fact that additional training was required and that there was not a 'bust' or a failed checkride, is a narrow distinction. Previously there was no such thing as a non-evaluative situation and such performance was simply considered a 'bust'. This is obviously a new approach to a difficult and unpleasant problem.

Checking is definitely a part of any LOE. It frequently is not easy for pilots in this process to consider the two sessions entirely separate. Even without considering the FAA's and each company's moral responsibility, there is no way to avoid the FAA requirement for periodic checking. The LOFT and LOS system provides training sessions before the required evaluations (LOEs). The system works well as long as the training is adequate for the pilots involved.

Training for Behavior in an Operational Context

While the industry is concerned with both technical and non-technical behavior, one of the goals of today's training is to deal effectively with what is called cognitive operational behavior.[4] The term at first sounds a bit like 'psycho-babble', but nevertheless is a useful concept used to help better understand the behavioral aspects of the operation of today's transport airplanes. Cognitive operational behavior can be divided into three categories—skill-based, rule-based, or knowledge-based behavior (Rasmussen, 1987). More than one behavior category can be involved in a single instance. These are important training considerations that vary depending upon the training objective and the characteristics of the individuals to be trained.

Skill-based Behavior

Skill-based behavior is characterized by smooth, automated and highly integrated patterns of mental and physical activities that take place without conscious control and with smooth, automated, and highly integrated patterns. There is almost complete knowledge of the activity and of the environment in

---

[4] Cognitive behavior denotes knowledge, perception, or awareness (or the lack of them) of the behavioral area involved.

which it is involved. With skill-based behavior, responses are usually straight-forward.

Rule-based Behavior

Rule-based behavior is typically guided or directed by a stored rule or proce-dure. Rules may be derived empirically through experience on previous occa-sions or a rule may be the product of conscious problem solving and a man-agement decision. Rules are usually communicated as an instruction or as a revision to a manual. A good example is in the use of Standard Operating Procedures (SOPs) which are stated in flight operations or equipment manu-als. Rules are usually communicated to pilots as either an instruction or as a revision to a manual. SOPs are often revised because of an accident or inci-dent.

As we have previously noted, it is the nature of an air transport opera-tion to be a rule-based operation to the extent possible. In most cases the operation is fairly fixed. When new situations arise or are anticipated, or when the airline flies into a new airport or area, flight operations management pro-mulgates the operating rules it believes will be necessary. At that point the operation should become simpler for all concerned. Rule-based operations are further discussed in Chapter 13—Crew Resource Management (CRM) and the Team Approach.

The industry practice to issue a new rule to deal with the stipulated causes of accidents or incidents satisfies the perceived needs of the operators, the regulators, and the investigators. In many cases, it also creates new training requirements. We have learned that the inevitable result of this rule prolifera-tion has been the growth and expansion of Flight Operation Manuals (FOMs). As the industry grew and as accidents or incidents occurred, Flight Opera-tions Manuals gradually grew in size. This is a reactive process and it in-creases an already strong tendency to make an air transport operation a rule-based operation. While training tries to cover the most important rules, nei-ther training, nor even the rules themselves, can cover all contingencies. A thick FOM does not eliminate the responsibility of the pilots to use common sense and good judgement.

Knowledge-based Behavior

Knowledge-based behavior is used when there are no rules or any established know-how on how to reach a well-defined goal. It involves general background and sometimes 'special' knowledge. Knowledge-based behavior is used in combination with skill-based or rule-based behavior. Frequently it is used in high-level decision making.

There are gray areas between the categories of rule-based, skill-based, and knowledge-based behavior. In specific instances it is not always easy to categorically decide which of the three are primary. Most airline flying assumes a high order of skill-based performance, and it is in this context that the relevant rules are applied. As we have noted, more than one category can be involved in a single instance. During training, control frequently moves from the knowledge-based or rule-based level towards the skill-based level as familiarity with a specific scenario is developed. Individuals are traditionally encouraged to stay within the rule-based area despite the fact that sometimes the actions required are routinely familiar and have transitioned into a skill. This is because a rule-based procedure is an acknowledged safe procedure and is a sure way to minimize the possibility of the pilot, or even the carrier, encountering legal or regulatory trouble.

Many events require combinations of all three categories of behavior. For example, a crosswind landing may involve both rule-based behavior (rules regarding limits and general manipulative rules) as well as a high level of skill-based and even knowledge-based behavior in order to successfully make difficult crosswind landings. In *New Technology and Human Error*, Professor Reason has related this classification to errors by explaining that skill-based errors are almost always a slip, and rule-based or knowledge based errors are almost always a mistake. Chapter 9—Man's Limitations, Human Errors, and Information Processing— discusses the difference between slips and mistakes. Readers who are interested in further ramifications of this technology are urged to study Professor Reason's latest book, *Managing the Risks of Organizational Accidents*.

Behavioral Combinations in the Real World

A very high order of a combination of knowledge-based, skill-based, and CRM trained behavior was used by Captain Al Haynes and his entire flight crew on United Airlines Flight 232 on 19 July 19 1989 in their Sioux City, Iowa accident. The flight, enroute to Chicago, made an emergency crash landing at Sioux City, Iowa after the fan section of the number two (tail) engine in their three-engined DC-10 failed in such a way that there was a total loss of hydraulic pressure in all three of the airplane's hydraulic systems. One result was an almost complete loss of flight control. As Capt. Haynes stated in an interview for *The Nuclear Professional,*

> Up there, we found ourselves in a whole new world. None of us had ever been in that type of situation. No simulator exercise had prepared us. Loss of hydraulics just wasn't supposed to happen. We had no procedures to follow.

Indeed, there were no formal rules or procedures that covered the difficulties encountered by the crew. The aircraft touched down with very little control at a speed of 215 knots (much greater than normal landing speeds) and broke apart as it impacted the runway. Of the 296 passengers who were onboard, 112 were killed. The knowledge- and skill-based behavior of the flight deck crew and cabin crew, and their crew resource management behaviors, made it possible for 184 passengers and crew to survive. The entire flightcrew was inducted into the Smithsonian Hall of Fame as a result of their performance.[5]

Stanley Stewart's *Emergency: Crisis on the Flight Deck* is devoted to examples that show exemplary pilot performance in situations that are not covered by the training pilots have received or by routine rule-based courses of action. Many of them are not well known outside of some esoteric aviation circles. Perhaps unfortunately, the industry makes very little effort to publicize such incidents. This may be to a considerable extent because they are not always reported accurately and in perspective.

Regulatory Aspects of Training

There has been a growing consensus among regulatory authorities and training experts, both within and outside the industry, that regulatory requirements and the training practices that must be based upon them, have not always kept up with the advanced technology found in the latest airplanes and with our changing environment. Regulatory authorities are particularly important because they set the minimum standards for all airlines that they regulate. In many countries, including the US, the regulatory authorities approve all training programs.

---

[5] The United Airlines Flight 232 crew consisted of Captain Al Haynes, First Officer Bill Records, Second Officer Dudley Dvorak, and Captain Dennis Fitch, a DC-10 flight instructor who left his seat in the cabin to assist in the cockpit after the malfunction occurred. The cabin crew consisted of First Flight Attendant Jan Brown and 8 other flight attendants. All of them performed superbly under very difficult circumstances.

Most regulatory authorities recognize the basic problem and the necessity to keep regulations up-to-date with changes in the operating environment and with advancing technology. Different authorities may take different approaches to specific problems, some of which have political ramifications, but virtually all of them recognize that required training should be sensitive to changing needs. The US FAA's Advanced Qualification Program (AQP), which is discussed below, is an example of one authority's approach.

Many of today's standard and required curriculum items did not exist in the previous decades. These include such things as Upset (sometimes called Advanced Maneuvers) training, CRM training, GPWS training, TCAS training, WSAS[6] training, etc. Advances in hardware and software require that pilots practice maneuvers and conditions that reinforce the lessons learned in training for this new equipment. Many of them have required changes in the regulations.

The FAA's Advanced Qualification Program (AQP)

In the US, an innovative and official advance in pilot training has been the FAA's Advanced Qualification Program (AQP). Details are specified in FAA Advisory Circular 120-54—Advanced Qualification Program. The AQP furnishes a regulatory attempt to both modernize training and to keep required training sensitive to changing needs. It represents a major regulatory agency's effort to encourage maximum flexibility in training in order to take full advantage of individual skills and knowledge, to recognize that traditional training may no longer be adequate or efficient, and also to meet the ever-increasing costs of airline pilot training.

Of course, all of this must be done without derogating the agency's basic responsibility for safe and efficient public transport by air. In order to furnish the FAA detailed proof of a new program's effectiveness, stringent rules and documentation of the results are required for any airline instituting an AQP. Each AQP is evaluated as a separate program. For example, an airline may have an advanced qualification program for its B747-400 training, but not for its B737-300 program. At present, participation in an AQP is voluntary and on a specific fleet basis on an individual airline. Considerable initial work is required by a carrier to generate an AQP. It requires detailed analysis, a great deal of documentation, and very close communication with appropriate FAA personnel. Today, nearly all major airlines and a growing number of regional airlines participate in the program.

---

[6] These acronyms in the text are discussed and described in the Glossary.

The AQP objectives have the solid support of the ATA Task Force on Human Factors, of ALPA and APA, and of many others. It is completely consistent with the broadly supported *National Plan to Enhance Aviation Safety Through Human Factors Improvements* (ATA, 1989, Foushee, 1990). While consensus has not yet been reached on all of the details required for an effective AQP, its importance in the field of airline pilot training can hardly be overstated. In 1990 the FAA promulgated Special Federal Aviation Regulation 58 (SFAR-58), which spelled out the details required to permit airlines to participate in this innovative program. An excellent overview of the entire AQP by Thomas Longridge is located on the US FAA's internet website at http:///www.faa.gov/avr/afs/tlpaper.htm. In addition, the FAA web site devoted to AQP is listed in Appendix Q.

Individual Needs

Inevitably, there will be a greater effort to meet individual needs as a part of the move toward the 'training to proficiency' concept that is being accepted in more and more operating circles. While these individual needs may or may not be job-related, they do affect pilot behavior and the efficiency of the training process.

Individual needs can vary by time. They also vary depending on recent and total experience, motivation, airline philosophy (particularly in cases of airline sales or mergers), individual capability, an elusive characteristic sometimes called 'computer literacy', and many other things. Effective training to meet these sometimes different needs can be important and economical.

In a similar fashion, accurate and specific diagnoses of the individual's need for special training and for the type of training needed can be very difficult to achieve. Everyone loses if an airline pilot's career is unnecessarily terminated. The FAA's AQP may provide more flexibility in meeting individual needs than is presently feasible under most conditions.

While the 'training to proficiency' concept deals with individual needs, it would be a mistake to believe that it will eliminate all proficiency problems, for it obviously will not. Unfortunately, there have always been a very few pilots who were unable to adapt to major changes in airline operations or who for some reason are unable to demonstrate acceptable performance. Because these pilots have sometimes had long and successful careers, it is not easy to find fair and responsible solutions to the dilemma they present to themselves, to the pilot union, and to airline management. The only answer may well be to put these very few difficult and apparently permanent proficiency problems back into the gray area of the collective bargaining arena where such problems can be best handled and probably belong.

Training Roles of the Manufacturer, the Airline, and Training Organizations

Manufacturers, airlines, and specialized training organizations all play an important role in airline pilot training. Their roles will be discussed in the following paragraphs.

The Manufacturer's Role

There is a growing recognition that the general training requirements and the skills and knowledge that will be required to operate new aircraft and systems safely and efficiently should be considered in the design process. The manufacturers of a new airplane or a new system are in a good position to know what will be required to operate their new airplane or system safely and efficiently in the environment in which it will be used. An important consideration for the manufacturer is the amount and kind of training necessary for the pilots who must fly the new airplane or system and for the mechanics who must maintain them. General training needs should be considered at the design level, even if training requirements need not be specific at that time. If the airplane can be established as simply a modification of an already certificated and approved type, substantial savings in certification and in training for the airlines are available. For completely understandable reasons, this also becomes an important airplane sales factor.

An expanded training role for manufacturers is required because the training capability of the buyers of advanced technology aircraft can vary considerably. Pilot and mechanic training has become even more complex and important. The training process provides the operational interface between the manufacturer, the regulator, and the operator. Some airlines may not have the facilities (trainers or training equipment) required to operate the new equipment safely and efficiently. One way that the manufacturer can be assured that the high-quality training necessary for a good interface between the manufacturer and the purchaser is always available is to provide it themselves. This capability solves a very real problem for some airlines.

Understandably, aircraft manufacturers provide initial training for all airlines that buy their transports. The larger airlines usually have the initial training for new airplanes done by the manufacturer for instructors and check pilots and then, because they have the requisite training facilities, continue by doing their own training.

The Airline's Role

Pilot training has implications for both pilots and managers because the functions of each group are absolutely crucial. The safety and efficiency of the airline's operation depends, to a considerable extent, on the efficiency and effectiveness of its training as well as the integrity of its flight operations management. The overall training aspect of airline operational management should be viewed in this context.

Smaller airlines do not always have the training capability and the expensive simulators that are required for today's training. Therefore, they frequently depend upon the manufacturer, another airline, or a separate training facility to train all of their pilots on new equipment. Some of these airlines even use the training and checking capability of these other facilities for their recurrent training. However, this does not reduce their basic responsibility for the safety and efficiency of their operations.

The Regional and Commuter Airlines

Because of an increased concern for commuter safety in the US based on a record that was somewhat distorted by several critics, it was announced, as we noted in Chapter 2—The Industry and its Safety Record, that after December of 1995 the US will have 'one level of safety' for both commuter and major airlines. Reduced to its simplest terms, the new regulation details the requirements necessary to bring 10-30 seat aircraft up to the same Part 121 standards used by airlines flying larger aircraft. While many commuter airlines have operated, in whole or in part, under Part 121 standards for many years, there is no question that the new regulation has serious training ramifications for many regional and commuter airlines (Regional Airline Association, 1996).

The Bottom Line

Airline operating philosophy, policies, procedures and practices (the 4 Ps which are discussed in Chapter 6—The Social Environment) are key ingredients in their day-to-day operation. The implementation of all four elements should be an integral part of training and, as in a good CRM program, this is an important part of total training. The teaching of the company's general policies and of its specific operating philosophy should be a normal part of training in all fleets.

Operational management, training officials and personnel, and the line pilots are all concerned with the specifics of company policy and of basic operating procedures as they are established by the manufacturer (and as they may be modified by the airline and approved by the FAA). All four—the manufacturers, the airline, the pilots, and the FAA—are equally concerned with the manner in which policies and approved operating practices and procedures are followed. These issues involve human factors problems that must be resolved for satisfactory implementation.

Operating practices and procedures include the 'monitored approach', the 'fail safe crew', 'the basic crew or team concept' and the development of a clear distinction between pilot-flying (PF) and pilot-not-flying (PNF) duties. They also include procedures for the handling of obvious and subtle incapacitation and should include use of the 'two communication rule'. The two communication rule is explained in the Incapacitation on the Job section of Chapter 15. All of these practices or procedures should be a part of pilot training.

The increasing emphasis on CRM principles is making the role of the flight instructor, the check pilot, and the new 'AQP' evaluator more complex. The roles of flight instructors and check pilots are changing, and both jobs are becoming more difficult. The teaching and evaluation (or assessment) of an airline pilot is not an easy task to do well and consistently.

Flight instructors and check pilots are keys to developing effective pilot training. Selection procedures for both instructors and check pilots now require emphasis on the ability to teach CRM and team performance in a LOFT environment. No longer will technical skills be enough. As this chapter is being written, methods for considering individuals in terms of team performance are being evaluated. The development of suitable criteria and the validation of workable methods deserve its high priority. Pioneering work has been done in the cooperative University of Texas/NASA/FAA (UT/NASA/FAA) project. Flight instructors and check pilots or evaluators require special training. A web site that details the University of Texas research efforts is given in Appendix Q.

A frequently ignored area that directly affects the effectiveness of training involves check airmen, training pilots, and other supervisory pilots, as well as the higher levels of flight operations management. It is the requirement for them to follow the rules and procedures that have been promulgated. They do not always do this. In the elegant language of the academicians, the 'theory espoused' must also be the 'theory expressed' whenever and wherever management is involved. While this sounds relatively simple and blatantly obvious, the theory practiced by some parts of management sometimes varies from the theory they espouse.

The reasons for non-compliance and deviations from procedures includes the belief by many line pilots (and some management pilots) that the standard operating procedures, which these management representatives may not be meticulously following, are less than optimum procedures, and do not reflect 'the way things are, or should be, done out on the line.' Regardless of the extent of this practice, it should be understood and corrected for the practice of not following SOPs can negate the effectiveness of good training and undermine otherwise good flight operations. Chapter 9—Man's Limitations, Human Errors, and Information Processing discussed some reasons that SOPs and procedures are not always followed. However, the main point here is that if for any reasons SOPs and procedures are not followed, the failure to follow them has additional ramifications.

Training Organizations

While most carriers conduct their own training, there is a growing trend to arrange for at least some of that training to be done by, the manufacturer, another airline, or a training organization. This arrangement is made usually because the airline does not have the facilities (simulators, instructors, etc.) to perform the required training in house. It has become increasingly undesirable to take an airplane out of line service to perform training exercises. First, it is obviously extremely expensive to take an airplane out of service for training. Secondly, however, training maneuvers can be accomplished in the simulator that would be extremely unwise to attempt in an airplane. A major and continuing problem is to ensure that the airline's philosophy, policies, and procedures are reflected in the training and evaluation furnished by the contractor.

The Development of Airline Specialized Training Centers

An increasing number of airlines have anticipated a continuing need for high-quality airline pilot training and have developed specialized training centers whose services are available for hire. These airlines attempt to maximize the use of expensive training equipment, and many of them are marketing their training capabilities. Several consider their flight academies as separate profit centers, and provide training ranging from ab initio, especially in Europe, to approved type training and evaluation.

There is no reason to anticipate a diminution of these efforts. Recently a major US airline announced a $130 million expansion of its training center not only in order to prepare for its increasing training needs but also to tap into the booming US and international third-party airline training market (Norris, 1997). Another major airline has just expanded its flight training center with a $340 million dollar upgrade that has 36 simulator bays, more than 600 instructors, and 104 class and briefing rooms. At present more than 200 national and international carriers train their flight crews at this flight training center.

Contracted Training to a Training Organization

Dedicated training organizations, which have been primarily concerned with corporate flight operations, are playing an increasing role in some trunk and regional or commuter airline operations. There seems little doubt that they will play a continuing and growing role in airline transport.

Recently a major US manufacturer and a major flight training organization have joined to create what has been termed a training 'superschool' which will combine training for pilots and mechanics and be the largest airline pilot training and maintenance personnel training company of its type in the world. It will not be restricted to products made by the manufacturing company. Predictions are that 'over the next 20 years, airlines are expected to order about 16,000 jetliners worldwide to meet air traffic growth rates averaging about five percent per year' (Harrison, 1997). It is also expected that more than 200,000 new pilots will require training in the next two decades. This organization (Flightsafety Boeing Training International) has just announced that it will establish a new $85-million European training hub in London. This will be the first training center hub outside of the US that the training organization has created and the first of a global network of such training centers it plans to set-up for pilots and mechanics.

Designated Training Centers

In the US, a new provision under the FAA's AQP is the formal establishment of designated 'training centers'. They have been defined in FAA Advisory Circular (AC 120-54) as an independent organization that provides training under contract or other arrangement to certificate holders. There is little doubt that already established training organizations will seek and secure FAA appointment as an AQP designated training center.

Under the AQP such training centers will be staffed with people who have been approved by the FAA. Approved training centers will be authorized to furnish air carriers with both flight instructors and evaluators for cockpit resource management and for training on specific airplanes for specific carriers. The carrier also has to approve the authorized training center's program. This will formalize a process that has grown over the past few years. The AQP Advisory Circular also deals with the problem of reinforcing the specific airline's philosophy, policies, and procedures with the training that is contracted.

Training Devices

Among the many training changes for the line pilot are the utilization of computer-based training (CBT) and its close relative, computer-assisted learning (CAL). Today, even the best training of this nature (sometimes called desktop training) seems to adapt better to needs in the technical training area than to areas requiring behavioral and other kinds of skills and knowledge.

Many early CBT training devices did not provide realistic training because they were essentially single-path trainers. These training devices had only one way to accomplish a specific goal. They became little more than a 'punch the right button so you can get to the next step' training device. They did not have the alternatives that were available in the airplane and therefore did not accurately simulate real-world operation. They were unpopular with many pilots and trainers. Later CBTs do not have this particular problem.

There have been substantial advances in virtually all CBT/CAL areas, and there is little doubt that the use of CBT/CALwill continue to increase. Among the advantages such training has are the following

- It is more economical than lecture-type instruction

- It ensures that all trainees receive the same and (correct) information

- It does not require large numbers of expert instructors, and

- It reduces undesirable pressure on trainees by permitting them to proceed at their own pace—an aspect of this training that is somewhat controversial.

Critics maintain that even the best of programs using CBT or CAL devices do not actually reduce the time pressure on trainees because there are usually expected and sometimes even published 'normal' time limits for each section. The critics also maintain that the classroom interface with other students and an instructor in the older classical ground school environment pro-

Figure 16.1 High Tech Classroom
Source: *ICAO Journal* April 1994, page 27

vides a valuable educational exchange that is impossible to achieve with CBT or CAL. The most successful training programs seem to blend a combination of these approaches.

Manufacturers have become very much involved in CBT/CAL training for aircraft maintenance technicians. For example, recently a major manufacturer, which trains more than 3,000 aircraft maintenance technicians at its \$108-million Customer Service Training Center, reported emphasizing interactive and troubleshooting skills utilizing CBT. Previously such training was as much as 95% instructor led (*Aviation Week & Space Technology*, 2 December 1996).

Flight Training Devices (FTDs) and Flight Simulators (FSSs)

From virtually the beginning of the advanced technology era, pilots have complained that they did not get enough basic system training, and in particular that they did not get enough training in the use of the flight management system. One answer seems to be the increased use of the part-task trainer, which is sometimes called a flight training device (FTD).[7] Flight training devices are used in training for pilots, mechanics, and also in more realistic training for flight attendants. The FAA has established seven different levels of complexity required in FTDs. In addition, the FAA certifies four levels of fidelity for advanced full flight simulators (FFSs).[8] In most countries, simulators must be approved by a regulatory authority before they can be used in pilot training. Unless a bi-lateral agreement or similar document approves it, separate simulator approval is usually required by each country that approves the training. This has been an expensive, laborious, and repetitive process.

One of the most promising developments in recent years, which was stimulated by recognition that the practice of each State making its own rules is burdensome and expensive, is a technical document developed by the representatives of 12 countries. This document sets out a common standard for the evaluation and certification of airplane flight simulators. The document

---

[7] An airplane flight training device is a full-scale replica of an airplane's instruments, equipment, panels and controls in an open flight-deck area or an enclosed airplane cockpit, including the assemblage of equipment and programs necessary to represent the airplane in ground and flight conditions to the extent of the systems installed in the device; and does not require a force (motion) cueing or visual system' (FAA).

[8] If a device does not incorporate at least three axes-of-motion and night visuals, it is not a flight simulator (FSS), but may be in the category of a flight training device.

Figure 16.2 Level D Simulator
Source: *ICAO Journal* April 1994, page 28

was adopted unanimously by 129 delegates in a January 1992 meeting in London. It is available from ICAO as the Manual of Criteria for the Qualification of Flight Simulators (Document 9625). The application of international criteria for the evaluation of flight simulators will reduce the considerable workload of aircraft manufacturers, simulator makers, and regulatory authorities involved in certifying the growing number of simulators that are being used for airline pilot training (Orlady, 1994).

The Question of Fidelity

The level of fidelity that is required for the most effective teaching and the best use of limited fidelity are both somewhat controversial topics. Inevitably, increased fidelity means increased costs so obviously there is no point in purchasing a higher level of fidelity than is needed. In addition, there is considerable evidence that in some cases unneeded fidelity actually decreases training effectiveness by causing unnecessary distractions. Several years ago an acknowledged training expert, Wallace Prophet, gave an outstanding paper on this subject. It was given at the Flight Safety Association's International Air Safety Seminar in Athens, Greece, 6 November 1963. Prophet very logically maintained that:

> A training device or simulator has one principal purpose—to provide an environment in which one can be trained. Since training and flight safety are so closely related, it is not surprising that safety personnel have considerable interest in training devices....My remarks are made in support of the following general thesis: Aviation personnel have concentrated too much on *devices*— simulators and trainers—and not enough on their *use*. We have not given the necessary attention to *what* is taught and *how* it is taught. We are *hardware* oriented, when we should be *training* oriented. Consequently, training devices often do not make the contribution of which they are capable to training and flight safety. (Prophet, 1963)

The other side of the fidelity coin is that the FAA and other regulatory authorities properly have required a very high level of fidelity if the simulator is used as a substitute for the actual airplane in training and checking. Therefore, most airlines buy simulators with that very high level of fidelity and on occasion do not use all of the fidelity they have purchased, or end up using the simulators for tasks that do not require such complexity. The use of part-task training devices can address some aspects of this dilemma. Thomas Longridge in an 'Overview of the Advanced Qualification Program' notes that the FAA's 'AQP encourages air carriers to utilize a suite of equipment matched on the basis of analysis to the training requirements at any given stage of a curriculum. Judicious analysis of their requirements can enable an AQP participant to significantly reduce the need for a full simulator (Longridge, 1998).

A Training Summary

It should be clear that training for airline pilots is a complex, expensive, and important undertaking. It is highly specialized and too important to not be considered a very high priority item. As we stated in the Preface, this book should not be considered an all-encompassing treatise on any subject and that point seems to apply particularly to this chapter. Training needs vary by airline and by the type of training being considered. In some cases, training needs may raise human factors questions that can be best answered by a human factors professional. Such areas as those involving the transfer of skills, potential negative training, determining the optimum use of part task training, determining the optimum use of very sophisticated simulators, and finally evaluating the overall effectiveness of the training system can raise difficult questions that are well beyond the purview of this book.

References

Air Transport Association of America's Human Factors Task Force (1989). *National plan to enhance aviation safety through human factors improvements,* Washington, D.C.

Aviation Week & Space Technology, (1996). 'Meeting the Training Challenge', Advertiser Sponsored Market Supplement in *Aviation Week & Space Technology,* 2 December 1996, McGraw-Hill Companies, New York.

Bartram, Dave and Baxter, Peter (1996). 'Validation of the Cathay Pacific Airways Pilot Selection Program', *The International Journal of Aviation Psychology,* 6(2), Lawrence Erlbaum Associates, Inc., Mahwah, New Jersey.

Billings, Charles E. (1997). *Aviation Automation: The Search for a Human-Centered Approach,* Lawrence Erlbaum Associates, Mahwah, New Jersey.

Buck, Robert N. (1995). *The Pilot's Burden,* Iowa State University, Ames, Iowa.

Chidester, Thomas R., Helmreich, Robert L., Gregorich, Steven E., and Geis, Craig E. (1991). 'Pilot Personality and Crew Coordination: Implications for Training and Selection', *The International Journal of Aviation Psychology,* 1(1), Lawrence Erlbaum Associates, Inc., Mahwah, New Jersey.

Cooper, George. E., White, Maurice D., and Lauber, John K., eds. (1979). *Resource Management on the Flight Deck,* NASA Conference Publication 2120, Proceedings of NASA/Industry Workshop, Ames Research Center, Moffett Field, California.

Damos, Diane (1995). 'Pilot Selection Batteries: A Critical Examination', in Johnson, N., Fuller, R., and McDonald, N., eds., *Aviation Psychology: Training and Selection,* Avebury Technical, Aldershot, United Kingdom.

Damos, Diane (1996). 'Pilot Selection Batteries: Shortcomings and Perspectives', *The International Journal of Aviation Psychology,* Lawrence Erlbaum Associates, Inc., Mahwah, New Jersey.

Dismukes, R. Key, Jobe, Kimberly K, and McDonnell (1997). *LOFT Debriefings: An Analysis of Instructor Techniques and Crew Participation,* NASA Technical Memorandum 110442, NASA Ames Research Center, Moffett Field, California.

Federal Aviation Administration (1991). *Line Operational Simulations,* Advisory Circular 120-35C, Washington, D.C.

Foushee, H.C. (1990). 'National Plan for Aviation Human Factors', presentation to the Society of Automotive Engineers G-10 Committee, Monterey, California.

Greene, Berk and Richmond, Jim (1993). 'Human Factors in Workload Certification', presented at Society of Automotive Engineers Aerotech '93, Federal Aviation Administration, Seattle, Washington.

Green, Roger G., Muir, Helen, James, Melanie, Gradwell, David, and Green, Roger L., (1991). *Human Factors for Pilots,* Avebury Technical, Aldershot, United Kingdom.

Harrison, Kirby J. (1997). 'Boeing/FSI Create Training 'Superschool', *Aviation International News,* 1 April 1997, Midland Park, New Jersey.

Johnston, A.N., (1992). 'The Development and Use of a Generic Nonnormal Checklist With Applications in *Ab Initio* and Introductory Advanced Qualifications Programs', *The International Journal of Aviation Psychology,* Lawrence Erlbaum Associates, Inc., Mahwah, New Jersey.

Lauber, John K. and Foushee, H.C. (1981). *Guidelines for Line-Oriented Flight Training, Vol. I and II.*, NASA Conference Publication 2120, Ames Research Center, Moffett Field, California.

Lauber, John K. (1993). In Foreword to *Cockpit Resource Management*, edited by Wiener, Earl L., Kanki, Barbara G., and Helmreich, Robert L., Academic Press, Inc., Harcourt Brace Jovanovich, Publishers, San Diego, California.

Longridge, Thomas M. (1998), 'Overview of the Advanced Qualification Program', http://www.faa.gov/avr/afa/tlpaper.htm, Federal Aviation Administration, Washington, D.C.

McDonnell, Lori K, Jobe, Kimberly K., and Dismukes (1997). *Facilitating LOS Debriefings: A Training Manual*, NASA Technical Memorandum 112192, NASA Ames Research Center, Moffett Field, California.

McFarland, Ross A. (1953). *Human Factors in Air Transportation*, McGraw Hill Book Company, Inc., New York.

Norris, Guy (1997). 'United Drives Hard To Gain A Place In The Training Market', *Flight International*, 2 - 8 April, 1997, Reed Business Publishing, Sutton, United Kingdom.

Odegard, John (1994). 'Expert Pilots Can Be Cultivated Through A Shift In The Way Students Are Taught', in *ICAO Journal*, May 1994, International Civil Aviation Organization, Montreal, Canada.

Orlady, Harry W. (1993). 'Airline Pilot Training Today and Tomorrow', chapter in *Cockpit Resource Management*, edited by Wiener, Earl L., Kanki, Barbara, G., and Helmreich, Robert, L., Academic Press, Inc., San Diego, California.

Orlady, Harry W. (1994). 'Airline Pilot Training Programmes Have Undergone Important and Necessary Changes in the Past Decades', *ICAO Journal*, April 1994, International Civil Aviation Organization, Montreal, Canada.

Orlady, Harry W. and Foushee, H. Clayton, eds. (1986). *Cockpit Resource Management Training*, NASA Conference Publication 2455, Ames Research Center, Moffett Field, California.

Orlady, Harry W. and Wheeler, William A. (1989). 'Training for Advanced Cockpit Technology Aircraft', *Proceedings of the Fifth International Symposium on Aviation Psychology,* Ohio State University, Columbus, Ohio.

Proctor, Paul (1996). 'Airlines Increase Hiring From Civil Ranks', *Aviation Week and Space Technology*, 8 April 1997, Hill Companies, New York.

Prophet, Wallace (1963). 'The Importance of Training Requirements Information in the Design and Use of Aviation Training Devices', presented at the Flight Safety Foundation International Air Safety Seminar, 6 November 1963, Flight Safety Foundation, New York (now Arlington, Virginia).

Rasmussen, Jens (1987). *New Technology and Human Error*, ed. by Rasmussen Jens, Duncan, Keith, and LePlat, Jacques, John Wiley & Sons, Chichester, United Kingdom.

Reason, James (1997). *Managing the Risks of Organizational Accidents*, Ashgate Publishing Limited, Aldershot, Hants, England.

Regional Airline Association (1996). *Annual Report of the Regional Airline Association*, Regional Airline Association, Washington, D.C.

Stewart, Stanley (1989). *Emergency: Crisis in the Flight Deck*, Airlife Publishing Ltd., Shrewsbury, England.

Vette, Gordon with Macdonald, John (1983). *Impact Erebus*, Hodder and Stoughton, Auckland, New Zealand.

17 The Challenging Role of the Flight Attendant

Flight Attendants have two basic responsibilities. One involves their important passenger service obligations and the second, which is even more important, involves the safety of their passengers and themselves under both normal and non-normal conditions. The flight attendant job requires hard work performed at all hours of the clock and the ability to deal with a wide variety of in-flight medical and other emergencies. Flight attendants are probably the most critical resource in emergency passenger evacuations.

Many air transport accidents have human factors ramifications that involve the cabin crew and their interface with the flight deck crew. The main focus of this chapter concerns the safety responsibilities of flight attendants. However, it is important to have an understanding of their history, their environment, and some of their particular challenges.

Historical Factors

The first cabin attendants were three 14-year old cabin boys, who were hired by Great Britain's Daimler Airways in 1922. They were dressed like bellboys and were sent aloft on key flights. Neither food nor drinks were served on these planes so that the young airborne bellboys were probably more ornamental than useful. The minimal duties of Daimler's bellboys were gradually upgraded but there was no question that their primary function was entirely oriented toward passenger service. European airlines continued to stress in-flight passenger service during the middle and late 1920s. Several used men as airborne stewards serving the passengers. These, plus Daimler's bellboys, were our first cabin attendants.

In 1930, United Airlines forever changed the nature of in-flight service. Because of the persuasion of Ellen Church, the world's first airline stewardess (see Figure 17.1), and the foresight of Steve Simpson, one of United's early executives, United hired attractive, single, registered nurses to be the world's first stewardesses on all United's flights. They had to be less than 25 years of age. Their duties were entirely to increase passenger comfort and well being and to increase confidence in the safety, comfort, and convenience of this new method of travel. Today, the nursing background requirement, as well as restrictions regarding age, gender, and marital status, have been abolished.

For a great many years the job of the flight attendants (stewardesses in the early days) was never considered a career job. A main reason was the strict requirement that the job was terminated immediately when these attractive, desirable, and non-unionized young ladies became married. Later cabin attendants became unionized, gradually pay was raised, and the other restrictions were formally lifted.

More of the so-called fringe benefits such as good vacations, pass travel, medical benefits and pensions have become a part of the job until today a great many cabin attendants of both genders consider their job a career. A great many flight attendants are married. In most cases, their benefits accrue to their spouses in exactly the same way as the benefits accrue to the spouses of pilots and other employees. Approximately 20% of flight attendants are males.

Primary concern with passenger service and passenger well being has always been a hallmark of the industry viewpoint toward flight attendants. It is still a high point in advertisements promoting air travel. Only recently have the safety ramifications of the flight attendants begun to receive attention. Their safety role has been led by the flight attendant's unions, by the pilot unions, and to an increasing extent by researchers and the FAA and NTSB. This has usually occurred in the aftermath of an accident or serious incident in which the role, training, and conduct of the flight attendants and their interface with the cockpit crew was considered a factor in the accident or incident. There is little question that the cabin attendant's job has grown considerably from the old days when the only real concern was passenger service. For example, one major airline has identified 23 safety-related instances that are mandatory reporting events. Those 23 mandatory reporting events, many of which must be dealt with on the spot, are listed in Appendix L.

Organizational Factors

A major organizational factor that began with the first stewardess service and continues to influence the flight deck crew/cabin crew interface is the prevailing airline view that the cabin crew's principal duties are concerned with passenger comfort and well-being, including the cabin crew's responsibilities during evacuation. The accompanying view that the flight deck crew has only a peripheral interest in passenger service has helped continue the industry practice of having a separate organization for pilot crewmembers and cabin crewmembers.

Figure 17.1 Ellen Church—the First Airline Stewardess
Source: Courtesy of United Airlines

Pilots are usually organizationally placed under a Vice-president of Flight Operations while the cabin crew is organizationally placed under a Vice-president of Sales or Marketing. This arrangement is fine from the standpoint of what the industry sees as the primary responsibility of each group.

Unfortunately, it has not facilitated a good interface between the two groups. Historically, both departments have given such an interface a very low priority, with the possible exception of the interaction involving the emergency evacuation responsibilities of the cabin crew. Because of the importance with which passengers view the flight attendants and their role in furnishing in-flight services, it is easy to see the marketing division's wish to keep control of the flight attendant division. In addition, there is almost certainly considerably greater flexibility in the airline's collective bargaining relationship with the flight attendants if they are kept separate. There is also little doubt that Flight Operations Departments are perfectly happy to have minimal responsibility for sales and marketing.

There are at least three schools of thought regarding the desirability of this type of organizational structure. One believes that the present arrangement is entirely logical. Its proponents believe that pilots, at best, have only a secondary interest in marketing considerations and that marketing is of such importance to an airline that it deserves undivided attention. A second school, which includes many academics and researchers, believes that the safety ramifications of a good interface between the two groups is so important that both groups should be responsible to a single individual. A third, and we believe unfortunately a smaller group, believes that it is the responsibility of each department to be aware of and sensitive to each other. The individual Vice-presidents of Marketing and of Flight Operations would then be responsible to a more senior Vice-president who has ultimate responsibility for both departments and who must ensure that both safety and the interface between the two departments is given proper attention. Members of this smaller group believe that the two departments should not be able to operate independently of each other as they so often do today.

Frank Hawkins, in *Human Factors in Flight* pointed out the need to consider the interrelationships of flight attendants with all aspects of their unique environment (Hawkins, 1993). For a discussion of this area, we highly recommend pages 286 to 325 of his book. The social environment is particularly important because the relevant social environment is a major factor in the interface flight attendants have both with their passengers and with the cockpit crew. A large part of the social environment stems from the organizational culture of the airline. The airline's organizational culture influences both the role of flight attendants in air safety and their relationship with their passengers and the cockpit crew. The concept of social environment is discussed in Chapter 6—The Social Environment.

Regulatory Factors

A long time goal of the cabin attendants has been to achieve some sort of FAA licensing. This goal has been opposed by the airlines for at least two reasons, one perhaps more honorable than the other. First, scheduling is much simpler if flight attendants are able to work all types of equipment, especially as compared to pilots who almost always are only qualified and current on a single aircraft type. This scheduling versatility can have benefits both for the flight attendants and for the company. The second reason, however, illustrates one reason that flight attendants, and in particular the flight attendants' union, have sought official licensing. If flight attendants are not licensed, it makes it much easier to secure temporary or even permanent replacements (usually other passenger service employees) during a labor management crisis such as a strike.

The FAA does have to approve the airline's syllabus for flight attendant training. Initial flight attendant training is quite rigorous. It typically requires about six weeks and includes learning all safety aspects as well as the specifics of the company's individualized cabin service. An example of the difficulties that the lack of licensing can cause for the flight attendants' union occurred in 1993 when a strike was threatened at a major carrier. This particular company quickly obtained FAA approval to reduce the required minimum flight attendant training from six weeks to eight days (Chute and Wiener (1995). The FAA's position was that eight days was enough time to satisfactorily train for the required safety items.

The Flight Attendant Today

Flight attendants are primarily concerned with three kinds of liveware—the other flight attendants, the cockpit crew, and their passengers. While a Boeing B-737 may typically have three flight attendants, a long-haul jumbo jet may have as many as twenty-one. All cabin crewmembers have to operate as a team, which is typically coordinated by a chief flight attendant, who is sometimes referred to as a chief purser or chief steward. The interface among the cabin crew and with the cockpit crew is obviously critical. As there are few restrictions on the type of passengers who are allowed to board an aircraft (other than those who are obviously inebriated) it is not surprising that the flight attendants have to routinely deal with a wide variety of passengers. The broad range of passengers that flight attendants serve include differences in age, gender, physical capability, sophistication, native language, and level of comfort with air travel. With predicted passenger loads numbering in the area of one and one-half billion passengers by the 21st century, a continued wide range of passengers is inevitable.

Flight Crew Communication and Coordination

One of the objectives stated in the 1985 FAA Human Factors Research Plan was to improve the effectiveness of communication and coordination between cockpit and cabin crews. 'Crew coordination is crucial not only in emergencies, but also during normal operations....The cockpit and cabin crewmembers must act as one cohesive crew, even though they are trained, scheduled and generally regarded as two independent crews' (Cardosi and Huntley, 1988).

Flight attendants are professionally concerned with communication between themselves, with the cockpit crew, and with their passengers. Because this book is concerned primarily with operational issues, the discussion in succeeding paragraphs will emphasize but not be limited to communication and coordination between and among members of the operating team. Many of the problems involve training related issues. Many are human factors issues because they directly affect the interface between the cockpit and cabin crews and also the interface among flight attendants, who must work together as a team. Safety issues are often involved. In an effort to better deal with the interface between the operational team of the pilots and the flight attendants, several airlines presently conduct a portion of their recurrent training with a combined group of pilots and flight attendants. Nearly all agree that this joint training is stimulating, positive, and educational training that should be encouraged.

Cabin-Cockpit and Intra-Cabin Crew Communications

Two categories of communication are involved in basic flight operations for flight attendants—flight deck-to-cabin communication and communication among cabin crewmembers. Prerequisites for good communication between cabin and cockpit as listed by Koan in 1985 are:

- respect and rapport among crew members,

- communication that will not fail in an emergency,

- an understanding of the other crew's duties, and

- the same information on specific topics (e.g., code words).

For example, code words or signals for hijacking or evacuation are virtually useless unless both the flight attendants and pilots are aware of their meaning and of the action that may be taken because of their use. A well-orchestrated preparation for an emergency evacuation, or the handling of any other emergency, requires well-coordinated training for both the cabin and flight deck crews.

The NTSB and the FAA have become increasingly concerned about differences in training between cabin and cockpit crews for identical safety items (NTSB, 1992). Chute and Wiener cite one example where for 9 years flight attendants were taught that in an emergency they could expect the flight crew to give them at least four critical pieces of information. The four pieces of information were the type of emergency, the signal to brace, the signal to evacuate, and the time available to prepare. None of a group of pilots from the same airline who were interviewed regarding it had ever heard of these 'four critical pieces of information' (Chute and Wiener, 1995). Virtually all serious observers of this problem agree that better coordination in the training and handling of safety items is needed on many airlines.

There are some obvious logistical problems in achieving good communication. For example, on some large airlines, the flight attendant employee group numbers in excess of twenty thousand. On a given trip, a flight attendant may know a few or none of the other flight attendants and none of the cockpit crew. The communication problem is complicated because as the number of cabin crew increases so does the complexity of the communication problem. Many airlines try to funnel all communications from flight attendants to the cockpit through the chief purser or chief flight attendant. Communication within small airlines is much less of a problem because cabin and cockpit crews fly together often and can know each other quite well.

A further complication results from the reluctance of some airlines to schedule cabin and flight deck crews as intact teams so that they fly multi-day trips together. As mentioned earlier, flight attendants are trained to be available to work all of a carrier's aircraft. It can be economically beneficial to use different scheduling practices for flight attendants than are used for pilots. This versatility can also mean better working conditions for the cabin attendants. However, this practice means that crews may not have the time to establish good lines of communication and to accomplish some of the teambuilding that was discussed in Chapter 13—CRM and the Team Approach.

Yet another challenge for crew coordination concerns the training given to the head of the cabin crew—to the chief flight attendant or chief purser. At some carriers, this position is bid by seniority and given little recognition of its additional responsibilities in terms of training and additional compensation. As a consequence, at some carriers, it is not unusual for this important position to be held by the most junior flight attendant. While this situation is not the same as having the most junior pilot in the cockpit fly as captain, it does suggest that safety considerations are given a fairly low priority in cabin crew scheduling.

Crew pairing problems are not limited to jumbo jets flying long segments. The captain of a medium-sized jet recently complained that on one of his normal schedules, four different cabin crews were regularly scheduled for a single day's flying. He had real difficulty knowing with whom he was flying. Flight attendants also operate under different work rules as defined by the FARs and by their union contracts. These differences can result, for example, with a cockpit crew being legal for a trip after a creeping delay, and the flight attendants not being legal. The converse can also happen.

It is clear that communication and crew coordination must be addressed in training. CRM training, whether done jointly or separately, should emphasize these points. Short lines in an operating manual are not enough. Pilots should be trained to solicit relevant information from flight attendants and should know that flight attendants need to have early and explicit information in any non-normal situation in which their performance or the actions of the passengers are important.

Sterile Cockpit Rule

The coordination problem has been enhanced by the 'sterile cockpit' rule. The sterile cockpit rule (FAR 121.542) states that during critical phases of flight and during all other flight operations (except cruise) conducted below 10,000 feet, no crewmember may engage in any activity or conversation that is not required for a safe operation. The regulation specifically excludes nonessential communication between the cabin and cockpit crews during the sterile period.

While the basic purpose of the rule is exemplary, its implementation has created some dilemmas. For example, one cockpit crew on approach below 10,000 received a succinct and precise communication on their cabin interphone. The flight attendant informed them that 'the fire is out now'. This communication was the first and only notification to the cockpit crew that anything was amiss, let alone the presence of a fire in the cabin. During a postflight debriefing, the flight attendants stated that they were hesitant to notify the cockpit because it was a two-pilot airplane without a flight engineer and because the fire had occurred during what they thought was the sterile cockpit period.

One problem with the sterile cockpit rule is defining what is a safety-related item and what is nonessential communication. A second problem is to reliably notify the cabin crew when the airplane has either descended below or has climbed above 10,000 feet. Some airlines use a particular PA announcement, others a sequenced aural signal, and yet others simply tell the flight attendant to regard the time period ten minutes after takeoff and ten minutes before landing as the sterile period.

Airline Differences in Coordination and Communication

The degree to which training departments emphasize coordination and communication between cabin and cockpit crews varies widely from airline to airline. Unfortunately accidents and incidents do not limit themselves to the airlines that are well prepared. This problem is exacerbated by the fact mentioned earlier, that cockpit crews and cabin crews usually operate under two different administrations, with each administration being responsible only for their specific crews. With two separate training departments, it is obvious that major efforts must be made to provide compatible programs, and programs that achieve common goals. Some airlines do this quite successfully, but it does take extra effort and consideration. Crew resource management programs provide an ideal way to emphasize the importance of good communication and coordination between cockpit and cabin crews.

Today's Coordination and Communication Challenges

Today, cabin crew/flight crew communication and coordination seem to be getting more of the emphasis they require—at least on some airlines. An important part of such an enlightened program is that each group understand the other group's duties and responsibilities during both normal and non-normal operations. While on trips, displays of common courtesy, such as pleasant introductions, foster an atmosphere of coordination in the achievement of common goals. If at all possible, a preflight briefing by the Captain that includes the in-flight weather, the estimated flight time, and any unusual circumstances involving the flight should always be made as well as a solicitation for communication and offer for assistance and support. However, and especially with the long range jets and multiple crews, it may be possible to give the preflight briefing to only the chief purser or first flight attendant. In any case, it is paramount that both the cabin and cockpit crews view each other as a resource with a common goal and feel comfortable communicating about all problems and irregularities.

A different sort of communication problem was highlighted in April of 1994 when the FAA proposed a rule that would require an Operator Flight Attendant English Language Program in order that flight attendants 'understand sufficient English language to communicate, coordinate, and perform all required safety related duties.' It would be comparable to regulatory requirements for other crewmembers and dispatchers. The Aviation Rulemaking Advisory Committee, which is comprised of aviation related organizations that advise the FAA on regulatory issues, has stated that 'it is inconsistent to assign flight attendants safety related duties aboard flights without ensuring that they have the ability to effectively communicate and coordinate these duties with other crewmembers.' To date, the FAA has not implemented such a rule, largely because of a lack of data or industry interest in the proposal.

Rebecca Chute and Earl Wiener have studied the problem of cockpit-cabin communication in research supported in part by grants from the NASA Ames Research Center to San Jose State and the University of Miami. Their reports in the *International Journal of Aviation Psychology* are recommended reading for anyone who wishes to further pursue this subject (Chute and Wiener, 1995 and 1996)

Medical Emergencies and Other In-flight Problems

Given the number and variety of passengers that are carried each day, it is not surprising that the airlines sometimes have medical emergencies in-flight The FAA has reported that 7,789 medical incidents were reported to it in the ten-year period between 1971 to 1980 (Poliafico, 1988). It is not possible to accurately estimate the frequency of such incidents because there was no mandatory requirement to report passenger medical incidents during this period, however there is no question that the number would be greater today because of the growth of the industry since 1980.

In another study, the medical services of Japan Airlines reported that the airline experienced 405 cases of medical emergencies within the 3-year period from April 1993 to March 1996. The frequency of these incidents was 3.6 per

1,000 flights or 1.4 per 100,000 passengers.

In some airlines, the aeromedical committee of the pilot association serves an important role in monitoring these incidents. One ALPA Aeromedical Committee reported that at a large US carrier, 2,400 in-flight medical emergencies were reported in 1997, an increase of 14 percent from the previous year. Two-thirds of these emergencies resulted from passenger illness, while the remainder resulted from passenger injury. This particular carrier, which operates well over 2,000 flights a day, averages 7 medical events per day. Of the 2,400 emergencies that occurred in 1997, only 100 resulted in a diversion from the planned destination. Twenty-five were on international flights.

One has to be careful about numbers in this area because there are no universally accepted definitions of what constitutes a medical emergency. Reporting of medical data varies considerably. However, there is little doubt that there are a great many in-flight medical events on an average day and that many flight attendants have received minimal effective training to deal with them.

A medical emergency in a passenger provides a difficult problem for flight attendants. Regardless of their first aid qualifications, if a medical emergency occurs, the flight attendant is not only forced to make a diagnosis of sorts but also to furnish effective assistance. A real disadvantage of the two-person flight deck crew is that the flight crew can be of little help since they do not have the additional resource of a flight engineer.

Most airlines are aware of and concerned with this problem. Some airlines have an in-house, full-time medical department which is an excellent resource. The pilots, working together with the flight attendants, are able to contact their own medical department using company communication channels. For those airlines without a regular medical department, at least one commercial establishment is available on an international basis to provide 24-hour medical consultation for any in-flight medical emergency or to supply whatever in-flight medical advice is deemed necessary. This process seems to work well. Unfortunately, some airlines have neither their own medical department nor any relationship set up with a contracted medical service provider. Some airlines provide little help to their flight crews in this area.

Most common causes for medical incidents in flight are gastrointestinal difficulties, shortness of breath, hyperventilation, chest pains, and loss of consciousness due to fainting or to some form of a heart attack. Hyperventilation was covered in Chapter 4—The Physical Environment and the Physiology of Flight. Other problems may require a sophisticated diagnosis.

If a medical emergency occurs in flight, the usual procedure is to inform the captain and then see if there is a medical person, such as a medical doctor or nurse, on board who is willing to treat the ill person. The captain may or may not decide to land as soon as possible. Some medically qualified individuals refuse to identify themselves because of the potential liability of a lawsuit, but fortunately, in most cases suitably qualified individuals make themselves available and help. Attempts have been made to secure a 'Good Samaritan' law, in order to eliminate the personal liability that could be associated with these cases, but so far the efforts, while continuing, have been unsuccessful.

In 1986 FAR 121.309 was promulgated by the FAA. It requires that a reasonably sophisticated medical kit be on board any US transport and specifies that only a qualified medical practitioner may use it. Many airlines outside the US have similar kits on their airplanes—some have had them for many years. These medical kits are not without their problems. For example, problems include identification of individuals who are qualified to use the kit, the basic security of the kit, specific responsibility for the freshness of some of the drugs in the kit, the liability (if any) of doctors who volunteer their services, language difficulties, etc.

Recently American Airlines has expanded the required kit by including a 'heart defibrillator' on all of its over water flights and has trained flight attendants in its use. The record of the use of the defibrillator and its effectiveness is being watched very closely by other airlines and the FAA. Defibrillators have been used selectively by several foreign airlines for some time and recently United, Delta, Northwest, and Virgin Atlantic have started installation of defibrillators on certain of their flights.

It is frequently suggested that all flight attendants be required to obtain a formal first aid certificate of some sort in order to handle the problem of the passenger who develops an acute medical problem while in flight. The FAA requires that approved first aid training be given to all flight attendants and that the training be checked periodically. Nearly all airlines throughout the world provide first aid training to their flight attendants in some form, but its quality varies considerably. Air New Zealand requires all flight attendants to hold a recognized first aid certificate and is among the airlines that provides annual recurrent training to ensure that the required first aid knowledge is kept current.

The most serious of the medical problems that can occur is, of course, death to the passenger. In-flight deaths do occur. During the seven-year period

1977 to 1984, 577 in-flight deaths were reported (Hawkins, 1987, 1993). The July/August 1997 issue of *The Airline Employee* states that there are about 15 medical emergencies in US airplanes every day and approximately 350 deaths per year, Most of the deaths are due to natural causes and 90 percent of them happen to elderly individuals. It has been estimated that approximately 10 percent of the in-flight deaths might be preventable if appropriate medical facilities are on board (Poliafico, 1988). Exactly what constitutes appropriate medical facilities and training suitable to use the appropriate medical facilities is not stipulated. One airline spokesman is quoted as asking, 'How far do you go in making aircraft into flying hospitals? A list of the medical in-flight emergencies resulting from diversions in 1996 is shown in Table 17.1.

Table 17.1   List of Medical In-Flight Emergencies Resulting in Diversions in 1996

Cardiovascular	183
Neurological	88
Seizure	51
Respiratory	37
Fainting or loss of consciousness	34
Other	164
TOTAL	**557**

Source: Air Transport Association of America

The Problem of the Unruly Passenger

The problem of the unruly passenger is on the increase. A major airline in the US reported that physical assaults to flight attendants rose from 33 in 1984 to 140 in 1995. Another major airline reported an increase from 53 incidents of criminal misconduct by passengers in 1994 to 88 incidents in 1996 by 12 December of that year. One factor is, obviously, simply growth of the industry. Japan Airlines has reported 16 violent episodes in their fiscal year 1994, 15 in 1995, 40 in 1996, and 15 in the first four and one/half months in the fiscal year of 1997. Sixty percent of these incidents have occurred on international flights, and since Japanese airlines do not serve liquor on domestic flights, some believe that the incidents may be alcohol related (*Aviation Week and Space Technology,* 8 September 1997).

A major US airline has reported that intoxication was responsible for approximately 25 percent of all abusive passenger behaviors. It would be incredibly naive to believe that alcohol, and in some cases non-prescription drugs, were not a contributing factor in many cases. British Airways has just reported that its crews had to deal with more than 260 disruptive passengers in 1997. Somewhat surprisingly, at least to us, was the fact that 70 percent of all incidents were smoking related. British Airways has no plans to revise its no smoking ban on all flights. It states that the vast majority of its passengers favors the ban.

Today, the airlines and the courts are seldom lenient to anyone responsible for incidents where a flight attendant or cockpit crewmember was injured while trying to pacify an unruly passenger. As Lamar C. Walter, an assistant US attorney, who has prosecuted such cases, has noted, 'an airplane is not a good place to pick a fight'. Unfortunately such incidents do happen and unfortunately again, the two-person cockpit crew can be of little help to the cabin crew in those situations.

In an effort to provide guidance on this serious question, the FAA has published an Advisory Circular[1] (to help airlines, flight and cabin crews, and law enforcement officers deal with increasingly frequent rowdy-passenger incidents that range from verbal abuse to actions that interfere with flight safety). Most airlines have on-board some sort of restraining device for truly unruly passengers. They have been used very infrequently. The problem is complicated on international flights where the law enforcement response is a responsibility of the destination government and where the incident can be further complicated by language difficulties.

American Airlines' President, Donald Carty, has been a leader in achieving a coordinated effort among airlines, flight and cabin crews, and law enforcement agencies. He has asked that enforcement action be a 'certainty' in each case of passenger misconduct that interferes with crew duties (*Aviation Daily*, 8 November 1996). Other leaders have been United Airlines, the Association of Flight Attendants, the Air Line Pilots Association, and the Federal Aviation Administration.

Cabin Hardware

Traditionally, human factors has received much more emphasis on the flight deck than in the passenger cabin. This is particularly true for the older airplanes.

---

[1] AC 120-65 AFS-200, Interference With Crewmembers In The Performance of Their Duties.

Because airplanes can be expected to be in service for up to 30 years, a wide variety of cabin design and equipment is flying somewhere. This situation forces flight attendants to cope with whatever is available on their flight and to be familiar with many different types of equipment.

Some cabin hardware considerations are fairly obvious. For example, delethalization is a major consideration. The human body can survive a surprising G-load, as long as the G-loads are not accompanied by sharp protrusions that penetrate the body.[2] Therefore, sharp objects or protrusions should be eliminated. In addition, equipment should be eliminated if it cannot be satisfactorily stowed. Seats and other cabin furnishings have come loose in accidents and caused causalities in otherwise survivable accidents. Flight attendants and passengers have perished, not because of the impact of a crash but because toxic gases from burning cabin materials killed them when they were perhaps shocked, but by no means fatally injured.

Seats and Rest Provisions

Seating for flight attendants is somewhat different than for passengers. Flight attendant seats have been controversial for years for many reasons. One of the reasons is that comfort is not a major design consideration. During flight, flight attendants are presumed to be up and about serving their passengers. Therefore, their seats were designed primarily to be used for restraint during takeoffs and landings and during turbulence. These seats are frequently called jump seats because they fold up when not in use. They must be located near emergency doors and are required to have shoulder harnesses in addition to the traditional lap belts. Many flight attendant seats are rearward facing because they take advantage of available space in front of bulkheads and because rearward seats enable the flight attendants to have a good view of their passengers.

The requirement for rest and relaxation on long range flights has significantly expanded initial seat requirements by requiring that a passenger seat or row of passenger seats be reserved as crew rest seats. This arrangement has been controversial, for flight attendants as well as for pilots, as the quality of rest can vary dramatically due to the proximity of the passengers and the ensuing cabin service. One of the latest transports (the Boeing 747-400) has a separate rest area in the aft of the aircraft with bunk beds and lounge chairs for the cabin crew. This aircraft also provides a separate bunk area aft of the flight

---

[2] 'G' is the basic unit of acceleration. 1 G is equal to 32 ft/sec^2.

deck for the pilots. Other long-range aircraft also have separate rest areas for the crews.

Carts, Galleys, and Control Panels

Carts and galleys often receive considerable well-deserved attention. They are important in terms of both safety and efficiency. Flight attendants are still injured from electrical appliances, oven and hot-cup burns, inadequately stowed equipment, latches that fail, and other things of that nature. A fully-loaded service cart is a heavy item, and must be handled carefully. There must be reliable methods of securing them during turbulence. The weight and mass of movable carts is an item that must be considered by both pilots and flight attendants during turbulence and during climbs and descents because of the sometimes steep angle of the aircraft and the cabin floor. Reliable braking devices are essential to maintain control of the carts. One has only to imagine a 110-pound person pushing a 250-pound cart uphill and encountering turbulence to see why food and liquor service carts are the single most common cause of flight attendant injuries.

In an extremely rare event, a flight attendant was killed when a galley service elevator malfunctioned during a flight from Baltimore/Washington to London. Blame was allocated to the service trolley retention and release system, a faulty switch design, and an inadequate pre-flight inspection (NTSB-AR-82-1). Clearly this accident showed deficiencies in the L-H (liveware-hardware) interface. The investigation revealed that in the previous ten years, there had been nine other injuries with like elevators. Injuries included contusions, lacerations, and broken bones. An unfortunate truism is that it has sometimes taken incidents such as this fatality to spur the development of safer and more efficient cabin hardware. In this case, the faulty design was corrected. From a positive side, it is good that the industry does take advantage of serious incidents and does make necessary and important interim improvements.

The number and location of galleys play a major role in the safety and efficiency of the multiple tasks required of flight attendants. Jumbo jets can have as many as 5 separate galleys and as many as 21 flight attendants. Control panel design, for both the galley service areas and for communication, for passenger service amenities, and for Public Address (PA) use, is finally getting the attention it deserves. Control panels now cover such areas as communication with other service stations, the cockpit and the passengers; galley equipment; lighting; passenger amenities like movies and audio presentations; and with some aircraft, relatively minor temperature control.

The latest airplanes truly represent state-of-the art in design and efficiency, some even to the point where cabin attendants can talk directly to ground operations without going through the flight crew. Flight attendants frequently need communication with the ground for such passenger service items as passenger medical problems, extra or special meals, wheelchairs, and gate and connecting flight information. Unfortunately, the need for such communication usually arises during periods that the flightcrew is busy with the operation of the airplane. All of these areas require human factors input at the design stage.

Emergency Equipment in the Cabin

Emergency equipment ranges from smoke masks and fire extinguishers to emergency oxygen. Passenger cabin considerations recently have been getting increased attention at all stages of design. For example, because we have seen accidents where passengers could not find the appropriate emergency exits, it is important that emergency exits be made both unobtrusive and conspicuous. There are many passenger human factors problems in the cabin.

Emergency equipment, while seldom used, is of direct interest to the flight deck crew, cabin crew, and passengers. Anthropometrical data that are relevant for the user population is obviously required, but the problem does not end there. Escape slides, which have sometimes been operated inadvertently and which have not always operated properly have been a continuing problem. The cabin environment at the time of an accident will inevitably be chaotic.

In an accident situation, a wide variety of passengers with highly charged emotions frequently find themselves in darkness, or nearly total darkness. Despite color-coded and placarded instructions, they often find it difficult to remember the instructions to follow the lighted guide markers, and seldom react with total objectivity. Significant water may be present and the airplane may or may not be right side up. Emergency exits and other equipment may be completely unfamiliar to individuals who are forced to use it to survive. It is extremely important that the handling and operation of any equipment that will be used by passengers be as self-evident as possible.

Theoretically, many of these problems are covered by pre-departure briefings. However, in spite of major efforts to make these briefings more effective, passengers seem to pay little attention to pre-departure briefings whether the briefings are given in a specially made movie or are given personally by the flight attendants. In 1983 and 1984 the FAA conducted a series of tests and found that approximately one in three passengers failed to don their life-vests properly, even after a demonstration. It is clear that the problem is not simply in the demonstration, and it is also clear that the industry has not been able to solve the problem of securing appropriate passenger behavior in an emergency (Hawkins, 1987).

It takes more than good design of emergency equipment for good performance in a real emergency. A recent study by Professor Helen Muir of Cranfield University in England entitled *Cabin Staff Behavior in Emergency Evacuations* found that emergency evacuation exercises in which there is assertive action by the cabin crew significantly increases the chance of survival after a crash landing or other land emergency. Assertive cabin crew behavior in the study included 'calling the passengers to exits and actively pushing them through exits as rapidly as possible in a highly active, but non-aggressive, manner' (Learmount, 1995). It is precisely this sort of behavior that enables cabin attendants to evacuate aircraft so quickly. In the early 1990s, a fully loaded DC-10 with 284 passengers was completely evacuated in 90 seconds after an aborted takeoff and ensuing fire. The role of the flight attendant in an evacuation cannot be underestimated and should not be undervalued.

Flammability of Cabin Materials and Toxic Smoke

The FAA has estimated that in 1980 between 30 and 40 percent of the fatalities in survivable accidents are related to fire and its effects. This figure applies to clothing as well as to cabin furnishings. It has resulted in considerable criticism to the Agency for failing to establish requirements for the utilization in civil aircraft of non-toxic and non-flammable materials. This is an important issue for the FAA.

A major hazard in airplane cabins is the toxic fumes formed when any polyurethane foam material burns. This foam material is used for many purposes in the cabin. The flight deck crew has protective smoke masks, but these are seldom available in the cabin. Additionally, the cockpit crew is specifically trained in the use of smoke masks and is kept familiar with their use in mandatory simulator training exercises. Such training and equipment is not available to passengers. Another device that has become a requirement in recent years is a protective breathing equipment device or PBE. The PBE is slipped over one's head, is relatively airtight and contains a chemical oxygen generator that provides about 15 minutes of oxygen. A typical airplane would have one PBE in the flight deck and two in the cabin. At one time, some aviation safety consumer advocate groups recommended, somewhat unrealistically, that each passenger should carry their own PBE in case of an emergency involving toxic smoke.

In response to criticism from major consumer advocates and air safety experts and as a result of a comprehensive 5-year study by the FAA, large aircraft built or operated in the US after 1987 must have fire-blocking seat cushion covers. Fire-blocking seat covers are also required in the UK by directives of the British CAA, and in several other countries by the directives of their regulatory authorities. These covers retard the access of a fire to the polyurethane cushions. When burning, these cushions generate toxic hydrogen cyanide and carbon monoxide. The covers provide at least an extra 40 seconds for evacuation. The extent to which other countries adopt such safety measures depends on their own certifying authorities. Many are covered in bi-lateral agreements.

Minor fires in the cabin are usually handled by flight attendants. They occur for any number of reasons, including the chafing of electrical wiring to overhead or other lighting. Pulling appropriate circuit breakers and the use of hand-held fire extinguishers usually handles such problems nicely. The prohibition of in-flight smoking has reduced the number of cabin fire incidents, although in spite of ample warnings and lavatory smoke detectors, passengers still manage to create fires in airplane lavatories by surreptitious prohibited smoking.

One of the worst examples of the toxicity of smoke in a cabin fire occurred in 1980 in Riyadh in which all 301 occupants died. The L-1011 airplane returned to Riyadh after cabin smoke, which originated in a cargo compartment, had been detected. The aircraft landed safely. After the airplane turned off the runway and stopped, the captain ordered an evacuation. Unfortunately, the airplane was still pressurized and the doors were held closed by the pressurization. No doors were opened from the inside and no doors were opened from the outside until about 26 minutes later. By then the entire interior of the cabin was engulfed in smoke and flames. Apparently no evacuation was attempted and 301 persons died as a result of toxic smoke and fire. Many of them were piled in front of the passenger exits and cockpit door. It was a tragic example of the speed with which toxic smoke can be created and of the importance of the prompt and efficient execution of well-established safety procedures. Almost certainly there were a large number of unnecessary deaths. Communication between the cockpit and the flight attendants was limited and ultimately proved to be fatal. Much can be learned from that fatal incident.

The inevitable chaos, which includes among other things the role of congestion, competitiveness and configuration, was demonstrated by Helen Muir et al. in a landmark study commissioned by the United Kingdom Civil Aviation Authority (1996). The report of this study notes that: 'The principal threat to life is the occurrence of fire and toxic fumes within the cabin; and it is clearly essential that passengers, assuming they survive the initial impact, are able to escape from this environment as quickly as possible.'

Another example of the lethal potential of the smoke and fumes that can accompany in-flight fires was the tragic case of a ValuJet DC9-30. It crashed in the Florida Everglades on 11 May 1996 and resulted in 110 fatalities. The carrying of some unauthorized time-expired chemical oxygen canisters apparently caused a fire in a cargo compartment. These canisters, when activated, can generate extreme heat. At least one of them activated and, among other things, set fire to aircraft tires that were being carried in a cargo hold. The ensuing fire caused considerable smoke and fumes that entered the cabin and cockpit. The aircraft ended in an out-of-control descent that culminated in the crash

Cabin Air

We have already discussed many aspects of the physical cabin/cockpit environment in Chapter 5—Those Magnificent Flying Machines and Their Internal Environment. The cabin crew is directly involved in all of the internal environmental considerations and more vulnerable in many of them than the flight deck crew for two reasons. The first reason is that cabin attendants have to be physically active during their duty periods. Cockpit crews normally are not physically active while on duty. The cabin temperature is usually set for sedentary passengers and is often higher than desirable for people being active. The second reason is that there is more control of recirculated air, temperature, and, in some airplanes, even humidity, in the cockpits than is available in the cabin.

Millions of passengers travel on US airlines annually and many millions more travel on foreign carriers. Because of the flight attendants inevitable contact with a very diverse population, concern has been raised regarding their exposure to disease organisms that might be present among those passengers.

Air quality surveys of airliner air have largely been concerned with environmental tobacco smoke and particulate matter and these are discussed in Chapter 5—Those Magnificent Flying Machines and Their Internal Environment, but there is also the question of microbiological disease transmission in recirculated cabin air. In one of the few studies of the normal internal aircraft microbiological climate while the aircraft was aloft, Doctors Wick and Irvine selected 36 domestic flights, four intercontinental flights, and two international flights in the Western Hemisphere that were operated by one of the largest US airlines.

The microbiological concentrations found with the airliners used in the Wick and Irvine study had levels which were typically much lower than the concentrations associated with ordinary activities such as the levels found in homes or in city buses and streets. One reason may be that bleed air from the jet engine's compressor stages provides cabin pressure. Compressor air is extremely hot—typically 250°C. This very hot air is then cooled to about 112°C and then further cooled by additional heat exchangers before it is ducted to the cabin. Doctors Wick and Irvine concluded: 'Because the high compressor temperatures effectively kill any living organism in the intake air, the air supply is virtually sterile as it enters the cabin air distribution system.... We, therefore, conclude that the risk of disease transmission as a result of microbial concentrations in airline cabin air is correspondingly low' (Wick and Irvine, 1995). Needless to say, this finding has not been received well by some of the flight attendant associations, and this remains a highly controversial area. From an observational and certainly unscientific view, exposure to long hours of low humidity seems to cause greater discomfort than does exposure to contaminated air.

Specialized Training

Today, flight attendants are required to be very flexible and versatile individuals. To achieve the required versatility, they receive several different types of specialized training. Their specialized training ranges for training from such things as the recognition and the handling of dangerous goods, fire fighting, first aid, and the handling of both normal and non-normal passengers to training in crew resource management, aviation oxygen requirements, and their responsibilities during an evacuation. Their profession is a demanding one and their contribution to the safe and efficient operation of the flight is often underestimated.

References

Cardosi, Kim M. and Huntley, M. Stephen, Jr. (1988). *Cockpit and Cabin Crew Coordination*, Report DOT/FAA/FS-88/1, US Department of Transportation, Transportation Systems Center, Cambridge, Massachusetts.

Chute, Rebecca D., and Wiener, Earl L. (1995). 'Cockpit-Cabin Communication: I. A Tale of Two Cultures', *The International Journal of Aviation Psychology.* Lawrence Erlbaum Associates, Inc., Mahwah, New Jersey.

Chute, Rebecca D., and Wiener, Earl L. (1996). 'Cockpit-Cabin Communication: II. Shall We Tell the Pilots?', *The International Journal of Aviation Psychology.* Lawrence Erlbaum Associates, Inc., Mahwah, New Jersey.

Hawkins, Frank. H., (1993). Chapter entitled 'The Aircraft and its Human Payload' in *Human Factors in Flight*, Second Edition ed. by Orlady, Harry W., Avebury Publishing Limited, Aldershot, England.

Howard, Benjamin (1954). 'The Attainment of Greater Safety', presented at the 1st Annual ALPA Air Safety Forum, and later reprinted for presentation at the Aircraft Accident Prevention Course, University of Southern California, July 1957.

Koan, Noreen, (1985). 'Cockpit and Cabin Crew Coordination and Communication'. Society of Automotive Engineers, Technical Paper 851918, Warrendale, Pennsylvania.

Learmount, David (1995). 'Assertive cabin crew save lives', *Flight International*, 24-30 May, 1995, Reed Business Information, Sutton, Surrey, United Kingdom.

Mohler, Stanley R., 1969. 'Crash Protection in Survivable Accidents'. Memorandum, Staff Study, FAA Office of Aviation Medicine, Washington, D.C.

Muir, Helen C. and Marrison, Claire (1991). 'The Effect on Aircraft Evacuations of Passenger Behavior and Smoke in the Cabin', *Proceedings of the Sixth International Symposium on Aviation Psychology*, 29 April-2 May 1991, The Ohio State University, Columbus, Ohio.

Muir, Helen C., Bottomley, David M., and Marrison, Claire, (1996). 'Effects of Motivation and Cabin Configuration on Emergency Aircraft Evacuation Behavior and Rates of Egress', *The International Journal of Aviation Psychology*, 6((1), 57-77, Mahwah, New Jersey.

National Transportation Safety Board (1992). *Special Investigative Report: Flight Attendant Training and Performance during Emergency Situations*, NTSB/SIR-92/02, Washington, D.C.

Poliafico, Frank J., 1988. 'Emergency Training for Flight Attendants: What and How', presented at the Fifth Annual Cabin Safety International Symposium, cosponsored by Federal Aviation Administration, Western Pacific Region, University of Southern California, and Southern California Safety Institute, Los Angeles, California.

Skogstad, Anders, Dyregrov, Atle, and Hellesøy, Odd H. (1995). 'Cockpit-Cabin Crew Interaction: Satisfaction With Communication and Information Exchange', *Aviation, Space, And Environmental Medicine*, September 1995, the Aerospace Medical Association, Alexandria, Virginia.

Vandermark, Michael J. (1991). 'Should Flight Attendants Be Included in CRM Training? A Discussion of a Major Air Carrier's Approach to Total Crew Training', *The International Journal of Aviation Psychology*, Lawrence Erlbaum Associates, Inc, Mahwah, New Jersey.

Wick, Robert L. Jr. and Irvine, Laurence A. (1995). 'The Microbiological Composition of Airliner Cabin Air', *Aviation, Space, and Environmental Medicine*, March 1995, Aerospace Medical Association, Alexandria, Virginia.

18 Non-punitive Incident Reporting

> The alarming thing is that we do not take advantage of our good fortune. Here we have a brush with disaster; a live crew and virtually intact aircraft ready to tell a story, and yet we never open the book.   (Bobbie R. Allen, 1966)

Non-punitive Incident Reporting

This chapter offers a brief description of non-punitive incident reporting. Incident reports tell us the way the aviation system works in the real world from the viewpoint of the people that have to operate it. It is a nearly ultimate test of aviation human factors and the aviation system itself, because it, along with the accident record, gives us an otherwise unobtainable picture of problems in the system. We have given considerable emphasis to the US experience because of our greater familiarity with the history of non-punitive incident reporting in the US and because of the availability of information regarding it.

The potential advantages of an effective non-punitive incident reporting system have been recognized for many years. Implementation of such a program, however, is not easy. One of the most successful is the US ASRS—The Aviation Safety Reporting System—that is administered by NASA. It will receive well over 32,000 reports in the current calendar year. In order to keep the significance of this very large volume of reports in perspective, it is well to remember the number of potential reporters that are a part of the US aviation system. We should also remember that ASRS has been successfully established in the US aviation system for 22 years. The history of the ASRS—illustrates very well the important contribution that can be made by individuals in the pursuit of recognized goals.

Developing an effective non-punitive incident reporting procedure is a complex process that involves many factors. Anything approaching a complete discussion of this process and all of the relevant factors, particularly on a worldwide basis, is beyond the purview of this book. Those interested in a comprehensive study are urged to read a monograph entitled *Incident Reporting Systems in Civil Aviation* by Dr. Charles E. Billings who can properly be called the conceptual founder and chief designer of the NASA/FAA's Aviation Safety Reporting System. The monograph is currently being written and will be published by the International Academy of Aviation and Space Medicine, which is presently headquartered in Toronto, Canada. Dr. Billings has been involved in non-punitive incident reporting since it began in the US and was instrumental in the design of the US ASRS. He was the NASA manager of ASRS during its first and formative years.

The History of Non-punitive Incident Reporting in the US

The first person in an official position to advocate strongly for non-punitive incident reporting in the US was the late Bobbie R. Allen, the Director of the Bureau of Safety of the Civil Aeronautics Board.[1] Mr. Allen firmly believed in the accident prevention potential of non-punitive incident reporting. In November of 1966, he gave a memorable paper to attendees at the Flight Safety Foundation's 19th International Air Safety Seminar. His paper was entitled 'Incident Investigation—The Sleeping Giant'.

Mr. Allen reminded delegates to the Seminar that: '...the thin line dividing an accident involving fatalities or major damage to the aircraft from an incident in which everyone got off scot-free is often nothing more than luck.' He also noted:

> The alarming thing is that we do not take advantage of our good fortune. Here we have a brush with disaster; a live crew and virtually intact aircraft ready to tell a story, and yet we never open the book. We must find a way of moving this raw material for accident prevention to the processing machine.
>
> What is it, then, that stands in the way of communicating this incident information to the appropriate governmental agency for processing? Repeatedly, when this question is asked, one hears the reply-FEAR: fear of litigation; fear of regulation; fear of punitive action. Admittedly such fear is not totally unfounded ...
>
> (Allen, 1966)

Unfortunately fear of litigation; fear of regulation; and fear of punitive action still impedes and sometimes prevents meaningful incident reporting in most parts of the world. The belief that punishment is indispensable and

---

[1] The US Civil Aeronautics Board was an independent federal agency that was established in 1940. Among its duties was the investigation of aviation accidents and determination of their probable causes. Its chief safety officer was the Director of its Bureau of Air Safety. The Board's investigative responsibilities were transferred to the newly-created National Transportation Safety Board (NTSB) with the passage of the Federal Aviation Act in 1958.

society's best protection against transgressions of any sort is an intrinsic part of many national, regulatory, and corporate cultures.

Reinforcing the importance of incident reporting, Mr. Allen noted that at an ICAO meeting on accident investigation, the Australian delegate told the attendees that 'a greater contribution to the safety of aircraft operations in Australia had been derived from the investigation of aircraft incidents than from the investigation of accidents'. These remarks were of particular interest because Australia has led all countries in the world in air transport safety for as long as records have been kept.

Australia had had an incident reporting system for several years. The incident reporting was required by the Australian Air Navigation Regulations and, because of superb and far from typical administration, was unique in the aviation world. Captain Arthur Lovell, Manager of Operational Safety for Ansett Airlines wrote eloquently of the Australian system and of its administration when he wrote:[2]

> We are very 'sold' on our system in that we are informed of incidents before they develop into accidents. The accent is very positively on fully investigating each report with the object of improving and strengthening airline operation for all concerned, and there is definitely no thought of using it punitively; in fact I cannot recollect an occasion on which this has happened. Punishment is a negative approach and it is doubtful that in itself, it achieves much, if anything.

Pilots are a critical part of any incident reporting system because, in virtually all cases, they are involved in the incidents and they usually furnish the incident reports. Clarence Sayen, former President of the Air Line Pilots Association, was an early proponent of non-punitive incident reporting. As early as 1954 he explained to the air transport industry 'that incident reporting systems aren't working because pilots fear disciplinary action by the carriers or the Government if they reflect dangerous occurrences'. Information regarding these occurrences, of course, was exactly the information needed to better control the risk to aviation safety that the occurrences represented. Unfortunately, within the aviation community there was an almost unanimous fear of the consequences of incident reporting. Fears of financial liability, personal incrimination, disciplinary consequences, and even license certification by the FAA kept preventing implementation of any meaningful incident reporting programs.

---

[2] This is an excerpt from a 19 February 1974 letter to Captain R.A. Stone of the United Airlines Department of Flight Safety.

United Airline's Flight Safety Awareness Program

Because of the fear issue, not much happened in non-punitive incident reporting in the US until Captain W. E. Dunkle, Senior Vice-President of Flight Operations of United Airlines, instituted United's Flight Safety Awareness Program (FSAP) late in 1973. Non-punitive incident reporting was an important part of the program. Capt. Dunkle made it clear that he (and United) would rather know about the incidents than punish the offenders.

In order to secure good incident reports, which were an inherent part of the Flight Safety Awareness Program, Capt. Dunkle, promised the United pilots:

- United Airlines will not take any punitive action against any pilot or flight dispatcher as a result of information procured through this program.

- We will not voluntarily divulge information secured in this program to any outside agency which will permit identification of any individual involved.

- We will vigorously protect individual anonymity in all aspects of this program unless this protection has been waived by the individual involved.

This was about as far as any company could go and still remain within existing US laws. For example, whether it agreed or not, there was no way a company could promise to ignore a valid subpoena from a court of law.

To get good data from the incident reports, a decision was made not to attempt a completely anonymous program. Because pilots could not be expected to be sensitive to all of the factors that were of interest, the FSAP Program would take reports from pilots only after they had been voluntarily interviewed. In order to get reasonably sophisticated interviewers, all of United's Flight Managers (over 100) attended two-day seminars at NASA's Ames Research Center. Not more than 15 were allowed at each seminar to be sure that all attendees could participate. NASA's Dr. Charles Billings and Dr. John Lauber participated in each seminar. Actually, they cooperated and participated in all aspects of the program. United and NASA had common aviation safety goals.

The Flight Safety Awareness Program seminars included a discussion of basic interview techniques and of the kind of human factors data that is too often not included in routine incident reporting. The seminars also included the reasons behind the Flight Safety Awareness Program, its premise and its goals; a detailed review of the airline industry safety record, and of United's safety record; and of the implications to both pilots and managers of the commitments made by Capt. Dunkle.

This program was a big step. Pilots were guaranteed anonymity and freedom from any punitive action. All reports of their incidents came from volunteered interviews that were given by whichever one of United's 100 flight managers that the pilot selected. The only stipulation was that the flight manager must have participated in a FSAP seminar. When the interview was finished, the flight manager, in the presence of the pilot, would telephone the results of the interview into a secure recording device where the report would be transcribed and the recording erased. At the conclusion of the telephoned report, the flight manager would destroy all records of the interview in the presence of the pilot. It was hoped that the ranks of the potential interviewers would include pilots who had been selected by ALPA and had attended a seminar identical to those given to the flight managers. Unfortunately the collective bargaining situation at the time precluded full ALPA participation. It is a considerable tribute to all of the parties involved—to the line pilots, to United's management, and to the ALPA representatives working during a stressful negotiation period—that the common goal of safety did not impede the progress of the program.

The reporting process used reflected both the technology of the time and a trusting relationship between line pilots and management that would be regarded as quite unusual today. In order to protect the line pilot who reported, the flight manager phoned in a synopsis of the interview regarding the incident to an automatic recording device that was kept in a locked cabinet in United's Flight Operations Department. No identifying data such as names, dates, trip numbers or the name or domicile of the Flight Manager making the report were included. The recorded tape was transcribed each morning by a secretary and immediately erased. With any one of over 100 flight managers as a possible source, it was virtually impossible for that source to be traced and identified.

It was a complex operation, but preservation of the anonymity of the individuals involved was considered crucial. The program was highly successful and the number of reports received did not change even when the program was expanded to include line-pilot interviewers. No attempt by anyone was ever made to breach its security. Unfortunately, this cannot be said of the management of all airlines that are participating in the FAA's newly instituted Flight Operations Quality Assurance (FOQA) program. (The FOQA program is discussed in Chapter 20—The Worldwide Safety Challenge.) There is no better way to sabotage a program based on confidentiality than to breach confidentiality promises that have been made by one of the involved parties.

Neither pilots, nor anyone else, will participate in a program that can cause serious personal jeopardy to those who participate. Beyond that, they will adamantly oppose any program that can cause such jeopardy. In the interests of overall safety, the authors completely agree with the position taken by Captain Dunkle when he said, 'I would rather know about the incident than punish the offender'. The NASA/FAA ASRS program, ICAO, and the Flight Safety Foundation take the same position.

Incident Reporting Can Help

The importance of non-punitive incident reporting received national attention when on 4 December 1974 Trans World Airlines (TWA) Flight 514 crashed while making an approach to Dulles Airport. The flight crew misinterpreted an approach chart and the clearance that they were given by Approach Control. The flight descended below the minimum safe altitude specified for the area in which it was flying and collided with a Virginia mountaintop called Round Hill.

The NTSB soon learned from United and from the FAA that only six weeks before the TWA crash, a United flight went through an almost exact duplicate of the events leading up to the TWA crash. A United flight had been at the same altitudes, had received an identical clearance, misunderstood the clearance in an identical fashion, and only through good fortune was able to avoid a similar fate.

Upon landing, the United crew realized that they had flown at a low altitude for an unusually long period of time during their approach and reported their experience through United's Flight Safety Awareness Program. The ambiguous nature of the charted procedures and the differences in the interpretation of the clearances delivered between the pilots and the controllers became clear during the interview the copilot gave following the incident. All United pilots were immediately made aware of the potential trap in clearance misinterpretation through a company emergency telemeter to all pilots.

The FAA was also notified. Unfortunately, there was no generally accepted method for promptly circulating such information throughout the industry. In its report on the TWA accident, the NTSB stated that it was 'encouraged that such safety experiences (as United's) had been initiated'. The report further stated: 'In retrospect, the Board finds it most unfortunate that an incident of this nature was not, at the time of its occurrence, subject to uninhibited reporting...which might have resulted in broad and timely dissemination of the safety message issued by the carrier to its own flight crews.' The TWA accident and the NTSB's report were almost certainly among the most crucial events that led to the US Aviation Safety Reporting System— NASA/FAA's ASRS.

NASA/FAA's Aviation Safety Reporting System (ASRS)

The FAA is the agency responsible for the promotion, regulation, and safety of aviation in the US. Following the TWA accident at Round Hill, it received considerable pressure to do something from Congress, the Board, and virtually all elements of the US aviation system. Some cynics believed that the proximity of Round Hill to Washington DC was a major factor in the high level of Congressional interest.

One of the FAA's actions was to institute its own declared non-punitive incident reporting system. This first FAA program was a failure because pilots simply would not trust the FAA. Properly or improperly, the FAA, as both the maker of the law and its enforcer, was not generally viewed as a properly disinterested party.

To its considerable credit the FAA did not give up. In August of 1975 it signed an interagency agreement with NASA in which NASA, as a neutral third party and 'honest broker' that had no rule-making or enforcing responsibilities, would collect, process, and analyze the voluntarily submitted reports received from the aviation community.

The result was the inauguration of the highly successful Aviation Safety Reporting System (ASRS). Through the ASRS, members of the aviation community were offered anonymity and limited immunity to individuals who reported incidents that identified deficiencies and discrepancies in the aviation system before they caused accidents or incidents. The major disclaimers were that the incident did not reveal a criminal action or a reportable accident. As mentioned at the start of this chapter, Dr. Charles Billings is a non-punitive incident reporting pioneer and was the chief architect of the present ASRS. He and Dr. John Lauber deserve much credit for the success of the FAA/ NASA program. Both were with NASA when the ASRS began.

The late William Reynard, who succeeded Dr. Billings as NASA's manager of the ASRS, was a lawyer. His legal training was considered essential for protection of the ASRS during the inauguration of an entirely non-punitive and confidential real-world incident reporting system in as litigious a society as exists in the US. During the initial part of the program, he, as well as Drs.

Billings and Lauber, read each report that was submitted. It takes very little imagination to see the damage that could be done to the program by even well intentioned legal intercession.

ASRS Today

ASRS report intake has grown from approximately 400 reports a month when the program was inaugurated in May of 1975 to the nearly 3,000 reports it now receives each month. Today, reflecting a total systems approach, the ASRS also has special reporting forms for cabin attendants and for mechanic technicians. If additional information is needed regarding a specific report, the process provides a very small telephone window for a trained ASRS analyst to contact the reporter. This window has never been abused. The system has remained truly anonymous.

To properly analyze the plethora of reports it receives, ASRS uses 15 trained full and part-time analysts. ASRS analysts include retired controllers who have ATC experience in all types of control facilities and pilots with certifications and ratings from single engine land to advanced 'glass cockpit' aircraft. The analysts have over 400 years of pilot and controller experience. With increased emphasis and recognition of cabin attendant and mechanic technician problems, expertise in these areas has been added to the analyst staff.

An obvious problem for ASRS is that it is inundated with pilot reports. Because it takes a variable and appreciable amount of time to properly analyze reports that are sometimes complex, the ASRS has developed a sophisticated processing method that assures that any significant problems reported will be detected. There is no way that each report can be completely analyzed within ASRS budgets, however the ASRS processing method assures that all critical reports are fully analyzed.

The ASRS uses several methods to make the information it receives available to the aviation community. Perhaps the most popular is the monthly newsletter, *CALLBACK*, which informs its readers of ASRS programs, report processing and research activity, and shares with the aviation community interesting and informative reports that ASRS receives. Extremely important are the alerting messages sent when a hazardous situation is reported. Alerting messages relay safety information to individuals in a position of authority so that they can investigate the allegation and take corrective actions if they are needed. From 1976 to 1996 ASRS issued 1,850 Alert Bulletins and FYI (for your information) notices to individuals in the aviation system who can deal with the problem reported. A majority of them were FYI notices. ASRS also publishes a quarterly safety bulletin called ASRS *Directline* that contains articles regarding data ASRS received. *Directline* was created to promptly meet the needs of operators and pilots of complex aircraft. It is distributed to operational managers, safety officers, training organizations, and publications departments.

Information in the ASRS is available to all interested parties through a simple request. To date over 5,770 search requests for data have been fulfilled. Primary requestors have been the FAA, the general public, the media, academic institutions, the aviation community, foreign organizations, research organizations, the NASA, and the NTSB. Over 103 Quick Response Data Analyses have been performed for the FAA, NTSB, and other Government Agencies.

The ASRS has published more than 55 Research Studies and Special Papers. They are published as NASA Contractor Reports (CRs), NASA Technical Memoranda (TMs), NASA Reference Publications, papers presented at the Ohio State University's Biennial International Aviation Psychology Symposia, and in special papers to the industry. To keep its research agenda focused on current problems, the research agenda is developed from consultation with the FAA, NASA, the ASRS Staff, and from ASRS Research Conferences.

A process that is used by ASRS, and that the authors believe is perhaps the most productive use of the ASRS, is called a 'structured callback'. In a structured callback a specially trained ASRS researcher calls back the reporter (who has specific experience relative to the research problem) and asks him/her an in-depth series of questions regarding the problem. The reporter's name and telephone number are available to ASRS during the very small window that is available from the reporter's incident form, after which all identifiable information is destroyed.

Invariably, the reporter knows considerably more than can be gained from material in the written report. While a Structured Callback is still a manpower intensive process, it is considerably less intensive and less expensive than the traditional personal interview. It also can produce data from a much broader sample.

To date the ASRS has conducted Structured Callback Analysis Studies on Wake Turbulence Encounters, Multi-Engine Turbojet Uncommanded Upsets, Runway Incursion Incidents, and EMS Operations for NASA, the FAA, and the NTSB. As this is being written a structured callback study on problems associated with the ATC system called FANS is just beginning. FANS is further discussed in Chapter 22— The Air Transport Future.

Finally ASRS provides a great deal of information through its web pages on the Internet. The ASRS home page address is http://olias.arc.nasa.gov/asrs. The information and services available on the website include ASRS publications, ASRS Operational Issues Bulletins, Reporting Forms, Immunity Policy, information regarding the ASRS Database, an ASRS Program Overview, a Program Briefing, and methods to contact various ASRS function managers via e-mail. This system makes ASRS reports and ASRS data and information easily available to any interested person.

Non-punitive Incident Reporting Programs in Other Countries

Incident reporting programs are not a new thought in Europe. Sporadic attempts began there in the early 1960s. The first programs were the product of informal agreements among certain airlines and later became more formal when several members of IATA agreed to exchange incidents among themselves. They were only partially successful for a number of reasons. Not surprisingly, the principal reason was the old bugaboo of fear of the personal liability from actions of either corporate or regulatory sources.

Jack Howell, Director of the Air Navigation Bureau of ICAO, has stated that one of ICAO's basic challenges in the development of human factors in aviation is to '…identify ways in which to address and demolish existing barriers to the implementation of confidential, voluntary incident reporting and information exchange systems, so that such systems can be implemented on a world-wide basis' (Howell, 1996). At the present time there are non-punitive incident reporting programs in existence, or at least developing, in 12 other countries and the number is growing

There are many reasons establishing such programs is difficult. One of the main reasons is that from past experience pilots are not inclined to trust either their regulators or their companies to make incident programs truly non-punitive. The authors recall an IATA Technical Conference held in Istanbul, Turkey in 1975 where the flight operations representative of a well-known European airline forcefully proclaimed: 'If the pilots make a mistake, they must be punished.'

While we believe that such thinking represents far from enlightened thinking at the end of the 20th century, it does not mean that those who hold such views are necessarily 'bad' people. Attitudes reflect societal norms of behavior for a particular time and geographic area. It is also important to remember that meaningful attitudinal change is very difficult for some individuals. Unfortunately, many still believe that punishment for either advertent or inadvertent transgressions is an indispensable part of society's protection against violation of established rules. This belief remains a basic component of many national, regulatory, and organizational cultures.

It is for this reason that aspects of a country's culture can make a non-punitive incident program very difficult to administer or even to conceive. Even if qualified people and the necessary funds to run a meaningful program are available, both companies and their pilots may well fear the potential financial jeopardy in which an incident program can place them, especially in a punishment-oriented litigious society. Despite all of these difficulties, the safety potential of an effective non-punitive incident reporting program makes its achievement a very worthwhile goal.

Fortunately today, the importance of anonymous non-punitive incident reporting programs is becoming well recognized and the development of such programs is a worldwide phenomena. There are active programs in countries ranging from Canada and New Zealand to Russia and China. In addition to the Aviation Safety Reporting System (ASRS) in the US, some of the most prominent are:

- the Confidential Human Incident Reporting Program in the United Kingdom (CHIRP);

- Canada's Aviation Safety Program (CASRP/SECURITAS);

- Australia's Confidential Aviation Incident Reporting System (CAIRS);

- Germany and the European Community's European Confidential Aviation Safety Reporting Network (EUCARE).

Others becoming prominent are the South African Aviation Safety Council (SAASCo), New Zealand's ICARUS program, and Russia's Voluntary Aviation Reporting System.

As one would expect, these programs differ in quality, effectiveness, and general acceptance. However, they all recognize the importance of getting realistic information about real-world incidents and the contribution that this information can make toward overall safety.

Some Basic Questions

Three basic questions that all non-punitive safety incident reporting programs face are the question of immunity for the persons submitting the report, the quality of the data such programs furnish, and the further question of whether or not these programs really increase aviation safety. These questions will be discussed in the following paragraphs.

Immunity

Immunity is a critical issue because pilots (or anyone else) will not report if reporting can put them in any sort of jeopardy. Therefore, all incident reporting programs have the immunity problem. It varies depending upon the regulatory and corporate cultures, and in some cases even upon the national culture of the parties involved. In the US, which has a very litigious society and one oriented toward punishment as a just consequence of transgression, the subject of immunity is still somewhat contentious. Much emphasis to date has been focused on securing participant immunity, for immunity is a problem for anyone involved in flight operations.

In 1979, the FAA modified the original ASRS immunity provisions and established the following:

> The filing of a report with NASA concerning an incident or occurrence involving a violation of the Act or the Federal Aviation Regulations is considered by the FAA to be indicative of a constructive attitude. Such an attitude will tend to prevent future violations. Accordingly, although a finding of a violation may be made, neither a civil penalty nor certificate suspension will be imposed if:
>
> 1. The violation was inadvertent and not deliberate.
>
> 2. The violation did not involve a criminal offense, or accident, or action under Section 609 of the Act which discloses a lack of qualification or competency, which are wholly excluded from the policy.
>
> 3. The person has not been found in any prior FAA enforcement action to have committed a violation of the Federal Aviation Act, or of any regulation promulgated under that Act for a period of 5 years prior to the date of the occurrence, and
>
> 4. The person proves that, within 10 days after the violation, he or she completed

and delivered or mailed a written report of the incident or occurrence to NASA under ASRS.

Two kinds of immunity are of particular interest. One is 'use immunity', which applies to the use of the data obtained. The other is 'transactional immunity', which results in the waiver of disciplinary action for the person submitting the report. The third point listed above describes a change in the original transactional immunity policy. It is neither as broad nor lenient as the immunity given to reporters as a part of the original ASRS. However, this 1979 modification was satisfactory to the ASRS's industry group and to the FAA, whose legal branch believed that the original transactional immunity was too broad. Our belief is that a period of two years prior to the date of the occurrence would have been satisfactory and still have furnished ample deterrence to any sort of frivolous disregard of the regulations.

Do Anonymous Reports Furnish Good Data?

By definition, information in reports that are truly anonymous is not subject to verification or to refutation. To require, or even permit, verification or refutation would destroy the anonymity that makes such programs successful. Therefore, such data and information is always unverified. Its validity in rigorous scientific circles can be inherently suspect, but by no means does this suggest that the data and information is not valid.

In virtually all cases, one can be reasonably sure that the incident reported actually occurred. Furthermore, all anonymous incident reporting requires some level of analysis and understanding by the reporting party. When several other reporters report the same occurrence, there is a reasonable suspicion that a valid problem is being reported. If the reporter also volunteers his or her reasons for the occurrence, the reason may be completely accurate only in the opinion of the reporter. However, even if the reporter is wrong, misconceptions are important. It is important that misconceptions be corrected. Misconception was the major cause of the December 1974 TWA accident at Round Hill that led directly to the formation of the US ASRS.

Without an effective incident reporting program, the information the reports contain is virtually unobtainable. A major value of these reports is that they can identify areas that deserve additional investigation (Hardy, 1990). It is one of the reasons that the availability to conduct data search requests makes an important contribution to aviation safety.

Do Anonymous Incident Programs Increase Safety?

Safety in flight is the result of many factors, including items such as training, pilot experience levels, the infrastructure, and all aspects of the safety culture. It is virtually impossible to assign a relative figure to the contribution to safety that is made by each factor. Seldom are reports as dramatic and potentially life saving as the initial United Airlines incident report that was an unfortunate dress rehearsal for the TWA accident at Round Hill. An exception to this general rule may be the effect of the FAA regulation requiring GPWS for Part 121 carriers in the United States. The regulation resulted in the virtual elimination of CFIT accidents among Part 121 carriers.

Because of the difficulty of precisely identifying the role of each safety factor, the industry is forced to rely upon good judgement and common sense. We must rely upon the judgement of safety experts such as Bobbie Allen and the Australian delegate who told the attendees at an ICAO Accident Investigation meeting that 'a greater contribution to the safety of aircraft operations in Australia had been derived from the investigation of aircraft incidents than from the investigation of accidents'.

The late Bill Reynard, who succeeded Dr. Billings as NASA's manager of the ASRS program, said it nicely when he observed that 'it is difficult to record a non-event'. Both incidents and accidents are indeed events. As an exception to normal operations, they are relatively easy to record although not always easy to analyze thoroughly and effectively. In contrast, a routine safe flight is simply uneventful. It is a non-event. Because routine safe flights are a non-event and not recorded in this context, it is particularly difficult to quantify the contribution to aviation safety that is made by incident reporting programs.

Data from the US ASRS is both readily available and helpful. Since 1976 ASRS has received and processed over 410,000 reports. It maintains a database that contains over 70,000 incidents which is the world's largest repository of human factors aviation data. Utilizing data reported to it and which in most cases would be otherwise unavailable, ASRS has issued over 2,300 Safety Alerting Messages on reported aviation hazards. In addition, the ASRS has accomplished over 5,770 database search requests for the government, public, academic institutions, research organizations, the media, foreign aviation organizations, and the aviation community. It has accomplished 103 Quick Response Data Analyses for the FAA, NTSB, and for other government agencies. And finally it has its regular publications such as *CALLBACK* and *Directline* that furnish the aviation community with current aviation safety material. While this does not prove the specific contribution of ASRS to aviation safety, there can be little doubt that it has been substantial.

Other Safety Considerations

Three other considerations are also important. One arises because of the inherent difficulty of securing reliable data on occurrences that happen in day-to-day line operations. An effective incident reporting program can provide this information. To be effective, the incident reporting system must be non-punitive. An enlightened management can obtain information that would be otherwise unobtainable by providing a non-punitive incident reporting system. It is impossible for anyone to fully supervise each and every flight. All of these considerations apply equally to the appropriate regulatory authority.

The second consideration arises because of the constructive and highly individualized training that happens to any individual that files an incident report. In virtually all cases, an incident is an unwanted occurrence to the individual involved in the incident. It is true that some reports are filed simply to try and avoid any punitive action that might occur—for example, deviating from an assigned altitude. But it is also true that reporting an incident requires reconstruction of and contemplation of the factors that were involved. Inevitably it also involves consideration of things that will prevent reoccurrence of the unwanted occurrence The result is an often timely, introspective, painless training dealing with the specific kind of unwanted occurrence that the crew has just experienced. The training is ideally timed to be truly effective.

A third consideration arises with questions of data sharing, confidentiality, and freedom from personal jeopardy for the participants of any voluntary program. At the moment these are very real problems with the FAA's Flight Operations Quality Assurance Programs (FOQA) which is discussed in Chapter 20—The Worldwide Safety Challenge. FOQA attempts to secure operational information that would be otherwise unavailable and promises airlines and their pilots freedom from jeopardy if they participate. Neither companies nor their pilots are so sure, and as we have seen, the administration of FOQA is still not without these basic problems. We will discuss some of these problems in Chapter 20—The Worldwide Safety Challenge.

This, of course, has been an overview of some of the contributions to aviation safety that have been made by the US ASRS. Similar contributions have been and will continue to be made by incident reporting programs in other countries throughout the world.

References

Allen, Bobbie R. (1966). 'Incident Investigation — The Sleeping Giant', *A Summary of the 19ᵗʰ Annual Air Safety Seminar*, Madrid, 15-18 November 1966, Flight Safety Foundation, Arlington, Virginia.

Ashby, Gus (1974). 'Task Force for Flight Safety Awareness Program', United Airlines, Chicago, Illinois.

Aviation Safety Reporting System (1997). *Program Overview*, NASA/Ames Research Center, Moffett Field, California.

Dunkle, William D. (1973). Letter to Harry Orlady re human factors activity, United Airlines, Chicago, Illinois.

Hardy, Rex (1990). *CALLBACK: NASA's Aviation Safety Reporting System*, The Smithsonian Institution, Washington, D.C.

Howell, Jack (1996). Opening address in *Proceedings of The Third Global Flight Safety and Human Factors Symposium*, International Civil Aviation Organization, Montreal, Canada.

ICAO (1996). *Human Factors Digest No. 13*, International Civil Aviation Organization, Montreal, Canada.

Kidera, George J. and Orlady, Harry W. (1975). 'Non-punitive Aviation Incident Reporting', presented at XXIII Congress of Aviation and Space Medicine, 30 September 1975, United Air Lines, Chicago, Illinois.

Orlady, Harry W. (1975). 'A Study of Turbo-Jet Transport Approach and Landing Accidents', Twentieth Technical Conference, International Air Transport Association, Montreal Canada.

Reynard, W. D., Billings, C. E., Cheaney, E. S., and Hardy, R. (1986), *The Development of the NASA Aviation Safety Reporting System*, NASA Reference Publication 1114, NASA/Ames Research Center, Moffett Field, California.

United Airlines Safety Awareness Program (1974). 'Guide for the Collection of Aviation Safety Data in Airline Operations', United Airlines, Chicago, Illinois.

United Airlines Safety Awareness Program (1975-1978). *Flight Safety Awareness Bulletins*, United Airlines, Chicago, Illinois.

19 Some Ramifications of Accident Analysis

> ...in general, errors are due to combinations of causes, rarely to a single cause, and this contributes to the difficulty of identifying and preventing them.
> (Senders and Moray, 1991)

The Assignment of Causes in Human Behavior

A major goal of human factors in aviation is to make the interfaces in the man-machine-environment more efficient and effective and to ensure that appropriate consideration is given to the long-term well being of the man. An important part of that goal is to identify the causal role of the factors that are involved, so those adverse factors can be eliminated, or if not eliminated, at least controlled. This is not an easy task. The difficulty is well illustrated with the example that Alphonse Chapanis used nearly 40 years ago to show the difficulty of determining a single causal role of human behavior in a relatively simple accident:

> The assignment of causes in human behavior is an extremely difficult thing to do. Take this example: A man has a protracted argument with his wife. He stamps out of the house to the nearest bar and drinks four highballs. He then decides to go for a ride. It is nighttime, there is a skim of snow on the ground, and the tires on our victim's car are smooth. In rounding a poorly banked curve at excessive speed, the right front tire blows out, the car leaves the road and is demolished. What was the cause of the accident? The argument? Drinking? Speed? The weather? The smooth tires? The blowout? The poorly designed highway? It is impossible to say, for if we had changed any one of these factors, perhaps the accident would not have happened. We have no way of assigning a "cause", even though we may have a complete description of the circumstances leading up to the accident. In fact, the chances are very good that a coroner, state policeman, minister, psychiatrist, and highway safety engineer would each find different causes in this event. (Alphonse Chapanis, 1959)

412

Traditional Analysis of Air Transport Accidents

Annex 13 of the Chicago Convention

Annex 13 of the Chicago Convention (see Appendix D) now states: 'The sole objective of investigation of an accident or incident shall be the *prevention* of accidents and incidents. *It shall not be the purpose of this activity to appor-tion blame or liability*' (Italics supplied). ICAO is, of course, a political orga-nization that is subjected to political pressures. The quoted sentence was not always there. It is an extremely important part of Annex 13 and was the result of a long and bitter struggle that extended over many years. In spite of the fact that the inclusion of this sentence to Annex 13 was signed by all 184 ICAO member States, living up to the statement is sometimes difficult in today's environment.

The 'Stop Rule' Problem

Air transport is a tightly coupled, complex system. In addition to inherent operational challenges, it is continually pushed to its limits by political and economic forces. As previously noted, one of the problems that was identified by Rasmussen, Duncan, and Leplat (1987) is that 'the identification of an event as a human error depends entirely upon the stop rule applied for the explanatory search after the facts'. They further state: 'How far back to seek is a rather open question; generally, the search will stop when one or more changes are found which are familiar and therefore acceptable as explana-tions, and to which something can be done for correction.' Their observation applies well to the traditional method of investigating air transport accidents.

There seems little question that in the past most air transport accident investigations have stopped too soon. Investigations have stopped when in-vestigators found what they considered the probable cause. Contributing and latent causes were ignored. It is possible that if they had been identified, mean-ingful corrections to prevent similar accidents could have been taken. Unfor-tunately, too often the things that were done for the correction of a single primary cause have not solved the problem or problems that created the acci-dent.

ICAO's Dan Maurino has attacked the culture of traditional accident investigators who, with few exceptions, have not considered it a part of their task to investigate the human factors aspect of their investigations. In some cases, one reason may have been the availability of qualified individuals.

Maurino tells us that until such individuals are seen as an integral part of any serious investigation '...we will continue to reinforce areas (of flight operations) which are already very good, and fail to deal with the weak aspects' (Maurino et al., 1995).

Many thoughtful critics maintain that traditional accident analysis in the US, and in many other parts of the world, seems more interested in finding fault or blame than in preventing accidents. This tendency to stop after finding fault may be due to the emphasis on finding a 'probable cause'. The law itself and the demand of the media, Congress, the public, and the attorneys for the unfortunate victims and their heirs to find quick and simple answers to what often are difficult and complex problems puts extraordinary pressure on the NTSB and on the FAA (Miller[1], 1998). This is unfortunate because most aviation experts (including those in the NTSB and the FAA) believe that the total anatomy of aircraft accidents virtually always reveals an extremely relevant sequence-of-events. This is very close to the 'error chain' theory, which is discussed in later paragraphs. Senders and Moray have noted that a basic difficulty in accident analysis always has been that 'in general, errors are due to combinations of causes rarely to a single cause, and this contributes to the difficulty of identifying and preventing them' (Senders and Moray, 1991).

In 'Tyranny of the "Probable Cause"', C.O. Miller states: 'A basic corollary of the sequence-of-events principle is that *the objective of a safety investigation is to break any possible link* of that accident-producing chain.' He further states: 'the hazard in accepting the "probable cause"' approach rests principally on its contravening the very fundamental sequence-of-events principle and leads to oversimplification of the analysis; hence, limited scope of possible corrective action' (Miller, 1998).

The Error Chain

A theory of air transport accident/incident causation that is widely supported by many experts is the concept of the 'error chain'. This concept holds that most accidents/incidents consist of a series of errors or chain links, any one of which, if interrupted, could have prevented the accident. Under this theory,

---

[1] C.O. (Chuck) Miller holds a BS in aeronautical engineering, a MS in systems management, and a JD. He has been professionally involved as a safety professional in the aviation industry, private consulting, research, and major universities. His government service included serving as director of the NTSB's Bureau of Aviation Safety from 1968 to 1974.

the prevention, or at least the control, of those errors by breaking the chain is a very effective way to improve the safety record. We should note that there are both internal and external causes of errors. Clear definition and precise identification of the elements in the 'error chain' and of the practical aspects of how the likelihood of the accident could have been prevented are not always a conspicuous part of many accident analyses.

In 1985, Richard Sears, a product safety engineer of the Boeing Commercial Airplane Company, advocated a similar, or perhaps tangential, approach (Sears, 1985). Accident reports covering accidents that occurred between 1977 and 1984 were critically reviewed. His study reaffirmed that the major accident rate (he defined a major accident as either a fatal accident or a hull loss) showed a nearly constant rate from 1967 through 1984. The study showed that the average worldwide rates (excluding those accidents that were incurred by either USSR or Mainland China aircraft or those accidents resulting from sabotage, hijacking, or military action) were approximately 20 major accidents per year. An average of five of them occurred to US operators. As Sears notes, a basic analytical problem is 'that careful study of each accident clearly brings out the random and unpredictable nature of commercial jet transport accidents....' The extent to which accidents during this period were completely analyzed was not discussed.

The 93 accidents that occurred to world airlines between 1977 and 1984 were further reviewed by Sears to discover the causal factors, which fulfilled the following criteria:

1. The accident might reasonably have been prevented had the factor not been present (this is essentially the 'error chain' concept), and

2. A definitive solution or remedy can be envisioned for the elimination of the factor.

Using these criteria, 182 causes were attributed to the 93 accidents. Their distribution is shown in Figure 19.1. Of these accidents, 54 percent had two factors; only 28 percent were single factor accidents.

There were 24 significant accident contributors that met Sears' criteria. These contributors ranged from 'pilot deviated from basic operational procedures' or 'inadequate crosscheck by 2nd crew member', to 'operational procedures did not require use of available approach aids' or 'Captain inexperienced in aircraft type.' Many of them involved human factors issues and many of them were categorized (properly or improperly) as flight crew

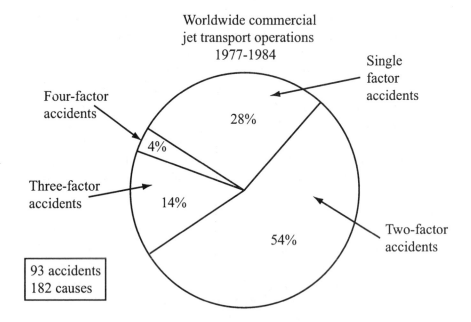

Figure 19.1 Distribution of the Number of Significant Contributors Per Accident

Source: 'A New Look at Accident Contributors and the Implications of Operational and Training Procedures', page 6, Sears, 1985

errors. Seventy-one percent of them directly involved compliance with standard operating procedures.

'The Significant Accident Causes and Their Percentage of Presence in 93 Major Accidents' and the 'Flight Crew Causes and Their Percentage of Presence in 93 Major Accidents' are given in Appendix M. They clearly illustrate the reasons that the industry is interested in flight crew errors and the flight crew training challenge to reduce (or hopefully eliminate) the crew-related accident. These data also illustrate the reason that while the time, effort, and money spent on flightcrew training is already enormous, it is money very well spent if it is spent on effective training.

Ultimately, the last chance for the detection of errors and for the prevention of an accident itself rests with the flight crew. This point was well made by Gerard Bruggink at a special meeting of ALPA pilots when he told them: 'As pilots you are at the receiving end of all the technical and attitudinal discrepancies in the system....You are often the last one who can interrupt the development of an unsatisfactory situation, **if** you are aware of it' (Bruggink, 1975).

The System-Generated Causes of Operational Errors

All factors are an important part of the accident/incident puzzle, yet many have been virtually ignored. A total system approach is needed. The public expects and the industry needs the analysis of accidents and incidents to identify and then correct all of the causes of air transport's unwanted events.

Fortunately, more of the industry (helped by ICAO) is now looking at world air transport safety from a system viewpoint. This includes both air transport's social environment and the system's infrastructure. While this new look does not suggest that the large question of traditional flight crew error be neglected, it does suggest that human error is not limited to the personnel directly involved in the accident or incident, or as several others have put it, to those that find themselves at 'the sharp end of the stick'.

We now know that limiting the search for the causes of air transport accidents to the performance of the individuals directly involved in those catastrophes and then stopping the search there invariably results in an incomplete analysis. A great deal of inertia remains in the old system, and we are still too much of a 'single-cause-oriented' society. The problem is exacerbated because, and this is certainly true in the US, we are also a litigious society.

It is important to recognize that, in too many cases, traditional analysis and the kinds of corrections usually made have often missed significant opportunities to increase air safety. Too often analyses of air transport accidents and incidents have ignored those causes that have their origin in national, regulatory, or organizational cultures (the social environment). Too often analyses have ignored the latent conditions, which are described later in this chapter.

The Canadian investigation of the 10 March 1989 Air Ontario accident at Dryden, Ontario is an exception. It took a total system approach and was thorough. A special Commission of Inquiry headed by The Honorable Virgil P. Moshansky conducted the accident investigation. He has written: 'The commission represented a rare opportunity to examine the entire aviation system for latent and active failures which might have contributed to the Captain's faulty decision.' A conscious decision was made to launch an in-depth search for such failures, and to investigate fully the impact of Human Factors throughout the aviation system upon the events at Dryden' (Moshansky, 1995). The result was a landmark accident investigation and report. Justice Moshansky later stated that, 'This accident was the result of failure in the air transportation system as a whole.' Emphasizing the importance of these findings, Professor James Reason of Manchester University noted: 'Those on the spot were more the inheritors than the instigators of the accident sequence.'

The Total System Approach

Professor Reason has pioneered a new organizational framework that involves a total system approach to accident investigation. It expands and refines the concept of the error chain. This approach has been endorsed by virtually all air transport organizations—by both domestic and international regulatory authorities, by both foreign and domestic airlines, by pilot organizations, and by the air traffic controllers. It extends incident or accident analysis from not only the individuals whose errors have directly played a part in these catastrophic events but also to all of the people or organizations that could have been involved. A major contribution has been the identification of weak spots or 'latent conditions' in the system that may have lain dormant for long periods until a combination of circumstances triggers the final and fatal human error. Passive latent errors are not only errors in themselves but they also increase the potential for later active human errors in the system. As Anchard Zeller, speaking of human limitations, told us many years ago: 'If this potential is repeated often enough, an accident will result' (Zeller, 1966). This principle also applies to latent conditions. If a latent condition exists for long enough, it will eventually lead to an accident.

This new organizational investigation framework has helped the approach toward a total systems analysis tremendously. It has been officially adopted by several diverse organizations—all of whom have the common goal of maximizing aviation safety. These diverse organizations include ICAO, the International Air Transport Association (IATA), the International Federation of Air Line Pilots Associations (IFALPA), the International Federation of Air Traffic Controllers Associations (IFATCA), the NTSB of the United States, the Bureau of Air Safety Investigation of Australia (BASI), Transport Canada, and many others, including US airlines and the Air Line Pilots Association (ALPA).

Active Failures and Latent Conditions

The total system approach extends incident or accident analysis from not only the individuals who directly played a part, to all of the people or organizations that could have been involved. A major contribution is recognition of the existence and relevance of latent conditions. Latent conditions can be defined as 'loopholes in the system's defenses...whose potential existed for some time prior to the onset of the accident sequence, though usually without any obvious bad effects.' Latent conditions may lie dormant for long periods, even years, before they combine with active failures and other local triggering events to breach the air transport system's defenses. While today latent conditions can achieve notoriety on rare occasions, they are important at all times.

Existing (latent) conditions are important because they create a state that facilitates the probability of an eventual active failure and an accident. Active failures, which are the type with which we are most familiar, have an immediate and direct impact upon the accident or incident. They receive immediate attention. If a system approach is used, and accidents and serious incidents are thoroughly investigated, latent conditions are also often implicated.

Figure 19.2 shows the way that active failures and latent conditions can combine to cause an undesired event.

Two of the most attractive concepts of this approach to accident analysis are the development of a process that includes (1) analyzing an accident or incident from a total system viewpoint, and (2) the identification of both *active failures* and *latent condition* pathways to an event. An event is defined as the 'breaching, absence or bypassing of some or all of the system's defenses and safeguards.' Considerations vary from 'managerial decisions to conditions in the workplaces involved (flight decks, hangars, etc.), and to personal and situational factors that lead to errors and violations. The event may result in either a catastrophe or simply result in an incident and a "free lesson", depending too often on mere chance and whether or not the system's defenses are operative and effective' (Maurino et al., 1995). Organizational or managerial decisions, as well as decisions made by the active participants, are all influenced by the social environment.

Active failures are caused by those in direct contact with the system such as pilots, air traffic controllers, mechanics, and the like. They are relatively easy to identify. In contrast, latent conditions are usually a part of the organizational culture, including the defenses, barriers, and safeguards that are a part of the total system. We have already seen that national and organizational cultural factors can be very slow to change. Nevertheless, the system must deal with them.

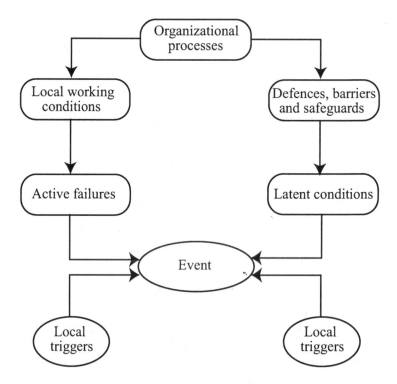

Figure 19.2 Active and Latent Failures Combining to Cause an Event
Source: *Beyond Aviation Human Factors*, page 24, Maurino, et al., 1995,
                 and used with permission of Ashgate Publishing Limited

Figure 19.3 shows the ways that active failures and latent conditions in the overall defense system can overlap or combine to result in an air transport accident. In the real world, the holes that represent elements in the defense system are not static. The position of the holes depicting active failures and latent failures can move depending upon time and conditions. An accident only occurs when the holes line up and none of the defenses within the aviation system and applicable to the particular situation are effective. The size of the limited window of accident opportunity shown at the top of the figure varies with effectiveness of total system management.

Management as a Causal Factor?

Some have argued that management is at least partially implicated every time air carrier personnel are identified as the 'primary cause' of an air transport

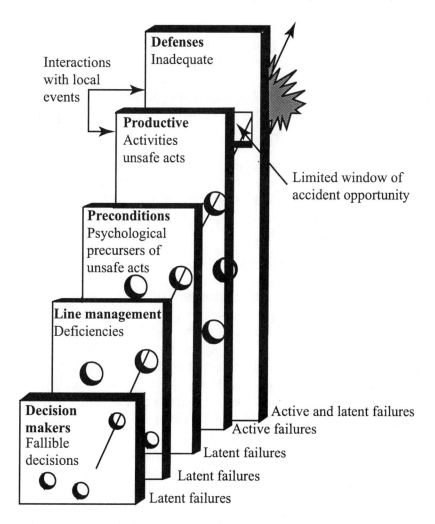

Interactions with local events

Defenses Inadequate

Productive Activities unsafe acts

Limited window of accident opportunity

Preconditions Psychological precursers of unsafe acts

Line management Deficiencies

Decision makers Fallible decisions

Active and latent failures
Active failures
Latent failures
Latent failures
Latent failures

Figure 19.3 Reason's Human Contribution to Accidents in Complex Systems

Source: Adapted from *ICAO Human Factors Digest No. 10*, page 19

accident. This is not a new thought (Prendal, 1974). Additionally, it is now clear that organizational factors are at least partially responsible for many air transport accidents or incidents (Johnston, 1991).

Managerial responsibility also has been linked in other industries. On 06 March 1987 a North Sea passenger and freight ferry, The Herald of Free Enterprise, capsized with a loss of 193 lives just four minutes after departing the inner harbor at Zeebrugge, Belgium. Managerial responsibility was directly involved. A paper, 'The Terrible Risk - It's Worth a Thought', dealing with the lessons that that accident might mean to air transport was written by a veteran pilot and manager, Captain B.S. Grieve.[2] Captain Grieve stated directly and bluntly: 'Managers must accept that the safety of an operation is entirely within their control.' He has written further and with considerable insight:

> The pilots who fly company aircraft are simply an extension of the care taken by a management team who aims to ensure that risk is reduced to an absolute minimum while recognizing the need for a delicate balance with commercial aspirations. (B.S. Grieve, 1990)

A Changing Accident Pattern

Historically, stereotyped revelations that the flight crew played a direct and primary role in air transport accidents have played an major part in the impression people in and out of the industry have of those accidents. In the US this perception was reinforced by a 1994 NTSB safety study (NTSB/SS-94/01), which stated: 'Actions or inaction of the flight crew have been cited in the majority of fatal air carrier accidents.' However, as a study by Gerard Bruggink has pointed out, the Board reached this conclusion by lumping together the data on 13 years (1978-1990) of major airline accidents.

In an effort to see if there had been any change in the accident pattern, Bruggink used NTSB data to examine US air carrier accidents that occurred in the 20 years between 1977 and 1996. He considered only those accidents in which 'the flight crew had operational control of the aircraft moving under its own power'. He excluded eight single-fatality ramp accidents, which carry the same weight in NTSB statistics as a catastrophic accident that kills all occupants of an aircraft.[3] Incidentally, in none of these ramp accidents did the NTSB attribute 'cause' to the flight crew.

In 'A Changing Accident Pattern', Bruggink reported that if the 20 years is divided into five year segments 'the percentage of accidents that were be-

---

[2] At the time of writing, Captain B. S. Grieve is the Operations Director of a major British airline.

[3] An interesting aspect of these eight cases is that they could have happened without the 'intention of flight' and therefore would not have fallen under the ICAO definition of an aircraft accident.

yond crew control were, according to the NTSB, 33% from 1977-1981, 22% from 1982-1986, 60% from 1987 to 1991, and 67% from 1992-96 (Bruggink, 1997). More dramatically, the percentage of accidents that were beyond crew control from 1977 to 1986 was 25.5%, from 1987 to 1996 the percentage was 63.5%. These data are shown graphically in Figure 19.4.

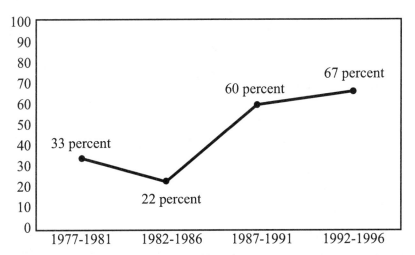

Figure 19.4 Percentage of Fatal Accidents Beyond Crew Control in Scheduled Operations Under Part 121

Source: Reprinted from *Air Line Pilot*, page 11, May 1997 with permission. Copyright ©*Air Line Pilot*

There is an obvious and striking difference in the accident pattern of air transport accidents the NTSB analyzed over the 20 years from 1977 to 1996. Bruggink's data illustrates the ease with which perspective errors can be made if data over too many years are lumped together. It is not as easy to identify the reasons for the obvious change. Bruggink listed two possibilities:

- The growing emphasis on better team performance of flight crews that began in earnest in the 1980s, and

- The gradual but consistent shift to regional carriers of short-haul passenger (operations)...

The NTSB grouped 13 years of data in its study. Its analysis cited flight crew error in most of the major air carrier accidents that occurred during that period. Now it has become clear that there has been a marked change in the distribution of the causes of these accidents. The NTSB conclusion is not true for the later years. In addition to the reasons for the pattern change that Bruggink has postulated, we believe another factor has been involved. That factor is the increased sophistication of airline training and the increased sophistication of accident analysis. It is important to remember that while the FAA regulation requiring GPWS in Part 121 and Part 125 air carriers substantially reduced, but did not eliminate, CFIT accidents, the regulation probably had minimal effect on the NTSB and Bruggink studies because the effectiveness of the regulation was well established before the change in the causes of the accidents that were examined.

Bruggink's study is a valuable and provocative report and there are lessons to be learned from it. The lessons may be quite different in the well-developed countries than the lessons that can be learned from those countries operating in significantly different environments.

A Final Thought

A purpose of this book has been to stimulate the type of awareness that will make all members of the operating team aware of their importance in accident/incident prevention and to help ensure that members of the operating team are not simply the victims of a series of situations that make an accident or incident eventually inevitable. When an accident does occur, it is clear that in virtually all cases a total system approach should be used for the analysis of the accident. A total system approach makes it possible for the aviation system to take advantage of the lessons that may be learned. It is clear that all aspects of the social environment must be considered.

References

Bruggink, Gerard B. (1975). 'The Last Line of Defense', presented to the special meeting of ALPA Pilots, 14 April 1975, New Orleans, Louisiana.

Bruggink, Gerard, B. (1997). 'A Changing Accident Pattern', *Air Line Pilot*, May 1997, The Air Line Pilots Association, Herndon, Virginia.

Chapanis, Alphonse (1959). *Research Techniques in Human Engineering*, John Hopkins Press, Baltimore, Maryland.

Grieve, B.S., (1990). 'The Terrible Risk', Britannia Airways Ltd., Bedfordshire, England.

Hawkins, Frank H. (1993). *Human Factors in Flight*, 2nd Edition, edited by Orlady, Harry W., Ashgate Publishing Ltd., Aldershot, England.

Howard, Benjamin (1954). 'The Attainment of Greater Safety', presented at the 1st Annual ALPA Air Safety Forum, and later reprinted for presentation at the Aircraft Accident Prevention Course, University of Southern California, July 1957.

Johnston, Neil (1991). 'Organizational Factors in Human Factors Accident Investigation', *Proceedings of the Sixth International Symposium on Aviation Psychology*, 29 April-2 May 1991, The Ohio State University, Columbus, Ohio.

Lautman, L.G. and Gallimore, P.L. (1987). 'Control of the Crew-Caused Accident', *Flight Safety Foundation Flight Safety Digest*, Flight Safety Foundation, Alexandria Virginia.

Lundberg, Bo O. K. (1966). *The 'Allotment-of Probability-Shares' - APS - Method, Memorandum PE-18,* The Aeronautical Research Institute of Sweden (FFA), Stockholm, Sweden.

Maurino, Daniel E., Reason, James, Johnston, Neil, and Lee, Rob B. (1995). *Beyond Aviation Human Factors*, Ashgate Publishing Limited, Aldershot, Hants, England.

Miller, C.O. (1998). 'Trapped by 'Probable Cause', *Air Line Pilot*, The Air Line Pilots Association, International, Herndon, Virginia.

Moshansky, Virgil P. (1995). From the Foreword to *Beyond Aviation Human Factors*, Maurino, Reason, Johnston, and Lee, Ashgate Publishing Limited, Aldershot, Hants, England.

North, David M. (1997). Editorial—'We Know the Safety Issues, Now Let's Push Solutions', *Aviation Week and Space Technology*, 1 December 1997, McGraw and Hill Inc., New York.

Prendal, Bjarne (1974). 'Management and Communication: Discipline and Motivation', Flight Safety Foundation's 27th International Aviation Safety Seminar, Flight Safety Foundation, New York.

Rasmussen, Jens, Duncan, Keith, and Leplat, Jacques, eds. (1987). *New Technology and Human Error*, John Wiley & Sons Ltd., Chichester, England.

Reason, James (1997). *Managing the Risks of Organizational Accidents*, Ashgate Publishing Limited, Aldershot, Hants, England.

Sears, Richard L. (1985). 'A New Look at Accident Contributors and the Implications of Operational and Training Procedures', presented at Flight Safety Foundation 38th International Air Safety Seminar, Flight Safety Foundation, Alexandria, Virginia.

Senders, John W. and Moray, Neville, P. (1991). *Human Error: Cause, Prediction, and Reduction*, Lawrence Erlbaum Associates, Hillsdale, New Jersey.

Wiener, E.L. (1989). *Human factors of advanced technology ("glass cockpit") transport aircraft*, NASA Technical Report 177528, Ames Research Center, Moffett Field, California.

Zeller, Anchard F. (1966). *Summary of Human Factors Session, 19th Annual International Air Safety Seminar*, Flight Safety Foundation, Alexandria, Virginia.

20 The Worldwide Safety Challenge

> Man's greatest sin is the unnecessary taking of human lives.
> (Benjamin O. Howard, 1954)

The Concept and Reality of Risk

There is an inherent risk in air transport as there is in any kind of travel. In 1903, Wilbur Wright noted that 'If you are looking for perfect safety, you would do well to sit on a fence and watch the birds.' Safety is inherently and inevitably relative, not absolute, for there is always an element of risk. Dr. John Lauber, a renowned expert and a leader in air transport safety, has stated rather bluntly that the 'talk of "zero risk" has done a disservice to the public....."zero risk" is not possible. Aviation operations are always going to be risk-intensive, and we should put that up front.'

The probably unattainable goal of '0' accidents and '0' incidents has been a goal for many years. In the US, the phrase, '0' accidents, was given recent attention by Frederico Pena, then the Secretary of Transportation, when he proclaimed that it was the goal of his administration. One of the first times either of the authors remember hearing this phrase was in early 1973 when Dr. George Kidera, then Medical Director of United Airlines, used it as a part of the closing statement of the training film, *Incapacitated Crew*. The film was made as one more step towards the ultimate goal of '0' accidents and '0' incidents.

It is important to recognize that '0' accidents and '0' incidents do not mean '0' risk. As we stated in Chapter 2—The Industry and its Safety Record, Benjamin Howard noted years ago: 'It will be found that the risks associated with flying are some 95 percent the result of the flight having been made and 5 percent the result of the flight time and distance involved' (Howard, 1953). Professor James Reason, utilizing the FSF's Icarus Committee's statistics wrote: '...in 1995, the risks to passengers (the probability of becoming involved in an accident with at least one fatality) varied by a factor of 42 across the world's air carriers, from 1 in 260,000 chance of death or injury in the worst cases to a 1 in 11,000,000 probability in the best cases' (Reason, 1997). The task of the industry is to control the risk that is inherent in air transport in order to make further progress toward the elusive goal of '0' accidents.

Accident Rates and Statistics

Progress in airline safety is sometimes not fully appreciated. Considerable progress has been made. As an example, Sir John Dent, when he was the United Kingdom's Civil Aviation Authority (CAA) chairman, noted that '... if the accident rate in 1985 had been the same as it was in 1950, we would have had 23,000 fatalities'. Instead, the industry had 2,000 fatalities. In 1997, despite a significantly larger base, the worldwide fatalities were down to 1,307.

Aviation as a worldwide industry has grown tremendously in the last two decades. Unfortunately, there has not been a significant reduction in either the total number of accidents or in the total number of fatalities. Virtually the only comparative statistics published by the popular media and press organizations are the number of accidents per year and the number of passengers killed. A perennial problem for the air transport industry is that facts and figures can be misleading if they are not kept in context.

This problem of a rapidly expanding industry and an essentially static safety rate has not suddenly appeared. Three decades ago, Bo K. O. Lundberg[1] warned the aviation world that: 'Public confidence in the safety of aviation *does not significantly depend upon the statistical risk level....* What really matters is how often catastrophic air accidents occur.' This is a fact of life. He noted that 'At a constant accident rate, the number of accidents will, of course, grow in direct proportion to the expansion of aviation' (Lundberg, 1966). The magnitude of this problem is shown in an estimate of industry progress that was given at an ALPA Technology and Flight Deck Symposium by David Hinson, when he was the US Federal Aviation Administrator. He told the attendees that there was a statistical probability that if the accident rate were to be held constant at the 1996 level of about one accident per million departures, the projected growth in traffic could result in a serious accident every week by 2015. As Mr. Hinson stated, '...the clock is running...time is short'.

---

[1] Bo O.K. Lundberg, a distinguished aeronautical engineer and aviation scientist, was Director General of the Aeronautical Research Institute of Sweden at the time this Memorandum (PE-18) was written. In 1964 he presented The Daniel and Florence Guggenheim Memorial Lecture at the Third International Congress of Aeronautical Sciences.

The Lautman and Gallimore Study

In 1986, Boeing's Les Lautman and Peter Gallimore reported on a study of 126 major air transport accidents that had occurred from 1977 through 1984. This is a major study and is particularly thought provoking. Among other things Lautman and Gallimore found that despite the beginning of increased attention to cockpit resource management programs and to human factors generally, over 70% of the major accidents—the fatal or hull-loss accidents––were still what have been called crew-caused accidents.

There is really nothing new in that statistic or that thought, although as we saw in the last chapter, the study by Gerard Bruggink, 'A Changing Accident Pattern', suggests that the old pattern is changing. What Bruggink's study shows us is that factors other than just the flightcrews should be considered in our accident analyses. It does not suggest that we should not be concerned about cockpit crew behavior. There is little question that aviation safety studies that attribute a high percentage of major air transport accidents to the flightcrew have been a frustrating fixture in air transport operations for a great many years.

The analysis of Lautman and Gallimore was based upon major accidents per million departures. We believe it is a much better measure than using accidents per million passenger miles or with using the large number of hours flown. If used selectively, accidents per million departures, which obviously also includes arrivals, can consider the meaningful exposure of each airline and each airplane.

We do know that the phase of flight is important and that arrivals and departures include approximately 94% of our accidents. In another study, that used a narrower definition of the time-span for departures and for approach and landing, Weener and Russell noted that 65% of the accidents occur in the first two and the last four minutes of the average flight, i.e., during the first two minutes after departure from an airport and the last four minutes of the approach and landing (Weener and Russell, 1993). These data are shown in Figure 20.1.

Safety—the Industry's Greatest Challenge

It is easy to underestimate the magnitude of the safety challenge the industry is facing. In the US alone, in the next decade the transport industry will be making as many as 35,000 takeoffs and landings each day. This means over

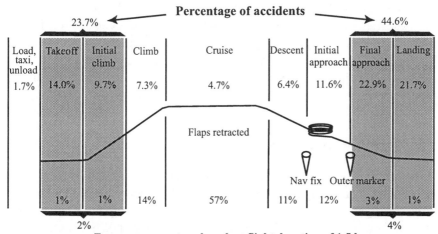

Figure 20.1 **Percentage of Flight Involving Hull Loss Accidents**
Source: Adapted from *Statistical Summary of Commercial Jet Aircraft Accidents—Worldwide Operations—1959-1996*, page 13, Boeing Commercial Aircraft Company

1 million each month and something like 12,600,000 per year—all with the goal of '0' accidents. While '0' accidents is certainly a worthwhile goal, unfortunately it is probably unobtainable.

Worldwide the numbers of forecast takeoffs and landings are significantly greater. The training, equipment, and infrastructure required to support operations such as these are complex and expensive. There is no question that safety costs money, but as an anonymous observer one time rather cryptically said: 'If you think safety is expensive, try having an accident.'

There are aviation safety experts who believe that at least some air transport accidents will be an inevitable part of the diversity and complexity of our multi-cultural, high-technology society (Perrow, 1984). Notwithstanding this pessimistic viewpoint, it is our belief that while an air transport system in which there is no risk and there are no incidents is not possible, an air transport system can be created in which the system's defenses can come very close to controlling or preventing the seemingly inevitable accident. Such a system obviously includes effective monitoring and habitual operational wariness on the part of all members of the flightcrew. If this can be done, so that we can control human error and then identify and eliminate what Professor Reason has labeled 'latent conditions', the inevitable accident will virtually disappear.

The goal of '0' accidents may well be unattainable. However, there is little question that the industry can, must, and will do better than it is presently doing. In the future, if the industry's own growth projections are realized and the safety rate is not improved, the number of accidents and the number of fatalities will be unacceptable. Herein lies a very great challenge for a complex and dynamic industry.

The Worldwide Safety Problem

Today, air transport is a world industry. Both air transport and its safety must be considered on a worldwide basis. This point is important for there are significant differences in safety in the various geographic regions. There are carriers with good records in all world geographic regions (Weener and Russell, 1993).

When the worldwide fleet of jet airplanes manufactured by the largest air transport aircraft company in the world (the Boeing Commercial Airplane Company) was examined, Lautman and Gallimore found that 16% of the operators had crew-caused accident rates higher than the fleet average. Of the operators, 80% had no crew-caused accidents at all. This was one of the first times that these subjects had been reported.

Particularly enlightening, and also disturbing, was the fact that 16% of the operators had 80% of the total accidents. With such data, it is difficult to avoid facing the fact that a relatively few number of operators are involved in the majority of air transport accidents.

The industrialized countries generate about 84% of all airline traffic and generally have the best safety records. Australia and New Zealand have led the world in air transport safety for as long as records have been kept. Although to keep their extraordinary record in perspective, it should be remembered that a large US airline has more exposure in a single year than Australian and New Zealand carriers have had for most of their existence.

The third world, including China and the CIS, furnish about 16% of the known traffic and have approximately 70% of all the known commercial jet hull losses. Table 20.1, which is taken from the Flight Safety Foundation's *Flight Safety Digest* of March 1996, illustrates the distribution of air transport accidents by ICAO Regions.

Table 20.1   Aircraft Accidents by ICAO Region

ICAO Region	# of Accidents	Movements	Rate/Mil Movements
Africa	17	562,734	30.21
Asia-Pacific	19	1,039,380	18.28
Eastern Europe	5	243,300	20.55
Europe	26	2,732,780	9.51
Latin America	34	1,050,632	32.56
Middle East	3	263,183	11.4
North America	28	6,680,700	4.08
Total	**132**	**12,752,709**	**10.35**

Source: 'Airport Safety: A Study of Accidents and Available Approach-
and-Landing Aids', page 9, *Flight Safety Digest*, March 1996,
Flight Safety Foundation

Using ICAO's regions, today's highest accident rates are in Africa, Asia-Pacific, Eastern Europe, and Latin America. According to the predictions of a major European manufacturer, if 'accident rates do not change, fatalities will increase by 250% in Asia, 190% in Europe, 150% in Latin America and 70% in Africa' (North, 1997). The percentages vary because the forecast growth and total air safety is different in each region.

It is important to realize that details of the safety challenge in each geographic region vary and that many factors are involved in determining an ultimate safety record. Factors such as infrastructure, regulatory and organizational culture, can play an important role. Political stability can be another potentially meaningful factor.

For example, the airlines that are in European countries that do not belong to the European Joint Aviation Authorities (JAA) have an accident rate that is 8 times greater than the record of the airlines in JAA countries. Twenty-four European countries belong to the JAA. In another example, Australia and New Zealand are sometimes classified in a very large Asia-Pacific region that has an overall high accident rate. Yet these two countries have been leaders in air transport safety for as long as records have been kept.

These examples by no means suggest that airlines in the countries that have the better records have no safety problems. Virtually all of them have incidents that are very close to a fatal accident. The safety problems of these airlines are of a different magnitude to the safety problems of those airlines with a poorer record, and they can be of a different kind. It would be a big mistake to assume that the safety problems in all countries have the same characteristics.

While there are regional differences in accident rates, there are carriers with good records in all world geographic regions (Weener and Russell, 1993). These carriers demonstrate that regional problems can be overcome, despite significant infrastructure and other deficiencies in some of the areas with the poorest records. Purely regional evaluations can be misleading if they are not fully understood.

Human Factors in Aviation Safety

For many years the air transport industry has tried to deal effectively with the 'most pressing—and depressing—matter' of aircraft accidents. An enlightened view was expressed by IATA's Secretary General, Knut Hammarskjöld when in 1975 he stated:

> The issues we now call 'human factors'...are the critical issues...I believe that at this point in time in the history of civil aviation there is enough knowledge available to launch a serious and effective attack on this—the 'Last Frontier'––of the airline safety problem. (Hammarskjöld, 1975)

Unfortunately it has taken nearly two decades 'to launch a serious and effective attack on this—the "Last Frontier"—of the airline safety problem.' But there is now a growing consensus in the worldwide industry, helped considerably by the leadership exhibited by ICAO that broadly based human factors is a core aviation technology. It is recognized as such in most progressive parts of the world. Human factors are now very much a part of any good accident or incident investigation. The US has been one of the leaders of this movement. Worldwide, the support and implementation of aviation human factors in other countries has ranged from very good to non-existent.

Recognizing the opportunity and the challenge represented by aviation human factors, the US created a 'National Plan For Civil Aviation Human Factors' in the early 1990s. The National Plan includes an Interagency Agreement between the FAA and NASA that shows great promise. One of its goals is to ensure that needed research is being done on basic questions in aviation human factors. Jane Garvey, at the time of writing, the FAA Administrator, has stated that human factors issues is one of the four areas regarding safety that the FAA will emphasize during her five-year term. The other areas that the FAA will emphasize are controlled flight into terrain accidents, loss of control accidents, and landing and approach accidents. She has also given full support to the Flight Operations Quality Assurance (FOQA) Program. Aviation human factors is an essential part of FOQA, which among other things plans to collect and monitor certain flight data parameters. The FOQA program is supported by the operators and is generally supported by pilot organizations. However, both pilot unions and many airlines have a few specific concerns and reservations regarding this program. FOQA is discussed in a later section of this chapter.

Aviation Safety and Human Factors in ICAO

ICAO is the most important of the international aviation organizations (see Chapter 6—The Social Environment, and Appendix D). It reaches all major and most minor countries. ICAO's influence extends to still developing countries and their airlines, which would otherwise be left to struggle on their own. While ICAO has other interests and responsibilities, increased flight safety has long been one of its major objectives. In 1986 the ICAO Assembly adopted Resolution A26-9 which emphasized human factors and reemphasized the importance of flight safety. As a follow-up to the ICAO Assembly resolution, the ICAO Air Navigation Commission further defined the objective of this resolution by stating that its purpose was:

> To improve safety in aviation by making States more aware and responsive to the importance of human factors in civil aviation operations through the provision of practical human factors material and measures developed on the basis of experience in States.

Three years following adoption of Resolution A26-9, the ICAO Council enacted a revision to ICAO's Annex I (Personnel Licensing) which required that all pilots in the future be familiar with 'human performance and limitations' as they relate to their flying activities and the formal privileges of their license. This was a major accomplishment for it is now required that formal licensing and certification procedures include human performance and limitations. Effective implementation of this provision is now the major problem. It is much easier to say than it is to accomplish effectively in all of ICAO's member States. Great Britain has been a world leader in the implementation of this program.

The ICAO Flight Safety and Human Factors Program

An active ICAO Flight Safety and Human Factors program was formed in accordance with the wishes of the ICAO Assembly and Council and to stimulate action regarding Resolution A26-9. The program is based on the belief that 'significant progress in Human Factors will only be achieved through the commitment of the user community worldwide'. The program further believes that meaningful widespread education is the first step in achieving progress in human factors. An advantage of such ICAO programs is that they can reach many airlines and countries that would otherwise be untouched.

Key segments of the ICAO program have been the organization of an active multi-cultural Human Factors Study Group, the production of 13 Human Factors Digests (see Appendix N), the organization and management of carefully selected International Flight Safety and Human Factors Symposiums, and the organization and management of Regional Flight Safety and Human Factors Workshops. The ICAO Human Factors and Safety Program also participates in events organized by other international organizations and publishes articles in the ICAO Journal and in other media. The following paragraphs will briefly discuss ICAO's Human Factors and Safety Symposiums and Workshops and its role in attacking a specific problem in international air traffic control.

ICAO's Global Symposiums

ICAO's four- to five-day Global Symposiums are part of a ten-year ICAO Flight Safety and Human Factors program. The Symposiums have been jointly sponsored by ICAO and by the host country's official aviation organization. Emphasizing the importance the world of transport aviation attaches to human factors, the Symposiums have been attended by not only officials of ICAO States, but also by representatives from airlines, operational employee groups, air transport and systems manufacturers, other concerned international organizations, and by representatives from academic, training, and research institutions.

The first Symposium was held in Leningrad (now St. Petersburg), Russia in April of 1990. The theme of the five-day Symposium was 'the application of Human Factors knowledge to aviation management, training and operations' with the objective: 'to improve safety in aviation by making States more aware and responsive to the importance of human factors in civil aviation operations through the provision of practical human factors materials and measures, developed on the basis of the experience in States'. The working languages of this Symposium were English, French, Russian, and Spanish. 'Working languages' means that headsets were provided to all participants that provided simultaneous translation of the presentations, similar to what is observed at the United Nations. It was attended by 230 participants from 30 States (nations) and several international organizations.

A second global symposium was held in Washington, DC, in April 1993. The theme of this symposium was 'Human Factors Training for Operational Personnel'. Its objectives were the same as stated for the Leningrad symposium. Again the working languages were English, French, Russian, and Spanish. The symposium was attended by 325 participants from 42 States and six International Organizations.

A third symposium was held in Auckland, New Zealand in April 1996. Its theme was 'Safety 2000: Integrating Human Factors Knowledge and Practice into Tomorrow's Aviation System.' The objective was to improve the safety and effectiveness of the aviation system through the integration and practical application of human factors into all aspects of aviation. Its working languages were English and French. This Symposium was attended by 533 participants from 52 States (nations) and 3 international organizations.

A fourth symposium—the last symposium held before publication of this book— took place in Santiago, Chile, from 12-15 April 1999. The symposium theme was 'Human Factors and CNS/ATM Systems Safety and Efficiency: Building the Future'. The working languages were English and Spanish.

ICAO's Regional Seminars

ICAO's four to five-day Regional Seminars are held in three-year cycles, two in each year. These seminars have been held in such cities as Cairo, Mexico City, Addis Ababa, Bangkok, and Rio de Janeiro. Participants are officials of the regional States and representatives from airlines of that region, manufacturers of airplanes used there, other concerned national and international organizations, and regional and sometimes worldwide academic, training, and research institutions. The faculty is composed of as many as 20 internationally recognized experts in aviation and human factors as well as local authorities and both local and international ICAO representatives.

The US ASAP and the ICAO Audit Program

The US FAA in the early 1990s developed a program called the International Aviation Safety Assessment Program (ASAP) in an attempt to increase world-wide safety and to protect US citizens. This program monitors the compliance of other countries with international aviation safety oversight rules and inci-dentally, reinforces the importance of ICAO's Standards and Recommended Practices (SARPS). The consequence for noncompliance with international oversight rules is that airlines from a country that does not provide the stipu-lated safety oversight are prohibited from entering the lucrative US market.

The FAA has made it clear that it is not assessing whether or not an individual airline is safe or less safe, but whether or not the country being monitored has a civil aviation authority in place that ensures that accepted operational and safety procedures are maintained by its air carriers. In a very real sense, it evaluates the safety effectiveness of the country's regulatory culture.

The ICAO Audit Program

Following the FAA initiative, but considerably expanding it, Dr. Assad Kotaite, President of the ICAO Council, told the Council on 24 February 1997 that audits of national air transport standards performed by ICAO should be the worldwide accepted norm. He called on ICAO's then 185 member States (na-tions) to give ICAO the necessary powers to oversee them. Dr. Kotaite sug-gests that the means to accomplish this: 'might well be found in the introduc-tion of international technical inspections, or safety and security audits, which call on states to rectify disclosed deficiencies.' He further stated that: 'ICAO, as an international body, should be empowered to check closely the imple-mentation of safety and security standards, and to carry out regular inspec-tions.' The ICAO Council agreed. 'ICAO would (then) become the recog-nized worldwide auditor of safety and security standards for international civil aviation' (Ott, 1997a). This obviously would be an expansion of the ICAO role in international air safety and eventually could well take the place of the FAA's ASAP program. 'Under current ICAO rules, safety oversight assess-ments are restricted to three areas in the ICAO annexes: licensing of pilots, aircraft technical operations and airworthiness. Kotaite has suggested that the audits be expanded to embrace air traffic services and airport infrastructures, facilities and services' (Ott, 1997b).

Traditionally, approval of similar projects has been a slow process, and very few ICAO observers expected rapid movement on any proposal with such far-reaching consequences as these proposed ICAO audits. However, the compulsory safety oversight initiative received considerable emphasis at a meeting held 10-12 November 1997. The ICAO Council's concept was supported by 148 of ICAO's 185 member States. One of the difficult areas discussed involved the requirement to preserve the safety oversight program's confidentiality. A compromise involving the issue of confidentiality might provide that a summary agreement of nonpunitive oversight reports would be made public only after providing enough time to correct major deficiencies. Dr. Kotaite stated: 'ICAO's ultimate goal is not to impose sanctions, but to make the airline industry safer.... Flight safety is no longer a taboo subject— a balance must be found between it and national sovereignty.' The ICAO Audit program (the ICAO Safety Oversight Program) has been supported by the US FAA, the European Civil Aviation Conference (ECAC), which represents 36 European nations, and by many others.

We have already seen that ICAO States take the safety of worldwide air transportation very seriously as does the entire air transport industry. An earlier ICAO Safety Oversight Program was inaugurated on a voluntary basis in 1995. ICAO received requests from 85 nations to conduct safety audits and has conducted 70 safety audits. In addition, there have been 37 follow-ups of the original programs. This level of response clearly illustrated the need for and general acceptance of such a program. The final mandate to carry out compulsory safety audits was given ICAO by the 32nd ICAO Assembly, which ended on 2 October 1998. As Dr. Kotaite stated '...it ushers ICAO into the 21st century'.

The ICAO Assembly also stated that the results of the future mandatory program can now be published. While this gives ICAO an enforcement mechanism based on publicity, Dr. Kotaite has stated forcefully, 'I do not like sanctions; boycotting a country will not make its airspace safe. If there are deficiencies in the airspace or on the ground, we will not get the result required if we are negative. We must exert pressure and provide help.' Dr Kotaite has also stated: 'States do not lack the political will (to follow international safety oversight standards)....They are either not well informed, or lack funds, or lack national legislation.' These three factors have emerged from ICAO's existing program of voluntary safety audits (Warwick, 1998).

A very positive advantage of the ICAO audits is that they can help the regulatory authorities of certain Third World, or what have also been called 'developing countries' achieve satisfactory compliance with recognized international standards. These countries may have special problems and specific needs. Many of them do not have individuals with the expertise that is required to develop a sound regulatory culture. The head of IATA's Safety Committee, Pakistani International Airline's corporate safety chief, Captain Amjad Faizi, recently very bluntly told an international conference that regulatory slackness is at the top of the list of Third World airline safety problems (Faizi, 1997).

IATA's 'Buddy System'

In another approach to world airline safety, the International Air Transport Association (IATA) has inaugurated a 'buddy system'. To achieve its laudable goal of halving hull-loss rates by 2004, will require a reduction of nine or ten crashes per year. IATA recognizes that '80% of its member's accidents are caused by approximately 20% of its airlines' and has made its 'buddy' concept a major plank in a seven-point program to achieve its admirable goal of reducing its accidents by half.

The innovative proposal by IATA is to have airlines with good safety records take airlines with poor safety records in their regions under their wings in a 'flight-safety buddy system' in an effort to improve their safety performance. Such a program is obviously fraught with difficulties. Difficulties include budgets (for such a program costs time and money for both 'buddies'), effective recognition of the possible difference in operational problems, effective commitment to the program by both 'buddies', the assignment of appropriate personnel, recognition of the need for such a program, implementation of an effective safety program, etc. Initially, the program will focus on Latin America and the Caribbean and at the time of writing was expected to begin in 1999. IATA has announced it is already matching partners for an experimental phase in South America. Several airlines have already expressed interest in the scheme (*Flight International,* 13-19 August 1997).

At its inception, the 'buddy' program envisions a senior flight-safety manager from an airline in the region with a good safety record spending up to five days with the designated 'buddy' airline. The flight-safety manager would provide comprehensive briefing and evaluation on such subjects as briefings on a flight safety organization, operational policies, procedures, flight safety reporting, training, crew resource management, and risk management. The program also includes the promotion of flight-data analysis, incident reporting, and the installation and maintenance of ground-proximity warning systems (*Flight International, 13-19 August 1997*). This is an ambitious program and even small parts of it, if effectively implemented, can help.

Data Sharing and the FAA's FOQA

The effective sharing of flight operations data has been a challenging area for many years. The voluntary collection, analysis and sharing of such data, primarily through the use of on-board quick-access recorders and ground-based analytical programs, has been used in parts of Europe with modifications for more than two decades. In the US operational data sharing is included in the FAA's Flight Operations Quality Assurance program (FOQA). Unfortunately FOQA has foundered on whether or not such programs surrender the FAA's authority to prosecute violations of safety regulations and also whether or not such data might be disclosed to the media and to plaintiffs in civil suits against an airline or its employees. It is a very difficult area. This is a critical issue for the pilots. Stuart Mathews, president of the Flight Safety Foundation, has warned that 'no FOQA program in the US would be successful if it were vulnerable to disclosure in court proceedings' (McKenna, 1997).

The problem is broader than just court proceedings for it includes vulnerability to FAA enforcement actions as well as to company discipline. (See the discussion regarding 'GAIN' in Chapter 22—The Air Transport Future.) On 7 February 1997, US District Court Judge Stanley Marcus supported the position of the airlines and the pilots on this issue. In a federal civil court decision, regarding the American Airlines December 1995 accident near Cali, Colombia, he wrote that data from the American Airlines' Airline Safety Action Partnership (ASAP) are 'entitled to qualified privilege and denied the plaintiffs access to them'. Judge Marcus further noted that, 'some observers view the ASAP program as a 'prototype for future partnership programs between the airlines and the FAA to promote safety'. In American's ASAP program, error reports are collected electronically, and deidentified reports are reviewed weekly by a joint committee comprising of representatives from American, FAA and APA—the Allied Pilots Association that represent the American pilots.

The Flight Safety Foundation, which strongly supported Judge Marcus's decision, reported it this way: 'A US District Court judge's recent decision upholding the right of an airline to withhold information collected in nonpunitive safety reporting programs is an important first step toward expanding the programs throughout the commercial aviation industry and reducing accidents. Fundamental to the success of such programs is the premise that the information disclosed is provided for the purpose of improving safety and is not to be used to penalize those making reports.' This approach has been endorsed by pilots, their unions, and airline operational management.

In a recent courageous and very welcome clarification of FAA policy FAA Administrator Garvey, at a November 1998 international GAIN meeting in Long Beach, California, stated that she would 'issue a policy directing agency inspectors not to penalize airlines and individuals based on data retrieved from aircraft quick-access recorders under Flight Operations Quality Assurance (FOQA) programs. The policy will be modeled on British Airways and NASA's confidential safety reporting systems. Garvey said she would consider expanding it to include problems reported by individuals under arrangements such as American Airline's Aviation Safety Action Program' (World News Roundup, 1998). Of course, the Administrator's most welcome announcement does not take care of, and cannot take care of, the civil legal problems remaining.

The Flight Safety Foundation is continuing its quest for what it calls the 'elimination of obstacles' that are impeding the implementation of FOQA to help reduce accidents. The FSF's Chairman, Stuart Mathews, has stated bluntly that the 'lack of codified protection of FOQA' remains the chief stumbling block to widespread implementation of the program. A number of airlines are participating in FOQA on a limited basis and with therefore limited success. Both airlines and airline unions are insisting on the inclusion of legal protection from potential litigation or FAA enforcement stemming from the use of such data before FOQA can be fully implemented. It is hoped that it will not take a tragedy to secure such protection as it did with nonpunitive incident reporting and the development of the ASRS.

Problems with the Misuse of FOQA Data

The importance of the security of this process cannot be overemphasized. For example, a major US airline is presently having a serious confrontation with its pilot association over alleged misuse of FOQA data. In this particular incident, a flight manager tried to track down a crew based on a stated exceedance from data obtained through the FAA's Flight Operations Quality Assurance (FOQA) program. In this case, the flight manager first attempted to work backwards from the deidentified FOQA information to identify the crew. The second violation of the program agreement occurred when the flight manager personally telephoned a pilot to further investigate the incident. Among other things, it appears that the flight manager identified the wrong month in which the exceedence occurred and called the wrong pilot.

The agreement at this particular airline is typical of many carriers in that a specific system is set up to handle any contact that might be made to follow up on a particular exceedence. As a result of this incident, ALPA wants the flight manager who was involved removed as a flight manager because he breached the anonymity and non-punitive understanding upon which the FOQA system was based. Typical language in the Agreement airlines have with their pilots regarding a FOQA program include clauses such as:

> The sole contact with any flight crew member associated with a specific exceedence event shall be through the designated 'FOQA Monitoring Team Member' and the crew member's LEC (Local Executive Council) representative or his designee.

> ...In the event any employee/agent divulges any identifying data to any individual other than the designated FOQA Monitoring Team member representative, such employee shall immediately be removed from participation in the FOQA program.

> Any violation of the requirement of the agreed-upon FOQA program, or the terms herein, shall cause the immediate termination of the FOQA program and destruction of all data.

Variations of this problem are not restricted to the US. The Airline Pilots Association of New Zealand has recently notified other Associations that a New Zealand appellate court has upheld a lower court ruling that, in the absence of any national statute to the contrary, cockpit voice recordings may be used against airline pilots in civil and criminal proceedings. In a related incident, the captain of a Garuda Indonesia DC-10-30 that crashed at Fukota Airport in June of 1996 is being criminally charged for 'professional negligence leading to death and injury' with a sentence that can vary from eight years in jail with forced labor to a fine of 20,000 yen. In another incident, the Japanese Aircraft Accident Investigation Commission blamed the captain's inadequate judgment when he opted to abort a take-off after the aircraft had passed the V_R rotation speed. In still another example, a French court has just sentenced the pilot of the A-320 that crashed in the 1988 Mulhouse-Habersheim airshow and killed three people, to ten years in jail and a $5,000 fine for involuntary manslaughter. The copilot was given a 12-month suspended sentence.

ICAO has stated that cockpit voice recorders (CVRs) are a necessary and effective tool for safety investigations, and that CVR recordings should be used solely for such purposes. In the US, the FAA is prohibited from using cockpit voice recordings for disciplinary purposes, and the NTSB is prohibited from releasing the portions of recordings that include any 'nonpertinent' conversations. Potential liability problems arising from data secured for safety purposes are a delicate issue. This is a very serious problem for the pilots and for the industry. Developing and successfully implementing a system for the effective sharing of flight operations data in the interest of increased safety is a very real challenge to both carriers and their pilots.

Some Final Thoughts on Airline Safety and the Industry Challenge

The priority given safety is not a new thought in the air transport community. Its importance has been discussed at some length in Chapter 6—The Social Environment. Unfortunately, one would be naive to believe that the prime importance of safety and the priorities it requires have always been well understood and effectively implemented by all pilots, airlines, or regulators. However, despite conflicting demands and sometimes inconsistent performance, the overriding importance of operational safety has been a central concept for most airlines for many years.

For example, United Airlines, one of the oldest and largest US airlines, began its original Flight Operations Manual by emphasizing its famous 'Rule of Three' in large bold black letters. It was simple and effective. The preface of that early Flight Manual stated that flight operations would always be conducted with the following priorities:

SAFETY
COMFORT OF PASSENGERS
SCHEDULE

This basic concept, which is not unique to United, is a core corporate philosophy that continues to this day although the 'Rule of Three' has been rewritten to read:

SAFETY
SERVICE
INTEGRITY AND RESPONSIBILITY

Safety remains the number one corporate requirement. A recent restate-ment of the company's corporate values by one of United's large unions—the Air Line Pilots Association—further emphasizes that concept by stating that: 'Operational Safety shall be the highest priority in all corporate decisions...Safety is a moral imperative that supersedes all other consider-ations. Reduced to its essence, the (airline's) historical stock in trade is to provide safe transportation without compromise.'

The Chairman of American Airlines, another top US airline,[2] has writ-ten:

> Quality means different things at different times to different people. Without denigrating the importance of any measure, I think all would agree that the most important measure of aviation quality is safety, and that a corporate com-mitment to operate at the highest possible level of safety is the sine qua non of a quality carrier. (Crandall, 1991)

In another example, the Chairman and Chief Executive Officer of Lufthansa,[3] the pioneering and still highly successful German airline was asked to describe his airline's culture to the employees of a US airline with whom it had just formed a code-sharing agreement. He replied emphatically:

> First, we are working very hard to be a safe airline. Safety is above everything for us.... (Weber)

Without question, a corporation must make a profit, provide good wages for its employees, and provide dividends and profits for its stockholders. Nei-ther is there any question that safety provisions frequently cost money that could be spent in other places. There is an obvious conflict between safety and profits, but without safety, air transport corporations cannot continue to exist. There is usually a delicate balance between safety and legitimate com-mercial aspirations. However, the bottom line is that to neglect safety for tem-porary profits is to create economic suicide.

John Kern, then vice president-aircraft operations and chief safety officer at Northwest Airlines, emphasized that point when he told the partici-

---

[2] Robert L. Crandall, Chairman and CEO of American Airlines in the May 1991 *American Way*.

[3] Juergen Weber, Chairman and CEO of Lufthansa in interview reported in United Airlines' *Our Times*. Mr. Weber was recently inducted into the Laureates Hall of Fame at the the National Air and Space Museum of the Smithsonian Institute.

pants at a Global Aviation Safety & Security Summit, 'You have to establish a safety culture, over and beyond safety requirements' (Shifrin, 1996).

Because safety is the major consideration in the long-term successful airline's day-to-day operations, it is an almost unwritten law that the preeminence of safety must be inherent in manuals, bulletins, and in all other company publications that involve flight or ground operations. The precept that the airline's primary consideration is safety must be recognized by all from the President, the CEO (or whoever is Number 1), and through the board of directors to the last employee. Unfortunately, in the real world this has not always happened.

In today's world, it is a fact of life that every airline must produce a product—passenger seat miles—that is first safe, and then efficient. For both moral and economic reasons, safety is always the paramount consideration. If any aviation product, airplane, airline, or system is even perceived to be unsafe, it cannot survive.

A Safety Philosophy

Several years ago, I. Irving Pinkle, a NASA scientist now retired, was quoted in the November, 1966 Newsletter of the Flight Safety Foundation as follows:

> Every activity is obliged to improve its safety record where it can. Those who insist on ignoring the smaller safety problems about which something can be done, pointing to the larger problems about which nothing can be done yet, are mostly evading the issue. Most safety measures adopted deal with small portions of the total hazard. Over the years, the steady improvement that results is significant. If each step is discouraged because it doesn't solve the whole problem, then nothing is accomplished. (Pinkle, 1966)

In the succeeding three decades, the air transport industry has made significant improvements in the safety of its operations. Most of the improvements were made because all parts of the industry made the improvements that were recognized. Many of the improvements helped solve relatively small safety problems, but each made a significant contribution to increased safety. In addition, there has even been at least one improvement that is far from a small addition. The development of GPWS and MSAWS (or its counterparts) has substantially reduced Controlled Flight Into Terrain (CFIT) accidents in the US and in other countries where GPWS is required by regulation. CFIT accidents have been the largest single category of air transport accidents and will be discussed in more detail in Chapter 21— Current Safety Problems.

References

Aviation Week and Space Technology, 9 November 1998. In World News Roundup, 'The Americas', McGraw and Hill, Inc., New York.

Degani, Asaf and Wiener, Earl L. (1990). *Human Factors of Flight Deck Checklists— The Normal Checklist*, NASA Contractor Report 177549, Ames Research Center, Moffett Field, California.

Degani, Asaf and Wiener, Earl L. (1991). 'Philosophy, Policies, and Procedures: The Three P's of Flight-Deck Operations', *Proceedings of the Sixth International Symposium on Aviation Psychology*, 29 April-2 May 1991, Columbus, Ohio.

Degani, Asaf and Wiener, Earl L (1993). 'Cockpit Checklists: Concepts, Design, and Use', *Human Factors*, Human Factors and Ergonomics Society, Inc., Santa Monica, California.

Degani, Asaf and Wiener, Earl L. (1994). *On the Design of Flight-Deck Procedures*, NASA Contractor Report 177642, June 1994, Ames Research Center, Moffett Field, California.

Faizi, Amjad, (1997). Quoted in 'The Last Challenge,' *Flight International*, 8 - 14 January 1997. Reed Business Publishing, Sutton, Surrey, United Kingdom.

Greene, Berk and Richmond, Jim (1993). 'Human Factors in Workload Certification', paper given at SAE Aerotech 93, Federal Aviation Administration, Seattle, Washington.

Grieve, B.S. (1990). 'The Terrible Risk', Britannia Airways Ltd., Bedfordshire, England.

Hammarskjöld, Knut (1975). Secretary-General of IATA at his Opening Address at IATA's 20th Technical Conference, 10-14 November 1975, Istanbul, International Air Transport Association, Montreal.

Hawkins, Frank H. (1993). *Human Factors in Flight*, Second Edition, edited by Orlady, Harry W., Ashgate Publishing Ltd., Aldershot, England.

Howard, Benjamin (1954). 'The Attainment of Greater Safety', presented at the 1st Annual ALPA Air Safety Forum, and later reprinted for presentation at the Aircraft Accident Prevention Course, University of Southern California, July 1957.

Lautman, L.G. and Gallimore, P.L. (1987). 'Control of the Crew-Caused Accident', *Flight Safety Foundation Flight Safety Digest*, Flight Safety Foundation, Alexandria Virginia.

Learmount, David (1997). 'Safety', *Flight International*, 1-7 January 1977, Reed Business Information, Sutton, Surrey, United Kingdom.

Lundberg, Bo O. K. (1966). *The "Allotment-of Probability-Shares" - APS - Method, Memorandum PE-18*, The Aeronautical Research Institute of Sweden (FFA), Stockholm, Sweden.

Maurino, Daniel E., Reason, James, Johnston, Neil, and Lee, Rob B. (1995). *Beyond Aviation Human Factors*, Ashgate Publishing Limited, Aldershot, Hants, England.

McKenna, James T. (1997). 'Garvey Commits FAA to Safety Partnerships,' *Aviation Week and Space Technology*, November 1997, McGraw and Hill, Inc., New York.

North, David M. (1997). Editorial—'We Know the Safety Issues, Now Let's Push Solutions', *Aviation Week and Space Technology*, 1 December 1997, McGraw and Hill Inc., New York.

Ott, James (1997a). 'ICAO Stresses Safety Compliance', *Aviation Week and Space Technology*, 2 June 1997, McGraw and Hill Inc., New York.

Ott, James (1997b). 'Civil Aviation Directors to Explore Expanded Safety Role for ICAO', *Aviation Week and Space Technology*, 18 August 1997, McGraw and Hill Inc., New York.

Perrow, Charles (1984). 'The Organizational Context of Human Factors Engineering', Department of Sociology, Yale University, New Haven, Connecticut.

Pinkle, Irving I. (1966). 'Something To Think About,' *Newsletter*, November1966, Flight Safety Foundation, Alexandria, Virginia.

Rasmussen, Jens, Duncan, Keith, and Leplat, Jacques, eds. (1987). *New Technology and Human Error*, John Wiley & Sons Ltd., Chichester, England.

Reason, James (1997). *Managing the Risks of Organizational Accidents*, Ashgate Publishing Limited, Aldershot, Hants, England.

Senders, John W. and Moray, Neville, P. (1991). *Human Error: Cause, Prediction, and Reduction*, Lawrence Erlbaum Associates, Hillsdale, New Jersey.

Shrifin, Carol A. (1996). 'Safety Experts Seek Data Sharing', *Aviation Week and Space Technology*, McGraw and Hill, Inc., New York.

Svàtek, Nicole (1997). 'Human Factors Training for Flight Crew, Cabin Crew, and Ground Maintenance', *Focus on Commercial Aviation Safety*, The United Kingdom Flight Safety Committee, Choham, Woking, United Kingdom.

Warwick, Graham (1998). 'Improving Safety', *Flight International*, 14-20 October 1998, Reed Publishing Company, Sutton, United Kingdom.

Weener, Earl F. and Russell, Paul D. (1993). 'Crew Factor Accidents: Regional Perspective', presented at the 22nd International Air Transport Association Technical Conference, Montreal, Canada.

Wiener, E.L. (1989). *Human Factors of Advanced Technology ('Glass Cockpit') Transport Aircraft*, NASA Technical Report 177528, Ames Research Center, Moffett Field, California.

Zeller, Anchard F. (1966). *Summary of Human Factors Session, 19th Annual International Air Safety Seminar*, Flight Safety Foundation, Alexandria, Virginia.

21 Current Safety Problems

Developing Priorities for Safety Problems

Within the industry, there is not always agreement on the priorities that should be given to efforts to increase safety. The identification of specific operational problems, the infrastructure, the operational and regulatory culture, and training are important areas that often differ considerably by region and country. Training is included as an area for not only is training very important in its own right but there are even differences in training between airlines within countries. Regardless of the geographic region, most pilot groups include fatigue as a factor that needs more study and human factors analysis, while several other industry groups do not consider fatigue a significant problem. Fatigue and stress are covered in greater detail in Chapter 14—Fatigue and Stress.

An example of differences in priorities in the US can be seen in the initial efforts of a recently formed industry team. This team, which is being led by the Air Transport Association (ATA), is the result of a coordinated effort by the Aerospace Industries Association (principal manufacturers of both airplanes and engines), the airlines, and the Air Line Pilots Association. This industry team is called the Commercial Aviation Safety Strategy Team (CASST). The CASST notes that in the past air safety initiatives have been '...all over the map, and often of marginal value'. It states that: 'It's time to focus on programmes that hold the greatest promise of meaningful progress in air safety' (Flight International, 18-24 February 1998). In an effort to ensure that independent views are considered in its activities, the CASST team will include three 'franchised' members who are not connected with any of the founding organizations.

CASST's approach is supported by the FAA, which also has its own separate and somewhat different safety agenda. CASST's goal is 'to come up with the critical few initiatives that will make a difference'. It is taking at least a limited systems approach and will look at flight safety, cabin safety, engineering and maintenance. Its initial efforts will be to look at CFIT accidents, including enhanced ground proximity warning systems; and the development of better windshear, ice, wake turbulence, and clear air turbulence detectors.

447

In addition, a CASST task force is looking at potential human factors errors in maintenance procedures, resource management programs in maintenance, and such maintenance issues as uncontained engine failures. Another ATA sponsored task force is investigating the widespread use of flight operations quality assurance safety data collection and distribution (McKenna, 1998a).

An example of the way that specific safety priorities can change as more data is secured and analyzed is the US NTSB's recent paring of its 1997 list of 'Most Wanted' safety recommendations from 21 items to 10 (Business and Commercial Aviation, September 1998). The four remaining aviation items are explosive mixtures in fuel tanks, human fatigue, airframe icing, and runway excursions. Flight data recorders, pilot background checks, and wake turbulence were removed from the Board's list of 'Most Wanted' safety recommendations from the previous year, sometimes because solutions satisfactory to the Board have been implemented.

The UK Civil Aviation Authority (CAA) pioneered a system over two decades ago that has achieved general acceptance throughout the world. It includes the computer analysis of data from sophisticated flight data recorders (FDRs) and quick access recorders (QARs) that is analyzed by specialized software to determine trends. While all segments of the industry recognize the positive advantages of similar programs, implementation in litigious countries can be a very real problem because of the liability concerns of both the pilots and the airlines.

At the FSF International Air Safety Seminar, which was held in Cape Town, South Africa from 16-19 November 1998, Elizabeth Erickson, the FAA's deputy director of air certification services, revealed that future efforts to increase safety by the agency will utilize a 'structured data-driven process to continually reduce accident rates'. The FAA is putting high hopes on data that will be received from the FOQA program. However, as we have seen, there are still major implementation problems associated with FOQA. FOQA was discussed in Chapter 20

Other interesting aspects of the FSF Seminar are that while the FAA and the JAA (two of the world's most influential aviation authorities and defined further in the Glossary) have common objectives there are certain differences in their other safety priorities. The differences are shown in Table 21.1.

Table 21.1   JAA and FAA Additional Action Items for Safety

JAA Action List

- approach and landing accidents

- loss of control accidents

- design related accidents

- weather related accidents

- occupant safety and survivability

The FAA's SASI (Safety Analysis and Strategic Intervention) List

- loss of control

- uncontained engine failure

- runway excursions

- approach and landing

- weather

Each agency drew up their list independently, and while they both agreed that CFIT was the primary problem, the differences in the two lists of other action items illustrate that safety problems vary by region and that action plans should be regionally based and oriented. The JAA list covers 89% of all European accidents, while the FAA list covers 80% of US accidents (Learmount, 1998b).

In an effort to effectively accept air transport's human factors challenge, which is an important but only a part of the total accident problem, the Flight Safety Foundation founded its ICARUS Committee.[1] The ICARUS Committee is supported worldwide by major aircraft and equipment manufacturers, airlines, pilot representing organizations, research organizations, and regulatory agencies. Its members are top-level representatives of those organizations and the related disciplines and it was formed to explore ways to reduce human

---

[1] The ICARUS Committee was formed in 1992 by former FSF President John Enders and FSF Board of Governors Member Jean Pinet. Its members have extensive backgrounds in the human aspects of design, manufacturing, flight and maintenance operations, operating environment, and research.

factors-related accidents. The first two years of the ICARUS Committee resulted in 18 findings and 10 recommendations for action. They were based on current human factor problems in air transport operations in the US and throughout the world.

A specific and localized problem is involved with the routine data sharing of safety information. It is a very serious problem in the US, in New Zealand, and in certain other countries. It is not a problem in those countries whose data sharing programs are essentially non-punitive. The quality of the implementation of the programs varies considerably. For at least the last decade, the United Kingdom and British Airways have been leaders in making maximum use of data secured during routine flight operations.

Controlled Flight Into Terrain (CFIT)

Controlled flight into terrain accidents still constitute the leading category of worldwide air transport accidents. Despite control of the CFIT hazard among the major airlines of the developed countries, the 640 fatalities in 1997's airline crashes identifies this category as the one that poses the greatest danger to worldwide loss of life. In 1998, the CFIT accident rate reversed a trend that had previously been decreasing. Many safety experts believe that a reason may be that some operators have ignored the proven CFIT aids that are discussed later in this section (McKenna, 1998b).

CFIT accident control began with the development of Ground Proximity Warning Systems (GPWS) in the early 1970s when several airlines introduced them. A spectacular reduction of CFIT accidents in the US followed the regulatory requirement that required Ground Proximity Warning Systems in all transport aircraft carrying 11 passengers or more. Since that time most major non-US airlines have seen a similar spectacular reduction in these accidents. GPWSs are now required by regulation in many other countries.

According to a widely quoted definition, a controlled flight into terrain accident 'is one in which an otherwise serviceable aircraft under the control of the crew, is flown (unintentionally) into terrain, obstacles or water, with no prior awareness on the part of the crew of the impending collision' (Wiener, 1977). It is a good, useful, and all-inclusive definition. It should be recognized that CFIT defines a 'condition'—a condition that can occur during any phase of flight. For example, a CFIT accident that occurs when an aircraft lands short of the runway is a CFIT accident that occurred during the approach and landing flight phase. The fact that an accident like this can be classified both as a CFIT accident and as an approach and landing accident can sometimes lead to misinterpretation of statistics.

Controlled flight into terrain (CFIT) accidents are truly tragic events. A perfectly good and flyable airplane that is under control is flown into the ground, a mountain, or some other obstruction with no prior awareness on the part of the crew of the impending disaster. CFIT accidents have been called the 'quintessential human error accident'. Virtually all CFIT accidents involve an error in navigation procedures, in effective monitoring, in not following SOPs, etc. There often have been additional contributing factors. There is little question that the airplane was not where it was thought to be when it crashed.

One of the most interesting things about virtually all CFIT accidents is that they have three characteristics in common. It seems almost certain that there were, or should have been, very clear indications within the cockpit that things were not going well. Second, in most cases there seems to have been plenty of time to make suitable corrections. And lastly, there seems to have been little awareness of the problem until the accident was inevitable.

CFIT accidents are accidents that should not have happened. Unfortunately, on a worldwide basis they continue to happen. While the role of latent conditions in air transport accidents is now being recognized and can be a factor, analysis still reveals that a high percentage of CFIT accidents involve an apparent failure by critical human beings in the operational loop. The critical human beings are mostly cockpit crews but in some cases are air traffic controllers. Fortunately, the industry has learned that GPWS and other electronic warning devices, better training, better and specific allocation of crew duties, and improving the basic aviation infrastructure can all make very real contributions in controlling this chronic risk.

A Perennial Problem

CFIT accidents are nothing new. The wreckage of the *Southern Cloud*, a tri-motor Fokker that crashed in March of 1931 was discovered 200 miles northwest of Melbourne in 1958 by a surveyor near a summit in the Snowy Mountains. Since that accident, over 30,000 passengers and crew have lost their lives in terrain-related accidents (Bateman, 1994). John H. Reed, then Chairman of the National Transportation Safety Board, gave a later example of the continued pervasiveness of CFIT accidents. On 18 September 1974, he told a Congressional Committee that: 'Since 1969, the Safety Board has been and continues to be concerned about the number of airworthy aircraft which are inadvertently flown into the terrain.' He further told the committee that in the years between 1968 and 1974 the Board investigated 15 air carrier accidents of this type that caused 490 fatalities. During the same period, 841 lives were lost in CFIT accidents that were investigated by other governments.

Worldwide, these accidents caused 1,331 fatalities, an average of slightly more than 220 fatalities for each of the six years. At the time Mr. Reed testified before the Congressional Committee, the US NTSB was investigating five accidents involving 179 fatalities, all of which had similar circumstances (Reed, 1974). It was the Safety Board's judgment that a glide path warning system 'might have provided the crew with information to facilitate safely clearing terrain during enroute descent and approach to landing'. Perhaps too narrowly, in 1974 much informed opinion centered upon the need for more and better ground-based aids without giving equal consideration to internal cockpit procedures. Twenty-five years later the need for glide slope guidance (either internal or external) is recognized throughout the industry.

CFIT accidents were then, and still are, a significant industry problem, in spite of the fact that today major airlines have learned how to prevent most classic CFIT accidents. In the period 1983 through 1992, the worldwide CFIT rate was 0.35 accidents per million departures. The corresponding CFIT accident rate for North America was 0.03 accidents per million/departures— less than one-tenth the world rate. Other major regions of the world had accident rates that were as much as 80 times higher than the rate in North America and in selected non-North American countries. Boeing data has shown us that hull loss insurance claims exceeded $334 million in this ten-year period. That figure does not include the millions of dollars in passenger and crew liability claims or lost revenue that are a part of an airline accident. It is a disturbing reality that if CFIT and approach and landing accidents could be eliminated, it would prevent more than 80 percent of total fatalities (Flight Safety Foundation, 1993).

The hard facts are that if the CFIT risk is not significantly reduced worldwide, the safety record of air transportation will not be controlled to an acceptable level. Fortunately, we now know how to do this. There are wide variations in the rate of CFIT accidents throughout the world. However, it has been demonstrated in North America, the developed countries in Europe, and in some other countries, that with the application of currently available technology and practices, CFIT accidents can be controlled and perhaps even eliminated. Table 21.2, which shows the ICAO geographical regions in which crew-caused CFIT accidents occurred, is important. It strongly suggests that

Table 21.2   CFIT Accident Rates per Million Flights by ICAO Region

ICAO region	(Accidents per million flights)		Risk multiplier FSF CFIT checklist
	This study	FSF CFIT task force	
Africa	0.70	2.40	8.0
Asia/Pacific	0.57	1.00	3.0
Europe	0.27	0.45	1.3
South America	0.63	1.14	5.0
Middle East	0.00	0.00	1.1
North America	0.00	0.03	1.0

Source: 'An Analysis of Controlled-flight-into-terrain (CFIT) Accidents of Commercial Operators, 1988 Through 1994', Khatwa and Roelen, *Flight Safety Digest,* April-May 1996

the aviation infrastructure and the social culture may be significant considerations. It further suggests that simply installing GPWS (or EGPWS) may not solve the entire problem.

Solving the Problem of CFIT

The industry had become reasonably frustrated with its inability to make any real progress with the CFIT problem until the introduction of the radio altimeter in the late 1960s. The radio altimeter made feasible the lowering of landings to Category II minimums[2] and it also made feasible the concept of a workable electronic ground-proximity warning system (GPWS). The concept of an electronic warning system was a new approach. It was started in Europe at Scandinavian Airlines (SAS) and by an electronics engineer, Don Bateman. Bateman was an avionics engineer with a predecessor company and is now chief engineer of Flight Safety Avionics for AlliedSignal Aerospace. By 1971 the electronic GPWS had been voluntarily installed by several airlines, led by SAS, Canadian Pacific Airlines, Maersk Air, Braniff, and the old Pan American Airways. By 1973, Boeing was offering GPWSs as a recommended safety device. By 1974, a GPWS was a basic part of all airplanes Boeing manufactured.

---

[2] The glossary defines Category I, Category II, and Category III minimums.

CFIT in the Decade of the 1980s

As we have mentioned, CFIT accidents vary by geographic region. Figure 21.1 shows the CFIT accident rate per million flights by region of the operator's home base for the ten-year period 1983 to 1992.

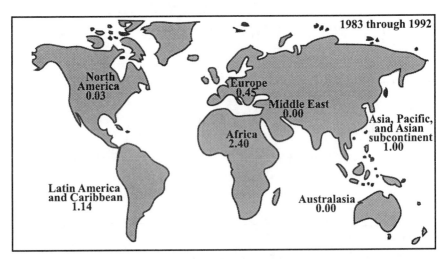

Figure 21.1 World Jet Transport Aircraft CFIT Accident Rates Per Million Flights by Region of the Operator's Home Base

Source: Adapted from 'Crew Factor Accidents: Regional Perspective', Figure 11, Weener and Russell, 1993

Figure 21.1 suggests several interesting things. It is not surprising that Australia and New Zealand and their airlines lead the pack in terms of the lowest CFIT accident rates. In the history of their jet operations, Australian airlines have never killed one of their passengers. Australia has had glide slope guidance in the form of either ILS glide slopes or T-VASI visual approach slope indicators at all of its jet approved runways from the beginning of the jet age. While Australia has very good regulations and a very good aviation in-frastructure, QANTAS, the Australian pioneering international airline always has had trips into countries that were not as fortunate. New Zealand's record is marred only by its 28 November 1979 tragic Mt. Erebus crash. To keep that superb record in perspective, one should remember that Australian or New Zealand airlines' jet transport exposure is considerably less than that of major airlines in the US and in some other parts of the world.

John Reed's testimony before a congressional committee was closely followed by the TWA B-727 CFIT accident that occurred 1 December 1974 at Round Hill, Maryland. The crash, which had 92 fatalities, occurred 50 feet below the last ridgeline and 20 NM from the destination runway at the Washington Dulles Airport. It very effectively dramatized the CFIT problem. Shortly after the crash, the FAA enacted the rule requiring all large transports to be fitted with a GPWS within one year (FAR 121.360). Some cynics believe that the proximity of the crash to Washington, D.C. was a major factor in the rapid official reaction to a known problem. The crash at Round Hill was also instrumental in the institution of non-punitive incident reporting in the US. (See Chapter 18—Non-punitive Incident Reporting.)

Figure 21.2 shows CFIT accidents per year for US and World Air Carriers. The outstanding feature in this figure is the dramatic decrease in CFIT accidents in large transports in the US following the FAA requirement, which was promulgated in 1975 shortly after the Round Hill accident, that all large jet transports be equipped with a GPWS.

During this same period FAA software experts also designed an ATC software package called the Minimum Safe Altitude Warning (MSAW). It was developed for use at ATC ARTS III facilities in the US as a follow–up to GPWS and is, in part, responsible for the good CFIT record in the US. MSAWs require ARTS III radar capability. MSAWs are installed in ATC installations and detect aircraft that are flying below minimum safe altitudes and enable the controller to warn the pilot. It has been described as the controllers' GPWS.

The installation of MSAW software at ATC ARTS III facilities continues to this day. MSAWs are also becoming available in a relatively few other non-US locations, specifically Israel, with limited applications in Japan, Italy, Switzerland, and soon Austria, Australia, Canada, Abu Dhabi, Bahrain and Hong Kong. Some countries, but certainly not all, have not felt it important to have MSAWs because they have essentially flat terrain. MSAWs have not been installed in other countries for a variety of other reasons.

The immediate drop-off in CFIT accidents for large transports in the US that followed the 1975 mandatory installation of GPWSs is significant because it clearly shows that CFIT accidents can be better controlled or even essentially eliminated. In the US, CFIT losses dropped from an average of eight hulls per year to about one every two years. The actual figures are 17 cases of fatal CFIT accidents involving commercial aircraft during the five years preceding the introduction of GPWS and only 2 in the following five years. Another graphic example of the effect of GPWS is the US experience with Part 135 (less than 30 seats) airlines During the period that Part 135 operators were not required to have GPWS, their record remained almost con-

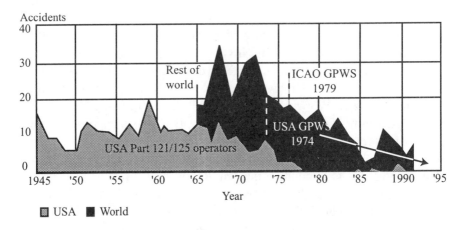

Figure 21.2 CFIT Accidents Per Year for US and World Air Carriers
Source: 'Past, Present and Future Efforts to Reduce Controlled Flight
            into Terrain (CFIT) Accidents', Bateman, 1990

stant. In the US in 1990, after adjusting for fleet sizes and departures, the risk
of a CFIT accident to a Part 135 commuter aircraft that does not have GPWS
is 100 times greater than that of a Part 121 aircraft with its mandatory GPWS
(Bateman, 1990). Some commuter airlines installed GPWS without being required to do so and their record shows that it was a good move to install it.

In the US, a major positive change has been the FAA's bringing Part 135
carriers using from 10 to 30 seats into the Part 121-Part 125 carriers' safety
umbrella. After March 1997, the lessor standards of Part 135 will apply only
to on-demand air taxis and scheduled service on aircraft with nine or fewer
seats. It is expected that the US commuter CFIT record will further improve
when all transports having more than 9 seats will be required to have GPWSs
and to meet other Part 121-Part 125 standards. While the practical problem
that limited previous installation of GPWS in many smaller, but still commercial
airplanes is frequently said to be the economic burden it placed on their
operators, the cost of GPWS is not really excessive. It has been stated that a
GPWS unit costs less than the exterior paint on a large transport aircraft. The
GPWS unit repays its original investment in from one to three years, based on
aircraft replacement costs and average settlement costs for the lives lost in air
transport accidents (Bateman, 1994).

A difficulty in the GPWS program has been the inevitable economic
replacement problem that is associated with virtually any advance. Several
improvements had been made in the original GPWS. The early GPWSs had

too many false warnings, late warnings, and on occasion even failed to give warnings when warnings were needed. The consequence was that pilots using the early GPWSs frequently delayed response to a valid GPWS warning. The result was an unnecessary crash or a serious aircraft incident. There have been several modifications or improvements of the original GPWS and they have eliminated virtually all of the original complaints. Unfortunately, the revised and advanced GPWSs have not always been installed and even today, some of the original Mark I systems are still in use.

In spite of all this, the contributions of the GPWS are now universally recognized. 'The demonstrated reduction in CFIT risk is about 20 times even when using early generation GPWS equipment. For the latest GPWS equipment the reduction is about 50 times' (Bateman, 1994). These are significant reductions.

Putting CFIT Accidents and GPWS in Perspective

Don Bateman, the father of GPWS, has written:

> It would be an overstatement to claim GPWS is the sole contributor to this significant reduction (in large US transports). The continual investment by the FAA in expanding and upgrading the ATC radar and tools, such as ARTS III Minimum Safe Altitude Warning Systems, approach lighting, VASI, ILS, DME and other navigation aids, along with improved procedures have all helped reduce the CFIT risk.

To this we would add that significant improvements in training, programs that increase pilot awareness of the importance of the CFIT hazard, and the growth of non-punitive incident reporting of CFIT warnings have also helped.

Unfortunately, CFIT accidents continue throughout the world. In an effort to better understand these unnecessary and tragic accidents, the Netherlands National Aerospace Laboratory (NLR) conducted *An Analysis of Controlled-flight-into-terrain (CFIT) Accidents of Commercial Operators, 1988 Through 1994*. This analysis of worldwide CFIT accidents was conducted by, and the subsequent report written by, Dr. Ratan Khatwa[3] and Alfred Roelen. The study was launched in association with the Flight Safety Foundation which has been leading a global effort to reduce CFIT accidents by 50%. The report stated that where data were known, 75% of the accident aircraft lacked a ground-proximity warning system (GPWS) and, perhaps surprisingly, that a significant

---

[3] Dr. Khatwa is now Manager of Flight Deck Design, Rockwell-Collins Inc., Cedar Rapids, Iowa.

percentage of the CFIT accidents occurred in areas without high terrain. The report further stated that the scheduled flights of major operators in North America and the Middle East (which has a very fine infrastructure) had the lowest CFIT rates.

The report by Khatwa and Roelen strongly recommends that the use of GPWS be mandated for domestic commercial operations, including those of all regional and air taxi operations. Seventy-one percent of the worldwide CFIT accidents experienced between 1988 to 1994 happened to smaller aircraft authorized to carry no more than nine passengers. To date, no appropriate rule or regulation applies to this category of air transport.

The following statement bears a chilling statistic: 'Currently, less than 5% of the world's commercial airplane fleet is not equipped with GPWS; however, these unequipped airplanes are involved in nearly 50% of CFIT accidents' (FAA/ICAO/Flight Safety Foundation Controlled Flight Into Terrain Education and Training Aid). While there can be more than a single reason for these accidents, it is clear that the data identifies a problem that must be effectively faced.

CFIT Internationally

In order to meet the CFIT problem on a worldwide basis, ICAO has issued standards, which are to be effective in 1998. The new standards require that GPWS equipment be installed in all commercial aircraft that fly internationally, have a maximum gross takeoff weight of 12,566 lb. or more, and that carry more than nine passengers. Unfortunately ICAO has no enforcement authority (its standards are recommended standards only), and its concerns are basically limited to international flying. This by no means implies that the ICAO regulation is not a step in the right direction. However at this time, there is no official mechanism, other than individual national regulations, that covers domestic commercial flying.

There are wide variations among ICAO nations. It will be interesting to see the effect of future ICAO audits on safety generally, and their effect on the CFIT problem specifically, as the ICAO audit program is implemented. A very real difficulty is that the new standards will not apply to most of the accident prone group, i.e., the smaller taxi operators flying domestically, including those in the US. In an effort to tackle this major safety problem, the Flight Safety Foundation (FSF) in the early 1990s took the lead in a worldwide industry effort to reduce CFIT accidents by organizing an international CFIT Task Force. The FSF Task Force included more than 150 representatives from

24 airlines, aircraft manufacturers, five equipment manufacturers, the principal pilot unions, and selected technical, research, and professional organizations.

The decreasing number of CFIT accidents, though still too high in the accident-prone group, seems to be moving in the right direction, although 1998 figures are creating a great deal of concern. The FSF CFIT Checklist, which is just one product of the FSF CFIT Task Force, is available in English, Spanish, Chinese, French, Russian and Arabic versions. Corporate underwriting has made possible the distribution of more than 30,000 copies of the Checklist worldwide at no cost to the recipients. The Foundation's final product is a comprehensive and highly recommended multimedia CFIT Education and Training Aid that in 1997 was developed by the Task Force and produced by Boeing. It includes a special video and the CFIT checklist. The Flight Safety Foundation received a *Flight International* Award for the work of its 'task force and training effort to combat CFIT incidents internationally'.

Figure 21.3 shows the worldwide decrease in CFIT accidents to large commercial jet airplanes in the past six years. This has been an impressive accomplishment in spite of a rise in 1998 that is not shown in the figure.

Overall the air transport industry has made significant improvements in

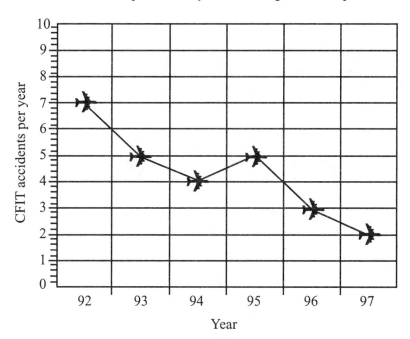

Figure 21.3 World CFIT Accidents in Large Commercial Jet Airplanes
Source: *Flight Safety Foundation News*, 27 February 1998

the safety of its operations. Such improvements as GPWS and MSAW, or its equivalent, along with improved procedures, training, and education have substantially reduced controlled flight into terrain accidents in the US for most transport operations and worldwide for most major airlines. However on a worldwide basis, CFIT is still the largest category of air transport accidents. Items like GPWS and MSAWs can help solve the specific problem but only when they are properly installed, utilized, and maintained and when they are accompanied by appropriate training, procedures and pilot education.

At present, MSAWs, which requires ARTS-3 radar capability, is not available in most foreign locations, in spite of the fact that many non-US ATC facilities do have ARTS-3 capability resident in their systems. Any readers interested in a more in-depth discussion of GPWS are urged to read Don Bateman's papers and the Khatwa and Roelen study discussed earlier. The FAA has recently started using the acronym, TAWS (Terrain Avoidance Warning System), in its regulations, apparently to avoid making a regulation using acronyms for systems such as GPWS or EGPWS that are closely associated with a particular vendor.

Loss of Control

'Loss of Control' is a relatively new category for air transport accidents and is particularly disturbing because at least some of the accidents attributed to loss of control seem to indicate that these accidents were caused by things the flight crew did or did not do to the airplane. Training and general pilot proficiency are frequently implicated. One of the latest loss of control accidents occurred on 6 November 1996 when the crash of an ADC Airlines Boeing 727-231 near Lagos, Nigeria killed 143 persons. In this case, the B-727 crew responded to a TCAS alert but lost control in an abrupt maneuver presumably caused by reaction to the TCAS alert. The aircraft then plunged into the earth from its altitude of 16,000 feet. The lack of effective flight training in the event of a TCAS alert may have been a factor in the accident. A TCAS alert should not cause a reaction so violent that a well-trained crew loses control of the airplane, particularly when it has nearly 16,000 feet to recover.

Other, and more difficult apparent loss of control accidents include the United Airlines B-737 crash, which occured on final approach at Colorado Springs and the B-737 crash of US Airways approaching Pittsburgh. In neither of these crashes has the NTSB been able to discover the cause for the loss of control that led to these tragic events. Recent training programs have been instituted at several airlines that are designed to deal with recovery from extreme positions, but there is no assurance that the new training would have prevented these accidents. Their cause is not known, and the industry is genuinely concerned. It is thought that aerobatic training may be important, but there is considerable disagreement on the details and on what parts of aerobatic training may be helpful.

Loss of Control accidents have a large number of fatalities—1,932 in ten years. It is exceeded only by the 2,806 fatalities that occurred in 36 CFIT accidents in the same period. A Joint Safety Analysis Team, organized from members of the Commercial Aviation Safety Strategy Group (CAAST), is studying 31 accidents involving in-flight loss of control fatal accidents that occurred in the ten years between 1988 and 1997.

In one analysis, these accidents were divided into three categories: Control Available, Control Compromised, and Other/Unknown. In 16 of the 31 accidents, control was thought to be available, and in 13, a known mechanical malfunction compromised control. In some cases the mechanical malfunction may have made the airplane unflyable. In two of these accidents, the cause of the accidents is listed as unknown. There is far from agreement in this classification. One member of the task force examining these accidents believes that only seven of the 31 accidents should truly be listed as loss of control. An interesting observation is that 25 of these 31 accidents happened outside of the US with non-US airlines (Dornheim, 1998).

Classification is a general problem that is certainly not limited to loss of control accidents. For example, the categorization and differentiation between a CFIT category and a loss of control category for an accident that occurs during an approach and landing seems to depend too often on the individual who is making the analysis. In many cases, the accidents are double classified. What is clear is that regardless of the classification, the industry is considering worldwide problems to a greater extent than in previous periods, and sincerely trying to do something about them.

As this chapter is being written, there may be an addition to this sorry list. Loss of control is considered by many to have been the direct cause of the China Airlines crash at Taipei Airport on Taiwan on 16 February 1998. The Taiwan Civil Aeronautics Administration has released flight data recorder information that shows the airplane had been cleared for runway 05L, was much too high, and that the crew apparently lost control while manually flying a go-around in which extreme pitch attitudes and speeds were allowed to develop (Learmount, 1998a). Unless there are additional complications, a manually flown go-around should not be a problem for well-trained pilots.

Approach and Landing Accidents

Most air transport accidents occur during the approach and landing phases of flight. Many of them are CFIT accidents. There are sound operational reasons for this distribution because the airplane must get close to the ground (terrain) during approach and landing. The operational CFIT risk is obviously higher during approach and landing than it is at cruise. It is hard to run into terrain at 35,000 feet.

One of the most comprehensive recent studies of the approach and landing risk in air transport was conducted by the Netherlands Directorate-General of Civil Aviation for the Flight Safety Foundation. The Study was named 'Airport Safety: a Study of Accidents and Available Approach Aids'. Like the CFIT Study, it was conducted by Ratan Khatwa and Alfred Roelen. The Study examined 557 airports that occurred around the world and analyzed 132 landing and approach accidents that occurred between 1984 and 1993. Forty percent of the accidents examined were CFIT accidents. It included a survey of flight crew training, cockpit procedures and operational documents. The Study found that 50 percent of all accidents occurred during approach and landing, which in this study included all accidents that occurred within 25 miles of the intended airport. A major conclusion was that 'properly executed precision approaches resulted in a five-fold risk advantage over non-precision approaches on a worldwide basis...'. In addition, when stratified by ICAO region, 'the risk associated with flying non-precision approaches compared with those flying precision approaches ranged from three-fold to almost eight-fold, depending on the region. Another finding was that 50 percent of CFIT accidents for jet aircraft for a five-year period involved non-precision approaches' (Enders et al., 1996). CFIT and non-precision approaches were very much involved in these accidents.

Not surprisingly, the study found that airport authorities could minimize the risk for approach and landing safety if they furnished precision approach and landing facilities. Underdeveloped regions had significantly greater problems in approach and landing accidents and underdeveloped regions also had a relatively high percentage of non-precision approaches. Additional factors that seemed important were the air traffic control operator's operating standards and practice and, in the Asia-Pacific area, the existence of surrounding unfavorable terrain and other obstacles.

The Study found that in North America there were virtually no ILS approaches conducted that did not have terminal approach radar. However, in Africa and Latin America, a significant number of airports offered precision approach facilities but did not have terminal approach radar. There seemed an obvious relationship between the approach and landing accidents and the lack of terminal approach radar, even at airports that had precision approach facilities. The lack of charted initial arrival procedures (STARs) and the effect of factors that were outside the direct control of the airport or authorities, such as light conditions and weather conditions were also important risk factors when they were combined with non-precision approaches. Finally, 16% of the accidents in this Study 'involved some type of mechanical failure that the crew was unable to manage'. Both pilot training and maintenance could well have been implicated.

Table 21.3 shows the accident categories that describe the 132 accidents analyzed in this study. The categories were mutually exclusive, so no single accident was placed in more than one category.

Almost certainly, pilots will always have to make non-precision

Table 21.3   Accident Categories in 132 Accidents

Accident Category	Number	Percent
CFIT, Unknown	1	0.8
CFIT, Land. Short	24	18.2
CFIT, Collision. High Terrain	22	16.7
CFIT, Collision. Object	4	3.0
CFIT, Water	2	1.5
Aircraft Collision on Ground	1	0.8
Landing Overrun	14	10.6
Runway Excursion	2	1.5
Landing Gear Problem	7	5.3
Wheel-up Landing	1	0.8
Unstable Approach	10	7.6
Loss of Control, Crew-caused	12	9.1
Wind Shear	3	2.3
Airframe Ice	1	0.8
Midair Collision	4	3.0
Loss of Power	7	5.3
Aircraft Structure	1	0.8
System Malfunction	6	4.5
Fuel Exhaustion	1	0.8
Unknown	9	6.8

Source: Airport Safety: A Study of Accidents and Available Approach-and-landing Aids (page 12)

approaches. One lesson from this Study is that these approaches increase the risk of an accident. The principles discussed in previous chapters become even more important. The successful completion of non-precision approaches is dependent on sound operational policies and practices. The use of the 'stabilized approach'[4] procedure is highly recommended.

Aircraft Collisions on the Ground

Aircraft collisions on the ground are a very real and a very old problem. They are usually associated with weather at or around minimums. When weather is at or around minimums, it is much more difficult or even impossible for tower personnel to know the location of taxiing aircraft. It also makes it much more difficult for the cockpit crew to know exactly where they are, especially if they happen to be on an unfamiliar airport. One of the worst ground collisions was the 1977 collision of a fully-loaded KLM B-747 that was making a takeoff and a fully-loaded PAA B-747 that had not cleared the takeoff runway at the mid-Atlantic island of Tenerife. In addition to the heavy fog that restricted visibility, many other problems were associated with this accident. The result of the collision was one of the world's worst crashes and 582 casualties.

The FAA tells us that the number of incidents where an airplane was not on the runway or taxiway it was expected to be on, increased nearly 19% in 1997. The fact that general aviation accounted for 59% of operations at towered airports but had 72% of the excursions, does not diminish the problem for the transport pilot. For example, a recent issue of the weekly computer web-based AVflash noted that:

> Speaking of 'runway excursions' NWA and CAL jets went nose to tail recently in Newark, N.J. Witnesses say ground radar was out when the NWA pilot stopped on a taxiway with part of his tail still hanging over the runway. The CAL pilot on his takeoff roll managed to stop about 1,200 feet from the NWA DC-9. On the west coast at Ontario, Calif., an SWA 737 turned onto the runway head-on with a UAL 737 already on its take-off roll. The UAL jet managed to swerve and stop, averting a tragedy. (AVflash, 1998)

Much of the effort to better control the problem of collision on the ground is centered on better ground radar for the towers, better maps of airport runways

---

[4] A stabilized approach is an approach flown with a constant rate of descent along an approximate 3 degree flight path with stable airspeed, power setting, and trim with the airplane configured for landing.

and taxiways, more readable and better maintained runway and taxiway signs, and specific procedures. There is also a need for better crew coordination and for accurate navigation while the airplane is on the ground. Good communications between all parties, which has sometimes been a weak link, is critical. Another approach has been to install directional lights on the runways and taxiways which are controlled by ground control at the tower and which reinforce voice radio instructions. They include red stop and green go lights at taxiway and runway entrances. This system has been used for several years with considerable success and with strong pilot approval at London's Heathrow Airport.

Go No-Go Decisions Close to Rotation During Takeoff

The problem of continuing or aborting the takeoff, particularly when the decision must be made close to the rotation point, is another old and one of the most difficult problems in air transport operations. There is an almost universal consensus that this decision must be made by the captain. There is also general recognition that the decision must be made very rapidly. There is no time for discussion with other crewmembers.

Unfortunately, even under optimum conditions, complex decisions cannot be made instantly. The take-off abort problem is often compounded because a takeoff abort from close to V_1,[5] especially for reasons other than an engine failure, involves not only a difficult decision that can involve multiple factors, but is also an unexpected event for the crew. If simulator training for rejected takeoffs is not given for other than an engine failure it is an obvious reflection on the adequacy of that training.

One of the first studies of problems associated with the accelerate-stop criteria that are a part of failures close to V_1 was made in 1969 by Foxworth and Marthinsen. There were obvious deficiencies in the then existing criteria, and the Foxworth and Marthinsen study was one of the first serious attempts to look at this problem (Foxworth, T.G. and Marthinsen, H.F., 1969). In another study, J. S. Clauzel found that when the ten year history of rejected takeoffs of the airplanes of a major US manufacturer were analyzed, approximately 75%

---

[5] V_1 was formerly called the 'critical engine speed' and is now defined as the 'takeoff decision speed'. It is the calculated speed below which an aircraft can lose an engine and still stop on the remaining runway. It can also be regarded as engine failure recognition speed. For real world application, it should include an increment due to pilot reaction time.

of the rejected takeoffs resulted in accidents or incidents that were initiated by either gear/tire problems or by engine problems. He also reported that those initiated by gear or tire problems were four times more likely to result in an accident or incident than those initiated because of engine problems (Clauzel, 1985). This result of Clauzel's study was not at all surprising. Virtually all airlines do a good job of training their crews for engine failures at or near V_1, but very few of them do an adequate job of training for the more common problem.

A later study by Harold Marthinsen was published in November 1993. It closely analyzes the go no-go process (Marthinsen, 1993). Marthinsen's study includes a review of the decision-making considerations that are involved in the very short time that is available. It reviews the effect of the following considerations:

- Runway positioning distance;
- Factors affecting the takeoff roll;
- Considerations of failures other than engine failures;
- Decision times;
- Factors affecting the stop distance;
- Factors affecting the continued takeoffs.

Marthinsen also lists 31 additional factors that can affect the original six. Each can involve the decision to reject or continue the takeoff. While obviously all of the factors are not relevant in each takeoff, each one can be a consideration.

The decision whether to continue the takeoff or to abort it must be made rapidly and usually under stressful conditions. There are many factors to consider and the ones that are relevant for the particular takeoff should be covered in the captain's briefing. All of these factors involve complex decisionmaking and are not simple issues. Most can be preconsidered in order to avoid having to evaluate a decision under stressful conditions with no time to reconsider. For example, a captain may brief that he/she will not abort the takeoff after V_1 for a door light (a warning that a door may not be completely secured). Manufacturers have assisted in this process also by designing warning systems on some of the later aircraft where certain warnings are inhibited during the takeoff roll and initial climbout.

Today, there are wide variations in the ways that airlines handle and train for this difficult problem. Solutions include those of at least one non-US airline that has all takeoffs made by the captain in order to eliminate change of control problems during this critical period. Many airlines have started calling out V_1 five knots early in order to allow for pilot reaction time. Other remedies are to have considerable training time spent on discussing this problem, including a discussion of the factors listed by Marthinsen. Considerable emphasis is being placed on the desirability of continuing takeoff even with a known failure because of the unknown stopping ability of the airplane, the hazards associated with a maximum stopping effort, and the demonstrated ability of the crew and the airplane to fly safely even with the airplane's degraded condition.

Perhaps the final word was best said by the captain of a Continental DC-10 who suffered an aborted takeoff on a wet runway at LAX because of a series of tire failures. In spite of rapid action by the captain, and the fact that the abort procedure was started well in advance of reaching the V_1 speed of 157 knots, the aircraft overran the 10,285 foot runway by 644 feet and was destroyed by fire. There were 2 passenger fatalities and 31 seriously injured passengers. When the captain, who was commended for his actions, was later asked if he would have done anything differently if he had known about the prospective abort, he replied, 'I would have called in sick'.

Other Rejected Takeoffs (RTOs)

Captain Roy Chamberlin studied the rejected takeoff incidents (RTOs) that were reported to the ASRS between 1 January 1989 and 30 November 1990 by pilots flying transport category aircraft in excess of 60,000 pounds. He found 168 reports that involved decision-making and procedural issues associated with the RTO's initiation and execution. Ninety-four of the study reports involved RTOs that were caused by flight crew errors. They included such things as unauthorized takeoffs, taxiway takeoffs, off-runway takeoffs (where the aircraft aligned with runway edge lights instead of the centerline lights), aircraft configuration anomalies, and ten incidents that involved loss of aircraft control. These flight crew procedural errors were often predisposed by frequency congestion, schedule pressure, the transfer of control to the first officer, and by environmental factors, such as visibility and the intensity of runway and taxiway lighting. Several other anomalies were caused by improper information transfer between flight crews and ATC, and deficiencies in task management and crew coordination (Chamberlin, 1991).

An important part of this study was its analysis of flight crew decisions following an RTO. Many RTOs do not involve critical go or no-go decisions made close to V_1 but occur early in the takeoff at relatively low speeds. As we have seen, RTOs can occur for many reasons. Decisions made after a RTO are dependent upon such things as warning system alerts, tactile sensing, engine instrument indications, aircraft-generated noises, or observations radioed by tower controllers or other airplanes. Many result in requests for emergency equipment, usually because of engine or other fires. RTOs may or may not include an aircraft evacuation, which never is just a routine operation.

In virtually all cases, the deceleration caused by heavy braking generates a great deal of heat in the tires and brake assemblies, and this can cause concern with further taxiing back to the ramp or during a second takeoff if one is attempted. Dangerously hot brakes can create a major hazard for any ground personnel who work in that area because of the danger of a tire or wheel hub explosion. A thermal fuse located in each wheel hub helps to protect against this hazard by releasing high pressure air in the tire before an explosion would occur. If a second takeoff attempt is made, dangerously hot brakes considerably increases the risk of brake or tire failure.

Brake energy charts help determine that brake kinetic energy capacity is not less than the brake kinetic energy absorption after a RTO. The charts specify cooling time required before another takeoff can be attempted, otherwise specific brake energy and RTO performance cannot be assured. To minimize problems in this area, latest airplanes have cockpit gauges that indicate brake temperature when an overheat condition exists.

Safety and Air Traffic Control in Africa

Air traffic control in Africa has been a particular problem. ICAO has declared all but seven African States to be 'critically deficient' in ATC capability. The deficient area includes approximately 75% of the continent, with only Egypt, Morocco, and Tunisia in the north and Botswana, Namibia, South Africa, and Zimbabwe in the south considered acceptable.

In October of 1996, the International Federation of Airline Pilots Association (IFALPA) published an extremely effective report which detailed the 'dangerous' state of enroute air traffic control (ATC) services over three-quarters of the African continent. It argued 'that the lack of mid-air collisions had less to do with ATC than with the region's relatively empty skies' (Learmount, 1997). The emerging of South Africa on the world stage following its rejection of apartheid and the growth of many of Africa's developing States has already significantly changed the air traffic pattern over Africa. Air traffic in Africa has been growing 6-8% per year. Air traffic over Africa grew 13.9%, to 5.1 million passengers in 1996-7 and has increased rapidly during 1998.

For several years, separation safety was provided by and pilots relied upon a procedure that was designed by the International Air Transport Association (IATA). The procedure depends upon the pilots to rely upon calls they hear on a common frequency as they report over navigational fixes. Pilots provided their own separation. The system, primitive as it was, worked fine for airplanes using the IATA system and because of the relatively limited traffic over the routes. The procedures could not, nor were they expected to, provide effective separation standards for all aircraft. As traffic grew so did the potential for disaster. It culminated in the 6 November 1996 crash of an ADC Boeing 727-231 near Lagos, Nigeria that killed 143 persons. The 727 crew responded to a TCAS alert but lost control of the airplane in an abrupt maneuver caused when the crew reacted to the TCAS alert. The airplane then plunged into the earth from its altitude of 16,000 feet. An additional factor could well have been inadequate training for the TCAS alert. This crash was also discussed under 'Loss of Control'.

It is hoped that an air navigation plan that has been endorsed by the ICAO Council will alleviate the very serious problem of ATC control over Africa. Actually, the ICAO plan is for the entire Africa-India region. It was approved by 350 participants from 56 nations at a meeting held in Abuja, Nigeria in May of 1996. The plan contains 128 separate recommendations that deal with the ATC concerns that have been expressed by IFALPA, IATA, ICAO, and the Air Line Pilots Association of South Africa.

The International Federation of Air Line Pilots (IFALPA) believes that TCAS should be required for transport flight over continental Africa. IFALPA states that for all practical purposes, much of the air over Africa is essentially uncontrolled airspace. The IFALPA position is supported in most operational circles.

Other Safety Problems in Africa

Africa has air safety problems other than just air traffic control. Its hull loss accident rate is 9.5 per million departures as compared with a world average of 1.4. The FSF Approach and Landing Accident Reduction Task Force revealed that Africa has by far the highest fatal approach and landing accident rate of 2.43 per million flights compared to the world rate of 0.43. The next worst was Latin America's 1.65, while the best was North America's 0.13. In a concentrated attempt to improve, Latin America is setting up its own structure for implementing and monitoring flight safety policies regionally. Safety experts in the US and in Europe have praised it for its efforts (Jones and Learmount, 1998).

In 1996 there were 17 fatal crashes in Africa, including one that resulted in 350 fatalities on the ground. The importance of the questions raised regarding safety over the African continent was illustrated at a recent ICAO meeting. US Secretary of Transportation, Rodney Slater, told the African Directors of Civil Aviation that no expansion of air services between the US and African nations would occur until air safety on the continent improves. He told the Directors that the US FAA was prepared to help in raising air navigation, airport security, and other areas to international levels. Starting 7 July 1998 Slater was scheduled to spend a week in Africa discussing safety and airport issues as part of the US $1.2 million program 'Safe Skies for Africa'. Presently only five African countries meet ICAO standards. Those countries are Egypt, Ethiopia, Ghana, Morocco and South Africa (Aviation Week and Space Technology, 1998).

The airlines have been paying the equivalent of approximately $6 million US dollars a month in overflight fees to these nations, and there is considerable doubt that very much of the $6 million per month has been used to improve the infrastructure of the individual States (*Aviation Week and Space Technology, 1997*). An anonymous (for obvious reasons) story from a previous decade tells of an airline employee who was sent down to an African State to talk about implementation of aid that would make air transport operations safer. One airline offered to provide crystals for additional NDBs. Another airline offered to install appropriate ILSs, maintain them, and ensure that they worked. In return the airlines expected to have their landing fees reduced. The response of a concerned native official was to go over to the window and point out the Rolls Royce parked in the spot reserved for the Director of Aviation. He threw up his hands in frustration and said, 'that's what your landing fees pay for'.

There are some bright spots. For example, ASECNA, which is a public air safety organization that represents the French-speaking states in west Africa, recently reported a noticeable improvement in reported air incidents. There were 14 reported air incidents in 1998 compared to 1997s 30, and there were only 17 near misses as compared to 26 the previous year. However, ASECNA warns that because there could be a number of unreported incidents and near misses, the data could be very misleading. ASECNA, which fortunately seems to be well-organized and has considerable funds, now has members in 14

African states and is responsible for managing air navigation over 24 international and 100 domestic airports.

There is a crying need for better interface and aeronautical coordination covering all of Africa, but unfortunately, there is no forum that considers the continent as a whole. Despite having many common goals and common objectives, there still is no aeronautical fixed telecommunications network between southern Africa's most cohesive infrastructural aviation forum, the South African Development Community (SADC) and ASECNA. Modern aeronautical communications is a major problem. Shortcomings in fixed and mobile communications, in pilot/controller communications, including extended VHF, in basic ATC tools and in personnel training are all very real problems in the developing states (Jones and Learmount, 1998).

Air Safety Problem Areas in Other Countries

Obviously many Third World countries have aeronautical safety problems. However, it should be clearly understood that safety problems are not restricted to only what have been called the 'developing countries'. While air transport is a world industry, the following paragraphs do not by any means discuss all of the industry's safety problems. Instead they offer discussion of several of them as they have been reported in the contemporary aviation press. Readers of this book should be aware of the limitations of the data reported. It is important to recognize that many of the inconsistencies in the data are due to problems inherent in the lack of agreed upon definitions of the terms and to the different reporting measures that are frequently used.

Problem areas that have been highlighted in other countries by IFALPA include the ATC facilities and operation in certain South American countries and in Greece. Problems vary by country. They include: very old or non-existent radar; poorly trained and compensated controllers that are often forced to have another job for subsistence; inadequate separation of civilian and military flight operations; and in some countries, a government policy that threatens pilots that complain about ATC inadequacies with license revocations or other adverse legal actions.

In the past decade, only five of the world's hundreds of airlines have had four or more fatal crashes. Four of them are Asian: Air India with seven; Korean Air with five; China Air with four; and Garuda Indonesian with four. Because Asian States recently have suffered a series of scheduled airline crashes, ICAO has publicly urged that they improve their airline safety oversight standards.[6]

Air transport has grown significantly in the region of the Asian States and this growth has been accompanied by virtually all of the problems we have discussed. Figures for the first half of 1998 confirm that South Asia and Asia-Pacific face serious safety challenges. In the first half of 1998, the region's scheduled charter and commuter airlines were involved in six of the 11 fatal accidents in worldwide commercial passenger-carrying services. While a six month period is much too short to draw meaningful conclusions regarding overall air safety, in this case familiar patterns are confirmed and there is little question that there are valid air safety problems in many countries in the South Asia and Asia-Pacific regions. In this same six-month period, none of air transport's serious accidents involved any of the world's major 'Western', Australasian or Middle Eastern carriers (*Flight International*, 22-28 July 1998).

There is general recognition that most of the Third World or developing nations have fatal accident rates that are much worse than the world average and many times higher than the best carriers. ICAO's mandatory audit program can be of significant help to those States with safety problems. The ICAO audit program was discussed in Chapter 20—The Worldwide Safety Challenge.

Automated Warnings

A perennial problem associated with at least some automated warnings is the tendency they create for some pilots to use them as primary alerting and warning devices, not just as the secondary warning device that they were designed to be. This is particularly true for some types of altitude warnings and for some configuration warnings such as those used to indicate takeoff flaps. With the exception of an alert tone associated with approaching a planned altitude, any time that one is seen or heard, it is an almost sure indication that a flight crew omission or error preceded it.

Ground Proximity Warning Systems (GPWS)

The latest GPWS warning tells the cockpit crew that it has made an operational

---

[6] One US airline made this infamous list. US Air—now US Airways—had five fatal accidents during the past decade. However while it has had incidents, since its reorganization approximately three years ago, its accident record has been spotless, except for an unexplained accident when a US Airways airplane went out of control from 6,000 feet while approaching Pittsburgh.

error, tells the crew what the problem is and that the airplane is either laterally or vertically not where it should be. The warning should not ordinarily be heard if the operation is conducted as planned. Unfortunately, there are still false warnings. Later versions of the original system have greatly reduced the number of false warnings, but they are still a problem. No pilot wants to unnecessarily perform the drastic maneuver that is called for if the GPWS warning is valid. Upon hearing a GPWS warning, most procedures call for the cockpit crew to take immediate aggressive action (full power and maximum climb attitude) unless there is a clear visual indication that the warning is not valid.

On very rare occasions, a difficult kind of an improper GPWS warning can be caused by an instrument approach procedure. In these cases, a valid warning is heard even if the procedure is followed meticulously. This is because the database has not been customized so that a particular terrain feature that does not affect safety unnecessarily triggers the GPWS. The newer versions of GPWSs can be programmed by the manufacturer for specific airfield approaches so these nuisance warnings are eliminated. These are rare conditions and safety is not adversely affected.

Primarily because of the early false warning problem, a delayed response to a GPWS warning has been an industry problem. One of the most graphic examples occurred in South America where an airplane installed with a GPWS had a tragic GPWS accident. The GPWS alarm sounded as it should have under the circumstances and included a voice that loudly and clearly said 'PULL UP, PULL UP'. Just before the airplane crashed into mountainous terrain, the Captain exclaimed in disgust at what he apparently thought was just another false warning, 'Yankee, shut up'. Forty-nine people were killed in an unnecessary crash.

Enhanced Ground Proximity Warning Systems (EGPWS)

A basic limitation with the original GPWS and its modifications was that it provided warnings only for terrain obstructions that were directly below the airplane with information provided by the already installed radar altimeter. Its warning gave little time for action by the pilots. It did not provide sufficient warning for steeply rising terrain that was directly ahead. American Airline's accident at Cali, Columbia made it clear that this was a meaningful deficiency.

A major expedited development following the Cali crash led to further refinement of the original GPWS concept and into the new Enhanced Ground Proximity Warning Systems (EGPWS). Two very big advantages of EGPWS are that it provides warning for rapidly rising terrain in front of the airplane, and it also increases the length of the warning. EGPWS compares an internal database of the world's terrain with the aircraft's location and altitude and generates a map-like display of the surrounding topography. It gives the flight crew a 60-second warning compared with the 10 to a maximum of 30-second warning provided by a traditional GPWS.

As this is being written several airlines have already started installation of the EGPWS in their fleets. More than 1,000 units are presently on order. The US ATA has stated that US airlines are already committed to installing EGPWSs in 6,300 airplanes by 2003 at an estimated cost of $600 million dollars. It has just been announced that in the US an EGPWS will be required equipment by 2001. In Great Britain, British Airways is spending 20 million pounds ($33.4 million) to install EGPWS in all of its aircraft. It is the first European carrier to adopt EGPWS and others are expected to follow. Additionally, Airbus has announced that it would install EGPWS in all airliners manufactured after 1999. Presently, it is an option on all Boeing aircraft and being specified by most airlines.

There is little question that EGPWS represents a significant improvement over GPWS and undoubtedly there will be improvements in the original EGPWSs as experience with them is gained. The EGPWS use of color has been questioned because it modifies existing crew alerting philosophy. Regardless of the merit in that question the only answer for the pilots that use EGPWS is for them to adapt to the specialized use of its color. There is general recognition that the EGPWS represents a significant advance in air transport safety.

Traffic Alert and Collision Avoidance (TCAS) and Windshear Advisory Systems (WSAS)

In the US, TCAS (Traffic Alert and Collision Avoidance System) and WSAS (Windshear Advisory System) are mandated for large transports, although at the time this is written, neither is required by regulation in cargo only airplanes. Despite this, at least one major air cargo airline is installing TCAS voluntarily in all its airplanes. GPWS has been required for some time. Early TCAS installations required that each airplane be equipped with TCAS for the TCAS to be effective. Later systems do not have that limitation.

Pilots believe in both TCAS and WSAS and state that TCAS has already prevented mid-air collisions. It is difficult to prove this claim for there is no organized method of securing that data. Current TCAS systems give vertical avoidance instructions only, not lateral. It is hoped that future systems will give both lateral and vertical advice. An interesting observation regarding WSAS is that the US Airways Charlotte, N.C. accident is the only large transport windshear-related accident to occur in the US since the promulgation of the regulation requiring the installation of a WSAS and a formal training program for its use.

The efficacy of WSAS is not a simple issue. In the Charlotte accident, the crew was going around and the flaps were in transit. The WSAS installation in the Charlotte airplane deactivated the WSAS when the flaps were in transit and this may have been a major factor This deficiency is now recognized and has since been corrected in many airplanes.

Another of the terminology differences that plague nearly everyone but the regulators and politicians, occurs because the US FAA terminology is called TCAS (traffic alert and collision avoidance system), while in ICAO it is called ACAS (airborne collision avoidance system). The FAA has required TCAS II on all transport aircraft having 30 or more seats since 31 December 1993, and TCAS I on aircraft having 10 to 30 seats since 10 February 1995. EUROCONTROL (a group of 33 European nations with the term further defined in the glossary) requires that all aircraft with at least 30 seats will be required to have TCAS/ACAS II by 1 January 2000. Cargo airplanes will be required to comply in EUROCONTROL nations. TCAS I, or its equivalent, which does not give any advisory information (TA), is not recognized by EUROCONTROL. Airplanes having from 19 to 30 seats will not be required to comply with this regulation until 2005.

In a good example of the continuing change of many electronic innovations made in order to improve and eliminate bugs in a system, Change 7 of TCAS/ACAS II (the latest version) is designed to ameliorate high vertical rate or bump-up encounters, phantom TAs, TAs that reverse the aircraft's vertical speed, and to clarify certain aural and visual TA annunciations. Change 7 will be made mandatory through a regulatory change in the US and in Europe and will allow pilots 20 to 48 seconds warning on traffic and 15 to 35 seconds on resolutions. However, resolutions still will be limited to vertical plane movement only.

To date, mid-air collisions have not been a significant problem for US air transport. However, the consequences of such an incident make a major catastrophe inevitable if one occurs. As traffic continues to increase, there is bound to be an increase in the risk of a mid-air collision. Many believe that ultimately more responsibility for maintaining adequate separation of aircraft will be transferred to the cockpit through TCAS, especially with general acceptance of the 'free flight' concept, which is discussed in Chapter 22— The Air Transport Future.

References

AVflash (1998). 'Airlines Get Up Close and Personal', 22 February 1998, AVflash@avweb.com.

Aviation Week and Space Technology (1998). 'Slater to Push African Safety', *Aviation Week and Space Technology*, 6 July 1998, McGraw Hill Inc., New York.

Aviation Week and Space Technology-Editorial (1997). 'Avoiding a Mid air Collision in Africa', *Aviation Week and Space Technology*, 7 April 1997, McGraw Hill Inc., New York.

Bateman, Don (1994). 'Development of Ground Proximity Warning Systems (GPWS)', presented at Royal Aeronautical Society Controlled Flight Into Terrain Conference, 8 November 1994, London, AlliedSignal Inc., Redmond, Washington.

Bateman, Don (1993). 'How to Terrain-Proof Corporate & Regional Aircraft', 5th Annual European Corporate and Regional Operators Safety Seminar, 3 March 1993, Amsterdam, AlliedSignal Inc., Redmond, Washington.

Bateman, Don (1991). 'How to Terrain Proof the World's Airline Fleet', 44th Annual International Air Safety Seminar, 14 November 1991, Singapore, AlliedSignal Inc, Redmond, Washington.

Bateman, Don (1990). 'Past, Present and Future Efforts to Reduce Controlled Flight into Terrain (CFIT) Accidents', Flight Safety Foundation 43rd International Aviation Safety Seminar, 1990, Rome, AlliedSignal Inc., Redmond, Washington.

Chamberlin, Roy W. (1991). 'Rejected Takeoffs: Causes, Problems, and Consequences', *Proceedings of the Sixth International Symposium on Aviation Psychology*, Aviation Safety Reporting System, Mountain View, California.

Clauzel, J.S. (1985). 'Should We Provide More Realistic Training for Airline Flight Crew?', presented at Flight Safety Foundation 38th International Air Safety Seminar, Boston, Massachusetts, 4-7 November 1985, Flight Safety Foundation, Alexandria, Virginia.

Dornheim, Michael A. (1998). '"Loss of Control" Under Scrutiny', *Aviation Week and Space Technology*, 27 April 1998, McGraw Hill, Inc., New York.

Enders, John H., Dodd, Robert, Tarrel, Rick, Khatwa, Roelen, Alfred L.C., and Karwal, Arun K. (1996). Referenced in 'Airport Safety: A Study of Accidents and Available Approach-and-landing Aids', *Flight Safety Digest*, March 1996, Flight Safety Foundation, Alexandria, Virginia.

Flight International (18-24 February 1998). 'Industry launches safety initiative', *Flight Safety International*, Reed Business Information, Sutton, Surrey, United Kingdom.

Flight Safety Foundation (1993). *Safety Alert*, 25 June 1993, Flight Safety Foundation, Alexandria, Virginia.

Foxworth, T.G. and Marthinsen, H.G. (1969). 'Another Look At Accelerate-Stop Criteria', AIAA paper No. 69-772, AIAA Aircraft Design and Operations Meeting, 14-16 July 1969, Los Angeles, California.

Hammarskjöld, Knut (1975). Secretary-General of IATA at his Opening Address at IATA's 20th Technical Conference, 10-14 November 1975, Istanbul, International Air Transport Association, Montreal.

Jones, Lois and Learmount, David (1998). 'African dawn', *Flight International*, 9-15 December 1998, Reed Business Information, Sutton, Surrey, United Kingdom.

Khatwa, Ratan and Roelen, Alfred (1996a). 'Airport Safety: A Study of Accidents and Available Approach-and-landing Aids', *Flight Safety Digest*, March 1996, Alexandria, Virginia.

Khatwa, Ratan and Roelen, Alfred (1996b). 'An Analysis of Controlled-flight-into-terrain (CFIT) Accidents of Commercial Operators, 1988 Through 1994', *Flight Safety Digest*, April-May 1996, Alexandria, Virginia.

Learmount, David (1997). 'Safety', *Flight International*, 1-7 January 1997, Reed Business Information, Sutton, Surrey, United Kingdom.

Learmount, David (1998a). 'Loss of control is key to China Airlines accident', *Flight International*, 17 March 1998, Reed Business Information, Sutton, Surrey, United Kingdom.

Learmount, David (1998b). 'The Precision Approach', *Flight International*, 9-15 December 1998, Reed Business Information, Sutton, Surrey, United Kingdom.

Lundberg, Bo O. K. (1966). *The 'Allotment-of Probability-Shares' - APS - Method. Memorandum PE-18*, The Aeronautical Research Institute of Sweden (FFA), Stockholm, Sweden.

Marthinsen, Harold F., (1993). 'The Decision-Making Process During Takeoff', the Air Line Pilots Association, Herndon, Virginia.

McKenna, James T. (1998b). 'Many Operators Ignore Vaunted CFIT Aids', *Aviation Week and Space Technology*, 9 November 1998, McGraw and Hill, Inc., New York.

McKenna, James T. (1998a). 'Industry Team Pushes Focused Safety Plan', *Aviation Week and Space Technology*, 16 February 1998, McGraw and Hill, Inc., New York.

McKenna, James T. (1997). 'Garvey Commits FAA to Safety Partnerships', *Aviation Week and Space Technology*, 3 November 1997, McGraw and Hill, Inc., New York.

Murphy, Frank B. (1994). 'Crew Centered Concepts Applied to Flight Deck Technology: the Boeing 777', presented to SAE Committee G-10, Boeing Commercial Airplane Co., Seattle, Washington.

North, David M. (1997). Editorial—'We Know the Safety Issues, Now Let's Push Solutions', *Aviation Week and Space Technology*, 1 December 1997, McGraw and Hill Inc., New York.

Ott, James (1997a). 'ICAO Stresses Safety Compliance', *Aviation Week and Space Technology*, 2 June 1997, McGraw and Hill Inc., New York.

Ott, James (1997b). 'Civil Aviation Directors to Explore Expanded Safety Role for ICAO', *Aviation Week and Space Technology*, 18 August 1997, McGraw and Hill Inc., New York.

Reed, John H. (1974). Statement of National Transportation Safety Board by its Chairman before the House of Representatives House Interstate and Foreign Commerce Subcommittee on Investigations Hearings on Ground Proximity Warning System, 18 September 1974, Washington D.C.

Sears, Richard L. (1985). 'A New Look at Accident Contributors and the Implications of Operational and Training Procedures', presented at Flight Safety Foundation 38th International Air Safety Seminar, Flight Safety Foundation, Alexandria, Virginia.

Senders, John W. and Moray, Neville, P. (1991). *Human Error: Cause, Prediction, and Reduction*, Lawrence Erlbaum Associates, Hillsdale, New Jersey.

Shifrin, Carol A. (1996). 'Safety Experts Seek Data Sharing', *Aviation Week and Space Technology*, 12 December 1996, McGraw and Hill, Inc., New York.

Taylor, Laurie (1988). *Air Travel: How Safe Is It?*, BSP Professional Books, London.

US Airways (1998). 'Controlled Flight Into Terrain', In *Flight Crew View*, July, August, September 1998, reprinted from FAA/ICAO//Boeing/Flight Safety Foundation Controlled Flight Into Terrain Education and Training Aid, US Airways, Pittsburgh, Pennsylvania.

Weener, E.F. and Russell, Paul D. (1993). 'Crew Factor Accidents: Regional Perspective', presented at 22nd IATA Technical Conference, 6-8 October 1993, Boeing Commercial Airplane Group, Seattle, Washington.

Wiener, E.L. (1989). *Human factors of advanced technology ("glass cockpit") transport aircraft*, NASA Technical Report 177528, Ames Research Center, Moffett Field, California.

Wiener, E. L. (1977). 'Controlled Flight Into Terrain: System Induced Accidents', *Human Factors Journal*, Volume 19, Human Factors and Ergonomics Society, Santa Monica, California.

22 The Air Transport Future

For I dipt into the future, far as human eye could see, Saw the Vision of the world, and all the wonder that would be... (Alfred, Lord Tennyson)[1]

Expected Growth for Today and Tomorrow

The predicted and continued growth of the air transport industry is a fact of life. Certainly developments in the future will include a host of human factor problems. While specific forecasts vary, depending frequently on who makes the forecast, there is unanimity that the air transport industry will continue its aggressive growth. Reasonably conservative forecasters predict that the world's airlines will be flying more that two billion international passengers annually in the next 12 years. These forecasts are important to those who are already in air transport and to those who are seriously considering becoming part of it. The human factors challenges and the developments that will be a part of this industry in the 21st century are an inherent part of these forecasts.

International Airline Forecast

ICAO's 185 member states carried nearly 1.5 billion passengers in 1997, and also carried in the vicinity of 100 billion metric ton kilometers (RTK) of cargo. International traffic, for both US and non-US airlines, is expected to continue to grow substantially, especially in the cargo area where a tripling of world cargo is expected by 2015. This will require approximately 850 additional large capacity freighter aircraft. The US airline share of world RTKs of cargo is expected to decline from its present about 30 percent to 27 percent by 2015.

Projected Asia-Pacific airline traffic has declined slightly in the past year and in some circles concern has been expressed over the future of Asian markets in the next decade. The concern arose because of fears resulting from 1997s financial and economic turbulence. Climatic variations and pollution concerns have at least temporarily reduced important tourist traffic in some Southeast Asian areas. Despite all this, there is general belief that the center of

---

[1] From 'Locksley Hall', 1842.

the world economy will move slowly to the Far East because of its very large populations. Far East airline traffic is expected to rebound and flourish (Shifrin and Thomas, 1998). Forecast growth rates, which can vary by country, are expected to be from five to six percent annually.

Unfortunately, in terms of worldwide fatalities and crashes, 1997 was very slightly worse than an average year. Internationally, there were 51 fatal crashes and 1,307 fatalities compared to an average of 49 fatal crashes and 1,243 fatalities per year over the past decade. In spite of the fact that traditionally safe Australia, Middle-Eastern, North American, and Western European airlines had no fatal accidents involving passenger carriers, other than the three fatalities mentioned under US Major Airlines, it is clear that improvement is still needed in world air transport safety. (Learmount, 1998). Improvement is especially needed in parts of Asia-Pacific, Latin America, and in the developing States in Africa. Improving air transport safety in these areas is one of the industry's major challenges.

The record in 1998 was mixed. Worldwide, CFIT was still a major problem in several international regions in both jet and non-jet aircraft. There were five CFIT jet accidents and eight involving propeller airplanes. While there was a continuing discrepancy in the level of safety demonstrated in various parts of the world, the areas with deficiencies have been identified and their problems are being attacked.

Regional and commuter airlines continued to show considerable improvement. The Asia-Pacific region was not as fortunate, and there is little question that it is a problem region. Safety areas that may need improvement include infrastructure, regulatory efficiency, training, the '4 P's', etc. It is important that operating problems are identified and then solved locally, for operating experiences are often unique in each region.

Overall figures are very close to the average for the decade. Nearly half (7940) of 1998's 1,244 fatalities occurred in scheduled jet accidents. Unfortunately, in 1998, western Europe's airlines had their first jet fatal accident since 1993. It was Swissair's MD-11 accident on 2 September 1998 off the coast of Nova Scotia, and all 229 people on board lost their lives. On a much smaller scale, Paukn Air of Spain also suffered a jet crash on 25 September during an approach to Melilla, Morocco; 38 lives were lost.

US Major Airlines

US scheduled passenger airlines carried 1.36 billion total passengers domestically and internationally in 1996 and are expected to carry 1.72 billion by the turn of the century. The FAA tells us that domestically, US airlines carried 546.2 million passengers in 1997 and at a projected increasing rate of 3.9% are expected to reach 827 million by 2008 (FAA, 1997). Carol Hallett, the Air Transport Association (ATA) president, has estimated that by 2015 passengers will double the 1996 figure and that cargo will grow even more.

Today, US airlines fly 30,000 flights per day and every day move about 1.5 million passengers virtually without a hitch. US airlines had a very good record in 1997 when there were only three fatalities, while in 1996 there were 342. None of the fatalities in 1997 were the result of a crash, and only two of the three were passengers. One of the passenger fatalities occurred when a passenger fell through an open catering door while boarding in Peru. The other occurred when a UAL B-747 encountered turbulence over the Pacific and an unrestrained passenger was killed. The seat belt sign was on. The third fatality was caused when a Delta ground crew member was crushed by a jet's nose wheel at La Guardia airport (AVflash, 3/2/98). The NTSB considers each of these incidents in the same category as an airline crash. Many, including experts at the Board, believe these categories are misleading and that the NTSB needs a classification system with figures that are not subject to misinterpretation and to misleading conclusions.

The year, 1998, was even better than 1997 in that there were no fatalities at all. Jerry Lederer, the well-known pioneering safety expert, reminds us that this was the third time that there were no US scheduled fatalities for an entire year. In 1923, the US Air Mail Service did not have a single fatal accident, although it did have several accidents and 477 forced landings. The record was remarkable, however, as the life expectancy of an air mail pilot in 1923 was four years. The second occurence of having a year with no fatalities was 1944. The US airlines, which were then flying DC-3's, set a new record when they had no fatalities for the 18 months preceding 1944. However, in the next year they had five fatal crashes (personal communication, 19 January 1999).

Impressive as the 1998 performance is, the lesson for all of us is that, as we have previously mentioned, short term records can be misleading. Accidents are rare events, and they are not programmed by the calendar. US airlines can be very proud of their record in 1998. The big challenge is to maintain that very good record.

Another very real problem that will be discussed later in the aviation infrastructure section of this chapter concerns the additional air traffic that will develop if the industry grows at the forecast levels. It is of some interest that North American airports, which account for close to half the world's airline traffic, handled an extra 70 million passengers in 1996 with virtually no increase in the level of air traffic. The answer, of course, is that the airlines are flying more large airplanes and they are also enjoying higher load factors. It is clear that the present national and international infrastructure is approaching its limits. The predicted additional traffic will create serious problems for airlines and regulators, not only in the US but also throughout the world. Some of those problems are discussed later in this chapter.

US Regional Airlines

Regional revenue passenger miles and passenger enplanements are forecast to continue to grow in much the same fashion that growth is expected to continue for the larger major airlines. Figure 22.1 is Saab Aircraft of America's forecast of the growth in the number of passengers it expects will utilize regional airlines through 2005. Table 22.1 compares aircraft fleet development in 1995 and the fleet development expected in 2005. Over the past three years, the regional/commuter industry has been the fastest-growing sector of the commercial aviation industry in the US. In the future worldwide, it is believed that the major expansion of regional airlines will be with airplanes in the 50 to 70 seat range, that there will be approximately 8,000 regional aircraft deliveries and that about one-half of the deliveries will be turbojets.

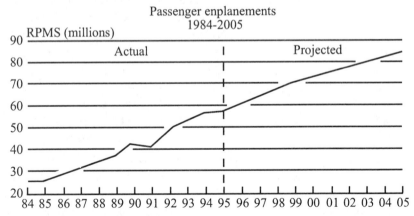

Figure 22.1 US Regional Traffic Passenger Enplanements 1984-2005
Source: *Regional Airline Association Annual Report*, 1996, page 57

By the year 2009, the FAA estimates that the regional airlines will carry 117 million passengers and account for 13.4 percent of all US domestic passenger enplanements. It is also forecasting a continued increase in the trip

Table 22.1 US Regional Airline Industry Aircraft Fleet Developments

Seat-class	Total fleet 1995	Total fleet 2005	Total new deliveries
15-19 seat	730	436	197
20-39 seat	697	1,481	795
40-59 seat	171	485	344
60-89 seat	174	311	201
Totals	**1,772**	**2,713**	**1,537**

Source: *Regional Airline Association Annual Report*, 1996, page 57

length of the regional carriers because of the continued integration of large numbers of high-speed turbo-props and regional jets into the regional/commuter fleets. These airplanes create opportunities for growth in non-traditional regional/commuter markets.

The Aviation Infrastructure

Today most experts believe that the aviation infrastructure will be under extreme pressure to adapt to ever-increasing needs. The aviation infrastructure includes the ATC system, fixed structures such as airport runways, taxiways, passenger and cargo terminals, fuel lines, control towers, and other ground installations, as well as installations for communications and radar. Infrastructure problems in some of the countries whose accident record places them in the high or medium risk categories include a lack of functioning instrument landing systems (ILS), poor air traffic control capability, lack of navigation aids, inadequate or no radar coverage, unreliable weather reporting, and substandard airport equipment.

Aircraft substantially larger than the transports of today will impose a severe burden on the airports that handle them and on the ground-side facilities that must serve them. It has been estimated that airplanes with passenger capacities of from 600 to 1,000 passengers can well be in regular service by the second or third decade of the new century. The only alternative to the ever-increasing size of airplanes of all categories, and the trend for all airlines to go to larger airplanes than they are now using, is to increase the number of airports, passenger gates, runways, taxiways, etc., or to further increase the number of aircraft operations without increasing the infrastructure. This can be very difficult if not virtually impossible in many locations. Very large funds are required, and it is a worldwide problem.

The World Infrastructure

While the magnitude and characteristics of the infrastructure problem vary, there is little question that it is a problem in most parts of the world. Infrastructure problems are not restricted to 'third world' or developing countries. A hopeful note for those countries with a developing aviation system is that many of them will be helped by satellite technology that can reduce or even eliminate the requirement for ground-based navigation and communication facilities.

Airports in several countries already have major construction projects underway. One of the most spectacular of these is at Hong Kong, which has recently completed the new Chek Lap Kok International airport (HKG). It was built into Kowloon Bay by flattening two islets and reclaiming a great deal of land for a cost of over 20 billion dollars. The new terminal, which occupies six million square feet and has eight levels, is nine times larger than the Kai Tak airport terminal it replaced. The new terminal is designed for 87 million passengers a year.

The Chek Lap Kok airport project included a 4,475-foot long suspension bridge (the longest of its type in the world), an express highway, and a railway to link the airport to downtown. Unlike Kai Tak, which is closed at night because of the noise caused by arriving and departing aircraft flying over residential areas, Chek Lap Kok will operate on a 24-hour basis because of its more distant location. Chek Lap Kok opened in July 1998.

In Malaysia, Kuala Lumpur has just completed the first phase of a new $11.5 billion airport to better handle anticipated freight and passenger loads and to compete more effectively with Singapore's Changi airport as the region's premier international airport. Similar development is happening throughout Europe. An example is the $416 million expansion planned for France's second busiest airport (Nice-Côte d'Azur). Because further expansion of that airport does not seem feasible, planning for a new additional airport is expected to start in 1999. If present forecasts are fulfilled, such expansion will be required not only in Europe, but also in the rest of the world.

The Infrastructure in the US

In the United States, as in many parts of Europe, virtually all elements of the infrastructure are presently being operated at or close to their capacity. Busy airports cannot handle any more traffic and already gate space is at a premium. Enroute air traffic routes are getting closer to their limits, and arrival and departures are being routinely limited. Because things like new or significantly expanded airports and ATC facilities have many difficulties, larger airplanes seem inevitable. That means that all segments of air transport will use larger airplanes than they are presently using. This trend has already started. North American airlines handled an extra 70 million passengers in 1996 but the volume of aircraft movements remained virtually unchanged. A consequence of larger heavier airplanes is that present airports will require modifications or additions to take care of the increased passenger volumes. A problem sometimes neglected in airport planning is that the heavier airplanes also will require substantially reinforced runways, taxiways, and ramps.

The National Civil Aviation Review Commission (NCARC) was formed as part of the Federal Aviation Reauthorization Act of 1996. It consists of 21 members from different aviation organizations and has recommended expenditures of $2 billion per year for the next 5 years to expedite capital development projects in its Airport Improvement Program. The Commission states that without additional funding, the number of airport improvements currently needed exceeds the money available to finance their construction.

Estimates by the NCARC are that the aviation system in the US will need added runways and ramp areas, especially on 'underused airports', at a cost which has been estimated to be approximately $5 to $7 billion per year. If the FAA's projection that the number of passengers will increase by 351 million in the next 12 years is correct, "then we need 10 new airports the size of Chicago O'Hare to meet such demands" (David Plavin, pres. of Airports Council International, North America).

The only choice for a large number of airports is to reinforce or upgrade existing facilities. While many present airports are presently, or already have expanded, unfortunately too many of them have no additional room. With the single exception of the new Denver, Colorado airport (DIA), no new major airports have been constructed in the US for several years and there is little hope that our society will construct more new airports in the near future. A very real problem has been the so-called urban sprawl around present airports. Airports are seldom considered desirable neighbors. As Charles Billings (1997) has noted, 'The NIMBY ("not in my back yard") syndrome is never more evident than in aviation matters, in part because airplanes, even relatively quiet ones, are perceived as significantly noisier than other machines in the environment.'

One US airport, whose managers are actively planning for the future, is Los Angeles International (LAX). At the present time, LAX airport managers are considering a 60% expansion for the 21st century at a cost of 12 billion dollars. Financing such a large sum is a major problem. Expansion of the airport and its facilities has become a requirement to keep abreast of a growing industry and its associated economy. Today, LAX is the 2nd busiest cargo airport and the fourth busiest airport in the world. Almost 25% of its travelers are on international flights. Ninety-six airlines use the LAX airport. They furnish clear evidence that air transport is a worldwide industry.

Capacity Enhancement of the Present System

Industry expansion and growth are generally considered desirable by all elements of the industry because expansion and growth invariably mean more and better jobs for employees, more profits for investors and stockholders, and better service for the public. For pilots it means more promotions and usually, but not always, better schedules because there are more trips to choose from. Unfortunately however, there are practical limits in the expansion and growth that can be attained without significant infrastructure expansion.

The very real difficulties inherent in expanding the present aviation infrastructure are forcing the industry to make major efforts to ensure that all aspects of present facilities are being fully utilized. A great deal of research is being conducted by the FAA, NASA, and other research institutions to see if any portions of the aviation system can be made more efficient by the use of better procedures, the latest technology, or by any other methods.

Today, there is little question that the industry must utilize present airports and other elements in the aviation system at higher levels if it is at all possible. However, capacity enhancement is not without its safety problems. Without significant infrastructure expansion and changes in basic procedures, there is no way that capacity can be enhanced without reducing lateral, vertical, or time separation. Reductions in any of these areas cannot be made without reducing safety margins unless they are accompanied by better procedures and advanced technology. Both must be thoroughly tested, for neither the airlines nor their pilots can afford the consequences of a dreaded mid-air collision.

The Air Line Pilots Association is particularly concerned with elements of such FAA programs as LASHO (Land and Hold Short—an outgrowth of the current SOIR or Simultaneous Operations to Intersecting Runway Program), PRM (Precision Radar Monitoring), CASTWIG (Converging Approach

Working Group's) proposals, Free Flight (which is discussed later in this chapter), and RNP (Required Navigational Performance). Each of them should enhance capacity; however unfortunately, at the moment there is not agreement between the FAA, the airlines, and the pilots as to the training and equipment required for their further implementation. Pilots consider this a serious problem for there is very little solace in being right if you are dead.

These are important and delicate issues. The air transport industry has a very large exposure because of the considerable number of operations it flies under a wide variety of conditions. If margins are drawn too tightly, at some point an accident is inevitable. At the same time, it is essential and in the best interests of the public, the pilots, the airlines, and the regulators that the system be used at its safe maximum. This is not an easy problem.

The ATC System

There is an overwhelming consensus within the industry that the ATC system must be modernized. The NCARC warns that by 2010 the US can become an international aviation backwater if it does not effectively modernize the National Airspace System (NAS). Critics state that the NAS is both underfunded and mismanaged. They believe that if air traffic in the next 10 years increases disproportionally to the funds spent on enhancing capacity and increasing safety, the result will be an air traffic jam that could kill the US economy at a time when global aviation is reaching a new level of profitability. Today, flight delays are costing the airlines billions of dollars each year and the crowded airways cause air safety concerns. In 1997 there were 225 near misses from aircraft that were flying too close together. This is an increase of 22 per cent from the near misses recorded in 1996.

As part of its program to meet these problems, the FAA has announced that it is initiating a plan to reorganize airspace throughout the country for efficiency, safety, and noise mitigation. It is expected to start in the Boston-to-Miami corridor, which includes the New York area and its three major commercial airports—La Guardia, John F. Kennedy International, and Newark International. At this writing the FAA is expected to announce a step-by-step series of improvements that will include the installation of new air-traffic-controller workstations at a cost of $2 billion. Development will continue on the 'Wide-Area Augmentation System (WAAS) that in some instances will permit airplanes to fly and land entirely independent of any ground systems. Details of these plans are discussed later in this chapter under Future Developments in the Air Transport System.

The San Francisco Bay Area

The San Francisco airport (SFO) is in the midst of a 2.4 billion expansion that has far-reaching consequences for transportation and parts of the environment of the entire Bay area. Severely squeezed by space, its airport designers are going up and not out. The 2,400 acres of the San Francisco airport is the smallest acreage of any major US airport. By contrast, Denver (DIA) has 35,000 acres. The challenge of the designers is to build a new airport essentially on top of the world's seventh busiest airport without disrupting the 40 million passengers per year that it serves. At the time this is being written, San Francisco International's \$2.4 billion expansion is the largest airport expansion project in the US. Others are the \$1.2 billion expansion of New York's John F. Kennedy International airport, and Detroit's \$1.6 billion expansion of its Municipal Airport. Both are scheduled to be completed by late 2001. SFO is aiming at completion one year earlier.

The economic importance of an effective air transport system is well illustrated in San Francisco. The airport is a money machine for the Bay Area. Last year it generated over \$7.5 billion in personal income while paying \$3.8 billion in federal, state, and local taxes. The total value of its airfreight was over \$76 billion in 1996. In addition, visitors are estimated to pump \$10.7 billion a year into the Bay Area. The economic stakes are high for maintaining a vigorous international airport in San Francisco and other cities.

One of the more controversial parts of the San Francisco airport expansion is the building of a new runway or runways out into the bay in order to accommodate the next generation of jets and supersonic transports. It would be the first major fill of the bay in 30 years and raises a great many environmental concerns. New runways are required to minimize the noise pollution and delays associated with Bay area weather. Presently the SFO airport has the third-highest number of flight delays among US airports, and there is evidence that the number of delays at SFO is increasing.

Limitations in the Industry's Future

Philip Condit, president and chief executive officer of the Boeing Company, while giving the Wright Brothers Lectureship in Aeronautics recently told the World Congress and Exposition that:

> ...I am certain that the real measure of their value (future technological breakthroughs) will be determined not by engineers—but by the airline customer. It

is the customer, and only the customer, who is the ultimate judge of value by deciding what products to buy at what price. As airplane designers, we can agonize over the right combination of technical performance, cost, and market timing to satisfy our customers—but the airline customer remains the final authority on whether we've done our homework correctly.

We can try to anticipate what customers will want five years from now, or ten years—but we'd better be sure that our technology is focused on solving real problems in ways that make good economic sense for the airlines that buy our airplanes. Ultimately, it is the customer who will determine the future direction of commercial aircraft design. (Condit, 1996)

The relevance of this comment was emphasized with the graph shown in Figure 22.2. The graph compares the yield per passenger versus the price per seat for jet air transports since the beginning of the jet era in 1960. Since 1960, jetliner prices on a per-seat basis have been rising faster than inflation while the yield per passenger-mile to the airline has been decreasing at about the same rate. Obviously, these trends cannot continue indefinitely. Today, a new B747-400 has a $150 million price tag, while back in 1970 a new B-747 cost only $23 million. In 1975, TWA sold six B-747s to Iran for $16.5 million each. Gary Behrens, Southwest Airline's chief operating officer, complains that the cost of a B-737 has increased to about $35 million this year from the $3 million that airplane cost in 1971. Those prices are typical of price increases for other airplanes. They clearly indicate one of the problems that the industry faces in the future.

On the day following Mr. Condit's lecture, NASA Administrator, Daniel S. Goldin, borrowed the Condit slide that showed the decreasing airline revenue per passenger mile and, while giving a major address at the World Aviation Congress and Exposition, told the industry: 'This is the only slide I think I've ever shown since becoming Administrator.... This, in my mind, is big-time trouble if we think evolutionary change will get us out of trouble.' Administrator Goldin believes that simply depending on aircraft derivatives, 'takes us down the wrong path' because 'it does not attack the curve'. Summing up, he asked his audience to remember two things. 'We cannot live with that curve ' and 'safety must improve by a factor of three just to stay even and by another factor of three to get ahead' (Ponticel, 1996).

A final limitation to the industry's growth may well be the willingness of governments throughout the world to improve and expand the infrastructure. Unfortunately, improving and expanding the infrastructure will require large amounts of money. The availability of these funds, and the discretion

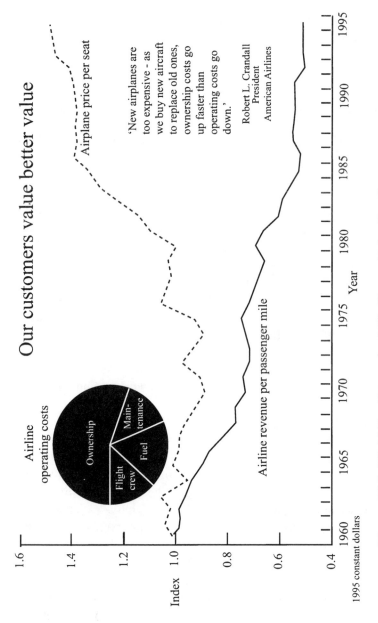

Figure 22.2   Yield Per Passenger Mile Versus Price Per Seat
Source: 'Performance, Process, and Value: Commercial Aircraft Design in the 21st
Century', page 8, Condit, Boeing Commercial Airplane Company, 1996

with which they are expended is bound to be an important factor in the expansion of the industry.

In the meantime all of us should remember that while the rate of traffic growth may periodically slow or even go negative, history tells us that traffic itself will rarely, if ever drop. The total industry, including the governments of most countries of the world, must be ready for significant air traffic growth in the 21st century.

Evolutionary Changes

It is of some interest that NASA Administrator Goldin has suggested that evolutionary changes will not be enough to 'get us out of trouble'. Evolutionary changes have been a part of this industry for many years. While this may be essentially a matter of definition, the only really revolutionary changes we can identify over the past half-century have been the change to pressurized airplanes and the change to the jet era. Virtually all of the other changes have been evolutionary. With the exception of the move to supersonic airplanes— the HSCTs that we will discuss later, we see nothing in the reasonably near future in the development of transport airplanes that we consider revolutionary. There is no chance that HSCTs will be a predominant part of the industry in the reasonably near future.

We are by no means suggesting that the airplanes built in the early 1970s are comparable to those being built 25 years later. Their only really equivalent features are their common designations. Changes in these airplanes have been gradual, continuous, and evolutionary. There have not been revolutionary changes.

An example of evolutionary, not revolutionary, change is the next generation of B-737s. They are aimed at a single-aisle market of something like 11,000 transports in the next 20 years. The changes are designed to include a family of airplanes that can be stretched again if desired, and can vary from 100 to 189 passengers. The new 737s will also have available such items as much more powerful jet engines, better fuel consumption, and innumerable other technological advances. Harold Arnold, chief project engineer and deputy program director for the new 737s, tells us that the last airplane to be certified (the B737-800) retains only about 10% of the technological features of the first aircraft. Appealing to the airlines is the fact that the latest 737s are expected to have about 15% less maintenance costs than their predecessors.

The new B737-800 is expected to share the same pilot type rating as all other 737s, although 'differences training' from 10 hours to eight days, depending upon the background of the trainees, is forecast. The 'differences training' reinforces the movement toward much more individualized training depending on the needs of the pilot. There is a definite move (led by Airbus) toward more common-type training whenever it is possible because of the substantial economies in initial certification, in the pilot training required, and in the scheduling flexibility gained. There are substantial economic advantages in these evolutionary changes and in the ability of the industry to take advantage of incremental technological changes. The newest Airbus transports compete with the new B-737s in all aspects.

The Standard Airplane

In an effort to reduce the cost of transport airplanes, the industry has asked Boeing and Airbus to consider building a 'Standard Airplane'. The SAE has scheduled a symposium in November of 1998 to publicly discuss the issues surrounding the Standard Airplane concept. Proponents maintain that if airliners were delivered in a basic standard design in which the number of changes requested by an airline customer were reduced to a bare minimum a great deal of money could be saved. The only options allowed under the concept of the Standard Airplane would be those things that were unique to the passenger in things the passenger could see, feel, or taste.

Boeing has stated that they have produced more than 14,000 separate changes on the five airplane models they offer and that 72% of those changes were at the request of a single customer. Airbus has had similar experiences. With a Standard Airplane, airlines would only specify things like seats, décor, food, entertainment, etc. There is little question that a Standard Airplane could save the airlines a considerable amount of money. Whether the development of a Standard Airplane would be an evolutionary or a revolutionary change seems relatively academic. The big question is whether Boeing and Airbus and the world's airlines and regulators will ever be able to agree on the details that would be required to implement the concept of the Standard Airplane. Customization costs money. It costs money in terms of engineering work-hours and special tools and in disruptions to the assembly process.

International Harmonization

Harmonization is the standardization between different countries of rules and regulations involving the manufacture, certification, and operation of air transports. It is important to all segments of the industry. Nearly all parties agree that it is an important concept that is well worth pursuing. ICAO is active in harmonization activities. One of the most successful harmonization efforts has occurred between the US FAA and the European Joint Aviation Authorities (JAA).

The JAA is made up of 27 European countries—15 of which belong to the growing European Union (EU). The JAA is a rule-making but not a law-making body. It is not a regulatory authority largely because individual countries are reluctant to give up their right to make the rules and regulations that govern aviation in their States. Member countries are required by law to adopt the JAA's regulations, but each has the right to change portions of them if they wish. The JAR-OPS (a set of rules governing aviation operations) have been adopted by many but not all of the JAA countries in spite of the fact that they have been accepted in principle by all of them.

Harmonization is a dynamic area that includes considerable political controversy. For example, the EU maintains that its member countries cannot legally adopt JAA regulations into law unless they are a part of EU law. This creates serious political problems for the EU holds that only it has the right to make aviation regulations that are binding on all members. Critics point out that the EU has no aviation expertise and is an entirely political body, while the JAA is recognized as the center of excellence in aviation matters in Europe. A far from minor problem is that the EU regulations require that everything be available in the 11 languages of EU members while international aviation (and the JAA) have traditionally used only four languages—English, French, Spanish, and Russian. (See Appendix O.)

Recently European Union transport ministers have established a European Aviation Safety Agency (EASA) that has been given power to 'develop, adapt and, if necessary, publish rules relating to the design, certification, manufacture, continuing airworthiness, maintenance and operation of civil aircraft and the certification of personnel engaged in those activities.' The new agency obviously conflicts with the JAA and unlike the JAA, the new agency will be backed by a legal framework that includes all EU members. The EASA hopes, to certify aircraft equipment and procedures based on international safety standards and to ensure that its standards are applied in all EU countries.

Harmonization of a lesser type is achieved through the bilateral agreement of individual countries and has been used for some time. The US, which presently has a commission considering its certification process, has in the order of 37 such bilateral agreements in which each country agrees to accept the regulations of the other. The ALPA and IFALPA, as well as IATA and ICAO believe strongly in the harmonization principle. It helps achieve recognized safety standards throughout the world.

Developments in the Air Transport System

As this book is being written, a great number of changes and innovations in the air transport industry are in the process of being put into effect. Some of them already have been partially implemented. A representative group of these changes and developments are discussed in the following paragraphs. Some of these changes are in every sense of the word revolutionary changes.

Historically, air transport navigation in three of the location dimensions has been concerned with keeping the airplane longitudinally, latitudinally, and in a vertical plane that all are within reasonable operational limits. The time dimension (the fourth location dimension) had considerable flexibility and has been mainly concerned with scheduled departure and arrival times, ETAs (estimated time of arrivals), and relatively rare curfews at some airports.

In the 21st century, the time dimension will probably lose a great deal of its flexibility and become as critical as the other ATC location parameters. Specific times over fixes will almost certainly become part of clearances, particularly for such ATC innovations as Free Flight. Free Flight is discussed later in this section. Current Flight Management Systems (FMSs) are able to adjust power and therefore airspeed to meet clearance requirements that are based on very accurate fix and destination arrivals.

Future Air Navigation System (FANS)

FANS is a future air navigation system whose specifications have been developed by ICAO. All of the critical functions of communications, navigation, surveillance, and air-traffic management (CNS/ATM)—the building blocks for the worldwide implementation of FANS—are now in place and are being introduced incrementally as they mature and as necessary modifications are adopted. The critical element now is the money required for implementation, especially in the developing nations. At a recent international conference, the Nepal delegate rather prophetically told the 800 plus delegates that: 'The success of global CNS/ATM implementation will not be measured by the fastest horse in the race, but by the slowest.'

ICAO and IATA have led increased development in the Asia-Pacific region. It has included cooperative efforts by Korean Airlines, Singapore Airlines, Japan Airlines, British Airways and KLM. IATA's Regional Coordinating Group (RCG) and ICAO have been able to bridge many of the political rifts existing in Asia and now new routes across Asia to Europe. CNS/ATM routes over the North Pole to the US East Coast are becoming an actuality. Long-standing plans to open a CNS/ATM route from cities in Southeast Asia to India, Pakistan, Iran, Turkey and eventually Europe are becoming closer to reality.

Aircraft and avionics manufacturers are forced to meet operators' present needs and at the same time fulfill their long-term requirements. The air traffic control problem is complicated because of the lack of a worldwide ATC system and the regional politics that can be involved in developing CNS/ATM standards that meet FANS requirements. Today a first step has been the inclusion of FANS-1 capability in the Boeing 747-400 and in the latest Airbus aircraft, where the navigation system is referred to as FANS-A. Fortunately, present traffic control and navigation systems can coexist with the satellite navigation systems. This makes the transition to satellite navigation much less stressful than it might otherwise be. The Air Transport Association has estimated that inadequate air traffic systems and procedures cost the domestic US air transport industry 3.5 billion dollars annually.

Data link is one of the most critical CNS/ATM building blocks. GPS (Global Position System or its European or Russian equivalents, GNSS or GLONASS) and Automatic Dependence Surveillance and its modifications (e.g., ADS-B) are key elements of FANS. They all furnish an airplane's position, altitude and velocity vector data automatically to ground control stations at frequent intervals.

One of the major problems the international community has with the FANS program is to reach agreement on how the data link that furnishes the automatic dependent surveillance-broadcast should be specified. The two major candidates are the Swedish-developed digital VHF data link and the US-developed and FAA sponsored Mode-S data link. The potential market is very large and the stakes for each country are high. Each would like to have its system adopted so that it would become the ICAO standard. Determination of the winning system is a lengthy technical and political process. It would be nice if the only considerations were the relative merits of the competing systems, but unfortunately in the real world, things are not that simple.

Global Positioning System (GPS), Europe's Global Navigation Satellite System (GNSS), and Russia's GLONASS

Global Positioning System (GPS), in which geographic location is obtained entirely from navigation satellites, has been called the next major step in the exploitation of global satellite technology especially when its navigation potential is combined with data link communication to ATC. While the project is still in an initial and exploratory stage, a carefully managed step forward has been over the Pacific with the use of FANS (including GPS and data link for CNS/ATM implementation) by four airlines. The four airlines are United Airlines, QANTAS, Air New Zealand, and Cathay Pacific. Their operation is coordinated by the US FAA, by Air Services Australia, by the Air Services Corporation of New Zealand, and by Honeywell and Boeing. The GPS uses US military satellites.

It is claimed by many that GPS meets the FAA requirements for a sole means of navigation for civil navigation. However, there is reluctance to depend upon GPS as the single source for instrument navigational data. The President's Commission on Critical Infrastructure Protection says: '...the most significant projected vulnerabilities are those associated with the modernization of the NAS and the plan to adopt the GPS as the sole basis for radio navigation in the USA by 2010.... Exclusive reliance on any single system creates inherent vulnerabilities, no single system can be guaranteed for 100% availability for 100% of the time.' Total reliance on the GPS for landing aircraft '...creates the potential for single-point failure and cascading effects'.

In a recent report to Congress, the FAA warns that 'technical uncertainties' relating to intentional jamming, unintentional interference and solar activity could require an independent backup to the global positioning system and the Wide Area Augmentation System (WAAS) that it has proposed. While no decision has been made at the time of this writing, it seems likely that the FAA will approve the GPS-based WAAS only as a primary, not as a sole means of navigation. A back-up navigation system, which the FAA says might be a combination of presently available systems, will then be required. The FAA has just given a contract to RTCA, Inc. to study this problem.

A very big advantage of GPS in many of the 'third world' or other countries that may not have well-developed navigation infrastructures (e.g., CIS, China, many African and South American States, etc.) is that GPS or other satellite navigation may be a way to open its airspace, bring in overflight revenue, and improve their internal air navigational capability. GPS can do all this without these countries having to invest in expensive ground-based navigation aids. Several developing nations have urged ICAO to establish an international monetary fund to help them finance implementation of CNS/ATM system programs. It is generally agreed that funding remains a key hurdle to be overcome before CNS/ATM can be implemented. Overall, it is believed that a coordinated CNS/ATM system can reduce costs by about $6 billion a year when compared with current operational procedures (Kelley, 1998a).

In Europe, the European Commission (EC), the European Space Agency, and EUROCONTROL have agreed to formalize their cooperation and to expand it to include a second-generation GNSS-2 that could be used instead of GPS and GLONASS (Russia's Global Navigation Satellite System). At present, the three organizations are already cooperating on GNSS-1, which is compatible with the US military's GPS, with the Russian military's GLONASS and with regional augmentation systems. GNSS-2 is designed to be used entirely under civilian control. Both the European Commission and the former Russian States are reluctant to depend upon a GPS that is entirely controlled by the US.

GPS, GLONASS, and Europe's EGNOS are commonly referred to as GNSS-1 and, particularly in Europe, are viewed with a fair amount of skepticism because both GPS and GLONASS are owned by the military in the US and Russia respectively. There is concern that either could be shut down for strategic purposes during a military conflict. Many European users favor going directly to a civilian owned and operated GNSS-2 in order to avoid the estimated $100 million costs that would be incurred in implementing an ultimately replaced GNSS-1 (*Global Airspace*, 1999).

As this chapter is being written, *Aviation Week and Space Technology* has reported that Continental Airlines has received the first differential-GPS instrument approach for a revenue-service aircraft in the US. The FAA has just certified Continental's MD-80s for the use of Honeywell/Pelorus Navigation Systems SLS-2000 for Category I landings at Newark and Minneapolis. This GLS (landing system) offers significant cost saving for airports compared with a conventional ILS. The single satellite system offers precision approaches to all of the airport's runways, while with a conventional ILS, separate localizer, glideslope antennas and marker beacons are needed for each runway. This is a first big step toward 21st century navigation using satellite navigation (GPS/GLS) for instrument approaches.

While all of the growth potentials for the new system have not yet been approved, the assumption is that it will provide multiple precision approaches to a runway and have the further growth possibility of reaching category II and category III minimums. The potential for the underdeveloped States that are developing modern aviation infrastructures is considerable.

Free Flight

Free Flight has been defined as '...a safe and efficient flight operating capability under instrument flight rules (IFR) in which operators have the freedom to select their paths and speed in real time. The concept moves the National Airspace System (NAS) from a centralized command-and-control system between pilots and air traffic controllers to a distributed system that allows pilots, whenever practical, to choose their own route and file a flight plan that follows the most efficient and economical route. Air traffic restrictions are only imposed to ensure separation, to preclude exceeding airport capacity, to prevent unauthorized flight through Special Use Airspace (SUA), and to ensure safety of flight. Restrictions are limited in extent and duration to correct identified problems. Any activity that removes restrictions represents a move toward Free Flight' (RTCA Task Force 3, 1995).

The Free Flight goal is to provide greater efficiency in flight operations by increasing airport arrival and departure rates, reducing enroute separations, and performing traffic flow management more efficiently. It is a highly worthwhile goal. Free Flight will attempt to take full advantage of the potential of GPS/ADS. GPS is a key element of CNS/ATM, the FAA's proposed Air Traffic Management System. In all probability the NAS' current operational concept will be replaced. Free Flight does limit pilot flexibility in certain situations, such as those when separation in congested airspace or at busy airports is required.

The general concept of Free Flight, which must also meet FANS requirements, has been approved by the managers of ATC in Europe, although a number of details need to be adopted and coordinated. As we mentioned in the discussion of FANS, there is presently an important difference of opinion between the US FAA and several European countries regarding the system that will ultimately furnish required automatic dependent surveillance (ADS). There is also some disagreement over requirements for the civilian use of satellite technology.

While the timing and ultimate future of Free Flight, both in the US and in other parts of the world still may be uncertain, it is clear that many of the people involved believe that Free Flight can significantly improve the present

ATC system. Without change, the present ATC system does not seem capable of meeting the future demands of a growing and dynamic industry. To equal or exceed the safety levels that have been achieved without restricting forecast growth will require the best efforts of everyone involved.

Advantages of Free Flight

One of the big advantages of the Free Flight concept is that it moves from the present rather rigid and procedural ATC system to a future collaborative system that provides increased flexibility for the users. It does this by utilizing space-based technologies such as global positioning system navigation satellites (GPS), two-way data links with navigation, and the latest automatic dependent surveillance-broadcast system (ADS-B). All of this puts much more flexibility and responsibility for enroute traffic control in the cockpit, although ultimate authority for separation remains with ATC. New displays, both in aircraft cockpits and at ATC stations, are real world requirements if the system is to develop an increasingly seamless form of air traffic and flow control management worldwide (Braune, Funk, Bittner, 1996).

With Free Flight, the pilot would choose the route, speed, and altitude. The Free Flight concept as envisioned by RTCA (Figure 22.3) is based upon two airspace zones—a protected zone and an alert zone. Size of the zones is based on the aircraft's speed, performance, characteristics, communication, navigation, and surveillance Aircraft can be maneuvered with autonomous freedom until an alert zone is breached, in which case an air traffic controller will give one or both of the aircraft instructions to ensure separation. Another concept for Free Flight currently being explored includes a four-zone concept that works much like current TCAS alerting.

The FAA's and Other Organization's Level of Commitment

The FAA made a major commitment to Free Flight in 1994. Its commitment is shared with world organizations, including ICAO and EUROCONTROL, industry and pilot organizations, NASA, and relevant academia. Already, there has been a marked elaboration of the original concept. It includes a large number of different, existing, or evolving hardware systems, software systems, procedures, training, and research programs that will require considerable operational and political work if Free Flight is to be successful.

George Donohue, former FAA associate administrator for research and acquisitions, has rightly stated: 'Anything of the magnitude of moving to free

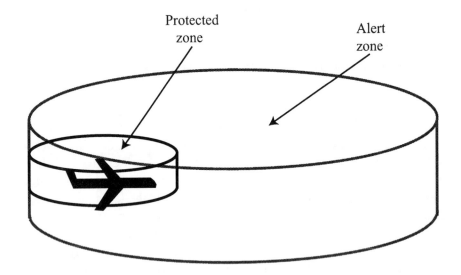

Figure 22.3 Computer-constructed Protected and Alert Zones
Source: *Report of RTCA Board of Directors Select Committee on Free Flight*. Reproduced by permission of the RTCA, Inc.

flight has got to have a significant demonstration of how it works before it will be accepted by the controller and pilot communities.'

In order to have a significant demonstration and to test the feasibility of the present concept of Free Flight, a two-year evaluation called Flight 2000 is planned between Alaska and Hawaii. Hawaii and Alaska were selected because of their environment and because of their affordable fleet size. It is hoped that any major operational bugs can be identified and corrected during this period. The two-year evaluation will include all categories of transport aircraft and several military aircraft. Approximately 2,000 airplanes are expected to be involved. General aviation participation is voluntary. In order to ensure that the evaluation covers all major user categories, an Ohio valley site may also be selected so that major air cargo carriers can participate in the evaluation. It is presently planned to have a complete evaluation and a report of Flight 2000 completed in 2005.

Modifications of Free Flight

The RTCA subcommittee on Free Flight has recently recommended that the Free Flight program be renamed the Flight Operational Enhancements Program. The subcommittee has identified nine operational enhancements it considers essential in the successful evolution of the US National Airspace System to Free Flight. The recommendations include: 'the use of the flight information system to improve cockpit situational awareness; the avoidance of controlled flight into terrain through the use of cost-effective terrain database and graphical display; improved terminal operations in low-visibility conditions; enhanced see and avoid; delegation of separation authority to the cockpit; improved surface navigation; enhanced airport surface surveillance for the controller; the use of automatic dependent surveillance-broadcast (ADS-B) in non-radar airspace; and the use of ADS-B-based separation standards.' (Kelly, 1998b). The FAA recently has announced that final plans for a revised Flight 2000 program will be announced in December of 1998.

There are many human factors, political, and economic questions that are directly involved with the implementation of worldwide Free Flight. It is not a simple process. A SAE G-10 Free Flight subcommittee is working on Free Flight problems and has developed what it calls eight critical issues, nine serious issues, and four desirable issues. The subcommittee believes that failure to resolve these issues, especially those that are critical, will affect the long-term viability of Free Flight.

Without in any way ignoring the fundamental problems associated with enroute Free Flight or ignoring its advantages over the present system, it seems that major problems that preclude unrestricted Free Flight for the foreseeable future will still exist in high density areas near busy airports and in already congested traffic areas. There are innumerable reasons that airline passengers want to arrive and depart from the same airports at essentially the same time. Sorting out the inevitable conflicts that will arise in Free Flight will require parts of a system fairly close (though perhaps significantly improved) to the system used today. The hub and spoke scheduling system, which is used at busy terminals throughout the world, exacerbates these problems. Free Flight can significantly improve the enroute portion of the ATC system, however it may well have considerably greater difficulty with congestion during arrivals and departures.

Air traffic controllers will continue to play a major role in the future development of Free Flight. Recently, Mike McNally, president of the National Air Traffic Controllers Association, said:

I know a lot of folks believe there is some planning to reduce personnel required as we put new technologies on board. The irony is that the opposite is true. You cannot have a controller looking at a radar scope and multiple screens and continue to do the job in the manner in which they do today. So we are going to need extra sets of eyes to work other types of equipment. We see more controllers being needed, not fewer. In fact, prior to his departure David Hinson [former FAA Administrator] indicated that more than a year ago. We'll need more controllers, not fewer.   (Lowe, 1998)

Free Flight and Pilot Workload

Considerable concern has been expressed that pilot's workload would be increased—perhaps to undesirable levels—with Free Flight. In a study by a Dutch aerospace laboratory (NLR) that was partially funded by the US FAA and undertaken in conjunction with NASA, the concern seems at least questionable. The study concluded that workload does not increase when a pilot is given responsibility for separation assurance in a 'free flight' environment. The project leader says that the conclusion surprised the research team. 'We anticipated a dramatic increase in workload.... The conclusions held even when traffic density was increased to levels that are never likely to be experienced' (Sheppard, 1998).

The results of this study seem counter-intuitive to most observers unless the study was confined to the enroute phases of flight and assumed a transfer to current procedures in terminal areas. One can expect considerable discussion regarding workload in the new environment. The critical question will be to determine if workload, even if increased, remains within acceptable limits. And, as was discussed in Chapter 10—Workload, the question of how to measure the workload in a reliable and valid fashion will certainly be raised

These are very complex issues. The chapter, 'Future Air Traffic Control and Management Automation', in Charles Billings' *Aviation Automation: The Search for a Human-Centered Approach* is highly recommended reading for anyone interested in automation-related and other human factors issues involving Free Flight.

FAA Wide Area Augmentation System (WAAS)

The Wide Area Augmentation System (WAAS) is the FAA's answer to air traffic control in the 21st century. WAAS consists of ground-based stations to provide near-pinpoint accuracy to satellite signals (such as GPS) that are broadcast from space. There is no direct connection between WAAS and Free Flight. WAAS is designed to provide navigation capability within the existing air traffic control system and, when it is introduced, Free Flight will utilize the existing WAAS system. It is completely consistent with ICAO's FANS.

European Geostationary Navigation Overlay System (EGNOS)

In an effort to do something about Europe's staggering air traffic control capacity problem, on 23 April 1999, EUROCONTROL's Provisional Council was slated to examine and approve a detailed plan to reduce vertical separation minimums (RVSM) in Europe's upper airspace. There is little question that there are significant obstacles to be overcome before the proposed plan can be implemented but the rewards are considerable. The proposed plan would make six new flight levels (30,000, 32,000, 34,000, 36,000, 38,000, and 40,000 feet) available to commercial traffic. Cost savings have been estimated by IATA to be in the vicinity of 2 billion eurodollars ($2.3 billion), most of which would result from reduced air traffic control delays. The rest of the cost savings is expected to be achieved by fuel savings achieved through improved access to upper flight levels and the ability to fly more economical flight levels.

Among the obstacles to be overcome is coordination with air traffic control among the 40 states that are participating and a part of Europe's air traffic system. Europe's air route structure is complex. It serves a wide variety of aircraft types, has a very high density of traffic, with a great deal more climbing and descending than is found at RVSM levels over the North Atlantic. Eric Sermijin of EUROCONTROL's Airspace and Navigation Division does not underestimate the problem within EUROCONTROL states but has stated that 'the main task has to be carried out at the (eastern) peripheries where the transitions are taking place' (Doyle, 1999).

Conversion to RVSM altitudes directly affects both the companies that operate as well as the pilots that fly the more than 8,000 civil aircraft in European upper level airspace and air traffic controllers in 38 countries. EUROCONTROL is taking responsibility for briefing national training managers from each of the countries involved. These managers will then be responsible for completion of the training of their air traffic controllers. The plan is to complete the required briefings for the trainers by 2000.

EUROCONTROL believes that the introduction of RVSM over Europe is the most cost effective way of finding badly needed extra capacity in Europe's upper flight levels and is committed to this program. It estimates that the cost benefit ratio up to the year 2015 will be eleven times over the large costs that will be incurred. Plans are for a pre-implementation safety assessment in July 2001 and a final go/delay decision in September of that year. IATA is in favor of earlier implementation anywhere that is possible. After implementation any aircraft not formally approved (must have MASPS altimetry) and without a suitably trained pilot will not be able to fly between 29,000 and 40,000 feet. Military, customs, and police aircraft are an exception to this rule.

Europe's answer to air traffic control in the 21st century is the European Geostationary Navigation Overlay System (EGNOS). It fulfills the same requirements as WAAS. It is essential that the two work together seamlessly in the interests of international aviation. Initial tests of WAAS/EGNOS interoperability were held in Iceland in October of 1998 and were highly successful. The deputy engineering manager of the UK's National Air Traffic Services has recently stated that 'interoperability between WAAS and EGNOS is no longer an issue' (*Global Airspace,* 1999).

Global Analysis and Information Network (GAIN)

The GAIN concept (Global Analysis and Information Network) was initiated by the US Federal Aviation Administration in May of 1996. Since that time it has gathered international support, including that of the United Kingdom's CAA, the US' NTSB, and the European Civil Aviation Conference. GAIN has been proposed as a system that would make aviation safety data instantly available worldwide. It is proposed that it be owned and operated by an international consortium. The hope is that it can avoid or minimize litigious problems in countries like the US.

The industry has held GAIN conferences in Boston, London, and San Francisco, which were sponsored by the FAA, the Royal Aeronautical Society, and by United Airlines respectively. An objective of the conference in San Francisco was to get greater participation from international carriers based on the Pacific Rim and Asian regions. Some individual airlines there have had outstanding individual programs but have not participated effectively. The latest GAIN conference was held 3-5 November 1998 in Long Beach, California. It was sponsored by United Airlines and included participation by officials from other US and international airlines. Representatives from the FAA, the British CAA (Civil Aviation Authority) and official representatives from other countries also attended.

In the US, three prototype-operating models of GAIN are under development. They include one by the Air Transport Association (ATA), which is based upon the very successful British Airways Safety Information System (BASIS). The ATA system is called the Aviation Safety Exchange System (ASES) and is intended chiefly for the free exchange of data between member airlines. A second operating model is being developed by the Flight Safety Foundation with the Battelle Corporation for the free exchange of airline safety data using the worldwide Internet as an electronic platform for disseminating information. The third model—the Operational Airport Safety Information System (OASIS) is under development by airport authorities in the Netherlands. It would share airline and airport safety information in a manner similar to that of ASES and BASIS (Phillips, 1997).

Ideally, GAIN would involve air carriers and other operators, manufacturers, insurers, pilots, mechanics, flight attendants, air traffic controllers, airport operators, and governments from all parts of the world, all linked together by a series of easily accessed computer networks. Key pieces of information would be cockpit voice recorder (CVR), flight data recordings, and Flight Operations Quality Assurance (FOQA) information.

A serious problem at this time is the fear that this data could be used for other than its intended purpose. Commenting upon a recent New Zealand appellate court ruling, Capt. Mike Glawe, the present chair of ALPA's Master Executive Council at United Airlines stated:

> Used appropriately, the cockpit voice recorder has been an invaluable tool; however, if it is used for anything else, it will rapidly lose its effectiveness and its support from professional aviators. Such shortsightedness sets the cause of aviation safety back for New Zealanders and international travelers in that country....We see this as a major setback in international efforts to improve aviation safety, and we will make our members aware that in New Zealand the cockpit voice recorder could be used for other than its intended purpose.
>
> (Lowe, 1997)

The concern for misuse is a major problem throughout the world. This problem was discussed in the Chapter 20 section, Data Sharing and Flight Operations Quality Assurance (FOQA).

The FAA's active promotion of FOQA and its attempt to keep safety data and other information confidential is a big help and so is the growing recognition of the contribution that a nonpunitive incident reporting system can make toward increased aviation safety. John Kern, vice-president of aircraft operations at Northwest Airlines has suggested the need for a working group to determine when safety matters require punitive or non-punitive action. The working group would define a 'box' inside which human errors could be reported without punitive action. Willful, negligent, or repeated actions would fall outside the 'box' and be subject to disciplinary action.

It takes little imagination to conceive of incidents in which determination of whether or not the incident belongs in the 'box' becomes very difficult. One can be assured that pilot groups will be extremely wary. A fundamental reason for an effective incident reporting program was well-stated by Captain W. E. Dunkle, former senior vice-president of operations at United Airlines, who inaugurated the first truly nonpunitive incident reporting program by simply saying, 'I'd rather know about the incidents than punish the offenders'. This is an important issue. Unfortunately, without another dramatic incident that illustrates the positive accident reducing potential of a non-punitive system, one can anticipate a great deal of debate before such a system can be implemented in the US. There is no question that TWA's unfortunate accident at Round Hill in December of 1974 made it possible to develop the NASA/ FAA Aviation Safety Reporting System.

NASA's Surface Development and Test Facility

In an effort to increase the effectiveness and productivity of an important part of Air Traffic Control, NASA is constructing a $9.3 million dollar simulator. The simulator is located at the Ames Research Center at Moffett Field, California, and will be housed in a two-story building inside a larger building. The new simulator will have the look and feel of an air tower control cab, with an image-generation system that will provide a realistic view of weather and environmental conditions, seasonal variations, and can portray up to 200 aircraft on the ground or in the air.

Steven Harke, the NASA simulator project manager has stated that: 'Airports are the bottlenecks of the whole damn system.' Researchers now will be able to simulate realistically virtually any major airport in the simulator's extensive database. Their task and goal will be to develop and test new technologies and procedures that will enable the movement of airplanes more safely and efficiently than the system is presently able to do (Sweeney, 1998). The human factors experts, who are being asked to develop improved ways to best handle the inevitable congested airport terminals, are assured of an excellent research facility

NASA's Taxiway Navigation and Situation Awareness Program (T-NASA)

The FAA reported that between 1900 and 1993 an average of 312,000 flights were delayed on the ground. These delays cost airline operations very large amounts of money—$3 billion in 1990 alone, while the cost impact of passenger delays was $6 billion for that year. To mitigate these delays NASA developed a total program called its Terminal Area Productivity program in

1995. The aim of the program is to safely increase capacities in the airport area in non-visual or instrument conditions until they are essentially equal to capacities in visual conditions. NASA is utilizing such advanced technologies as satellite navigation systems, digital data communications, information presentation technology, and advanced ground surveillance systems that must be integrated into the flight deck in order to reach this particular goal.

In an effort to improve the ground operations of aircraft, particularly during periods of poor visibility, NASA developed a Taxiway Navigation and Situation Awareness program. It contemplates providing route/taxi information projected on a HUD display to provide guidance during low visibility periods as well as providing an audio ground collision avoidance warning system (Foyle et al., 1996).

The complete implementation of such a system is not on the immediate horizon for a great many reasons. This brief discussion is included simply to illustrate the complexity of a truly all-weather operation and also to illustrate the fact that technology, while wonderful, is sometimes not without considerable cost and complexity.

Aircraft of the Future

Very Large Transports

The aircraft industry is actively planning a next generation of transports with up to 650 to 700 passengers. It also has been discussing a very large airplane (VLA) with a capacity of up to 1,000 passengers (Air Transport Medicine Committee, 1997). Recently, eight European countries have combined to request proposals for a Future Large Aircraft (FLA) stating that they have a requirement for nearly 300 such transports. Russia, which already has the world's largest cargo airplane (the Antonov AN-124), has just announced that it is beginning design on an 860 to 1,000 passenger transport. Russia is also proposing to build a stretched AN-124 cargoliner, which would be called an AN-125 and carry up to 250,000 kg. (551,150 pounds) either internally or externally.

Airbus Industrie has estimated that there will be a market for 1,442 aircraft that are 'larger than anything flying today'. A tangential but very real problem is the concern that aviation insurers have for their potential liability in the event of the crash of a super-jumbo passenger jet. For example, if one of the planned 650 passenger jets crashes, the losses could exceed two billion dollars. Individual aircraft values have increased steadily to the current high of $200 million for each airplane, and the norm for passenger deaths in the US exceeds $2.5 million per passenger. There seems little question that insurance premiums will continue to rise. The potential liability costs are astronomical.

High Speed Civil Transport (HSCT)

The commercial transport industry in the US is presently studying the feasibility of a new generation supersonic transport called the HSCT (high-speed civil transport). The presently planned HSCT is a vehicle with a gross takeoff weight of over 700,000 pounds, a length of over 300 feet, and a supersonic cruise speed in the area of Mach 2.4. It would be capable of carrying 300 passengers with a range of approximately 5,800 miles. The HSCT would cut Trans-Pacific flight times by approximately 60 per cent. Primary advantages are the number of additional trips an HSCT can make per year. This assumes that schedules can be planned that will attract international passengers.

When it flies above Mach 1, HSCT airplanes will fly at higher altitudes than do the subsonic transports. For example, the Concorde's supersonic cruise is usually begun at about 49,000 feet, and as fuel is burned and the airplane becomes lighter it drifts up to about 59,000 feet. The normal cruising altitude of NASA's HSCT is 60,000 feet. An interesting meteorological phenomenon is that at above 45,000 feet winds are generally considered relatively constant and insignificant. Optimum flight planning for the HSCT will probably be based on temperature profiles. Cooler temperatures translate to higher thrust, which in turn translates to a reduced fuel burn at constant speed. There will be a need to develop up-to-date wind, temperature, and weather forecasting for SST supersonic flight levels.

If the operating environment permits, such scheduling would permit airlines to take advantage of the HSCTs capabilities in potential trip time saving and operating economies. It is believed that within the CNS/ATM environment and the 4-D capability (longitudinal, latitudinal vertical, and time ATC geographical dimensions) of present FMCs any kind of existing or anticipated delays could be communicated in real time. Theoretically, it would permit an airplane to manage its flight in such a way that today's hold procedures can be reduced significantly or even eliminated, permitting economies even for those flights that have a portion of their cruise at sub-sonic speeds.

Development of an Economically Viable HSCT

The development of an economically viable HSCT is fraught with human factors problems. For example, aerodynamic considerations and the ability to eliminate the mechanism and the extra weight of the droop-nose, which is needed for takeoffs and landings with the Concorde, make it seem probable that the HSCT will have no forward cockpit windows (Regal, Hofer, and Pfaff, 1996). They would be replaced with new forward-looking all-weather video that is still being developed and which will then have to be tested and approved. While the cockpits need not be more complex than cockpits in today's airplanes, additional automation is inevitable. A principle reason is that for maximum aerodynamic efficiency, HSCT airplanes will be flown under a wide range of CG conditions that will require advanced control laws.

A NASA team, cooperating with manufacturers, is actively involved in a high-speed research program. Its mission statement is:

> To provide validated flight deck control concepts, technologies, and guidelines which enable, with acceptable risk, the development, certification, and operation of a safe, efficient, and economically viable HSCT aircraft.

The NASA HSCT parameters are comparable.. They include an aircraft that is 320 feet long, has a wingspan of 130 feet, and a takeoff weight of 750,000 pounds. A problem for the engine manufacturers is the high emission goals that have been set for any HSCT. NASA, which has been using a Russian Tu-144 in its research, recently announced that its development date for a viable HSCT transport is indefinite because present noise targets are unreachable with present technology.

The Ultimate HCST Economic Problem

Finally, any HSCT will face daunting economic problems. Manufacturers will have to be reasonably assured that there is a market large enough to offset the large development costs any HSCT is bound to incur. The big asset that any HSCT has is the time that it can save the passenger. However, with the popularity of the 'hub' concept and the increasingly long ranges available in subsonic transports, passengers may prefer a non-stop sub-sonic flight from cities like Philadelphia or Boston to a variety of cities in Europe to a HSCT flight that is available only from cities like London or New York. In one example, a passenger wishing to fly from Philadelphia to Frankfurt using a HSCT might be required to take a subsonic flight from Philadelphia to New York and change planes for the HSCT flight to London. Another plane change would then be required for the subsonic flight to Frankfurt. Very little, if any, time would be saved over a non-stop subsonic flight from Philadelphia to Frankfurt The same problem arises on the West Coast where the logical HSCT airports are Los Angeles, San Francisco, and possibly Seattle with flights to Narita for the Tokyo market in Japan and then sub-sonic flights to other Asian cities. A critical factor is the number of so-called 'hub' airports that will support the HSCT. Boeing has recently announced that it is postponing development of an HSCT indefinitely. Several experts believe that only a hypersonic airplane (speeds above Mach 5) is economically feasible.

HSCT in Other Countries

On the other side of the Atlantic, the European Supersonic Research Program (ESRP) is actively pursuing the development of a supersonic transport (SST). It currently is looking at a 250 seat aircraft that would fly at Mach 2, and have a 5,500 nautical mile range. The ESRP believes that a Mach 2 aircraft would have direct operating cost that would be significantly lower than one designed to fly at Mach 2.4, and that on transatlantic routes it would suffer only a 20 minute time disadvantage.

A recent issue of *Flight International tells* us that Japan's SST-related research now constitutes the single largest line item in the Ministry of International Trade and Industry's aerospace budget. While a majority of its funding so far has been spent on experimental propulsion work, its goals for the next generation SST closely mirror those of European and US manufacturers, i.e., for a 300 passenger airplane with a cruise range of around 11,000 kilometers or 6,800 miles (Paul, 1998). Because of the very large sums of money required for a SST program, Japan apparently sees its role as that of an international cooperator.

Vertical/Short Takeoff and Landing Aircraft (V/STOL)

An early kind of aircraft in this category was known as a Vertical Short Takeoff and Landing aircraft (the V/STOL). C. W. Harper of NASA (now retired) described some of its early problems using this analogy:

> The V/STOL aircraft has been to the transport industry just as girls are to a young boy. In both cases very attractive features can be recognized in this new

object of interest but the way in which advantage could be taken of them is not at all clear. Just as the boy learns eventually that success is achieved through a sophisticated and often expensive approach to the problem, so the V/STOL user has finally realized that a simple, cheap approach will not lead to success. In both cases substantial satisfaction should follow successful solution to the problem.　　　　　　　　　　(Flight Safety Foundation Newsletter, November 1966)

In the US there is a very active project designed to develop a viable STOL aircraft. It utilizes tilt-engine technology in which the engines provide primarily vertical thrust for takeoffs and landings and lateral thrust for cruise. At the time this chapter is being written there are approximately 60 options for a certificated STOL. The initial STOL will carry about 9 passengers so that interest has been mainly in the corporate sector. While it seems a little visionary at the moment, we believe that in the future successive models will be larger and attract airline interest for short length travel between adjacent population centers.

Operational Innovations

The preceding paragraphs have discussed broad-based operational concepts that indicate a range of future developments for the air transport industry. The next section deals with future developments in the cockpit. Human factors problems are inherent in their development and implementation.

Electronic Flight Bags

A long-range objective of the air transport operating industry has been to achieve a 'paperless cockpit'—to make available on CRTs or flat panels all of the information pilots need on any given trip. This would mean that all of the information and data currently in flight operation manuals, equipment manuals and on take-off, approach, departure, and enroute navigation charts would be resident within the FMS. It would be made available simply by asking for the information required and then having it displayed on a specified CRT or flat panel on the forward instrument panel of transports with that capability. It has also been proposed that such data be made available on a cockpit printer.

　·　A great many problems are associated with this objective. One of them concerns the basic reliability of the equipment. Electronic failures are not unknown and the best protection against electronic failures, particularly in case of maps or charts, seems to be the present plethora of paper versions. To date printers have been susceptible to paper jams, running out of paper, and outright printer failures. Printer output is at best relatively coarse. While printers have fault lights to help monitor their output, the only reliable backup is a paper copy. There is no way to be assured that the printer is producing the correct information.

Much of the data and information needed is specific to the airline or route being considered. Enroute, take-off, departure, approach, and landing data change frequently. A reliable and verified source of this kind of data is an absolute requirement. Air New Zealand, whose magnificent record is marred only by the fatal accident on Mt. Erebus, had significant on-board FMS data problems that were associated with that accident.

While on-board electronic displays may seem better than a printer to display things like approach plates, even with electronic displays there are problems in addition to that of basic reliability. One of them involves the display of the information that must be shown. A very fundamental problem is display resolution. Screens cannot handle the detail, much of it important, that can be found on a printed page or graphic, and a lot of approach charts carry a large amount of very fine detail. As would be expected, on-board printers also have resolution problems. See Appendix F for the detailed information now required on approach charts.

It seems reasonable to expect that the first information in an electronic flight bag will be furnished by the manufacturers on a disk that goes with the airplane. Revisions to equipment manuals can be furnished to the airlines for distribution as required for their airplanes. The navigation material can be more difficult for it will have to be updated and placed on a disk by a contractor and then distributed in an acceptable fashion. Jeppesen, which has competitors in Europe, will be the logical source for navigation material for most carriers and others using a 'paperless cockpit' and is working very hard to perfect the electronic flight bag concept.

Pilots have had mixed reactions to the idea of an electronic flight bag. While many relish the thought of no longer having to keep up with chart revisions, many are also skeptical of the reliability of such a system at this time. An obvious advantage of the electronic flight bag is that it will ensure that latest revisions are installed and available in the cockpit. While most observers believe that the trend toward a 'paperless cockpit' will continue, there is bound to be a long and perhaps permanent period during which electronic flight bags are both feasible and economic for some users, while others find it feasible and economical to continue with the present system

Head-Up Displays in Civil Transport

Traditional head-up displays (HUDs) could well have been discussed in an earlier chapter because they have been used for some time. We have included them in this Chapter—The Air Transport Future—because there have been significant improvements in HUDs. Only recently has their been a pronounced movement to include HUDs in all transports although some manufacturers and others have expressed doubts regarding their necessity. Essential head-up displays have been described as:

> A method of presenting images to the pilot of an aircraft while he/she is looking forward through the windscreen. These images are generated from a device out of the pilot's field of view and reflected from a transparent surface in front of the pilot. Thus, the pilot looks through this transparent surface and sees the information superimposed on his/her vision of the real world.[2]
>
> (Koonce and Allen, 1988)

All HUD displays utilize an overhead projector and combiner assembly that can be pulled down just in front of a forward window to provide basically the same navigational information that is displayed on the regular instrument panel. They are not useable from the opposite seat. The information HUDs furnish is focused at infinity rather than at the instrument panel distance. This means that the pilot looking through the display sees all of the information he/she needs to fly manually or to monitor the aircraft's position during an automatic landing. At the same time the pilot can see the approach and runway lights in the real world without having to switch from head-down to head-up in the period immediately before the landing. It takes seconds for eye adaptation and information interpretation in this process of focal and cognitive switching. This process is required at a time when the pilots are very busy and can consume a significant amount of time.

Late model HUDs furnish additional information by displaying computer derived pictures of the runway, the projected touchdown point, velocity vectors (which can be air mass or inertia derived), and aircraft total energy symbols. They furnish either an air mass flight path based on information from air data information or produce an inertial flight path that is based upon

---

[2] From *Aerospace Glossary for Human Factors Engineers* (1988). SAE ARP 4107, Society of Automotive Engineers, Warrendale, Pennsylvania.

an airborne inertial system. The former tells the pilot where he/she is going with respect to the earth but is only completely accurate in still air.

The information regarding the flight path that is derived from air mass data shows the movement of the aircraft in a moving air mass. This can result in an erroneous flight path relative to anything on the ground. In contrast, the inertial flight path instantaneously tells the pilot where the aircraft is going with respect to the earth (runway). It is a significant improvement.

An international airline pilot, J.E. Hutchinson, listed what he regards as the following significant advantages to an inertial-based HUD:

- It provides profile guidance for visual approaches when ground-based aids are lacking; also profile and track guidance for non-precision approaches by flying the FPV (flight path vector) at the desired azimuth and elevation from the approach fix;

- it provides directional guidance during take-off, so that the take-off minima can be matched to Cat IIIB landing minima;

- the FPV provides a direct and immediate indication of wind shear, and can be used to follow a flight path command programmed for an optimum escape trajectory;

- during climb the slope shown by the FPV can be used in conjunction with a total energy symbol to fly max gradient when this is needed, for example after engine failure on takeoff.

(Hutchinson, 1989)

An inertial-based HUD combined with an autopilot capable of automatic landings has been called a hybrid landing system. A hybrid landing system is described in the ICAO All Weather Operations Manual. The manual describes a primary fail-passive automatic landing system combined with a secondary independent guidance system that is generally understood as a Head-Up Display. US ALPA's All Weather Flying Committee believes that 'the hybrid concept is far superior in both reliability and performance to a simple fail-passive autoland'.

The Use of HUDs in Air Transport

The French airline, Air Inter (now part of Air France), pioneered the use of HUDs in 1974 and continues to use them on all airplane types. It uses them for monitoring when automatic landing systems are installed. Aer Inter now has over two decades of successful CAT IIIA operations eventually operating down to minimums of 20 feet and a runway visual range (RVR) of 150m (Hutchinson, 1989). In the US, Alaska Airlines has used a Head-up Guidance System (HGS) since 1984 and has achieved manual landing minima of 50 feet decision height and 210m RVR. Its takeoff minimums have been reduced from 210m to 90m.

The movement to incorporate HUDs in more airline aircraft is growing in the US and throughout the world. It is stimulated by a continuing movement to further lower minimums and to take advantage of HUD's other uses, particularly its enhanced vision potentials and its claimed ability to increase situational awareness for both members of the flight crew. HUDs furnish a highly competitive market in spite of the fact that there is still not a consensus on the symbology that should be used to display the very useful information a HUD can provide. Avionics manufacturers in the US, Great Britain, Italy, and France are among the forerunners. A UK company predicts a market for 10,000 civil airliner HUDs. Another has predicted that 'in 20 year's time, HUDs will be like headlights on cars: you won't be able to see without them'.

A HUD is claimed to increase safety for all landings, whether visual or normal IFR. Delta, Southwest, Lufthansa, Alitalia, and other airlines are installing HUDs in several of their airplanes and the movement toward greater use of HUDs is continuing. Recently American Airlines has ordered 75 wide-angle HUDs and has taken an option for an additional 400 wide-angle HUDs (Fitzsimons, 1998).

Flight Dynamics of Portland, Oregon, a leading manufacturer of HUDs, is stressing what it calls the core benefits of a conformal head-up display. It says its HUD helps every day on every landing (Phelps, 1998). Present HUDs are particularly useful if no other glide slope information is available or if the aircraft is making a non-precision approach (NPA). It is clear that many airlines are installing today's HUDs fleet-wide for more than a reduction of minimums.

Typical costs of the hardware suggest that, at least in some cases, HUDs may be provided for the left-seat pilot only. Obviously, this raises several other human factor questions, including the effect on crosschecking, operational monitoring and the problem of potential incapacitation. Some proponents of HUD believe that it can be considered a primary flight display. Others, while acknowledging the increasing utility of current HUDs, believe that there are still substantial human factors issues involving the at least partial blocking of real-world visual cues in a low visibility setting. These critics believe that these practical human factors issues keep HUDs from certification as a primary flight display (Greene and Richmond, 1993). As long as FAR Appendix D requires consideration of pilot incapacitation and for other reasons, we believe the installation of dual HUDs is inevitable, and that HUDs only can be considered a primary flight display if the human factors problems we have indicated can be resolved.

Airbus Industrie, which currently makes HUDs available at its customer's option, remains skeptical about the need or desirability of HUDs in their airplanes. Airbus uses its CAT IIIb autoland, with triple redundancy autopilots, to handle the low visibility precision approach situation. Because GPS now is being certified as a primary means of navigation, the Airbus FPA/TRK (flight path angle and track) mode means that all approaches can be flown with vertical guidance. Information is displayed on the PFD in a normal fashion. It should not be considered a conformal display. Vertical guidance is an important attribute because it eliminates the need to fly any non-precision approaches (NPAs), which, of course, are associated with a large number of CFIT accidents. US Airways has just removed the ADFs[3] from its Airbus fleet and will not have NPAs in its operational specifications for Airbus airplanes. The Airbus competitive position is that there is no need to incur the additional cost and complexity of a HUD. This is only one of the reasons that the entire HUD area is complex and very competitive.

Enhanced Vision Systems

Enhanced Vision Systems (EVSs) are in the research design phase and not yet operational. They can be used in advanced HUD systems and in regular cockpit CRTs or flat panels. Many EVSs use digital terrain data that matches the terrain profile measured by radar altimeters against database profiles and they also use Forward Looking Infrared (FLIR) or millimeter wavelength airborne radar. Both FLIRs and millimeter wave have advantages in restricted visibility depending on the specific weather conditions involved. It is conceivable that both may be required to provide an always available video of the runway in use. Proponents believe that EVS eventually will make airplanes independent of ground-based electronic aids. EVSs are also able to detect ground obstacles, thus permitting taxiing in periods of very restricted visibility. Northwest and other major US airlines are actively pursuing the acquisition of EVS

---

[3] An ADF is an automatic direction finder—an airborne radio navaid that is tuned to an NDB (non-directional beacon) or other suitable broadcast source.

on at least portions of their fleets.

Taylor has noted that: 'An ultimate form of EVS might be to provide Synthetic Vision Systems (SVS) using a terrain data base and sensors for back up, eliminating the need to see the "real world"' (Taylor, 1998). This is a concept that attracts designers of future supersonic airliners wishing to avoid the complicated and heavy nose droop system of the Concorde (see HCST discussion in this chapter).

TCAS with Both Vertical and Lateral Advisories

Presently, even the most advanced TCAS (or for ICAO, ACAS) give only vertical traffic avoidance advisories. However, there is little doubt that in the reasonably near future they will be capable of giving traffic avoidance advisories both vertically and laterally. Such a development will undoubtedly increase a tendency to make the flight crews more responsible for collision avoidance under all conditions. It is an inevitable modification of the traditional responsibility of ATC to maintain safe separation for all IFR aircraft. An equally inevitable consequence will be an increase in the workload of the cockpit crew. This increase in workload affects both the certification of airplanes and individual systems and also the allocation of cockpit duties and of procedures in general. While workload will almost certainly be increased, the perennial problem of whether the increased workload adversely affects air safety has not yet been determined.

Integrated Hazard Avoidance System

One of the latest attempts to put some order into what has been called a proliferation of warnings is called an Integrated Hazard Avoidance System (IHAS). It is an integrated system for external hazard detection and avoidance. An IHAS is designed to take the place of EGPWS, TCAS, Mode-S ATC Transponders, Weather/Windshear Radar and their aircraft warning systems and place them in a completely integrated system configuration.

IHAS also claims to have the potential to give pilots wake-vortex prediction, wing-ice detection, clear air turbulence detection, volcanic ash detection, and an enhanced vision system (EVS) as technology develops an effective and feasible EVS. While IHAS is being developed by several hard-working scientists and engineers and will undergo rigorous certification testing, one should not be surprised to see an initial 'shakedown period' for the IHAS. As we have seen, there were several modifications to the original GPWS, even before EGPWS. EGPWS can be considered as still another modification and a substantial expansion of the original GPWS.

To our knowledge, very few airline transports have installed an IHAS. Because of the economics involved, retrofitting of existing airplanes seems unlikely without a regulation requiring IHAS or its equivalent unless there are major advances in its additional potential use areas. It is an advancement that will probably have to wait for the next generation of aircraft before widespread installation without a regulatory requirement. Of course, a major air transport accident or even incident in which an IHAS may have helped can change all that.

An exception to the previous paragraph occurred in late 1998 when the FAA certified the modification of an IHAS that is scheduled to be installed in a British Airways B-777. Boeing has announced that a similar system will now be available for at least some of its new airplanes. The IHAS was developed by Boeing and the AlliedSignal engineers who developed the EGPWS. A similar system can be expected to be developed by engineers for other avionics companies.

Prioritization of IHAS Warnings

The general alarm prioritization in IHAS places 'rocks and weather' threats ahead of lesser problems such as a TCAS advisory. Its complexity is illustrated by the fact that warnings from six different initiating systems have been given a prioritized order to display 25 different warnings. The six initiating systems are GPWS or EGPWS, PWS (predictive windshear), and TCAS, with both TA (Traffic Advisory) and TR[4] (Traffic Resolution) capability. The new system includes an 'aural declutter' that reduces aural warnings where considered appropriate, while leaving the visual system untouched. The IHAS interleaves its alerts and warnings so that if the system senses that a recovery maneuver is underway, it cancels follow-on aural warnings (Proctor, 1998). Tongue-in-the-cheek pilots are waiting to see the extent that such automation reduces their cockpit workload, minimizes conflicting information, and truly increases air safety. The complexity of an integrated cockpit safety system is illustrated in Table 22.2, which shows one company's aural alert prioritization.

---

[4]These acronyms are all explained in the Glossary.

Table 22.2   Integrated Cockpit Safety System

Priority	Initiating System	Description	Ground	Air	Continuous?
1	GPWS	Reactive wind shear warning	Warning	Warning	
2	GPWS	Sink Rate Pull-up Warning		Warning	Yes
3	GPWS	Terrain Closure Pull-up Warning		Warning	Yes
4	GPWS	Terrain Terrain		Warning	
5	GPWS	V1 Callout (777 only)	Info		Transitional
6	GPWS	Engine Fail Callout (777 only)	Warning		
7	EGPWS	Terrain Awareness Pull-up Warning		Warning	
8	PWS	PWS Warning	Warning	Warning	
9	GPWS	Terrain Caution		Caution	Yes
10	GPWS	Minimums		Info	Transitional
11	EGPWS	Terrain Awareness Caution		Caution	7-sec
12	GPWS	Too Low Terrain		Caution	
13	EGPWS	Too Low Terrain Caution		Caution	
14	GPWS	Altitude Callouts		Info	Transitional
15	GPWS	Too Low Gear		Caution	
16	GPWS	Too Low Flaps		Caution	
17	GPWS	Sink Rate		Caution	
18	GPWS	Don't Sink		Caution	
19	GPWS	Glideslope		Caution	3-sec
20	PWS	PWS Caution	Caution	Caution	
21	GPWS	Approaching Decision Height		Info	Transitional
22	GPWS	Bank Angle		Caution	
23	GPWS	Reactive Wind Shear Caution		Caution	
--	GPWS	TCAS RA: 'Climb', 'Descend', etc.		Warning	Yes
--	GPWS	TCAS TA: 'Traffic'		Caution	Yes

GPWS: Ground Proximity Warning System
EGPWS: Enhanced Ground Proximity Warning System
PWS: Predictive Wind Shear
TCAS: Traffic Alert and Collision Avoidance System
RA: Resolution Advisory
TA: Traffic Advisory

Source: Reproduced by permission of AlliedSignal Aerospace

Safety Innovations for the Future

A somewhat esoteric development is a research study in Great Britain that is examining the usefulness of video camera systems that pilots can use to monitor the physical condition of their aircraft.

A quite different system that is here already is a ground maneuvering camera system that has been developed because of the length of the latest Boeing transport. The B-777-300 is 33 feet longer than a B-747. Because of the length of the aircraft, Boeing has provided a ground-maneuvering camera that provides a display of the nose gear and the two main landing gears to help the pilots maneuver the airplane on the ground. Three cameras are used, one under the forward fuselage and two on the leading edge of the horizontal stabilizers. The cameras are heated to avoid ice buildup and have lights for nighttime use. Camera images are displayed on the lower multifunction display as selected by a pilot.

Jerome Lederer, the World's 'Mr. Aviation Safety', has recently proposed that present FDRs be replaced with today's telemetry that would make operational data available immediately in the case of any crash. He cites the example of Apollo 6, an unmanned space flight that was lost and never seen again. However, within three weeks, by using telemetry and simulation, NASA was able to identify the cause of the loss to be a malfunctioning oxygen line. It would have been virtually impossible to have found the cause if it were not for telemetry. There are, of course, many very real problems, including cost, associated with implementation of this suggestion

Two observations, quoted by Frank Hawkins in *Human Factors in Flight,* are worth remembering. The philosopher and mathematician Bertrand Russell once said: 'Not only will men of science have to grapple with sciences that deal with man but – and this is a far more difficult matter – they will have to persuade the world to listen to what they have discovered.' Somewhat earlier, the physicist Max Planck, observed, slightly more pessimistically: 'An important scientific innovation rarely makes its way by gradually winning over and converting its opponents...what does happen is that its opponents gradually die out and the growing generation is familiarized with it from the beginning.' Practitioners of air transport operations and air transport safety will recognize that there is substantial truth in these two statements when they are applied to recommended innovations in air transport. However, observers must also recognize that while the almost innate conservatism of much of the operational airline industry may well be one of the reasons for its extraordinary success, the industry has shown that it can operate promptly and without regulation if safety is involved.

The air transport industry has grown extraordinarily over the past years. While challenges abound, there is no reason to think that this growth and continued technological developments will not continue. Advanced and sophisticated human factors have become an integral part of the operational industry. They can be expected to make a substantial contribution toward reaching the ultimate and probably unreachable goal of '0' accidents and '0' incidents in the future.

References

Air Transport Medicine Committee (1997). 'The Very Large Airplane: Safety, Health, and Comfort Considerations', The Aerospace Medical Association, Alexandria, Virginia.

Bateman, Don (1990). 'Past, Present and Future Efforts to Reduce Controlled Flight Into Terrain (CFIT) Accidents', Presented at Flight Safety Foundation 43rd International Air Safety Seminar, Flight Safety Foundation, Alexandria, Virginia.

Billings, Charles E. (1997). *Aviation Automation: The Search for a Human-Centered Approach*, Lawrence Erlbaum Associates, Mahwah, New Jersey.

Boyle, Andrew (1999), 'New Levels', *Flight International*, 3-9 February 1999, Reed Publishing, Sutton, Surrey, United Kingdom.

Braune, Rolf J., Funk, Kenneth H. II, Bittner, Alvah C. (1996). *Human Engineering Process in Systems Design and Integration*, SAE 965533, Society of Automotive Engineers, Warrendale, Pennsylvania.

Bruggink, Gerard B. (1975). 'The Last Line of Defense', presented to the Special LEC meeting of ALPA Pilots, 14 April 1975, New Orleans, Louisiana, USA, New Orleans.

Condit, Phillip M. (1996). *'Performance, Process, and Value: Commercial Aircraft Design in the 21st Century'*, the Wright Brothers Lectureship in Aeronautics presented at the World Aviation Congress and Exposition, Los Angeles, California, The Boeing Company, Seattle, Washington.

FAA (1997). FAA Commercial Aviation Forecast Conference, first week of March, 1997, Federal Aviation Administration, Washington, D.C.

Fitzsimons, Bernard (1998). 'Wide-angle GEC-Marconi HUD to debut with American Airlines,' *Aviation International News*, 26 February 1998, Midland Park, New Jersey.

Foyle, David C., Andre, Anthony D., McCann, Robert S., Wenzel, Elizabeth M., Begault, Durand R., Battiste, Vernol (1996). *Taxiway Navigation and Situation Awareness (T-NASA) System: Problem, Design Philosophy, and Description of an Integrated Display Suite for Low-Visibility Airport Surface Operations*, presented at 1996 World Aviation Congress, 21-24 October 1996, SAE International, Warrendale, Pennsylvania.

Global Airspace (1999). 'WAAS & EGNOS Working Together', *Global Airspace,* January 1999, Phillips Information, Inc., Potomac, Maryland.

Greene, Berk and Richmond, Jim (1993). 'Human Factors in Workload Certification', paper presented at SAE Aerotech 93, Federal Aviation Administration, Seattle, Washington.

Hutchinson, J.E. (1989). *The Hybrid Landing System - An Airline Pilot's View*, SAE Technical Paper 892377, Society of Automotive Engineers, Warrendale, Pennsylvania.

Kelly, Emma (1998a). 'Developing nations ask ICAO to help with CNS/ATM funding', *Flight International,* 20-26 May 1998, Reed Publishing, Sutton, Surrey, United Kingdom.

Kelly, Emma (1998b). 'FAA close to finalising Flight 2000', *Flight International*, 23-29 September 1998, Reed Publishing, Sutton, Surrey, United Kingdom.

Koonce, Jefferson M. and Allen, Jeremiah M. (1988). *Aerospace Glossary for Human Factors Engineers*, SAE ARP 4107, Society of Automotive Engineers, Warrendale, Pennsylvania.

Learmount, David (1998). 'Safety Perceptions', *Flight International*, Reed Business Publishing, Sutton, Surrey, United Kingdom.

Lewis, Paul (1998). 'Supersonic Rising Sun', *Flight International*, 17 – 23 June 1998, Reed Publishing, Sutton, Surrey, United Kingdom.

Lowe, Paul (1998a). From 'Voice behind the mike at controllers' union', *Aviation International News*, 1 January 1998, Midland Park, New Jersey.

Lowe, Paul (1997b). 'Washington report', *Aviation International News*, 1 August 1997, Midland Park, New Jersey.

Lundberg, Bo O.K. (1966). *The 'Allotment-of Probability-Shares' - APS - Method. Memorandum PE-18*, The Aeronautical Research Institute of Sweden (FFA), Stockholm, Sweden.

Newman, R. L. (1995). *Head-Up Displays: Designing the Way Ahead,* Ashgate Publishing Ltd., Aldershot, England.

Phelps, Mark, 1998. 'Head-up displays gaining favor in airline and bizav operations', Aviation International News, 1 February 1998, Midland Park, New Jersey.

Phillips, Edward H. (1997). '"GAIN" Committee Seeks Third Airline Safety Conference', *Aviation Week and Space Technology,* 7 July 1997, McGraw and Hill, Inc., New York.

Ponticel, Patrick, (1996). 'First World Aviation Congress pitches a curve', *Aerospace Engineering*, December 1996, the American Institute of Aeronautics and Astronautics, Washington, D.C.

Proctor, Paul (1998). 'Integrated Cockpit System Certified', *Aviation Week and Space Technology*, 6 April 1998, McGraw and Hill, Inc., New York.

Regal, David, Hofer, Elfie, and Pfaff, Thomas (1996). *Integration of Primary Flight Symbology and the External Vision System of the High Speed Civil Transport*, presented at the 1996 World Aviation Congress, Society of Automotive Engineers, Warrendale, Pennsylvania, and the American Institute of Aeronautics and Astronautics, Washington, D.C.

RTCA (1995). From 'Report of the RTCA Board of Directors' Select Committee on Free Flight', RTCA Inc., Washington, D.C.

Sears, Richard L. (1985). 'A New Look at Accident Contributors and the Implications of Operational and Training Procedures', presented at Flight Safety Foundation 38th International Air Safety Seminar, Flight Safety Foundation, Alexandria, Virginia.

Sheppard, Ian (1998). 'Free flight study finds pilots' workload is not increased', *Flight International*, 20-26 May 1998, Reed Business Publishing, Sutton, Surrey, United Kingdom.

Shifrin, Carole A. and Thomas, Geoffrey (1998). 'Asia/Pacific Market Healthy in Long-Term', *Aviation Week and Space Technology*, 23 March 1998, McGraw and Hill Inc., New York.

Stokes, A., Wickens, C. D., and Kite, K (1990). *Display Technology: Human Factors Concepts*, Society of Automotive Engineers, Warrendale, Pennsylvania.

Sweeney, Frank (1998). 'Simulated Skies: Ames to make airport testing "as real as it gets"', *San Jose Mercury News*, 21 April 1998, San Jose, California.

Taylor, Laurie (1998). *Air Travel: How Safe Is It?*, Second Edition, Blackwell Science Ltd., London, England.

Epilogue

From its beginning, air transport has been a challenging, growing and dynamic industry. There is no reason to think that similar challenges, growth and dynamism will not continue. We should never forget that all of powered flight began in the 20th century and that in the 21st, we can anticipate developments in air transport that are hard to imagine today. Two cartoons that have been favorites of ours for a long time help develop a perspective regarding aviation and the extraordinary industry that is truly changing our world.

Just how far we have come is well-illustrated by comparing today with the 1921 experience of pilot W. D'Courcey-Watney in his Nieuport Scout.

In clearing the pilot of culpabilty, the court decided "... that there was absolutely no lift in the air."

April 1921

Pilot W. D'Courcey-Watney's Nieuport Scout was extensively damaged when it failed to become airborne during take-off. The original Court of Inquiry found that the primary cause of the accident was carelessness and poor airmanship on the part of a very experienced pilot.

The Commandant General, however, not being wholly convinced that D'Courcey-Watney could be guilty of so culpable a mistake, ordered that the Court should be reconvened. After exhaustive and extensive inquiries and lengthy discussions with the Meteorological Officer, the Court came to the conclusion

523

that the pilot unfortunately was authorized to fly his aircraft on a day when there was absolutely no lift in the air and that he therefore could not be held responsible for the accident.

The court wishes to take this opportunity to extend its congratulations to Pilot W. D'Courcey-Watney on his reprieve and also on his engagement to the Commandant General's daughter, which had been announced shortly before the accident. (Courtesy ICAO)

The second cartoon was drawn by Don Hart for the May, 1961 issue of the United Airlines' flight publication, *The COCKPIT*. It reminds us that the marvelous and extraordinary developments that we see in all of aviation truly did have a very humble origin.

Everything we have been discussing really did start in a bicycle shop.

Appendix A

Safety in the Commonwealth of Independent States (CIS) and in the Republic of China

It is difficult to get accurate statistics regarding air transport safety in either the Commonwealth of Independent States (CIS, formerly known as the Soviet Union) or Republic of China. Until comparatively recently, reliable information was simply not available in the West. The major reasons seem political. However, the political view seems to be changing. With air transport becoming an increasingly global industry and with major political and attitudinal changes in these countries, formerly unavailable information is becoming available to interested parties and the aviation press. The following articles or condensations of relevant recent articles should be of interest to anyone following worldwide safety developments.

The Commonwealth of Independent States

1. The following short article is taken directly from the 17-23 January 1996 *Flight International*. It represents a major change in available air transport safety information.

Russia Follows International Statistics Line

'RUSSIA WILL need to amend its statistical base to bring it into line with international norms. Traditionally, with no private or business aviation in Soviet times, the accident statistics included all civil-aviation aero clubs; aircraft carrying personnel or cargo for industry and operated by that industry; helicopters and aircraft of Antonov An-2 category undertaking crop spraying and other aerial-work assignments or on Arctic duties — work which would not be included in statistics on airline safety in other countries.

All is not a picture of gloom. Overall, Russian aircraft have had reasonable safety records, considering the conditions of Russian airports and their weather. Of 108 Ilyushin Il-86s built, only one has been lost — it was parked at Delhi Airport when it was hit by a landing Boeing 737 in 1994. An accident in

December at Khabarovsk, when a Khabarovsk Air Tupolev Tu-154B disappeared, was the 21st fatal accident suffered by the type, of which over 900 have been built. An Azerbaijan Airway Tupolev Tu-134B which crashed shortly after take-off from Nahicevan, Azerbaijan was the 23rd of the type to be lost of the 852 built since 1963. Of just over 275 Ilyushin Il-62s built, 14 have been lost in fatal accidents. Bearing in mind their years of service and the number of flight hours, these figures compare reasonably well with aircraft of a similar age and category in the West.'

Flight International, 17 - 23 January 1996

2. In an article in *Flight International* 4 -10 June 1997, Paul Duffy wrote:

'THE CREATION OF THE MAK (Interstate Aviation Committee) in December 1999, as the first intergovernmental body to be formed by the then-new CIS, was an acknowledgment of the need to present a common approach to major questions of aviation among the countries of the former Soviet Union.'

The article was entitled 'MAKing aviation work in the CIS and noted in its subhead that, 'In the CIS, the Interstate Aviation Committee (MAK) is working to adapt standards to international requirements'. The quotation is included here simply as an example of CIS intentions.

3. Another view of present conditions can be seen in these quotations from selected paragraphs in Aviation International News, 19 June 1997. In an article entitled 'Russian airline industry grapples with free market', Reuben Johnson wrote:

'These two carriers [Aeroflot Russian International Airlines (ARIA) and Transaero] and a few other airlines that have emerged from the breakup of the former Soviet Union have done so well that several of them have taken delivery of leased Western aircraft, such as Boeing 737, 757 and 767s; Airbus A310s; and MD DC-10s.

On March 13, Aeroflot announced that Russian President Boris Yeltsin's son-in-law Valery Okulov, had been appointed the new general director. Clearly, the upper levels of the Russian government want the nation to retain a prominent place in the world of aviation. Putting Okulov in this position assures the Russian carrier that it will continue to enjoy priority status in any major government decisions affecting the aviation community.

The paradox for Russia, however, is that while the airline business appears to be good for some of the country's major carriers, the remaining sectors of the civil aviation industry seem to be deteriorating at an alarming rate.

Negative publicity about the safety of Russian airspace is another problem. A few years ago, an air traveler flying over the former Soviet Union was rated as ten times more likely to be killed in an air accident there than in any other part of the world. In 1996 alone, Russian officials said there were 13 crashes of civilian aircraft in all of Russia, for a total of 219 fatalities.....

The Russian FAS, which was formed over a year ago to create a regulatory function for commercial aviation in Russia similar to what the FAA does in the U.S., has tried to address the many 'holes' that still remain in the civil aviation system. The FAS remains hampered by the nonpayment-of-salaries issue, plus the fact that there are far too many commercial and charter airlines—about 400 at last count—operating in Russia with 108 of them certified to operate international flights. The latest proposal to address this problem outlines the consolidation of these airlines into about 100 or so larger companies that would be capable of more (of the) self-regulation that airlines in most other nations perform.'

4. As an important part of a concerted effort to improve overall air transport safety, the CIS's Interstate Aviation Committee (MA) has recently released figures that the region has just had its worst-ever fatal airline accident rate. The MA states that this gives a fatal accident rate of 3.5 per 100,000 flights, which is more than double the previous worst rate, which occurred in 1994. The chairman of its Commission of Flight Safety attributes the 13 major fatal accidents encountered in 1996 primarily to four factors: industry deregulation, aging aircraft, poor maintenance standards, and poor loading, including overloading.' The MA also expressed concern that 10 of the 13 fatal accidents involved CIS-registered freighters and was particularly concerned with the number of 'crew' on-board some of the cargo flights. In three of the accidents to Ilyushin II-76s which are supposed to have a maximum crew of six, they found that 21, 29, and 37 people were on board. (Duffy, 1997). Since its deregulation of air transport activities, over 400 airlines have begun operation in CIS territories.

5. The following appeared in the 19 January 1998 *AVflash* newsletter illustrating that major changes are continuing to take place in Russian aviation:

'RUSSIAN CRACKDOWN: Russia will have only about eight federal air carriers and 40-50 regional airlines by the year 2000 — down sharply from the current 315 carriers. Ivan Valov, first deputy chief of the Russian Federal Aviation Service, said the reductions will come through stricter licensing, certification, and air safety standards.'

6.   The following appeared in *The Aviation International News*, 1 February 1998:

Deaths in Russian Air Crashes Down in 1997

'Deaths were down in air crashes involving Russian civil passenger aircraft last year—from 219 deaths in 13 fatal crashes in 1996 to 80 dead in ten fatal crashes in 1997. The worst year in Russian civil aviation in this decade was 1994, with 16 fatal crashes killing 310 people.'

7.   Russia's concern and major problems regarding its airline industry is reflected in the following short article in the *Wall Street Journal* on 20 March 1998:

Russia Unveils Plan to Overhaul Airline Industry

'After years of neglect, Moscow apparently is ready to tackle a badly-needed overhaul of its airline industry. With only a handful of its 315 carriers on solid financial ground, Russia unveiled a plan to reduce the number of domestic airlines operating in its skies by two-thirds through mergers and alliances over the next few years. While **Aeroflot** had its troubles in the old days, its breakup after the collapse of the Soviet Union into hundreds of smaller carriers—one of which still bears the Aeroflot name—has done little for the Russian aviation industry.

Many of the smaller companies that evolved out of what had been the world's largest airline have been operating at a loss and their safety records have kept many potential travelers close to home. The plan unveiled yesterday foresees about 100 airlines, including five to eight national carriers, and a 4% to 5% increase in passenger traffic by the year 2000. The government expects the consolidation to allow the industry to purchase and lease new aircraft as well. Only a fraction—about 1%—of planes being flown by Russian airlines are new.'

8.   AVFlash of 19 April 1998 reported that:

'RUSSIA DEVELOPS SUPER-JUMBO: Russia's Sukhoi design bureau is reportedly developing a 1,000-seat airliner to compete against super-jumbos on the drawing boards at Boeing and Airbus. Russia can already lay claim to the world's largest cargo plane — the Antonov AN-124 — but the proposed KR-860 will be designed to carry between 860 and 1,000 passengers over a range of 14,000 km.'

9.   The following excerpt from Paul Duffy's article in *Flight International*, 19-25 August 1998 was entitled 'Russia puts civil airline safety oversight into military hands'.

'RUSSIA'S MILITARY has been charged with taking overall control of civil aviation safety oversight and accident investigation in a new presidential decree.

President Boris Yeltsin signed a decree on 10 August, transferring responsibility for the flight safety of Russia's civil airlines to the Ministry of Defense.

The reasoning behind Yeltsin's action is that Russia is said to need a single centre for all aviation oversight, given the overlaps between civil and military flying, and that the role could not be fulfilled by a civilian body.

The move appears to be a setback for the Federal Aviation Service (FAS), the administration which has responsibility for civil aviation regulation and control, but which will now have to report to the defense ministry....

The new decree is intended to provide a supervisory structure to oversee all aviation safety matters. It also calls for accident and incident investigations to be completed within four months.'

The Republic of China

1.   *Flight International* recently awarded Shen Yuankang, Vice Minister of General Administration of Civil Aviation of China for 'helping to strengthen air safety oversight in China's fast-growing civil aviation sector.' This is a prestigious award and Mr. Yuankang was selected by a well-qualified international selection committee. The following two paragraphs from the *Flight International* 18 -24 June 1997 announcing this award is illustrative.

'The improvement in China's levels of air safety has been dramatic over the past three years. The country's civil-aviation regime has undergone a major overhaul, with a sharper focus on safety procedures and oversight within China's fast growing airline industry.

In particular, the country has ended a spiraling rate of airline crashes. Over the last three years, China has had one of its best ever spells for air safety.'

2.   From a short article in *Flight International* 2 -8 July 1997 entitled:

JAA/CHINA HARMONY

'Europe's Joint Aviation Authorities (JAA) and China's civil-aviation authority have signed a memorandum of understanding pledging gradual harmonization of airworthiness standards. JAA says that the first area to be covered will be "...regulations and procedures covered in Joint Aviation Requirement No 21, on the certification of aircraft and related products and parts". The JAA has welcomed the move as simplifying the eventual certification of the planned Airbus Industrie/Aviation Industries of China AE3IX regional-jet family.'

3. The following excerpts from an article in the 2 October 1995 *Aviation Week and Space Technology* illustrate some of China's air transport problems:

China Tackles Critical Aviation Safety Issues
Michael Mecham/Beijing

'The world's most populous country has been collaborating with the US government in establishing a regulatory system to provide a sound basis for safer flight operations. Aviation's largest manufacturers, airlines, and flight safety specialists have become involved in training the Chinese, at home and abroad.

....Two sets of statistics illustrate the paradox of Chinese aviation as the 20th century closes. According to government data, China's annual air traffic growth has averaged 23.6% for the first half of the decade. The second half is predicted to taper off some—to an average 13% per year—but growth is forecast to remain in double digits until at least 2010.

Passenger growth jumped from 3.7 million in 1980 to 40 million last year. Growth in domestic air routes has been especially large, rising from 159 in 1980 to 630 last year. Total air routes have climbed from 180-727 in the same period. When Airbus took a look at this in terms of seat capacity, it forecast growth of six times today's levels by the year 2014, or an increase from 55,000 seats to 349,000.

BUT ANOTHER SET OF STATISTICS is also important. In 1992, the country suffered five air crashes that killed 295 persons. Chinese accident reports are abbreviated, but they showed that cockpit crews are often unfamiliar with their equipment, that English air-to-ground transmissions confused them, that poor maintenance, overloading of aircraft, and lack of flight and safety discipline are too often present.

To make matters worse, China led the world in aircraft hijackings in 1993 with 13.

....BY 1993, CHINA HAD become so important to aircraft manufacturers that it accounted for one out of every six sales of major transports. But over the past 21 months, the CAAC (Civil Aviation Administration of China) has kept its ban (prohibiting further purchases of Russian or Western-built aircraft) with few exceptions.

.... Improving safety standards over the long run will remain an enormous task for China. Among the hurdles:

- The lack of regulations.

- Cultural and institutional barriers.

- Inadequate management skills.

- Poor English standards.

....During the past two years, the CAAC has implemented FAR Part 61 concerning pilot's working conditions, Part 21 for modifying aircraft, Part 23 for airworthiness of small aircraft, Part 25 for major transports, Part 43 for mechanic's standards and Part 145 for repair stations.

Still to come is Part 121 for major transport operating licenses, but the US system does not translate well into China's structure, so that is proving a difficult job.'

4.   The following is from *Flight International* 8 - 14 October 1997.

Classroom Revolution

China aims to have 10,000 qualified pilots in five years time. The training task is immense.
Paul Lewis/Beijing

'CHINESE CIVIL AVIATION has been transformed since the country began to open up in the early 1980s. China's monolithic state carrier and its antiquated Soviet hardware have gone, replaced by a proliferation of international, regional and provincial airlines, operating the latest in Western designs. More recently, there has been a quieter and less-visible software revolution in the cockpit.

Over the past four years, pilot training in China has undergone an extensive overhaul and a fourfold expansion in an effort to bring the country's flying schools and technical institutes in line with the demands of a modern airline industry. The qualitative and quantitative refinements coming to the fore result from a major injection of money and effort by the Civil Aviation Administration of China (CAAC).

The single most valuable return on this investment has been a turnaround in China's once-abysmal safety record. The country has suffered only one accident and 35 deaths in the three years since CAAC vice-minister Shen Yankang announce a safety crackdown in July 1994, In the two years before that, seven crashes killed more than 500 people.

KEEPING UP WITH TRAFFIC

Chinese civil aviation is not completely in the clear yet, however. To keep pace with projected traffic growth, China will have to enlarge its pool of qualified pilots to more than 10,000 within the next five years....'

Appendix B

Academic Diversity in Human Factors Personnel

Good illustrations of the diversity of specialties in human factors can be found in the following lists of the educational background of the over 5,000 members of the Human Factors and Ergonomic Society and in the 83 graduate degrees, 41 of which are PhDs. that are found in the members of the Aviation Human Factors Division of the NASA-Ames Research Center. The NASA-Ames group specializes in Aviation, the Human Factors and Ergonomic Society is more broadly based.

HUMAN FACTORS AND ERGONOMICS SOCIETY (1995-1996)

Academic Specialty	Total	% PhDs
Psychology	38.85%	
General		11.72
Experimental		6.30
Industrial		1.12
Engineering		0.92
Cognitive		1.33
Other		1.42
Engineering	23.58%	
General		0.59
Industrial		5.73
Mechanical		0.55
Electrical		0.15
Aeronautical-Astronautical		0.13
Other		0.96
Human Factors/Ergonomics	9.83%	
General		1.70
Psychology		0.70
Engineering		0.31
Industrial Design	1.60%	0.02
Medicine/Physiology/Life Sciences	4.88%	2.04

Academic Specialty	Total	% PhDs
Education	1.57%	0.78
Business Administration	2.25%	0.11
Computer Science	1.07%	0.35
Safety	8.27%	0.16

NASA-AMES RESEARCH CENTER

Advanced Degree	Number	Total
Psychology		55
General Psychology	25	
Experimental Psychology	15	
Cognitive Psychology	7	
Quantitative Psychology	2	
Developmental Psychology	1	
Engineering Psychology	1	
Cognitive/Mathematic Psychology	1	
Experimental/Engineering Psychology	1	
Psycho/Physiology	2	
Engineering		11
Electrical Engineering	4	
Aeronautical and Astronautical Eng.	2	
Mechanical Engineering	1	
Engineering Mathematics	1	
General Engineering	1	
Aeronautical Engineering	1	
Industrial Engineering	1	
Physics		2
Computer Science		2
Mathematics		1
Aeronautics and Mathematics		1
Computer Music		1
Intelligent Systems Programing		1
Human Factors		1
Neuroscience		1
Biophysics and Neuroscience		1
Behavioral Sciences		1
Social Sciences		1
Management		1

Appendix C

Classification of Illusions

Draft two of the Air Force's *Medical Board Member Investigation Guide gives* the following classification of illusions that are relevant to aviation:

1. Kinesthetic Illusions: An erroneous perception of somatosensory stimuli to the ligaments, muscles, or joints of the body.

    a. G-Adaptation Illusion: An erroneous perception that motion has ceased after continued exposure to a sustained velocity. For example, movement in an elevator is only perceived at the beginning and end of the ascent or descent.

    b. G-Differential Illusion: An erroneous perception of aircraft attitude based on 'seat of the pants' sensations. For example, without other sensory inputs, a 30-deg-bank level turn feels the same as a 60-deg-bank turn.

2. Vection Illusions: Visual Illusions of motion, erroneously detected peripherally, in which a person perceives that he/she is moving when in fact an external object is moving.

    a. Circularvection: An erroneous sensation of rotation due to movement detected in the visual field, especially peripherally.

    b. Linearvection: An erroneous perception of linear movement due to motion detected in the visual field, especially peripherally.

3. Vestibular Illusion: Erroneous perceptions of orienting stimuli to the semicircular ducts or otolith organs of the vestibular apparatus.

    a. Coriolis Illusion: An erroneous sensation of rotation due to the movement of the head into a plane of angular or linear acceleration which induces fluid movement in the semicircular ducts.

b. Elevator Illusion: An erroneous sensation of pitch-up after level off from a steep climb or when in turbulence.

c. Giant-Hand Illusion: The erroneous sensation that controls will not respond to inputs, even with seemingly great effort, when the source of resistance is in fact the operator himself/herself attempting to respond to conflicting sensory cues.

d. Leans: An illusion of angular displacement (bank) due to an undetected, subthreshold angular acceleration followed by a detected transthreshold angular acceleration.

e. Somatogravic Illusion: An erroneous sensation of tilt in the vertical plane due to linear acceleration. This illusion is most common during rapid acceleration or deceleration.

f. Somatogyral Illusion: An erroneous perception that rotation has ceased because the semicircular canal fluid has stabilized after angular acceleration. The graveyard spin and graveyard spiral are results of the somatogyral illusion.

4. Visual Illusions: Erroneous perceptions of stimulus to the visual system.

a. Autokinesis: An erroneous perception of movement of a light when stared at for a length of time in a dark visual field.

b. Chain-Link-Fence Illusion: The blending into the foreground of nearby objects when focusing on a distant object.

c. Empty Field Myopia: The tendency for the eyes to focus at a distance of about one meter when viewing a visually non-stimulating field.

d. Flicker Vertigo: The disruptive psychological effects of cyclic visual stimulation of about 10 to 15 cycles per second.

e. False Horizon Illusion: An illusion created by sloping cloud formations, an obscure horizon, a dark scene with ground lights and stars or certain geometric patterns of ground light, which results in the pilot placing the aircraft in a dangerous attitude because of the perception of not being aligned properly with the actual horizon (AIM).

f. Geometric-Perspective Illusion: An erroneous perception of being nearer to, or farther away from, an object than one actually is, due to equating retinal image size to distance or angular displacement of familiar objects. For example, an 8,000-ft runway viewed from 1000 ft up may appear the same size as a 10,000-ft runway viewed from 1500 ft up; another example is the tendency to flare high on a wider than usual runway.

This classification of illusions appears in the *Aerospace Glossary for Human Factors Engineers*, SAE ARP 4107.

Appendix D

The Development of ICAO and the Six Freedoms of the Air

The following discussion of ICAO and the politically important six freedoms of the air are admittedly beyond conventional human factors considerations. However, the discussion of ICAO is included here because of ICAO's importance as the universally accepted primary international institution concerned with the orderly growth and development of worldwide civil air transport. The need to develop accredited worldwide standards rather than individual standards developed by sovereign states has not always been recognized among the nations of the world. The purpose of the next few paragraphs is simply to provide the background necessary to understand the development of ICAO as an important international organization and to better understand the importance of the implementation of aviation human factors on a worldwide basis.

The International Civil Aviation Organization (ICAO) was formed in 1944 when the US government invited 55 allied and neutral States in World War II to meet in Chicago to consider anticipated problems in postwar aviation. Fifty-two States sent representatives and after five weeks of deliberations, they adopted what is known as the Chicago Convention. The Chicago Convention included 96 articles. These 96 Articles established the privileges and restrictions of all contracting states, provided for the adoption of International Standards and Recommended Practices affecting air navigation (SARPS), recommended the installation of navigation facilities by member states, and suggested the facilitation of air transport by the reduction of customs and immigration formalities.

One of the results of the Chicago Convention was the formation of ICAO, which has become a specialist agency of the United Nations. It performs an important safety role in international aviation and directly affects the regulatory culture of virtually all of ICAO's individual States. Each ICAO State is a sovereign political entity and quite different from a US state.

ICAO has a sovereign body called the Assembly, and a governing body called the ICAO Council, which is a permanent body responsible to the Assembly and whose members are elected for three year terms. The Assembly meets at least once every three years. All work of the organization is reviewed at the Assembly's sessions. An important work of the Council is the creation, adoption, and amendment of air navigation specifications known as International Standards and Recommended Practices (SARPS). A Standard is defined as 'a specification the uniform application of which is necessary for the safety of regularity of international civil air navigation, while a Recommended Practice is one agreed to be desirable but not essential. There are nineteen sets of international SARPS and eighteen of them are within the technical field' (Taylor, 1997).

ICAO attempts to minimize differences among its member States with its SARPS. However, two or perhaps three, provisions of ICAO's rules hinder these efforts. The first, and probably the most important, is that ICAO itself has no supervisory or enforcement capability. While States generally agree with the ICAO SARPS, the implementation of SARPS is a very real problem. A much lesser problem is that while exemptions to ICAO SARPS are relatively rare, a State must only declare an exemption if it does not intend to comply with a given SARP. Another ICAO provision, which was a part of the original Chicago Convention and is universally accepted, still can create valid air safety problems. This provision is the right of each State to remain entirely sovereign for all operations within and above its national boundaries.

The Six Freedoms of the Air

The six freedoms of the air are an important State consideration and are shown in Figure D.1. The six freedoms are jealously guarded by States and cannot be taken for granted. They are important in international aviation. Each has been the subject of rigorous diplomatic, and usually bilateral, negotiations. The relatively recently constituted European Union feels strongly that it should have the right to negotiate these items for its members and that present bilaterals should be declared null and void. The economic and political consequences are high. This is a contentious and political area. Negotiations for either operating specifications or economic rights have not always been successful.

- The right of an airline to takeoff from A and overfly B to get to C.
- The rights of an airline to land in B for fuel or maintenance, but not to pick up or discharge traffic.
- The right of an airline to discharge traffic from A to B.
- The right of an airline to carry traffic back to A from B.
- The right of an airline to collect traffic in B and fly the traffic to C.
- The right of an airline to pick up in X traffic that is bound for C and route it through A. This traffic normally belongs to the airlines of X and C.

Six Freedoms of the Air

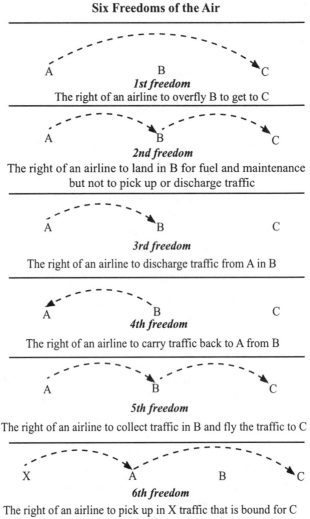

1st freedom
The right of an airline to overfly B to get to C

2nd freedom
The right of an airline to land in B for fuel and maintenance but not to pick up or discharge traffic

3rd freedom
The right of an airline to discharge traffic from A in B

4th freedom
The right of an airline to carry traffic back to A from B

5th freedom
The right of an airline to collect traffic in B and fly the traffic to C

6th freedom
The right of an airline to pick up in X traffic that is bound for C and route it through A. This traffic normally belongs to the airlines of X and C.

Figure D.1   The Six Freedoms of the Air

References

Taylor, Laurie (1998). *Air Travel: How Safe Is It?*, Second Edition, Blackwell Science Ltd., London, England.

Appendix E

The ICAO Standard Alphabet (ICAO Annex 10)

A	alfa	H	hotel	O	oscar	V	victor
B	bravo	I	india	P	papa	W	whiskey
C	charlie	J	juliet	Q	quebec	X	x-ray
D	delta	K	kilo	R	romeo	Y	yankee
E	echo	L	lima	S	sierra	Z	zulu
F	fox trot	M	mike	T	tango		
G	golf	N	november	U	uniform		

ICAO Manual of Radiotelephony

Letter	Word	Pronunciation
A	Alpha	**AL** FAH
B	Bravo	**BRAH VOH**
C	Charlie	**CHAR** LEE or **SHAR** LEE
D	Delta	**DELL** TAH
E	Echo	**ECK** OH
F	Foxtrot	**FOKS** TROT
G	Golf	GOLF
H	Hotel	HOH **TELL**
I	India	**IN** DEE AH
J	Juliet	**JEW** LEE ETT
K	Kilo	**KEY** LOH
L	Lima	**LEE** MAH
M	Mike	MIKE
N	November	NO **VEM** BER
O	Oscar	**OSS** CAH

Letter	Word	Pronunciation
P	Papa	PAH **PAH**
Q	Quebec	KEH BECK
R	Romeo	**ROW** ME OH
S	Sierra	SEE **AIRRAH**
T	Tango	**TAN** GO
U	Uniform	**YOU** NEE FORM
		or **OO** NEE FORM
V	Victor	**VIK** TAH
W	Whiskey	**WISS** KEY
X	X-ray	**ECKS** RAY
Y	Yankee	**YANG** KEY
Z	Zulu	**ZOO** LOO

Transmission of Numbers

Numeral or numeral elements	Pronunciation
0	ZE-RO
1	WUN
2	TOO
3	TREE
4	FOW-er
5	FIFE
6	SIX
7	SEV-en
8	AIT
9	NIN-er
Decimal	DAY - SEE - MAL
Thousand	TOUSAND

Note: The sylables printed in capital letters in the above list are to be stressed: for example, the two syllables in ZE-RO are given equal emphasis, whereas the first syllable of FOW-er is given primary emphasis

All numbers except whole thousands shall be transmitted by pronouncing each digit separately. Whole thousands shall be transmitted by pronouncing each digit in the number of thousands followed by the word THOUSAND.

Number	Transmitted as	Pronounced as
10	ONE ZERO	WUN ZE-RO
75	SEVEN FIVE	SEV-en FIVE
100	ONE ZERO ZERO	WUN ZE-RO ZE-RO
583	FIVE EIGHT THREE	FIFE AIT TREE
2 500	TWO FIVE ZERO ZERO	TOO FIFE ZE-RO ZE-RO
5 000	FIVE THOUSAND	FIFE TOUSAND
11 000	ONE ONE THOUSAND	WUN WUN TOUSAND
25 000	TWO FIVE THOUSAND	TOO FIFE TOUSAND
38 143	THREE EIGHT ONE FOUR THREE	TREE AIT WUN FOW-er TREE

Numbers containing a decimal point shall be transmitted with the decimal point in appropriate sequence being indicated by the word DECIMAL.

Number	Transmitted as	Pronounced as
118.1	ONE ONE EIGHT DECIMAL ONE	WUN WUN AIT DAY SEE-MAL WUN
120.37	ONE TWO ZERO DECIMAL THREE SEVEN	WUN TOO ZE-RO DAY SEE-MAL TREE SEV-en

When transmitting time, only the minutes of the hour are normally required. However, the hour should be included if there is any possibility of confusion. Coordinated universal time (UTC) shall be used.

Appendix F

Jeppesen Briefing Bulletin – The New Approach Charts

The following material is a Jeppesen Briefing Bulletin. It is slightly enlarged and reproduced with permission of Jeppesen Sanderson Inc. It is not to be used for navigation. Copyright 1997 Jeppesen Sanderson, Inc.

BRIEFING BULLETIN

DEN 97-M

INTRODUCING the
NEW JEPPESEN APPROACH CHART
FORMAT
19 September 97

Starting with the 19 September 97 revision a new Jeppesen approach chart format will begin to appear in your revisions. The new format is based upon the "briefing strip" concept, as described below. The new format will first appear at selected locations. Initially, not every chart in the revision will be converted to the new format, however over time there will be an increased rate of conversion. Therefore, you may not see a new format chart in every revision. For compatibility purposes airport chart headings will also be modified in a similar way. Eventually all Jeppesen approach and airport charts worldwide will be converted to the new format within a few years.

The new Jeppesen "briefing strip" approach chart format is a significant part of an ongoing effort to upgrade chart usability and readability. Jeppesen worked closely with many of our airline, corporate and general aviation customers, and a wide range of professional pilots and aviation organizations to develop the new format. The result is a chart that presents basic approach information in the sequence in which pilots would normally brief or review the procedure prior to flying it. The new format incorporates:

> Human factors evaluations.
> A standard pre-approach briefing sequence of information.
> Crew resource management, or C R M techniques.
> An emphasis on usability and legibility.

DEN 97-M4

NEW "BRIEFING STRIP" APPROACH CHART FORMAT
(Continued)

On the new charts, information is essentially the same, except that it is arranged in a manner based on human factors studies about how charts are reviewed and used in the cockpit. The top of the chart contains the location name, procedure title, identifier and airport name, positioned so they may be visible even when clipped to most control column clips. The next row contains all the communication frequencies in the order of normal use. The third row includes the pre-approach briefing information consisting of the primary navigation aid, final approach course, and altitudes. Below this row is the complete missed approach text. The Minimum Safe Altitude (MSA) is located to the right. Applicable procedural notes are located in the last row of the heading section.

The plan view and the profile view are unchanged except for some minor changes in appearance intended to reduce visual congestion and to help visually identify specific information.

Other noticeable changes are below the profile view. Conversion tables are placed here based on relative use. Additional features include a new graphic depiction of the Approach Lighting System (ALS) for the straight-in runway, as well as a new symbolic depiction of the initial "Up and Out" maneuvers (ICONs) for the missed approach procedure.

As a result of many requests, as approach charts are converted and reissued they will be resequenced according to runway number. This simplified method is easier to understand and will make it easier to retrieve a particular chart.

This new format represents the culmination of a dedicated effort by several airlines and organizations, as well as the input of and cooperation of thousands of professional pilots. Customers surveys have shown enthusiastic support for the new approach chart format by pilots of varied backgrounds and from different parts of the world. We extend our gratitude to everyone who contributed to the development effort, in each and every way.

22 August 97   (Continued)

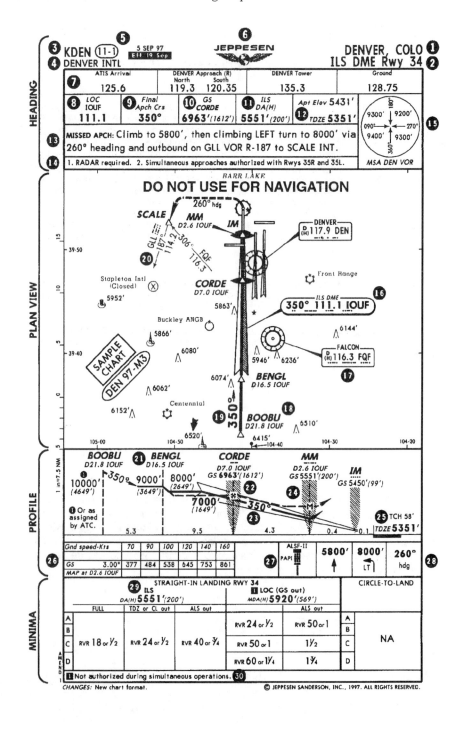

DEN 97-M2

NEW "BRIEFING STRIP" APPROACH CHART FORMAT
(Continued)

The following pages provide a sample chart and an explanation of the new format features.

❶	City/Location and State/Country names.
❷	Procedure identifier.
❸	Jeppesen NavData/ICAO airport identifier.
❹	Airport name.
❺	Index number, revision and effective dates. Charts sequenced by runway number, lowest to highest.
❻	Heading data arranged to avoid coverage by control column clip.
❼	Communications frequencies arranged horizontally.
❽	Primary navigation aid.
❾	Final approach course bearing.
❿	Glide Slope altitude at OM position (or equivalent) for Precision approaches or, minimum altitude at Final Approach Fix (FAF) for Non-Precision approaches.
⓫	Decision Altitude DA(H) or Minimum Descent Altitude MDA(H) for straight-in landing.
⓬	Airport and touchdown zone/runway end elevation.
⓭	Complete instructions for missed approach procedure.
⓮	Common placement of notes applicable to the procedure.
⓯	Minimum Safe Altitude (MSA) graphic placed in a consistent location.

Plan View
⓰	Primary navaid information enlarged and made bold.
⓱	New style for all navaid box outlines and thinner leader arrows.
⓲	Names and idents of airspace fixes associated with the approach procedure are enlarged and made bold.
⓳	Final approach course bearing is enlarged and made bold.
⓴	Formation radials and secondary airports are screened to reduce visual congestion.

Profile View
㉑	Names and idents of airspace fixes associated with the approach procedure are enlarged and made bold.
㉒	Glide Slope altitude at OM position (or equivalent) or minimum altitude at Final Approach Fix (FAF) is enlarged and made bold.
㉓	Final approach course bearing is enlarged and made bold.
㉔	Symbols for navaids and fixes are screened to reduce visual congestion.
㉕	Touchdown zone/runway end elevation is enlarged and made bold.
㉖	Conversion table positioned below profile view for improved usability.
㉗	Graphic depiction of applicable approach light system (ALS) and/or visual descent lighting aid.
㉘	Initial pilot actions ("up and out") for missed approach are symbolized (also referred to as "ICONs"). Complete missed approach instructions are located in the heading.

Minima
㉙	Decision Altitude DA(H) or Minimum Descent Altitude MDA(H) for straight-in landing is enlarged and made bold.
㉚	Notes applicable to landing minimums commonly located below minimum band.

22 August 97 (Continued)

Appendix G

ICAO Domains for CRM Training

Communications
- cultural influence
- role (age, crew positions, etc.)
- assertiveness
- participation
- listening
- feedback

Situation Awareness
- total awareness of surrounding environment
- reality vs. perception of reality
- fixation
- monitoring
- incapacitation (partial/total, physical/psychological)

Problem-Solving/Decision-Making/Judgement
- conflict resolution
- review (time-constrained)

Leadership/Followership
- team building
- managerial and supervisory skills
- authority
- assertiveness
- barriers
- cultural influence
- roles
- professionalism
- credibility
- team responsibility

ICAO Domains for CRM Training continued

Stress Management
- fitness to fly
- fatigue
- mental state

Critique
- preflight analysis and planning
- ongoing review
- postflight

FAA Domains for CRM Training

Communication Processes and Decision Behavior
- briefings
- inquiry/advocacy/assertion
- crew self-critique (decisions and actions)
- conflict resolution
- communications and decisionmaking

Team Building and Maintenance
- leadership/followership/concern for task
- interpersonal relationships/group climate
- workload management and situational awareness
- preparation/planning/vigilance
- workload distribution/distraction avoidance
- individual factors/stress reduction

Appendix H

Ten Ways to Kill a Good CRM Program

The top ten ways to kill off a CRM program was hatched at an Australian Aviation Psychology Association meeting at Manly Harbor/Sydney, Australia in 1992. The original list looked something like this.

1. Not Integrating CRM into LOFT, PT, and Other Operational Training.

2. Failing to recognize the unique needs of your own airline's culture.

3. Allowing the CRM Zealots to run the show.

4. Bypassing Research and Data Gathering Steps.

5. Ignoring the Checking and Standards Pilots.

6. Having Lots of Diagrams, Boxes, and Acronyms.

7. Making the Program a One Shot Affair.

8. Using Pop Psychology and Psycho babble.

9. Turning CRM training into a Therapy Session.

10. Redefining the "C" to mean Charismatic.

(Taggart, 1993)

Appendix I

Alertness Management

The following copy from the United Airlines Flight Operations Manual shows a part of what a major airline is doing to deal with the area of alertness and fatigue and is reprinted with their permission.

'Alertness Management' (Fatigue countermeasures)

'It is of utmost importance that flight operations be conducted by alert crews. The nature of flight operations is such that maintaining alertness is both necessary and difficult. Flight operations occur during all hours of the day and across widely separated time zones. While there are FAA regulations and contractual provisions that limit flight and duty time and protect certain rest periods, maintaining a high level of alertness requires more than rules. It requires a commitment from each and every pilot.

'**BEFORE FLIGHT** Numerous factors impact crew alertness before a flight, both at home and during layovers: the quantity and quality of sleep, exercise, nutrition, general health and fitness, consumption of alcohol or drugs, general mental and emotional state, etc. Crewmembers should discover what works best and practice those things found to help in maintaining alertness during flight. While all the above factors play a definite role in alertness, the importance of sleep cannot be overstated. Sleep is a physiological need just as food and water are, and the only way to satisfy the need for sleep is to sleep!

'**DURING FLIGHT** Certain long flights provide an opportunity for sleep en route by way of a relief pilot. In those circumstances, crewmembers are encouraged to take advantage of the opportunity to sleep. It is widely known, however, that crewmembers of augmented crews are not the only ones that face fatigue during flight. The FAA recognized this in commissioning NASA research and in establishing a government/industry working group 'to determine the feasibility of pre-planned rest in the cockpit...and, if feasible, determine the criteria for the establishment of such rest periods.' That working group reviewed the NASA study results and developed a draft Advisory Circular. The FAA has not yet published the advisory circular.'

Appendix J

Flying in the Early Days

Reporting in the Early Thirties

This is the transcript of an actual radio transcript circa 1931. Edgar Edson was the dispatcher and the late Frank Crismon, the pilot.

Edson:	Cheyenne to Crismon, plane 264, Report!
Crismon:	I'm 73 miles east of Rock Springs.
Edson:	How do you verify 73 miles?
Crismon:	Account of a snow squall. I've landed on the Lincoln Highway. I'm at an intersection and a highway sign says Rock Springs 73 miles. As soon as the squall passes, I'll takeoff.

A Very Eventful Trip

The following letter was written in 1975 by the late Captain Larry Letson to his friend, the late Captain Joe Hutchinson. Both flew for United Airlines until they retired, and both were early transport pilots.

May 30, 1934. Boeing 247

Concerning that trip when we went down in Connecticut. I'll try to give you some of the details as we lived them on that trip. First, it was not my scheduled trip. I was displaced on my schedule by Jack Herlihy who flew copilot to Cleveland on my trip and I was pushed back to the next trip an hour later. That was a nightmare of a trip as it turned out.

We left Chicago at 4:00 or 5:00 P.M. on May 29, landed at Cleveland. We were supposed to go on to Newark but the weather there was stinky and had been all day. Since it was the copilot's duty to check

552

the gas before departure - stick the tanks - and thinking we might need all the gas we could get, I filled the tanks - ran them over - to be sure they were full (286 gals.). Out of Cleveland we were cleared only to Kylertown, PA. Night had fallen by the time we left Cleveland. I took off at Cleveland and John Wolf, the pilot, requested clearance to Albany, N.Y. for better train connections for the passengers to New York. Then I headed for the Cleveland to Albany airway over to my left to follow the (airway) beacon lights to Albany. Johnny went back in the cabin and stayed quite a while. At a point up the line to Albany Johnny came up to listen to the weather broadcast. We were near the north-south airway, which crossed our route about 50 miles west of Newark. The weather at Newark on that broadcast was good, 600-1/2. Johnny signaled me to head for Newark.

When we got down to our own airway Johnny reported over that range station (I forget the name of that emergency field). That surprised everyone, for at that time we should have been nearing Albany. Johnny took the airplane and when we got to Newark it was down again. Newark had centerline runway lights - and I think they were 200 ft. apart. Johnny did a good job and on each approach would let her down to the ground (but on each try was) on the left side of the lights. I had my head out the side window and could see only one light - dimly - at a time. Also we could not stay down there too long because hangars were close to the side of the runway at the other end, and on each pull-out the red hazard light on our hangar showed up right off my wing tip. After the fourth attempt we had to give up and go back up on top. The top was 1200 ft., clear above with the stars and moon out. The Empire State building sticking out like a sore thumb. It was beautiful up there.

We were now on our last tank of gas with 36 gals. left. I had pumped the other two tanks dry. Remember those engines used about a gallon a minute, (Boeing 247, NC13334) so we had 36 minutes to do something. At about the 15 gallon mark Johnny started letting down slowly, hoping to get underneath, and look for a flat area - an apple orchard or corn field - we couldn't be fussy about an airport. I had my head out my side window, looking for breaks or a field or anything, when I noticed what appeared to be "white caps" *behind* the prop on my side! I thought we were out over the Atlantic, running out of gas, and I couldn't swim. I checked the altimeters and they showed 900 ft. It then dawned on me that the "white caps" were the undersides of tree leaves (you know, the undersides of large tree leaves are lighter color that the top sides that the prop blast was turning over. I horsed back on the wheel and we busted out on top again at 1200 feet. That was a narrow escape - but we had more coming. I then suggested to Johnny that we turn 90 degrees to the coast and maybe we would run off (the edge of) the overcast and find an open field. We headed northwest but as far as I could see it was overcast. Now we were down to 4 - 5 gallons. Johnny started letting down slowly again - we didn't know what the hell was under us. Finally I saw lights below - we were over a town. Johnny took a quick look and told me to kick out a flare. In just seconds the flare landed among a lot of houses. We went ahead for a minute and Johnny asked for the other flare. It wouldn't release. We had hit something which had partially closed the tube the flare slides out through. (We found out later we damned near knocked over a church steeple in this little town -which was Bethel, Conn. 70 miles north east of Newark). By then we were down to 1 or 2 gallons of gas - nothing to do but level off - go straight ahead and get away from this town. Finally, just after a few seconds, the fuel pressure lights came on. I pulled my head back in - might as well hang on to it as long as . possible. We said so-long to each other - Johnny slowed her down as much as possible and the last thing I remember was seeing tree branches going by the right landing light which was turned on.

When I 'came to' it seemed as quiet as a vacuum. My first thought was, 'This trip is over'. I took inventory of myself -left arm was OK, I could move it. Then my right arm, I could move it. My left leg - NO -then my right leg -NO -.Then I called the Stewardess and asked if she was OK? All were outside and on the ground, and she was OK. "Are you sure? - I'm sure - are you OK? - I'm OK - may have a couple of broken legs".

We had crashed 18 minutes after midnight, May 30, 1934. The tail section broke off behind the cabin door. It had whipped around, turned upside down and the end of the stabilizer leaned right up to the cabin door, so the passengers could slide right down it to the ground. We woke up this little town and a lot of people came over to the wreck and hauled the people over to Danbury, Conn. Hospital, 3 or 4 miles away.

In the meantime men were trying to remove wreckage from around me but couldn't get me out through the broken-up nose section. They had to drag me through the cabin and down the tail section to the ground. Then they sort of wadded me up and put me in the back seat of a car and took me to the hospital. That was a busy sight at the hospital. None of the passengers was injured seriously. One lady had a broken finger - some of the passengers got a black eye when they hit the back of the seat in front of them. John Wolf got a long scalp cut from his forehead to the back of his head. The Stewardess, a nurse, discovered that somehow his scalp had fallen down over his ear. She pushed that back in place, pulled his cap down over his head, to slow the bleeding, and had him sit by a tree until someone took him to the hospital. The Stewardess had a bruised shoulder and a bruised leg. We were all in the hospital about 45 minutes after the crash. From there the Stewardess called the Company.

The passengers were all released from the hospital in a couple of days or so. John Wolf had to stick around for a week or ten days. I stayed there for 8 1/2 months. Both my legs were broken, the skin peeled off my left heel and I had dozens of cuts and bruises. After the doctors were sure I had no internal injuries they put a steel plate in my left femur and put me in a body cast from my chest down to the soles of my feet. After about 3 1/2 months they cut me out of that cast - took out the steel plate and I'll be damned if I didn't wake up from the operation in another cast - that one for 3 1/2 months. Then when out of that one I had to learn to walk again. Also had to gain some weight. I looked like Mahatma Gandhi - you could hang your hat on my hip bones.

When it was all over, it didn't seem too bad. That was a good hospital, good doctors and nurses, the people were nice. I learned a few lessons - one can get accustomed to most anything if absolutely necessary - like "concrete pants". That wreck too, I think germinated a few ideas - like having an alternate before takeoff -reserve fuel-to get there and when landing. If I remember correctly, we had no minimums - if you could get in, OK ...no questions. Also I think that might have been the beginning of thinking about approach lights, etc. Don't believe we had any of these things in '34. Anyway, a lot of improvements in operations and facilities were thought up after each crash. As you know Joe, everything was pretty crude back in those days. You and I, and especially those before us paved the way and pioneered improvements in facilities and methods of operation to make it easier for the present generation of pilots. We wish all of them well.

Joe Wolf (quite possibly due to his head injury) never flew again as a United pilot. Seventeen months later, on Dec.15, 1936, while riding as passengers, Capt. Wolf and his wife were killed in a Western Air Lines B-247 crash.

Appendix K

Selected Incapacitation Incidents in Air Transport

1. One of the first and most famous pilot incapacitations on record involved a DC-4 crash at Brisbane, Australia in the 1940s. Dr. Lloyd Buley identified pilot incapacitation as the probable cause of the crash. Somewhat later, a Constellation crashed at Oklahoma City in which Lloyd Pittman, the pilot, was found to be taking contraindicated cardiovascular medicines and was found to have a cardiovascular condition that would have prevented him from flying if he had not kept his condition from the FAA.

2. Recognition and control of the incapacitation hazard started with a study conducted by Captain Harry Orlady, Dr. Richard Harper and Dr. George Kidera for United Airlines in the mid 1960s. They developed the Two-Communication Rule for the detection of subtle or obvious incapacitations. The Two-Communication Rule states that one should have a high index of suspicion of an incapacitation at any time a cockpit crewmember does not respond appropriately to two verbal communications, or at any time a crew member does not respond appropriately to any verbal communication associated with a significant deviation from either an SOP or a standard flight profile.

   They also developed the following four rules for the operational control of any incapacitation rule:

   1. Maintain control of the airplane.
   2. Take care of the incapacitated pilot.
   3. Reorganize the cockpit and land the airplane.
   4. Plan actions to be taken after the landing.

3. The problem of pilot incapacitation was given official notice when Dr. Lloyd Buley, as the Chief Medical Officer of ICAO and Secretary of the ICAO

Medical Study Group, reviewed the progress made in a collaborative ICAO-IATA-IFALPA study of pilot malfunction. Dr. Buley concluded his 'interim review' by stating:

> ...it is suggested that acknowledgment of pilot on-duty incapacitation...as a permanent part of the air transport industry scene in the foreseeable future constitutes a constructive rather than a defeatist medical position. Further, it appears essential that the design, management, operational, training, and licensing disciplines should recognize that pilot incapacitation must be given due weight...in the overall judgment of what level of safety is practically attainable. It is suggested that only through such recognition will we achieve satisfactory control over all aspects of this unpalatable but not intractable problem.

These were prophetic words. At the time the review was issued, this was a revolutionary statement. It eventually resulting in a major change in pilot licensing standards. No longer did the industry and the regulators need to consider that the incapacitation of a pilot constituted an emergency condition of any greater magnitude than the other emergencies which the aircraft and its systems were designed to control. Recognizing the inherent logic of this statement, today in the United Kingdom stricter medical standards are applied to pilots who fly single crew aircraft than are applied to pilots who fly multi-crew aircraft and whose airline provides incapacitation training.

4. Obvious incapacitations are those immediately apparent to the remaining crewmembers. They can occur suddenly; are usually prolonged; and can result in a complete loss of function. Death in the cockpit is the most dramatic form of obvious incapacitation. In the US there were 15 inflight pilot deaths between 1956 and 1966. There were 3 deaths in 1982—the latest year in which data are easily available. The youngest of these pilots was 47 when he died.

5. 'Subtle' incapacitations occur more frequently than the 'obvious' type. They are frequently unreported, partial in nature and usually transient—lasting for from a few seconds to several minutes. They are insidious because the affected pilot may look well and continue to operate but have only a partially functioning brain. The pilot may not be aware of his/her problem, *nor be capable of rationally evaluating it*. There are many potential causes of subtle incapacitation.

6. The British Airways Trident crash at Staines on 18 May 1972 killed 118 people. The official report stated: 'The abnormal heart condition of Captain Key led to lack of concentration and impaired judgement sufficient to account for his toleration of the speed errors and to his retraction of, or order to retract, the droops in mistake for the flaps,' and 'Lack of training directed at the possibility of "subtle incapacitation."' Subtle incapacitations can create severe operational problems.

7. Dr. Peter Chapman, then medical director of British Caledonian airline, studied the inflight experience of 48 IATA carriers in the 17-year period (1965-1981). He found that reported incapacitations ranged from 4 incapacitations in 1965 to 27 in 1977. He also conducted a comprehensive study simulating incapacitation at the most critical periods of flight and confirmed that the incapacitation problem could be effectively controlled with adequate training.

8. When asked if there was any possibility of a one-man crew in airline operation, veteran FAA test pilots Berk Greene and Jim Richmond replied:

> As long as FAR 25 Appendix D requires consideration of pilot incapacitation or removal, there will never be a one-pilot cockpit approved or an operation predicated on only one pilot. Our experience with pilot health tells us that the current requirement is correct. Because of the incapacitation requirement, all current cockpit designs allow basic, normal aircraft operation with only one pilot onboard, but the design requirement will continue to require complete accommodation for two pilots.

Appendix L

23 Reasons for Required Flight Attendant Reports

1. When an act of aggression (e.g. **BOMB THREAT** or **HIJACKING**) occurs.

2. When **SECURITY** procedures are breached.

3. When the cabin is **PREPARED** for an **EMERGENCY LANDING**.

4. When a **COMMUNICATION SYSTEM** (e.g. PA, Video Equipment, or Call Bells) fails or becomes impaired.

5. When there is a **DECOMPRESSION** of the aircraft.

6. When a **DISRUPTIVE PASSENGER** is confronted.

7. When **EMERGENCY EQUIPMENT** is **NON OPERATIONAL** or **NOT PRESENT**, (also notify Captain).

8. When an **EMERGENCY LANDING** is performed.

9. When the aircraft is **EVACUATED**.

10. When **FIRE/SMOKE/FUMES** are present in the passenger compartment.

11. When a **HAZARDOUS MATERIAL** is present in the passenger compartment.

12. When an **INTOXICATED PASSENGER** is confronted.

13. When the jumpseat is **BROKEN** or **INOPERABLE**.

14. When an **OVERFLOW** of the **LAVATORY WATER** occurs.

15. When there is a **POTENTIAL HAZARD** which may cause injury to a customer or Flight Attendant. For example: torn carpet or broken cart.

16. When there is a **PROBLEM ENFORCING FARs**.

17. When there is a **SAFETY RELATED INTERRUPTION DURING**

STERILE COCKPIT.

18. When significant **TURBULENCE** is encountered.

19. When a **SLIDE** is inadvertently **DEPLOYED**.

20. When a **LAVATORY SMOKE DETECTOR** is activated or vandalized.

21. When there is a passenger **SMOKING** incident. (Section 17 of the CSR must be completed).

22. **ANY EVENT WHERE SAFETY STANDARDS MAY HAVE BEEN COMPROMISED.**

23. **ANY EVENT WHICH MAY PROVIDE USEFUL INFORMATION FOR THE ENHANCEMENT OF CABIN SAFETY.**

From US Airways *Safety Hotline*, June 1998

Twenty-three mandatory reporting events have been identified for the Flight Attendants. 'Although in some situations an ASR [Air Safety Report] will also be filed, the Flight Attendant perspective is essential to ensure a complete look at the problem.'

Appendix M

Significant Accident Causes and Their Presence in 93 Major Accidents

33%	Pilot deviated from basic operational procedures.
26%	Inadequate crosscheck by 2nd crewmember.
13%	Design faults.
12%	Maintenance and inspection deficiencies.
10%	Complete absence of approach guidance.
10%	Captain did not respond to crew inputs.
9%	ATC failures or errors.
9%	Crews not conditioned for proper response during abnormal conditions.
9%	Other.
8%	Weather information insufficient or in error.
7%	Runway hazards.
6%	ATC/CREW communication deficiencies.
6%	Pilot did not recognize the need for go-around.
5%	No GPWS installed.
5%	Weight or center-of-gravity in error.
4%	Deficiencies in accepted navigation procedures.
4%	Pilot incapacitation.
4%	Inadequate piloting skills.
3%	Pilot used improper procedure during go-around.
3%	Crew errors during training flights.
3%	Pilot not trained to respond promptly to GPWS command.
3%	Pilot unable to execute safe landing or go-around when runway sighting is lost below MDA or DH.
3%	Operational procedures did not require use of available approach aids.
3%	Captain inexperienced in aircraft type.

Flight Crew Causes and Their Percentage of Presence in 93 Major Accidents

33%	Pilot deviated from basic operational procedures.
26%	Inadequate crosscheck by 2nd crewmember.
10%	Captain did not respond to crew inputs.
9%	Crews not conditioned for proper response during abnormal conditions.
6%	Pilot did not recognize the need for go-around.
4%	Deficiencies in accepted navigation procedures.
4%	Pilot incapacitation.
4%	Inadequate piloting skills.
3%	Pilot used improper procedure during go-around.
3%	Crew errors during training flights.
3%	Pilot not trained to respond promptly to GPWS command.
3%	Pilot unable to execute safe landing or go-around when runway sighting is lost below MDA or DH.
3%	Operational procedures did not require use of available approach aids.
3%	Captain inexperienced in type.

Percentage of Accident Contributors Bearing Directly on Compliance with Standard Operating Procedures

33%	Pilot deviated from basic operational procedures.
26%	Inadequate crosscheck by 2nd crewmember.
6%	Pilot did not recognize the need for go-around.
3%	Pilot not trained to respond promptly to GPWS.
3%	Operating Procedures did not require use of available approach aids.

These tables are taken from a paper, *'A New Look At Accident Contributors and the Implications of Operational and Training Procedures'*, which was given by Richard L. Sears of the Boeing Commercial Airplane Co. at the 38th International Flight Safety Foundation Air Safety Symposium held in November, 1985 in Boston, Massachusetts.

Appendix N

List of 13 *ICAO Human Factors Digests*

Human Factors Digest No. 1   *Fundamental Human Factors Concepts*

Human Factors Digest No. 2   *Cockpit Resource Management (CRM) and Line-Oriented Flight Training (LOFT)*

Human Factors Digest No. 3   *Training of Operational Personnel in Human Factors*

Human Factors Digest No. 4   *Proceedings of the ICAO Human Factors Seminar* (Leningrad, April 1990)

Human Factors Digest No. 5   *Operational Implications of Automation In Advanced Technology Flight Decks*

Human Factors Digest No. 6   *Ergonomics*

Human Factors Digest No. 7   *Investigation of Human Factors in Accidents and Incidents*

Human Factors Digest No. 8   *Human Factors in Air Traffic Control*

Human Factors Digest No. 9   *Proceedings of the Second ICAO Flight Safety and Human Factors Symposium* (Washington, D.C., April 1993)

Human Factors Digest No. 10   *Human Factors, Management and Organization*

Human Factors Digest No. 11   *Human Factors in CNS/ATM Systems*

Human Factors Digest No. 12   *Human Factors in Aircraft Maintenance and Inspection*

Human Factors Digest No. 13   *Proceedings of the Third ICAO Global Flight Safety and Human Factors Symposium* (Auckland, April 1996)

These Human Factors Digests are published in separate English, French, Russian and Spanish editions by the International Civil Aviation Organization. Orders should be sent to one of the following addresses, together with appropriate remittance in US dollars or in the currency of the country in which the order is placed.

> Document Sales Unit
> International Civil Aviation Organization
> 1000 Sherbrooke Street West, Suite 400
> Montreal, Quebec
> Canada H3A 2R2

Egypt	ICAO Representative, Middle East Office, 9 Shagaret El Door Street, Zamalek 11211, Cairo.
France	Représentant de l'OACI, Bureau Europe et Atlantique Nord, 3 bis, villa Emile-Bergerat, 92522 Neuilly-sur-Seine (Cedex).
India	Oxford Book and Stationery Co., Scindia House, New Delhi, or 17 Park Street, Calcutta.
Japan	Japan Civil Aviation Promotion Foundation, 15-12, 1-chome, Toranomon, Minato-Ku, Tokyo.
Kenya	ICAO Representative, Eastern and Southern Office, United Nations Accommodations, PO Box 46294, Nairobi.
Mexico	Representante de la OACI. Oficina Norteamérica, Centroamérica y Caribe, Apartado postal 5-377, C.P. 06500, Mexico, D.F.
Peru	Representante de la OACI, Oficina Sudamérica, Apartado 4127, Lima 100.
Senegal	Reprèsentant de l'OACI, Bureau Afreque occidentale et centrale, Boîte postale 2356, Dakar.

Spain   Pilot's, Suministros Aeronáuticos, S.A., C/Ulises, 5-
                 Oficina Núm. 2, 238043 Madrid.
Thailand   ICAO Representative, Asia and Pacific Office, P.O. Box
                 11, Samyack Ladprao, Bangkok 10901.
United Kingdom   Civil Aviation Authority, Printing and Publications
                 Services, Greville House, 37 Gratton Road, Cheltenham,
                 Glos., GL50 2BN.

Appendix O

Human Factors in Multi-Cultural Certification
by Harry W. Orlady and Robert B. Barnes

Worldwide there is increased recognition that modern operational human factors needs considerably more emphasis if the air transport industry is to accomplish the growth that is foreseen for the industry in virtually all quarters of the globe. It is essential to further improve the already good safety record of the certificated air transport industry. Operational safety is an inherent part of certification.

Historically, it has been highly desirable that new airplanes or aviation systems, if they are designed for a worldwide market, be designed to meet appropriate US, British, or European regulations and achieve certification under those regulations. Countries who are not a part of the original certification do not always agree. Thus certification, even by a major certification country, does not guarantee that a given airplane or system will be accepted universally without modification. Other countries have their own and sometimes additional requirements that they often consider an important part of their national sovereignty

An outstanding example of reciprocal standardization has been made by the 27 European countries that have formed the Joint Aviation Authorities (JAA). The JAA's achievements include the promulgation of a several Joint Aviation Requirements (JARs). ICAO is leading the movement to eliminate multiple certification that is always extremely expensive. It is at the forefront of a growing movement (sometimes called harmonization) to simplify the certification process.

Increasing the operational safety of air transport is unquestionably a global challenge. Certification, and the training that is associated with it, is a meaningful part of that challenge. Bilateral or multi-cultural certification, which is highly desirable, complicates the certification and the safety process for several reasons.

One of the major reasons for this complication is that the safety issues in certification must be viewed within the context of each state's and airline's operational realities. These contexts can differ dramatically between sovereign states. They do not change the goal.

Whenever an airplane or component is certified and then purchased by an individual operator, the new airplane or component must be operated safely and efficiently in a day-to-day routine in the State in which it is operated according to the regulations of that State. In some cases, it must be operated in an environment that may be very different from those of its designers or manufacturers.

Nations, and airlines within nations, may not operate in the same manner. Many states have different regulations and different infrastructures. None of these often legitimate differences makes the bi- or multi-cultural certification problem any easier.

Bi- or multi-lateral certification is a growing phenomenon. For example, at the present time the US is involved in 27 bi-lateral certifications and additional bi-lateral certifications are under development. These not only include countries with different national cultures and different regulatory provisions, but they also ultimately involve airlines with differing operating philosophies.

A recent complication in this process is occurring in Europe, where the European Union (a growing political organization of European States) recently stated that the bi-lateral agreements which had been signed by EU members were no longer valid—that only the European Union could make such agreements. The issues involved go well beyond certification. Fifteen of EU member States are among the 27 European States that are members of the JAA. While this is largely a political problem, its ramifications could very well cause serious operational and certification problems.

Air Transport Certification Today

Today, both language and cultural differences play an increasing role in the worldwide operation of air transports. These differences must be recognized during the certification and the training process if the operation of new airplanes or components is to be successful. Dr. Mica Endsley, who is concerned with these issues, has written that:

> Most design guidelines apply to components—a gauge, a lever, a chair. Many human factors issues, though have to do with how those components are brought together and interact with each other in the context of particular tasks and with particular user populations. The acceptability of a given system design

configuration cannot be adequately assessed in a vacuum. Certification must ultimately take into account how the system as a whole is used, how various components interact with each other in producing the system's performance and who is using it. This requires that a complete understanding of the proposed (and possible) uses of the system be determined. (Endsley, 1994)

Present day certification involves both traditional engineering concepts regarding measurement and standards, and also the more complex human behavioral issues. In spite of the fact that the technical aspects of certification are reasonably straightforward, this is not true of many of the behavioral issues. There can be valid differences between authorities of various states.

Our increasingly interdependent world, and our industry's significant role in this interdependence, has made it vitally important that all parties to a certification fully understand not only the certification process within their own country, but also the certification process in other countries. The cultural basis each country has for its certification process, and its rationale may impact harmonization. Cross-cultural cooperation is a critical factor.

Importance of Operational Procedures and Training

Operational procedures and training are an important part of the certification process. They must fit the operational culture and environment of the user. We should avoid having a good airplane or component used either with less than optimum procedures, or by individuals who have inadequate training to operate that equipment safely and efficiently. Unfortunately this has happened. The inevitable result is a less than desired performance, which can lead to a determination that the new equipment is poorly designed. This may not be the true problem. The true problem may be simply inadequate training or a combination of both.

Regardless of where an airplane or component is designed, it is designed to accomplish specific tasks. The design engineers involved are highly trained professionals and have an intimate understanding of the skills and knowledge required to use their designs efficiently. However, that is no longer enough. The design engineers must also consider the skill, knowledge, and experience of the pilots who will be required to operate the new designs safely and efficiently during day-to-day operations in the operational environment in which they will be used.

This latter point is extremely critical. No one involved in the air transport design and certification process can any longer assume that all pilots have the same skill and knowledge base that are found in pilots of the designer's country. A training program that is completely acceptable in country 'A' may not be acceptable in country 'B', or in country 'C', or in country 'D', etc. Serious consideration must be given to the prospective user's cultural norms and experience, and to the environment in which the new design will be operated. The training required to resolve differences and to ensure operational safety can vary considerably between States.

Language is important, for words, phrases, and concepts often do not translate in a functionally effective manner. If the airplane or component is to be used in countries other than in its origin, then the national, regulatory, and organizational cultures that effect line operations in that country must be recognized.

Simply considering Aviation English to be the language of international aviation does not solve these problems. We have all seen situations where an instructor, lecturing in a language which was not the mother tongue of the students, received many affirmative head nods that had very little to do with an understanding of the lecture material.

An Actual Cross-Cultural Training Challenge

The authors recently participated in an effort to address these cultural issues. The United States Federal Aviation Administration had been asked by the Russian Federation to help it achieve US certification of a four-engine jet transport—the Ilyushin IL-96T. This certification included the reduction of the original flight deck crew of three to a flight deck crew of two.

This apparently straightforward task was actually a complicated challenge because it involved two dramatically different yet highly sophisticated technical cultures and two very different languages.

We were asked to conduct a two-week workshop for a mixed group of Russian aviation certification officials, design engineers, and test pilots who would be directly involved in accomplishing the certification. The Russian participants were very highly qualified in their country's aviation industry and they were equally highly qualified by US standards. However, they spoke almost no English and had little understanding of American (or Western) engineering or air transport operation cultures.

Conducting a Cross-Cultural Workshop

The primary goal of the workshop, which was designed to address this cross-cultural challenge, was to enable the participants to actually demonstrate that they could apply the regulatory aspects of minimum crew and workload certification in the US (and from the US perspective) at the conclusion of the workshop.

We believe that such a narrow focus and demonstration of proficiency is imperative since it makes it possible to design specific training objectives which then can be measured incrementally as the program progresses. This is a dramatically different approach than that of the unfortunately more prevalent technique of straight lectures with little opportunity for participant interaction with new and different concepts. Another important advantage of the approach we used was that the responses we got from the daily assignments given the students enabled us to more fully understand the reasons for any problems or difficulties that they had.

A basic premise of the workshop design was that people are more comfortable and tend to think much better if they can work in their native language than they do if they are forced to struggle with the usual straight translation from one language to the other. In the same fashion, if the results of their discussions and deliberations can be given in their mother tongue, the quality of their conclusions is not degraded by their language skills.

A critical support element to the workshop was a comprehensive bi-lingual participant workbook. The workbook included the entire course outline and supplementary information for all of the topics that would be covered during the two-week workshop. The workbook included daily assignments and additional references. One page was written in English, which was the author's native tongue, while the opposite page had the same information in Russian. Each page had numbered paragraphs so that there would be no misunderstanding of the basic information being discussed. In addition, all lectures and classroom discussions were translated as they occurred. This was done by two accomplished aviation-oriented professional interpreters who spoke Russian as their native tongue. We found that having highly qualified interpreters, who were also familiar with aviation topics and terminology in both countries, was extremely important to the success of the workshop. The interpreters worked side by side with the discussion leaders or lecturers and they were a part of the daily teaching/facilitation team.

The workshop was organized to be fully participatory rather than merely a note-taking exercise. At its completion, the goal of the workshop was for each participant to be able to:

- Demonstrate an understanding of the steps required for minimum crew certification in the United States;

- Identify the key issues in US minimum crew certification from both the FAA's and the applicant's perspective;

- Find and use appropriate US regulations, advisory circulars, research processes; and

- Design and conduct a basic crew workload analysis plan.

The workshop required eighty classroom hours over a two-week period. Each of the participants had a full schedule and worked very hard. During the first week they were introduced to the evolution of the US crew concept which is now an integral part of certification. Historical references were provided to help each participant understand how current US certification policy has evolved and why it has evolved in the way in which it has.

An important part of this background information was an introduction to what we, at least in the United States, refer to as Crew Resource Management (CRM) Training. While CRM sometimes is called by different names in other countries, we believe that CRM principles are universally accepted. CRM consideration is now an essential element in any US certification activity. We used discussions of basic CRM principles to assist participant understanding of how inter-personal and inter-group communications affect the design, certification, and operation of our commercial transports.

Throughout the two-week workshop, the Russian participants were divided into small sub-groups and were given several practical problems to solve. Each sub-group then reported its proposed solution to the full group in their native Russian language. The presentation was interpreted to us for comment as the presentation was given. This made it possible for all Russian participants and their American facilitators (which included FAA personnel involved in this bi-lateral certification and our workshop) to have a meaningful interactive discussion on the implications of the various proposed solutions.

Lessons Learned

The time and effort spent in developing the cross-cultural workbook with opposite pages in English and then in Russian was time and effort well spent.

It quickly became obvious that it was very helpful for the participant to be able to discuss their assignments and reactions to the various alternative solutions in their mother tongue. This proved to be an effective method for addressing complex culture-based issues without further complicating the situation. There were no multiple interpretation steps. It was particularly important that an intimate interaction with fundamental certification concepts take place since the participants would be the key people adapting these concepts to their own culture when they returned home.

The Russian participants responded very well to both the challenges we gave them and to the process used. There is no question that the ability to work in their own language was a big help. In a post-workshop critique (also conducted in Russian), the participant comments indicated that the workshop had accomplished its objective, that it was a very useful experience, and that the workshop would greatly facilitate their important work at home.

The fundamental lessons that the authors learned from conducting this workshop include that:

- Present day certification should include both traditional engineering concepts and a much-expanded human factors involvement.

- 'Expanded' human factors includes the total social culture (i. e., the national culture, the regulatory culture, and the corporate or organizational culture) and that each of these should be an integral part of certification;

- The adequacy of training and procedures for the pilot population and the specific environment in which they will operate the airplane or component must be considered an integral part of the certification process.

These three precepts are important concepts in a rapidly changing world and a world that is becoming increasingly dependent upon safe, reliable, and efficient air transportation.

If the training procedures suggested by the manufacturer or original certifying authority do not fully meet the needs of the purchaser of the new airplane or equipment, individual airlines, from states in which the airplane was not manufactured, can deal with bi- or multi-cultural problems by developing a one-time special workshop for their key operating and training people. Such a process would enable the key people to fully understand and implement the operating premises that were behind the new airplane or component and permit them to ensure understanding and implementation of those premises in the pilots who will operate the new equipment. If, for any number of reasons, such a procedure is not desirable or possible for the State in question, it becomes an absolute necessity for the operator of the new airplane or system to accomplish these objectives. The result must be that the pilots who will have to operate the new airplane or system have received training that will permit them to operate it safely and efficiently in routine day-to-day line flight operations.

Certification

An important part of regulatory provisions is certification. Certification is directly affected by the State's regulatory culture. Airplanes are not allowed to take-off or land within individual states unless they have been certificated, or at least officially approved, by that state. This concept has significant political considerations because transport aviation is a global industry that covers nearly all countries in the world.

Historically, it has been highly desirable that new airplanes or aviation systems be designed to achieve US or British certification if they are designed for a worldwide market. While there are some differences between US and British requirements that are being slowly and laboriously resolved, certification in both includes meaningful recognition of modern human factors. Even such certification does not guarantee that the new airplane or system will be accepted universally without further modification because other countries have their own and sometimes additional requirements.

A simple fact of certification life is that many countries do not have the resources or expertise to deal with the complex issues that are now a part of certification. This is another compelling reason for making certification by a State with a mature certification capability acceptable in other states or in using such organizations as the JAA. Today, this is done with reciprocal agreements. Attaining a reciprocal agreement can be a very painstaking process, particularly if the countries have language or other cultural differences. When we consider these issues, we should always remember that individual countries have sovereign control over the airspace within their own boundaries and that many of them jealously guard that sovereign control.

References

Adam, C.F., Barnes, R., Orlady, H., and Ostrovsky, Y. (1996). 'Human Factors in the International Certification of Transport Category Aircraft', presented at ICAO Third Global Flight Safety and Human Factors Symposium, April 1996, Auckland.

Barnes, Robert B., Orlady, Harry W., and Orlady, Linda M. (1996). 'Multi-Cultural Training in Human Factors For Transport Aircraft Certification', presented at ICAO Third Global Flight Safety and Human Factors Symposium, April 1996, Auckland.

Endsley, Mica R. (1994). 'Aviation System Certification: Challenges and Opportunities', in *Human Factors Certification of Advanced Aviation Technologies*, Embry-Riddle Aeronautical University Press, Daytona Beach, Florida.

Maurino, D.L. and Galotti (1994). 'The Point of View of the International Civil Aviation Organization (ICAO)', in *Human Factors Certification of Advanced Aviation Technologies*, Embry-Riddle Aeronautical University Press, Daytona Beach, Florida.

Orlady, Harry W., (1994). 'Airline Pilot Training Programmes Have Undergone Important and Necessary Changes in the Past Decade', *ICAO Journal*, April 1994, Montreal.

Appendix P

FAA Medical Standards - Effective September 16, 1995
(from *The Guide for Aviation Medical Examiners*)

Medical Certificate	First Class	Second Class	Third Class
Pilot Type	Airline Transport Pilot	Commercial Pilot	Private Pilot
Distant Vision	20/20 or better in each eye separately with or without correction		20/40 or better in each eye separately, with or without correction
Near Vision	20/40 or better in each eye separately (Snellen equivalent), with or without correction as measured at 16 inches		
Intermediate Vision	20/40 or better in each eye separately (Snellen equivalent) with or without correction as measured at 16 inches		No requirement
Color Vision	Ability to perceive those colors necessary for safe performance of airman duties		
Hearing	Demonstrate hearing of an average conversational voice in a quiet room, using both ears at 6 feet, with the back turned OR pass one of the audiometric tests below.		
Audiology	Audiometric speech discrimination test: Score at least 70% discrimination in one ear		

Pure tone audiometric test: Unaided with thresholds no worse than:

	500Hz	1000Hz	2000Hz	3000Hz
Better ear	35Db	30Db	30Db	40Db
Worst ear	35Db	50Db	50Db	60Db

Medical Certificate	First Class	Second Class	Third Class
Pilot Type	Airline Transport Pilot	Commercial Pilot	Private Pilot
ENT	No ear disease or condition manifested by, or that may be reasonably be expected to be manifested by, vertigo or a disturbance of speech or equilibrium		
Pulse	Not disqualifying per se. Used to determine cardiac system status and responsiveness.		
Blood Pressure	No specified values stated in the standards. Hypertension covered under general medical standard and in the *Guide for Aviation Medical Examiners*.		
Electrocardiogram	At age 35 and annually after age 40	Not routinely required	
Mental	No diagnosis of psychosis, or bipolar disorder, or severe personality disorders		
Substance Dependence and Substance Abuse	A diagnosis or medical history of 'substance dependence' is disqualifying unless there is established clinical evidence, satisfactory to the Federal Air Surgeon of recovery, including sustained total abstinence from the substance(s) for not less than the preceding 2 years. A history of 'substance abuse' within the preceding 2 years is disqualifying. 'Substance' includes alcohol and other drugs (i.e., PCP, sedatives and hynoptics, narcolytics, marijuana, cocaine, opoids, amphetamines, hallucinogens, and other psychoactive drugs or chemicals).		

Medical Certificate	First Class	Second Class	Third Class
Pilot Type	Airline Transport Pilot	Commercial Pilot	Private Pilot
Disqualifying Conditions	Examiner must disqualify if the applicant has a history of		
	1. Diabetes mellitus requiring hypoglycemic medication;		
	2. Angina pectoris;		
	3. Coronary heart disease that has been treated or, if untreated, that has been symptomatic or clinically significant;		
	4. Myocardial infarction;		
	5. Cardiac valve replacement;		
	6. Permanent cardiac pacermaker		
	7. Heart replacement;		
	8. Psychosis;		
	9. Bipolar disorder;		
	10. Personality disorder that is severe enough to have repeatedly manifested itself by overt acts;		
	11. Substance dependence;		
	12. Substance abuse;		
	13. Epilepsy;		
	14. Disturbance of consciousness without satisfactory explanation of cause; and		
	15. Transient loss of control of nervous system function(s) without satisfactory explanation of cause.		

Appendix Q

Selected Aviation Websites of Specific Interest

The following websites are some that the authors have found particularly useful. It is not intended to be any kind of exhaustive list, but rather to give the reader some areas to explore and an idea of some of the resources available. The information on the websites is frequently updated and often represents some of the latest information available. Related links to other useful sites can be particularly helpful. As with a fair amount of information available today on the Web, two cautions are necessary. First, some of the data contained on the website may represent the opinion of the particular authors and should be treated as such. Secondly, on occasion information is posted before it has been scientifically researched and/or verified.

I. **Government Organizations and Federal Agencies**

    A.   FAA human factors: http://www.hf.faa.gov/

    B.   Fedworld Information Network: http://www.fedworld.gov

    C.   German National Aerospace Research Center (DLR):
        http://www.dlr.de

    D.   IATA: http://www.iata.org

    E.   ICAO: http://www.icao.org

    F.   NASA: http://www.nasa.gov

    G.   NASA's ASRS: http://www-afo.arc.nasa.gov/asrs/

    H.   NTSB: http//www.ntsb.gov

II. **News Sources and Publishers**

    A.   Ashgate: http://www.ashgate.com

    B.   Aviation Week & Space Technology:
        http:/www.aviationweek.com

    C.   Aviation Week Safety Resource: http://www.awgnet.com

    D.   Aviation International News: http://www:ainonline.com

    E.   News and database source: http://www.avweb.com

    F.   CRM Industry Developers Group:
        http://www.caar.db.erau.edu/crm

    G.   News and database course: http://www.landings.com

    H.   Professional Pilot magazine: http://www.flightdata.com/propilot

III. Manufacturers and Vendors

    A.   Boeing Airplane Company: http://www.boeing.com

    B.   Airbus Industrie: http://www.airbus.com

    C.   Jeppesen: http://www.jeppesen.com

IV. Associations

    A.   Air Line Pilots Association (ALPA): http://www.alpa.org

    B.   Airport Operators and Pilots Association (AOPA):
        http://www.aopa.org

    C.   Australian Aviation Psychology Association (AAvPa):
        http//vicnet.au~aavpa/

    D.   European Association for Aviation Psychology (EEAP):
        http://www.eaap.com

    E.   EUROCONTROL: http://www.eurocontrol.fr

    F.   Flight Safety Foundation: http://www.flightsafety.org

    G.   Human Factors and Ergonomic Society: http://hfes.org

    H.   International Society of Air Safety Investigators (ISASI):
        http://awgnet.com/safety/isasi.htm

    I.   National Business Aircraft Association (NBAA):http://nbaa.org

    J.   Royal Aeronautical Society: http://raes.org.uk

    K.   Society of Automotive Engineers: http://www.sae.org

V. Universities with Aviation Sites

A. Cranfield University: http://www.cranfield.ac.uk/coa

B. Embry-Riddle University: http://www.db.erau.edu/

C. University of Illinois: http://aviation.uiuc.edu

D. Massachusetts Institute of Technology (MIT):
    http://www.mit.edu

E. Ohio State University: http://aviation.eng.ohio-state.edu

F. Purdue University: http://www.tech.purdue.edu/at/

G. University of North Dakota:
    http://www.aero.und.edu/Academics/Aviation

H. University of Texas:
    http://www.psy.utexas.edu/psy/helmreich/nasaut.htm

Selective Glossary of Acronyms and Abbreviations

The following acronyms, abbreviations, and descriptive words or phrases have been selected and modified from a variety of sources. The sources include the ATA's *The Airline Handbook*, Charles Billing's *Automation: The Search For A Human-Centered Approach*, Jane's *Aerospace Dictionary* edited by Bill Gunston, Boeing's *A Glossary of Terms: Human Errors in Aviation*, and the SAE G-10 FAA-sponsored *Aerospace Glossary for Human Factors Engineers* (ARP 4107).

The purpose of this Glossary is to help make this text more understandable to readers who may not be familiar with these terms. It does not include many terms that are defined in the text, is not all-inclusive and should not be considered a definitive source.

AC   **Advisory Circular. ACs** are sent out by the FAA, to give guidance to the industry on specific subjects. AC material is not mandatory and does not constitute a regulation. It is for guidance purposes only.

ACARS   **Aircraft Communications Addressing and Reporting System**.

ACAS   **Airborne Collision Avoidance System**. ICAO terminology for a system installed in commercial jets to search for and alert pilots to the presence of other aircraft. It is virtually identical to the US TCAS. ACAS is also used by EUROCONTROL and by many other countries. See TCAS.

ADM   **Aeronautical Decision Making**.

ADS   **Automatic Dependence Surveillance**. Automatically provides aircraft position, altitude, and velocity vector data to ground stations at frequent intervals.

ADS-B   Automatically broadcasts the aircraft's position, altitude, and vector information for display by other aircraft or ground users at frequent intervals. It is an advanced ADS and requires a data link to support its performance requirements.

ADVTECH   **Advanced Technology Aircraft**. Also called Glass Cockpit Aircraft.

AGARD   **Advisory Group for Aerospace Research and Development**. AGARD was formed in 1952. Its purpose is to 'foster and improve the interchange of information relating to aerospace research and development between the NATO (North Atlantic Treaty Organizations) nations in order to ensure that the advances made by one nation are available to the others'.

AIM   **Airman's Information Manual**. A FAA publication for practicing pilots. AIM 'is designed to provide basic flight information and ATC procedures for use in the National Airspace System (NAS) of the United States. The information parallels the U.S. Aeronautical Publication (AIP) distributed internationally'.

ALPA   **Air Line Pilots Association**. ALPA is the oldest and largest professional pilot union in the world. It includes members of the pilots of most U.S. airlines. A significant exception is the pilots of American Airlines (about 8,000) who are members of the APA (the Allied Pilots Association) and there are other smaller pilot unions. ALPA has a very active Engineering and Safety Department and has played an important role in all US certification and other safety-related activities. The Canadian Air Line Pilots Association (CALPA) merged with ALPA in February of 1997. Since that merger, ALPA is the only North American member of IFALPA, the International Federation of Air Line Pilots Associations.

APA   **Allied Pilots Association**. The union representing the pilots of American Airlines.

AQP	**The FAA's Advanced Qualification Program**. An innovative program designed to increase aviation safety through improved training and evaluation. AQP is designed to be responsive to changes in aircraft technology, operations, and training methodology.
ARINC	**Aeronautical Radio, Incorporated**. A corporate organization that provides international and domestic data transmission, receiving and forwarding services for air carriers and other subscribers, and that provides specialized research on request. Does considerable work for the US FAA.
ARTCC	**Air Route Traffic Control Center (US)**. Also known as an **En Route Center**. It houses the air traffic controllers and equipment needed to identify and direct aircraft, primarily during the enroute portion of their flights.
ARTS III	**Automatic Radar Terminal System III**. A advanced (third development) of an automatic radar terminal system that is capable of using MSAWs ARTS III was developed by the FAA and is used principally in the US.
ASAP	**Airlines Safety Action Partnership**. Under the ASAP, error reports from American Airline's pilots, and other operational safety data, are collected electronically, deidentified and the data reviewed weekly by a joint committee composed of representatives of the FAA, American Airlines, and the Allied Pilots Association.
ASCENA	**ASCENA** is a public safety organization that represents French-speaking States in West Africa. It is responsible for air navigation over 24 international and 100 domestic airports.
ASES	**Aviation Safety Exchange System**. An ATA sponsored system for the exchange of air safety data, potentially useful under GAIN.

ASRS   **Aviation Safety Reporting System**. The US non-punitive incident reporting system. It has been successfully administered by NASA for the past 23 years and presently receives in the area of 32,000 reports (incidents and other items of safety interest) each year. A majority of ASRS funding is provided by the FAA. Historically, the FAA has not interfered with ASRS operation enabling it to continue to remain, and be perceived by the aviation community, as a productive and non-punitive aviation safety resource.

ATA   **Air Transport Association of America**. An association of airlines in the US which was formed in 1936 and is concerned with air transport safety, with the government rules and regulations, that affect the airlines, and with any other matters that jointly affect them. It is the air carrier industry organization.

ATC   **Air Traffic Control**. Provides tactical control of air movements by towers and air route traffic control centers. The FAA manages air traffic control in the US.

ATM   **Air Traffic Management**. FAA's proposed system for the management of air traffic in the US.

BASI   **Bureau of Air Safety Investigation**. The Australian agency that investigates and reports on aviation accidents in Australia.

BASIS   **British Airways Safety Information System**.
Developed by British Airways in 1990. It is an open penalty-free system that has become one of the world's most popular aviation management tools. BASIS is owned and operated by British Airways but licensed to other carriers and organizations. Today the system is used by over 150 organizations, regulatory authorities, and aircraft manufacturers.

CAA   1. **Civil Aviation Administration (US)**. The CAA was eliminated in the Deregulation Act of 1978. In previous years it was responsible for the general operation of all aspects of the Aviation System in the US. Its former responsibilities have now been given to the FAA.

2. **Civil Aviation Authority**. The United Kingdom Civil Aviation Authority, the UK equivalent of the FAA. The CAA designation is also used in other countries.

3. **Civil Aviation Administration**. The governmental civil aviation authority in several designated countries.

CAB   **Civil Aeronautics Board**. The US CAB was eliminated in the Deregulation Act of 1978. The CAB was responsible for the certification of aircraft, the determination of minimum crew complement etc. These responsibilities are now given to the FAA.

CAIRS   **Confidential Aviation Incident Reporting System**. The Australian confidential aviation incident reporting system (CAIRS) should not be confused with the long-standing incident reporting requirement of the Australian Air Navigation Regulations.

CASRP/SECURITAS   **Canadian Aviation Safety Reporting Program**.

CASST   **Commercial Aviation Safety Strategy Team**. This team, which is being led by the ATA, is the result of a coordinated effort by the Aerospace Industries Association, principal manufacturers of both airplanes and engines, the airlines, and the Air Line Pilots Association. The FAA and NASA are active participants.

CAT I, CAT II, and CAT III - Categories for Landing Minimums. These involve weather minimums and combinations of decision height (DH) and runway visual range (RVR) while using an ILS. They have been agreed to by ICAO and are in general use throughout the world.

    Category I   200 ft. decision height (DH) and a runway visual range RVR) of not less than 1800 feet.

Category II	200-100 ft DH and a RVR of not less than 1200 ft.
Category IIIa	no DH, and a RVR of not less than 700 ft.
Category IIIb	no DH, and a RVR of not less than 150 ft.
Category IIIc	no DH and no RVR (i.e. no external visibility).

CDU   **Control and Display Unit**. Provides much of the human-system interface in the flight management systems of many of today's aircraft. On Boeing aircraft, allows the pilot to program the FMC, and therefore, the FMS. On Airbus, the CDU is referred to as the MCDU (the 'M' standing for multifunction). It provides the interface with the FMGC (Flight Management and Guidance Computers).

CFIT   **Controlled Flight Into Terrain**. An accident 'in which an otherwise serviceable aircraft under the control of the crew, is flown (unintentionally) into terrain, obstacles or water, with no prior awareness on the part of the crew of the impending collision'.

CHIRP   **Confidential Human Incident Reporting Program**. The confidential incident reporting program of the United Kingdom.

CIS   **Commonwealth of Independent States**. States of the former Soviet Union.

CNS/ATM   Communications, navigation, surveillance and air traffic management. See FANS.

Code Sharing   A marketing practice in which two airlines share the same two-letter code used to identify carriers in the computer reservation systems used by travel agents.

CRM   **Crew Resource Management**. A concept to utilize and improve the resource management skills of pilots, flight attendants and others in the aviation system. The term was originally coined by John Lauber as Cockpit Resource Management but did have several antecedents. It arose as a result of a general realization that the safe and efficient operation of modern transport required more than technical skills, that a total team concept was required, and that the requisite skills and philosophy could and should be taught.

CRT   **Cathode Ray Tube**. Used to display information on the instrument panel of glass cockpit aircraft.

CVR   **Cockpit Voice Recorder**. Standard equipment that records cockpit sounds, including voice comments and radio transmissions from the cockpit for a period of 30 minutes. Future developments are expected to increase that time frame. CVRs are designed to withstand the forces of a crash so that its information may be used to reconstruct the circumstances leading up to a crash.

DFDR   **Digital Flight Data Recorder**. See FDR.

DH   **Decision Height**. Used in determination of landing minimums.

Dirigible or Airship
A lighter-than-air aircraft that has its own motive power and can be steered in any direction by its crew. See Zeppelin.

DME   **Distance Measuring Equipment**. Navigation equipment installed in an airplane that measures slant range distance.

Doppler   An airborne navigation system that uses Doppler radar to sense the rate of change of position of the aircraft and also a ground-based radar system that detects the wind shears or microbursts associated with thunderstorms.

DOT   **Department of Transportation.** Headed by the Secretary of Transportation and is responsible for all commercial transportation in the United States. The FAA a part of the DOT organization.

EASA   **European Aviation Safety Agency.** An aviation agency recently formed by the European Union that is expected to become the cornerstone agency that EU plans to have full legal power to issue and enforce aviation regulations throughout the EU. Full legal and operational status is expected to be reached by 2003.

ECAC   **European Civil Aviation Conference.** At the present time the ECAC has 36 member nations.

EEC   **EUROCONTROL Experimental Centre.** Established in 1962 as part of EUROCONTROL. Primarily supplies support for the study, design, development and improvement of air traffic control systems.

EGNOS   **European Geostationary Navigation Overlay System.** EGNOS is the satellite navigation system that is the European equivalent of the US WAAS. It is essential that the two systems work seamlessly together. Initial coordination tests of the two systems in Iceland have been successful.

EGPWS   **Enhanced Ground Proximity Warning System.** It is a considerably advanced GPWS.

ESRD   **European Supersonic Research Program.**

EU   **European Union.** A growing political organization composed of 27 European States.

EUCARE   **European Confidential Aviation Safety Reporting Network.** A newly established and independent European safety reporting system. Reports to EUCARE may be submitted by any member of the aviation community and are treated with absolute confidentiality. EUCARE is associated with the Technical University Berlin.

EUROCONTROL
The European Organization for the Safety of Air Navigation. EUROCONTROL is still growing and in 1995 had 33 member nations.

EVS
Enhanced Visual Systems. Advanced HUD systems that are not yet operational. Expected to make airplanes independent of ground aids. EVS displays can be shown on CRTs, flat panels, or advanced HUDs.

FAA
Federal Aviation Administration. Headed by the Administrator and organizationally structured under the DOT. Responsible for air safety and operation of the air traffic control system. Since aviation regulation in the US began in 1926 with the Aeronautics Branch of the Department of Commerce, a governmental agency has been responsible for both regulating safety in aeronautics and simultaneously, promoting the efficient growth of the aviation industry.

FADEC
Full-Authority Digital Engine Control. Automatically determines engine parameters and can control engine power for all flight phases, including takeoff.

Fail Active
System or group of systems that remain correctly operational after any failure.

Fail Operational
A system whose failure has no (or limited) effect on operation despite the failure of one or more of its components.

Fail Passive
System or systems that become inoperative after failure requiring operation to be terminated. A failure inactivates the system thus preventing a dangerous spurious operation or a hardover output.

FANS	**Future Air Navigation System**. FANS is a planned for worldwide air-navigation system, formulated by ICAO that is being implemented incrementally. A first step is to reach agreement on standards for the future communication, navigation, surveillance, and air- traffic management (CNS/ATM) system. Compliance with requirements for FANS-1 or FANS-A (Europe) is being included in new airplanes in the US and abroad.
FARs	**Federal Aviation Regulations**. The FARs are composed of 'Parts' such as Part 23, Part 25, Part 121 etc. These Parts are specific sections of the US regulations.
FCU	**Flight Control Unit**. The tactical mode and input data control panel for the Airbus autopilot system. See FMS.
FDR	**Flight Data Recorder**. Records pertinent technical information about a flight such as aircraft performance, including speed, altitude, heading, and other flight parameters as well as selected aircraft systems data. Modern FDRs, known as digital flight data recorders (DFDRs) record up to 88 parameters. Both are designed to withstand forces encountered in a crash.
FEIA	**Flight Engineers International Association**. Just after World War II a particularly important part of the conflict in the US over the number and qualifications of crew required in US airplanes. FEIA was a major contributor to the US President's Task Force on Aircraft Crew Complement.
FLA	**Future Large Aircraft**. The European equivalent of the VLA—the US designation of very large aircraft.
Flat Panel	Flat panel displays provide the same information as the earlier CRTs. Flat panels are considered more efficient than CRTs because they require considerably less overall space and have other advantages.

FLIR   **Forward Looking Infrared Radiation**. Infrared radiation is an electromagnetic radiation having wave lengths greater than those of visible light and shorter than microwaves. It has significant viewing advantages in certain kinds of weather, and disadvantages in other. FLIR are considered an important EVS.

Fly-by-wire   Conventional flight controls use cables from the control column in the cockpit to hydraulic actuators near the control services. Fly-by-wire places computers between the pilot's controls and the control surfaces, essentially inserting a microchip between the pilot and the controls. A simpler definition is to simply call fly-by-wire a flight control system with electric signaling.

FMA   **Flight Mode Annunciation Panel**. In older aircraft it is a dedicated panel usually located above or near the attitude indicator; in glass cockpit airplanes, it contains a display of flight modes that is located at the top of the primary flight display.

FMS   **Flight Management System**. FMSs are becoming increasingly versatile and complex. They are an essential part of the effective operation of today's advanced technology airplanes. In Airbus airplanes, FMSs are called FMGCs for Flight Management and Guidance Computers.

FOQA   **Flight Operations Quality Assurance**. A controversial and only partially implemented FAA and industry program. FOQA includes the AQP and the routine collection of flight operations data that will be automatically recorded on quick access recorders on each flight when and if it is fully implemented. It is an important part of a fully implemented GAIN.

FPV   **Flight Path Vector**. Prediction of future flight path. Replaces traditional flight director in advanced EFIS. A FPV is especially useful to detect and deal with windshear.

Free Flight   An advanced air traffic control concept which has been defined as, '…a safe and efficient flight operating capability under instrument flight rules in which operators have the freedom to select their paths and speed in real time…. Limitations are limited in extent and duration to correct the identified problem. Any activity which removes restrictions represent a move toward Free Flight' (RTCA Task Force 3, 1995).

FSB   **Flight Standardization Board**. An internal FAA Board.

FSF   **Flight Safety Foundation**. An apolitical, independent, nonprofit, and international organization that has more than 660 member organizations in 77 countries. The FSF provides an information and collection function that many lesser-developed aviation industries rely on for aviation safety information. It also provides confidential safety audits of worldwide corporate and airline operations.

G   'G' represents the downward acceleration of all objects on earth due to gravity. While G varies slightly with latitude it is usually considered as 32 ft/sec^2 (981cm/sec^2).

GAIN   **Global Analysis and Information Network**. An international information structure to collect, analyze and disseminate aviation safety information and thus to provide a significantly improved operational early-warning safety capability. It is envisioned as an information system run by the aviation community with ultimate cost savings through the sharing of data, analytical methodologies, and results.

GLONASS   **Global Navigation Satellite System**. A satellite navigation system deployed by the Russian Federation and operated by its defense department. GLONASS has disavowed any plans to degrade its signals availability for civilian use. Like GPS, it offers precise global and continuous position-fixing capability.

GNSS   **Global Navigation Satellite System**. European satellite-based navigation system that is compatible with GPS and

GLONASS.

GPS	**Global Positioning System**. A US satellite-based navigation system owned and operated by the US defense department. Offers precise, global, and continuous position-fixing capability.
GPS/ADS	**Global Positioning System and Automatic Dependent Surveillance**. See FANS.
GPWS	**Ground Proximity Warning System**.
Harmonization	Standardization between different countries of rules and regulations involving the manufacture, certification, and operation of air transport.
HGS	**Head-Up Guidance System**. See HUD.
Hub and Spoke	A system for utilizing aircraft efficiently by increasing service options. It uses the hub airport (the hub) to serve as an exchange point for flights to and from outlying towns and cities (the spokes).
HUD	**Head-Up Display**. Provides head-up display of flight path information to the pilot by projecting information on a transparent surface in front of the pilot that is superimposed on his/her vision of the real world.
Hypersonic	An aircraft having speed above Mach 5.
IATA	**International Air Transport Association**. IATA is an industry association of 235 international airlines. It is involved with all aspects of international air transportation, including both economic and technical issues.
ICAO	**International Civil Aviation Organization**. This is the specialized aviation agency of the United Nations. It currently has 186 member States. ICAO was formed in 1947 and has a sovereign body, the Assembly, a governing body, the Council, and an international staff.

IFALPA **International Federation of Airline Pilots Associations**. IFALPA has a membership of pilot associations representing 120,000 pilots from 90 nations. When the United Nations formed ICAO, it also granted official observer status to IFALPA as the single legitimate voice of the world's airline pilots. IFALPA provides a professional observer to ICAO and professional representation on 28 ICAO Committees, Panels, and Secretariat Study Groups. IFALPA is an active and influential contributor in international air safety forums.

IHAS **Integrated Hazard Avoidance System**.

Infrastructure That part of the Aviation System that includes the ATC system, fixed structures such as airport runways, taxiways, ramps, passenger and cargo terminals, fuel lines, control towers, and other ground installations, as well as radar and communications facilities.

INS **Inertial Navigation System**. An airborne system that keeps track of aircraft speed and position by means of gyroscopes and accelerometers.

JAA **Joint Aviation Authorities**. The JAA consists of 24 European countries—24 of which belong to the European Union (EU). It is responsible for rulemaking and the standardization of all areas of European aviation. The EU believes that it should be responsible for JAA functions.

JAR **Joint Aviation Requirement**. The JARs, formed by the JAA, are the product of a serious attempt of European authorities to reach a consensus on airworthiness issues. JARs contain both requirement and advisory material—Advisory Circulars Joint (ACJ); Advisory Material Joint (AMJ); Acceptable Means of Compliance (AMC); and Interpretative and Explanatory Material, as well as General Documents regarding Maintenance, Certification, Operations, and Licensing.

KUFAC **KLM Human Factors Awareness Course**.

LAAS	**Local Area Augmentation System**. GPS augmentation to augment WAAS in order to meet stringent Cat. 2 and Cat. 3 approach and landing requirements.
LCD	**Liquid Crystal Display**. LCDs are used in flat panels which seem to be gradually replacing CRTs in the latest airplanes.
LORAN	**Long-Range Navigation** system that uses ground-based low-frequency radio aids. In the US, the LORAN system is operated by the US Coast Guard.
MASPS	**Minimum Aircraft System Performance Specifications**. Operational system criteria specifying digital air data computers and associated systems that enable height keeping within 50 feet. MASPS are required for operation in RVSM airspace.
MEL	**Minimum Equipment List**. Aircraft equipment that must be in good working order for an aircraft to legally take off with passengers. The FAA may defer items not considered essential to an aircraft's airworthiness for limited and specific times.
Mode S	An enhanced transponder system that permits data link information to be communicated between ATC and aircraft. Mode S transponders can also communicate potential collision avoidance information between aircraft.
MSAW	**Minimum Safe Altitude Warning System**. A ground based system in ATC ARTS III facilities that automatically warns if an aircraft is below the minimum safe altitude.
NAS	**National Aerospace System (US)**.
NASA	**National Aeronautics and Space Administration (US)**.
NASA/UT	**A University of Texas research group with funding from NASA** (sometimes includes work with or sponsored by the FAA and is then designated as NASA/FAA/UT).

NCARC	**National Civil Aviation Review Commission.** The Commission was formed as part of the Federal Aviation Reauthorization Act of 1996, and consists of 21 members from different aviation organizations.
NPA	**Non-precision Approach.** An instrument approach with lateral guidance only from the final approach fix (FAP) to the runway touchdown zone. NPA is also defined as a standard instrument approach in which no electronic glide slope is provided.
NPRM	**Notice of Proposed Rule Making.** Issued by the FAA to allow maximum knowledge of its intentions and as part of its process of developing regulations.
NTSB	**The National Transportation Safety Board.** The official US Board that investigates aircraft accidents.
OASIS	**Operational Airport Safety Information System.** Joint effort of FSF and Battelle to develop a safety information system under GAIN.
PF	**Pilot Flying.**
PNF	**Pilot Not-flying.**
POI	**Principal Operations Inspector.** FAA's chief operations inspector for each airline.

Precision Approach

An instrument approach that furnishes lateral and vertical guidance from the final approach fix (FAP) to the runway touchdown zone. Instrument landing system (ILS), microwave landing systems (MLS), and precision approach radar (PAR) are considered precision approaches.

PWS	**Predictive Wind Shear.**
QAR	**Quick Access Recorder.**

Regional Air Carrier or Regional

An airline with annual revenues of less than $100 million and whose service is frequently limited to a particular geographic region.

RTCA
Radio Technical Commission for Aeronautics Inc. Formed in 1935 as the Radio Technical Commission for Aeronautics, it became incorporated in 1991 as simply RTCA, Inc. Its stated mission is 'to advance the art and science of aviation and aviation electronic systems for the benefit of the public'. It is funded primarily by dues from its over 200 members. Members include approximately 150 US government and business entities as well as academic and international associates. RTCA functions as a federal advisory committee and has been called a consensus builder.

RTK
Metric Ton Kilometer. A measure frequently used in measuring international cargo.

RTO
Rejected Takeoff.

RVR
Runway Visual Range.

RVSM
Reduced Vertical Separation Minimum. This is reduced vertical separation minimum above 29,000 feet. One thousand feet is still indicated barometrically, but less than 1,000 actual feet is provided because of the decreased air density at higher altitudes. Special controller training is required for all controllers handling transitions to and from RVSM airspace. Special equipment and aircraft operational approval are required for all aircraft utilizing RVSM airspace.

SADEC
South African Development Community.

Social Environment

The social environment includes the national environment, the regulatory environment and the organizational or corporate environment.

Stabilized Approach

An approach along the extended runway centerline with a constant, in-flight verifiable, descent gradient from the final approach altitude to the runway touchdown zone. An ILS approach is essentially a stabilized procedure except for offset-localizer approaches.

Stage 2

Term used to describe jets which meet specified noise parameters on takeoff and landing. Jets in this category include the Boeing 727 and the McDonnell Douglas DC-9.

Stage 3

Term used to describe jets which meet specified noise parameters on takeoff and landing. Stage 3 describes the quietest jets in service today, including the B-757 and the MD-80s.

SST

Supersonic Transport. A transport capable of flying at speed greater than the speed of sound, which varies with altitude but which is more than 700 miles per hour at sea level.

STC

Supplemental Type Certificate. Certificate for an airplane that is a derivative of an already certificated airplane which has a Type Certificate.

STCA

Short Term Conflict Alert. Software developed for controllers to give controllers warning if aircraft loss of separation is imminent. Developed in and used mainly in Europe.

STOL

Short Takeoff or Landing Aircraft. An aircraft that has the capability of operating from a short runway in accordance with applicable airworthiness and operating regulations. See VTOL.

SVS

Synthetic Vision System.

TA

Traffic Advisory. From a TCAS warning, usually before a TR.

TAWS	**Terrain Avoidance and Warning System**. The acronym used by the FAA in its regulations. GPWS and EGPWS are both TAWS.
TC	**Type Certificate**. Refers to a specific aircraft certified by a regulatory authority to be airworthy.
TCAS	**Traffic Alert and Collision Avoidance System**. A system installed in commercial jets to search for and alert pilots to the presence of other aircraft. More advanced versions of TCAS also give TAs (Traffic Advisories) and TRs (Traffic Resolutions) that furnish lateral avoidance instructions. Future versions are expected to give vertical avoidance instructions also. ICAO and EU use ACAS for the same function.
TR	**Traffic Resolution**. From a TCAS warning, usually after a TA.
Transponder	An electronic device that 'responds' to interrogation by ground-based radar with a special four-digit code that specifically identifies the aircraft on which it is located. Later versions have the ability to transmit the altitude of the aircraft automatically.
Transport Canada	The governmental regulator for civil aviation in Canada. The Canadian equivalent of the US FAA.
TSU	**Time of Safe Consciousness**. The time a person can remain unconscious from lack of oxygen without incurring risk of brain damage.
TUC	**Time of Useful Consciousness**. The time in which a person without normal oxygen can be expected to take effective preventive measures.

UTC	**Coordinated Universal Time**. The standard of time kept by atomic clocks around the world. UTC was internationally adopted in 1964 and has now largely replaced Greenwich mean time (GMT) as the universal standard time.
V_1	Formerly called the 'critical engine speed' and now defined as the 'takeoff decision speed'. It is the speed below which an aircraft can lose an engine and still stop on the remaining runway.
V_2	Takeoff safety speed—the normal takeoff speed for a multi engine aircraft. It is the lowest airspeed at which the airplane complies with the climb criteria following one engine out after takeoff. If an engine is lost after V_1 the pilot should continue to accelerate to V_2 and continue the takeoff.
V_R	Rotation speed—the airspeed during takeoff roll at which the airplane is rotated up to attain its takeoff attitude.
VHF	**Very High Frequency**. A portion of the electromagnetic spectrum whose characteristics limit its aeronautical use to line-of-sight communication and navigation.
VLA	**Very Large Aircraft**. The US equivalent of the European FLA—Future Large Aircraft.
VTOL	**Vertical Takeoff and Landing (Aircraft)**. Aircraft capable of vertical climbs and descents and of using very short runways or small areas for takeoff and landings. See STOL.
WAAS	**Wide Area Augmentation System**. A part of the ATC system proposed by the FAA that takes advantage of ground-based stations that permit near pinpoint accuracy from satellite signals that are broadcast from space. It is designed to provide navigation capability within the existing air traffic control system, to permit free flight, and to utilize satellite systems that are approved. The FAA has also developed a LAAS (Local Area Augmentation System) which

is designed to meet the rigorous Category 2 and Category 3 approach and landing conditions.

WSAS   **Wind Shear Advisory System**. A system that provides warning of wind shear to pilots. The system may be passive, reacting to accelerational forces on the airplane detected by airborne sensors, or active where it searches the environment for evidence of shears. Systems may be located in the airplane or on the ground.

Zeppelin   A rigid airship or dirigible named after the inventor, a German engineer and brigadier of cavalry, Count Ferdinand von Zeppelin. See Dirigible or Airship.

Index

A

ACAS 475, 517 *See also TCAS*
Acceleration 85
  'G' loads 86
    long duration 86
    short duration 86
Accident causes
  active failures 418-20
  Annex 13 of the Chicago Convention
    413
  assignment of causes 412
  compliance with SOPs 563
  contributing and latent causes 418
  Dryden, Ontario 417
  error chain 414, 418
  flight crew causes 56
  Howard, Benjamin 29, 40, 198, 426
  latent conditions 418-420
  management as a causal factor 420
  probable cause 414
  significant causes 562
  'stop rule' 413
  system-generated causes 417
  total system approach 418
  traditional analysis 413
  tyranny of the probable cause 414
    See also Miller, C.O.
Accident factors
  cultural 32, 114-115, 117-119, 124,
    178, 407
  data sharing 439
  human factors 432
  infrastructure 33, 470, 483, 485, 491
  organizational 32
  percentage of contributors 563
  regulatory 32, 33, 36

  risk 28, 38, 426, 427
  training 32 *See also Training*
Accident pattern 422-424
  Bruggink, Gerard 186, 422
  changing pattern 1997 and 1998 480
  Commonwealth of Independent
    States 525
  fatalities 29, 30, 38-39, 42, 391, 393,
    397, 427, 430, 431, 450, 452, 455,
    461, 467, 470, 480-481
  hull losses 30-34, 429
  human error 186-190
  IATA airlines 438
  Lautman and Gallimore 124, 428,
    430
  NTSB safety study 442
  regional rates 37
    statistics 426, 427
  Republic of China 529
  Russia 525
  worldwide safety 430-432
Advanced Qualification Program (AQP)
  193 *See also FAA AQP*
Airbus 25, 96, 109-110, 214, 233, 241,
  253, 492, 507, 516
Airline specialized training centers 366
  contracted training 366
  designated training centers 367
Airmail Days 10
  Brown, Walter Folger 11-12
  cancellation of the airmail contracts
    12-13
  Kelly Act (Airmail Act of 1925) 10
  Watres-McNary Act of 1930 11-12
Airplane inventors 1

G

T